Advances in Optical and Photonic Devices

Edited by

Ki Young Kim

Advances in Optical and Photonic Devices
Edited by Ki Young Kim

Republished by InTech
Janeza Trdine 9, 51000 Rijeka, Croatia

Technical Editor Teodora Smiljanic

Additional hard copies can be obtained from orders@intechopen.com

Advances in Optical and Photonic Devices, Edited by Ki Young Kim
p. cm.
ISBN 978-953-7619-76-3

For EU product safety concerns:
IN TECH d.o.o., Prolaz Marije Krucifikse Kozulić 3, 51000 Rijeka, Croatia,
info@intechopen.com or visit our website at intechopen.com.

Preface

The title of this book, *Advances in Optical and Photonic Devices*, encompasses a broad range of theory and applications which are of interest for diverse classes of optical and photonic devices. Unquestionably, recent successful achievements in modern optical communications and multifunctional systems have been accomplished based on composing "building blocks" of a variety of optical and photonic devices. Thus, the grasp of current trends and needs in device technology would be useful for further development of such a range of relative applications.

The book is going to be a collection of the contemporary researches and developments of various devices and structures in the area of optics and photonics. It is composed of 17 excellent chapters covering fundamental theory, physical operation mechanisms, fabrication and measurement techniques, application examples. Besides, it contains comprehensive reviews of recent trends and advancements in the field. First six chapters are especially focused on diverse aspects of recent developments of lasers and related technologies, while the later chapters deal with various optical and photonic devices including waveguides, filters, oscillators, isolators, photodiodes, photomultipliers, microcavities, and so on. Although the book is a collected edition of specific technological issues, I strongly believe that the readers can obtain generous and overall ideas and knowledge of the state-of-the-art technologies in optical and photonic devices.

Lastly, special words of thanks should go to all the scientists and engineers who have devoted a great deal of time to writing excellent chapters in this book.

January 2010

Editor

Ki Young Kim

Department of Physics
National Cheng Kung University
Tainan, Taiwan
E-mail: kykim1994@gmail.com

Contents

Lasers

1

Broadband Emission in Quantum-Dash Semiconductor Laser

Chee L. Tan[1], Hery S. Djie[1] and Boon S. Ooi[1,2]
[1]Lehigh University,
[2]King Abdullah University of Science & Technology,
[1]United States
[2]Saudi Arabia

1. Introduction

A new type of semiconductor laser is studied, in which injected carriers in the active region are quantum mechanically confined in localized finite self-assembled wire-like quantum-dash (Qdash) structures that are varied in sizes and compositions. Effects of such carrier distribution and quasi three-dimensional density of states contribute to a quasi-supercontinuum interband lasing characteristics, which is a new laser design platform as compared to continuous broad emission spectrum generated by nonlinear media pumped with ultrashort laser pulse. The wavelength profile of quasi-supercontinuum emission is tunable at near-infrared wavelength spanning across several optical communication bands at around ~1500 nm in addition to the ability of operating condition at room temperature as opposed to that previously obtained only at operating temperature below 100 K. In this chapter, a thorough analysis of the Qdash material system, device physics and the establishment of ultrabroad stimulated emission behavior will be presented and discussed.

2. Background

The generation of white light by laser radiation was first reported by Alfano and Shapiro (Alfano & Shapiro, 1970) who observed a spectral broadening of a picosecond second-harmonic output of a neodymium garnet laser (400-700 nm) with an energy of about 5 mJ in the bulk of borosilicate glass. Experiments performed with these bulk samples were then followed by studies on waveguide white-light generation in air-silica microstructure optical fibers to date (Ranka, 2000). The generation of the artificial white light is mainly due to the effective nonlinear-optical transformations of ultrashort laser pulses. Owing to its broad and continuous output spectrum, such radiation is called supercontinuum. Supercontinuum generation is an interesting physical phenomenon and the relevant technology is gaining in practical implications - it offers novel solutions for optical communications and control of ultrashort laser pulses (Nisoli et al., 1996), helps to achieve an unprecedented precision in optical metrology (Lin & Stolen, 1976), serves to probe the atmosphere of the Earth (Zheltikov, 2003), and suggests new strategies for the creation of compact multiplex light sources (Morioka et al., 1993) for nonlinear spectroscopy, microscopy, and laser biomedicine.

The first mid-infrared broadband semiconductor laser was demonstrated in an intersubband structure by adopting a quantum cascade configuration (Gmachl et al., 2002). Laser action with a Fabry-Perot spectrum covering all wavelengths from 6 to 8 µm simultaneously is demonstrated with a number of dissimilar intersubband optical transitions. Recently, similar unique spectral properties in the form of quasi-supercontinuum lasing characteristics have been demonstrated on semiconductor quantum-dot (Qdot) and Qdash platforms in different near-infrared wavelength regime without the need of ultrashort pulse laser excitation or the engineering of the intersubband transition level. ([a]Djie et al., 2007; Kovsh et al., 2007 ; [b]Djie et al., 2007; Tan et al., 2008). Most important, this unique feature of quasi-supercontinuum emission occurs at high temperature, i.e. room temperature in addition to identical operating conditions of a conventional semiconductor lasers. Owing to its broad and continuous spectrum emitted via only a single semiconductor laser diode, such device is called broadband laser. The broadband laser technology utilizes largely inhomogeneous quantum nanostructures active medium such as quantum-dot (Qdot) structures for wavelength emission in ~1200 nm ([a]Djie et al., 2007, Kovsh et al., 2007); and Qdash medium for emission in center wavelength of ~ 1600 nm ([b]Djie et al., 2007; Ooi et al., 2008). Bandgap tuning with emission width widening is possible and can be realized in Qdash materials via postgrowth lattice interdiffusion technique (Tan et al., 2008; Tan et al. 2009). Furthermore, interband optical transition in quantum confined heterostructures will contribute to a highly efficient broadband laser action as compared to other emitter technologies.

A brief review of current state-of-art of Qdot/Qdash technology is necessary to comprehend the origin and progress of semiconductor broadband laser. To date, conventional self-assembled Qdot/Qdash semiconductor nanostructures have attracted considerable interest in the fabrication of semiconductor lasers and optical amplifiers due to the unprecedented potential offered by three-dimensional energy levels quantification that lead to vastly improved optoelectronic characteristics as compared to conventional quantum-well (QW) structures and bulk materials (Bimberg et al., 1997; Wang et al., 2001; Ooi et al., 2008). Apart from its predominant applications in optoelectronics industry, self assembled Qdot/Qdash demonstrate a number of unique features as compared to QW materials. In particular, self-assembled Qdot lasers have been shown to emit unique lasing spectral characteristics, where the laser emission spectra are broadened with modulated non-lasing spectral regions and the number of lasing modes increases above threshold (Harris et al., 1998). Furthermore, early experiments showed an extraordinary wideband lasing coverage of 50 nm, in the absence of modulated non-lasing spectral regions, but only at a cryogenic temperature (60 K) from Qdot gain medium (Shoji et al., 1997; Jiang & Singh, 1999). These phenomena have been attributed to the carrier localization in noninteracting or spatially isolated dot ensembles and nonequilibrium carrier distribution among highly inhomogeneous Qdots (Harris et al., 1998; Jiang & Singh, 1999). The most recent study reveals that a low-ripple (<3 dB) non-modulated broad interband lasing coverage of ~40-75 nm from GaAs-based Qdot lasers can be achieved at room temperature with center wavelength of ~1160-1240 nm by employing a highly inhomogeneous InGaAs Qdot structures (Djie et al., 2007) and chirp InGaAs Qdot structures (Kovsh et al., 2007). These novel semiconductor light emitters are particularly attractive for many practical imaging and sensor applications due to their compactness and relatively low energy requirement in comparison to other state-of-art broad spectrum light sources.

The effort of achieving this interesting broadband lasing action in a longer wavelength region (1.5-1.6 μm) will thus be critical for broader applications relevant to the important low-loss transmission window of fiber optic system such as multichannel optical communication system, fiber-based optical coherent tomography, interferometric fiber optic gyroscopes, optical measurement systems, etc. Intensive research in the advanced growth of InP-based self-assembled Qdash had enabled the realization of high quality lasers and optical amplifiers that potentially cover the wavelength operation on both 1.31 μm and 1.55 μm of optical communication bands (Wang et al., 2001). Qdash assembly is essentially comprised of isotropic Qdots and finite quantum-wires (Qwires), whose cross section is similar to that of a typical Qdot, 3-4 nm (height) × 10-20 nm (base), while its length is varied from tens to hundreds of nanometers, as pictured by scanning electron microscopy (Dery et al., 2004) and atomic force microscopy (Popescu & Malloy, 2006). Due to quasi-three-dimensional carrier confinement and intrinsic properties, Qdash enable several interesting laser diode characteristics such as improved temperature insensitivity, optical feedback resistance, wide spectral tunability, and broad stimulated emission (Sek et al., 2007; Lelarge et al. 2007; [b]Djie et al., 2007). In addition, the gain properties of a Qdash amplifier bearing the distinct fingerprint of a quantum-wire-like density of states (Dery et al., 2004) while gain recovery characteristics and recovery time constants resembling Qdot characteristics (van der Poel et al., 2006). More so, it has been proposed that the role of optical gain broadening (Tan et al., 2007) that results in broadband emission from Qdot lasers is also inherent in Qdash lasers. These unique features will help to overcome the challenges in the nanoscaled epitaxial engineering of highly inhomogeneous Qdot for broadband laser applications.

All the interesting features of broad interband lasing actions from self-organized, spatially-isolated semiconductor nanostructure technology can be widely applied in optical telecommunications, various optical sensors detecting chemical agents, atmospheric or planetary gases, high-precision optical metrology and spectroscopy, and biomedical imaging (Ooi et al., 2008). In addition, it is natural to expect that narrow pulses can be generated by locking the phases of modes in this quasi-supercontinuum interband laser spectrum under mode-locked operation (Xing & Avrutin, 2005) due to the fast carrier dynamics and the broad optical gain bandwidth (Lelarge et al., 2007). Furthermore, the high power emission capability of ~1 W per device from these ultrabroadband Qdash lasers at room temperature (Tan et al., 2008) can be potentially employed as a high efficiency resonant pumping source (Garbuzov et al., 2005) for eye-safe Er-doped amplifiers and solid-state lasers.

In this chapter, we will present the generation of ultrabroad stimulated emission at room temperature operating condition in the InP-based broadband laser with wide wavelength coverage. For the first time, the InP-based unique dash quantum confined heterostructure properties is exploited to generate a broad lasing spectrum, following the prior success of short wavelength GaAs-based Qdot broadband laser. The fabricated Qdash laser diode emits at ~1.64 μm center wavelength with wide wavelength coverage of 76 nm. Unlike conventional diode lasers, the rule changing broadband lasing is obtained from the quasi-continuous interband transition by the inhomogeneous Qdash ensembles.

To further enhance the broad spectrum emission and fine tune the lasing wavelength coverage, we further engineer the bandgap energy of Qdash material with postgrowth lattice intermixing process utilizing impurity-free vacancy induced disordering (IFVD) technique. We successfully demonstrated a 100 nm wavelength blue-shifted Qdash lasers exhibiting a room-temperature broad lasing spectral coverage of ~85 nm at a center wavelength of ~1.55 µm with enhanced total emission power of ~1 W from a single as-cleaved broad area laser structure (50 x 500 µm²). The peculiar broad lasing spectra from fabricated diodes with different cavity lengths related to the effect of nonequilibrium carrier distribution in these highly inhomogeneous dashes will also be discussed.

3. Experiments and theoretical modelling

3.1 Materials and laser structure

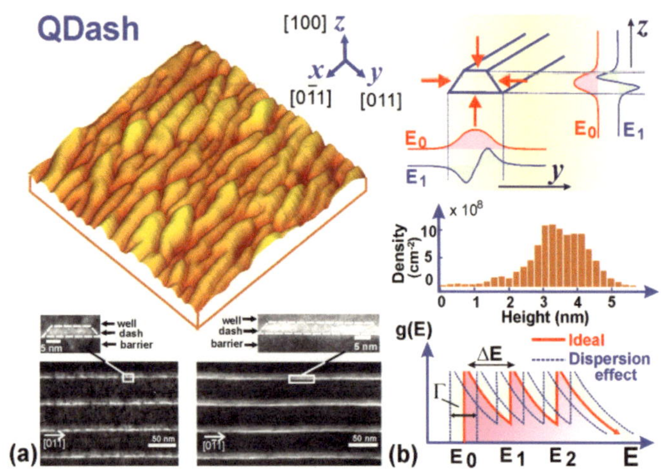

Fig. 1. (a) The plane-view AFM image (area of 0.5×0.5 µm²; height contrast of 8 nm) and the cross-sectional TEM images across [110] and $[1\bar{1}0]$ directions. (b) The illustration of carrier confinement in Qdash structure (top). Only the first two energy levels (E_1 and E_2) are shown for clarity. The height distribution profile of dash islands from AFM image (middle), that results in the density of states (DOS) spreading over the energy and forms the quasi-continuous interband transition (bottom).

The InAs/InAlGaAs Qdash material used in this study was grown by molecular beam epitaxy (MBE) on (1 0 0) oriented InP substrate. The laser is a *p-i-n* structure with active region consisting of four-sheet of InAs Qdashes, and each Qdash layer is embedded in an asymmetric InAlGaAs QW. The QWs are then sandwiched between two sets of SCHs. The Qdash-in-well structure consists of a 1.3 nm thick compressively strained $In_{0.64}Ga_{0.16}Al_{0.2}As$ layer, a five-monolayer (ML) thick InAs dash layer, and a 6.3 nm thick compressively strained $In_{0.64}Ga_{0.16}Al_{0.2}As$ layer. Each dash-in-well stack is separated by a 30 nm thick tensile strained layer of $In_{0.50}Ga_{0.32}Al_{0.18}As$ that acts as the strain compensating barrier. The lower cladding consists of a 200 nm thick $In_{0.52}Al_{0.48}As$ layer doped with Si at 1×10^{18} cm^{-3}, which is lattice matched to the InP substrate. The upper cladding and contact layers are 1700 nm thick $In_{0.52}Al_{0.48}As$ and 150 nm thick $In_{0.53}Ga_{0.47}As$, respectively. Both layers are doped with Be at 2×10^{18} cm^{-3} (Djie et al., 2006; Wang et al., 2006). Fig. 1(a) shows the plane-view atomic force microscopy (AFM) of the surface Qdash and the cross-sectional transmission electron microscopy (TEM) images of the laser structure. The Qdash structure comprises three-dimensional elongated nanostructure preferentially aligned along $[1\bar{1}0]$ direction with an average height of 3.2 nm, an average width of 18 nm, and the base or length varied from 20 to 75 nm. The individual Qdash provides strong carrier confinement along y- and z-directions and weaker confinement along the x-direction [Fig. 1(b-top)]. The AFM image reveals the nanostructure networks composed of dot-like and finite wire-like quantum confined structures with a bimodal height distribution profile [Fig. 1(b-middle)]. The isotropic, dot-like structure with a comparable width over the length have a relatively larger height than the wire-like structure suggesting that the elongated islands are formed by the coalescence of two or more dot-like islands. Considering the large dispersion in shape, size and composition, the inhomogeneous Qdash gives a wide energy spreading in the confining potentials. This effect leads to the broadened optical gain characteristics [Fig. 1(b-bottom)] suitable for the wideband optical devices such as superluminescent diodes (SLD) (Djie et al., 2006).

For the purpose of further enhancement of the broad spectrum emission and fine tune of the lasing wavelength coverage, we performed the dielectric cap annealing technique to induce selective intermixing using 475 nm thick SiO_2 layer as an vacancy source deposited using plasma enhanced chemical vapor deposition system. During the annealing, the SiO_2 cap will enhance the preferential atomic outdiffusion hence enhancing the group-III atomic interdiffusion in the Qdash active region and resulting in the effective bandgap modification of Qdash material (Tan et al., 2008; Wang et al., 2006). The dielectric cap also serves to protect the surface quality during annealing from the thermal induced decomposition. Following the dielectric cap removal, state-filling PL spectroscopy using a 980 nm diode laser as an excitation source was performed at 77 K to assess the bandgap modification from the interdiffusion effect on the laser structure. The IFVD process is performed by annealing the SiO_2 capped sample in nitrogen ambient for one minute in a rapid thermal processor (RTP). Fig. 2 gives the summary of PL peak shift and the linewidth as the annealing temperature increases from 600°C to 850°C. At the temperature of 750°C, the PL peak shifts towards a shorter wavelength emission while the linewidth is the broadest. Further increase in annealing temperature initiates more intermixing, and therefore improves the uniformity in shape, size and composition of Qdash leading to reduction in PL linewidth. The result points out the linewidth broadening at intermediate stage of intermixing due to non-uniform interdiffusion, which will be further selected to broaden the Qdash laser emission.

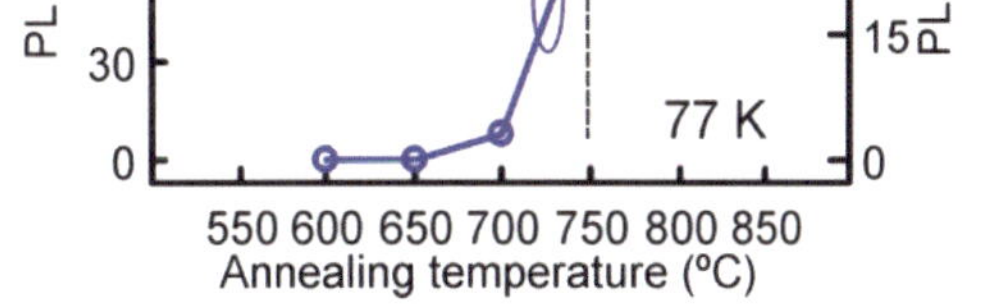

Fig. 2. The evolution of PL peak shift and linewidth measured at 77 K with varying annealing temperature of rapid thermal processor from SiO_2 capped Qdash samples. The inset depicts the normalized PL spectra for selected temperatures clearly showing the broadening of PL linewidth at the intermediate degree of intermixing.

Broad area lasers with 50 µm wide oxide stripes with no facet coating were then fabricated from both the as-grown and intermixed Qdash samples (under annealing temperature of 750 °C) with SiO_2 capped layer. In order to maximize the gain (Ukhanov et al., 2002), the optical cavity of the laser is aligned along the [011] orientation and is perpendicular to the dash direction. Current injection was performed to the non-facet-coated Qdash lasers under pulsed operation at 0.2% duty cycle with a 2 µs pulse width.

3.2 Simulation model of group-III interdiffusion

The understanding of diffusion processes is important to the interpretation of interdiffusion induced compositional change and the band structure modification related to the experimental works and selected postgrowth operating conditions presented in the previous sections. In IFVD process, majority vacancies are injected vertically from the dielectric cap during thermal treatment and therefore the interdiffusion will occur more effectively in the transverse direction that corresponds to the dash height (Tan et al., 2008; Wei & Chan, 2005). This diffusion effect becomes more pronounced at very thin Qdash layer when the dash height to base ratio is of ~ 0.1 or less (Wang et al., 2006). At intermediate stage of intermixing, the partial intermixing might occur, which the thick dash family will experience a larger degree of wavelength blue-shift due to the larger concentration (Crank, 1975) of active medium composition and hence its interdiffusion length is larger than the thin dash family. The solution of the diffusion problem (Crank, 1975) in the Qdash can be estimated by an equivalent one dimensional quantum-confined model (transverse direction of vacancies interdiffusion) by assuming a substance of concentration C_0, confined in a region of n repeating well and barrier of width w and b, respectively, centered at zero (Gontijo et al., 1994) that is given by

$$C_n = \frac{C_0}{2}[2 - \sum_{n=1}^{k}[erf(\frac{x-(n-1)(w+b)}{2L_D}) + erf(\frac{x-[nw+(n-1)b]}{2L_D})] \quad (1)$$

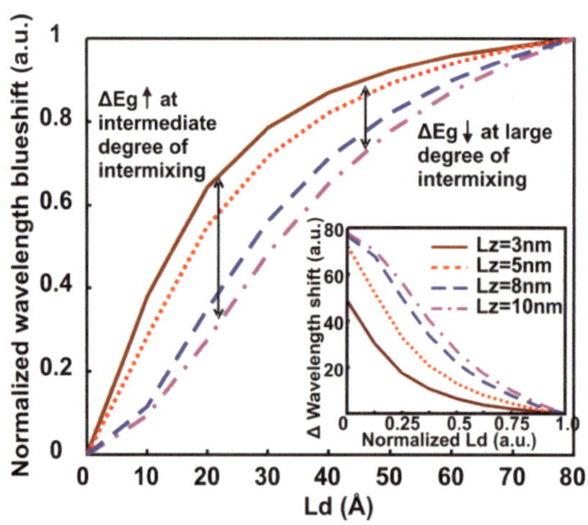

Fig. 3. The blueshift of normalized transition wavelength when diffusion length increases in one-dimensional quantum confined structure with different well widths (L_z). The inset shows the corresponding change of the transition wavelength shift with normalized diffusion length to QW/Qdash.

The origin of the *x* coordinate is at the left barrier of the first well. The one-dimensional quantum-confined structure of four repeating wells and barriers with arbitrary width are used in the simulation model to calculate the confined ground state energy level. The chosen material system is not critical because it serves only as a reference for the change of transition energy states with diffusion length (L_d) and well thickness. The quantum confined energy levels can be obtained by solving the one-dimensional time-independent Schrodinger equation and the results are shown in Fig. 3 and its inset. Different well widths (3, 4, 5 and 8 nm) with varied L_d are used in the calculation model to represent the different dash heights in the real Qdash assembly. As L_d increases, the wavelength shift to shorter wavelength due to group-III interdiffusion. The blue-shift rate is faster at same L_d for thin nanostructure than the thick nanostructure, as stated in Fig. 3. At intermediate stage of intermixing, the disparity in wavelength blueshift is notable. As intermixing proceeds further, the variation rate in wavelength blueshift becomes less until the nanostructure becomes fully intermixed and the wavelength blueshift converges. Noting a high dispersion in Qdash structure used in the experiments [16], widened gain characteristics can be practically achieved by selecting suitable degree of intermixing to the Qdash structure. Thereafter, the broadened linewidth can be attributed to the different intermixing results from inhomogeneous nanostructure in Qdash assembly at a medium degree of intermixing. Experimentally, this can be obtained by a given dielectric film properties heat-treated under the suitable annealing temperature and/or duration. Furthermore, variation in transition energy state is more sensitive to the L_d as compared to the well width. The thick dash family that tends to induce larger L_d contribute to larger blueshift of peak emission is as shown in the inset of Fig. 3. Hence, there is a smaller peak emission blueshift in the intermixed samples as compared to the as-grown samples under high excitation.

4. Results and discussion

4.1 Optical properties of Qdash – As-Grown and intermixed materials

State filling photoluminescence (PL) spectroscopy of as-grown Qdash samples were performed at 77 K by varying optical excitation density. As comparison, InAs Qdot embedded in InP matrix was grown and also characterized. The ground state PL peak emission is longer in Qdot as the InP matrix has a larger bandgap energy than InAlGaAs confining layers in the Qdash. Qdot structure shows well-resolved quantized states (E_0 to E_4) with a large energy separation between E_0 and E_1 (ΔE = 34 meV) compared to Qdash characteristics (up to E_4 with ΔE = 30 meV). At similar excitation density, more states are excited in Qdot than Qdashes as a manifestation of the enhanced DOS in Qdot subbands [Fig. 4]. At high excitation (1500 W/cm^2), a large number of minima in Qdash spectra are populated, resulting a broad emission line while in Qdot spectra, the individual minima is more apparent. These properties corroborate the quasi-continuous interband transition characteristics in Qdash over a wide wavelength range. The discrepancies between Qdot and Qdash are due to the shape of DOS [Fig. 1(b)]. Qdash with size and composition fluctuations have overlapping states with nearly identical transition energies in the high-energy portion that contributes to the gain broadening and thus produces less resolved confined state recombination in PL spectra. However, this is not the case in the Qdot assembly due to its delta function DOS leading to the atomic-line luminescence spectra.

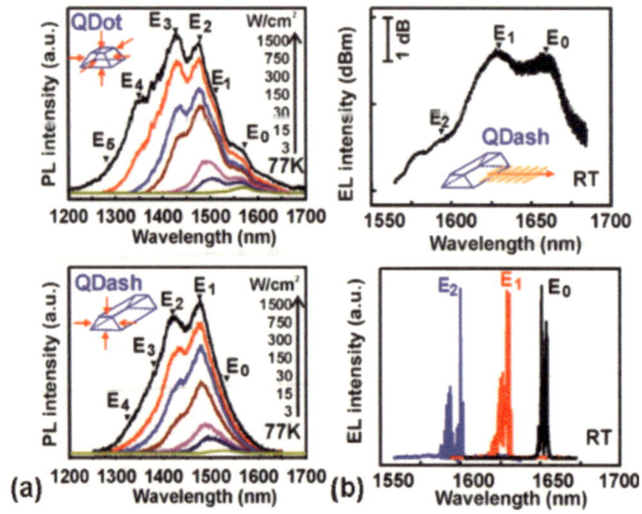

Fig. 4. (a) PL spectra at 77 K with varying optical pumping level taken from InAs Qdots within InP matrix (above) and InAs Qdashes within InAlGaAs matrix (below). The confined energy subbands are indicated with the arrow, after the deconvolution with the multi-Gaussian spectra. (b) EL spectra at RT showing the spontaneous emission (top) at J=0.8×J_{th} from a 300 µm device and the lasing emission spectra (bottom) from E_0, E_1 and E_2 states. These individual lasing lines are obtained from laser with cavity length L of 1000, 300, and 150 µm, respectively, at 1.1×J_{th}.

Broad area as-grown Qdash lasers with 50 µm wide oxide stripes without facet coating were fabricated and characterized. Fig. 4(b) shows the electroluminescence (EL) spectrum of the Qdash samples revealing fine structures of amplified spontaneous emission from different energy subbands in correlation to different lasing peaks. Up to three distinct laser emissions (1.65, 1.62, and 1.59 µm) from E_0, E_1 and E_2 energy transitions were obtained at J = 1.1×J_{th} from lasers with cavities L of 1000, 300, and 150 µm, respectively. The distinct lasing wavelength peak is attributed to the finite modal gain of each quantized state in Qdash assembly.

Carriers localized in different dots/dashes, resulting in a system without a global Fermi function and exhibiting an inhomogeneously broadened gain spectrum, have shown an interesting phenomena of lasing spectra (Harris et al., 1998; [a]Djie et al., 2007; Tan et al., 2007; Matthews et al., 2002). This unique feature of dot/dash can be well studied after postgrowth interdiffusion technique, from the evolution of state-filling spectroscopy from intermixed Qdash structures at 77 K, as shown in Fig. 5 and its inset. At low excitation below 3 W/cm², the ground state emissions of 1.57 µm and 1.50 µm are dominant in the as-grown and the intermixed samples, respectively. The PL spectra are gradually broadened in both samples with increasing optical excitation densities. An increase in the excitation power density leads to the filling of lower-energy states, allowing recombination from higher energy levels of Qdash structure. Under the same excitation density, the PL linewidth of intermixed sample is wider than the as-grown sample. At the power excitation density of 1500 W/cm², the PL linewidth increases by 11 nm (from 94 nm to 111 nm) after intermixing process. The phenomenon of carrier localization in Qdash becomes more evident when the intermixed sample shows a larger variation of full-width-half-maximum (ΔFWHM up to 47 nm) than the as-grown sample (ΔFWHM up to 18 nm) under various power excitation densities relative to the FWHM obtained at the optical excitation of 3 W/cm^2, as shown in the inset of Fig. 3.

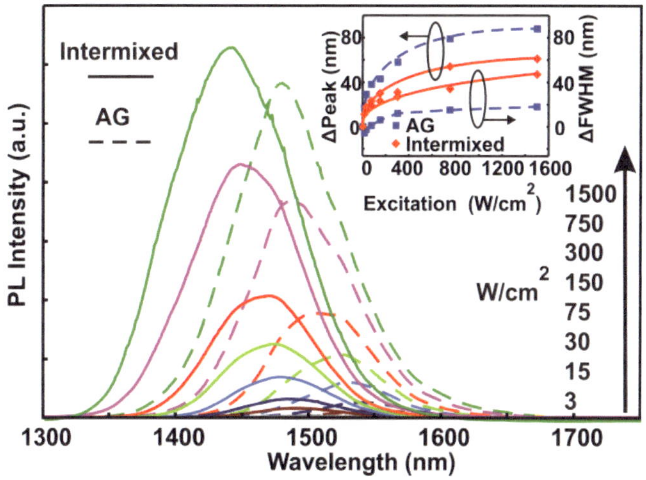

Fig. 5. The PL spectra of both as-grown and intermixed samples, with varying optical pumping levels, show global blueshift after intermixing. The inset shows the corresponding changes of FWHM and PL peak wavelength as compared to those obtained under optical excitation of 3 W/cm^2.

These enormously large broadening of the PL spectra from both the as-grown and intermixed samples is attributed to the contribution of multiple transition states ([a]Djie et al., 2007) or large inhomogeneous broadening of the non-interacting Qdash ensembles (Tan et al., 2007; Van der Poel et al., 2006). This observation is also clearly different from that of both conventional QW (Ooi et al., 1997) and Qdot structures (Wang et al., 2006). The IFVD technique is generally well-known to improve the size homogeneity of a highly inhomogeneous semiconductor nanostructure system and thus will contribute to smaller variation in energy transition after intermixing. For instance, at the power excitation density of 1500 W/cm^2, the PL linewidth decreases by 6 nm (from 94 nm to 88 nm) after the IFVD process is performed by annealing the SiO_2 capped sample at 750°C for two minutes (Djie et al., 2008). However, the opposite observations in the Qdash, i.e. larger PL linewidth after intermediate intermixing, suggests the presence of different interdiffusion rates at a given intermixing degree in the Qdash nanostructures as a consequence of wide variation in surface to volume ratio in Qdash ensembles. The presence of more non-interacting Qdash with wider distribution of energy levels will contribute to radiative recombination emission over larger wavelength coverage and thus a larger FWHM in PL spectra. In other words, carrier localization is more prominent in an isolated Qdash, which affects the optical properties of these material systems. Nevertheless, both intermixed and as-grown Qdash samples showing saturation of ΔFWHM at excitation power over 400 W/cm^2 indicates that large degeneracy levels in highly confined energy states of Qdash is still preserved as can be seen in Qdot nanostructures (Hadass et al., 2004).

The nearly symmetric Qdash PL spectra in Fig. 5 are broadened with increasing optical excitation densities. Furthermore, an increase in integrated PL intensity after intermixing occurs. All these observations are contrary to the conventional quantum-confined nanostructures. These can be attributed to the continuous PL wavelength blue-shift observed in both as-grown and intermixed samples, as shown in Fig. 5, with increasing optical excitation densities. The continuous blue-shift of the PL peak wavelength up to 88 nm in the as-grown sample and 61 nm in the intermixed sample at the optical excitation density of 1500 W/cm^2, relative to those obtained at the excitation of 3 W/cm^2, are shown in the inset of Fig. 5. The effect of band-filling is insufficient to explain the large degree of blue-shift observed from sample excited under high density excitation. Hence, it is reasonably ascribed this to the postulation of continuum states (Van der Poel et al., 2006) in the Qdash nanostructures, although spectral widening at a shorter wavelength is expected in an inhomogeneous Qdash structure (Hadass et al., 2004). Continuum states serve as an effective medium for exciton scattering and thus change the dephasing rate (Tan et al., 2007) at each energy level within the highly inhomogeneous ensembles and the radiative recombination profile will be different from that of conventional QW. The wide distribution of energy levels due to the nature of Qdash inhomogeneous (FWHM of 76 nm from PL measurement of as-grown sample at low excitation of 3 W/cm^2) will further serve as the radiative recombination states or "sink" for the scattered excitons from the dense continuum states. Consequently, quasi-supercontinuum lasing spectra of the diode laser fabricated from these samples are observed, which will be discussed in the later section. Nevertheless, smaller blue-shift of PL peak wavelength in the intermixed sample, as depicted in the inset of Fig. 5, indicates that IFVD enhances the Qdash inhomogeneity more so in larger sizes of Qdashes, which emit at longer wavelengths. Assuming a uniform injection of group-III vacancies from the surface during the IFVD process, the interdiffusion in the vertical direction will affect the dash height more than other directions (Djie et al., 2008; Wei et al., 2005). At an intermediate stage of intermixing, the thick dash family, where the quantized energy level located closer to the conduction band minima, will experience a larger degree of intermixing as the effective height or thickness of the dash decreases, as depicted in the inset of Fig. 3. In addition, the local effective concentration for the thick dash family is higher than the thin dashes. Under uniform annealing temperature, the thick Qdash family that has larger interdiffusion length will yield larger degree of intermixing. As a result, largest degree of wavelength blue-shift (~65 nm) is observed at low excitation of 3 W/cm^2 (dominant emission from thick dashes) as compared to the smaller wavelength blue-shift (~38 nm) at high excitation of 1500 W/cm^2 (dominant emission from thin dashes).

4.2 Effect of nonequilibrium carrier distribution from intermixed lasers

Broad area laser characterization of the intermixed samples further provides evidence of a multi-state emission as shown in Fig. 6. A spectral widening is apparent as the bias increases (Hadass et al., 2004). The emission spectra show multi-state lasing emission as injection increases to current density J of 1.5 x J_{th} (threshold current density) and above as opposed to a series of well-defined groups of longitudinal modes (Harris et al., 1998) emission above threshold in highly inhomogeneous Qdot. This implies the preservation of 3-dimensional energy confinement of the Qdash in addition to the emission from multiple sizes of Qdash ensembles as shown in Fig. 7 and Fig. 8 for fabricated lasers with different cavity lengths. The localized active region of the device can be treated as a large number of Qdot or Qdash, which can be further treated as a broad distribution of discrete energy levels (Shoji et al., 1997). This is owed to the inhomogeneous broadening nature of Qdash ensembles and the dash variation from different dash stacks. The light-current (*L-I*) curve of the short cavity Qdash laser (L = 300µm) yields a J_{th} and slope efficiency of 2.3 kA/cm^2 and 0.46 W/A, respectively, as depicted in Fig. 7(a). Measuring the temperature dependent J_{th} over a range of 10-50 °C, reveals the temperature characteristic (T_o) of 41.3 K. On the other hand, the long cavity Qdash laser (L = 1000µm) yields J_{th} = 1.18 kA/cm^2, slope efficiency of 0.215 W/A, and T_o of 46.7 K over the same temperature range, as shown in Fig. 8(a).

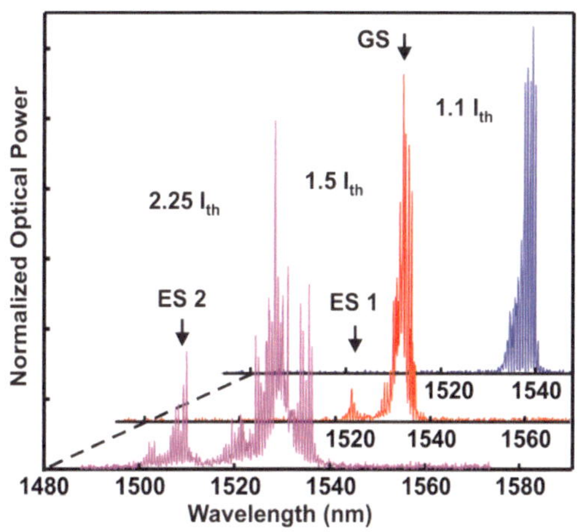

Fig. 6. The lasing spectra show the changes of multi-state emission, from ground state (GS), first excited state (ES 1) and second excited state (ES 2) of the 50 x 500 µm^2 broad area Qdash intermixed laser, under different current injection of 1.1 x I_{th}, 1.5 x I_{th} and 2.25 x I_{th}.

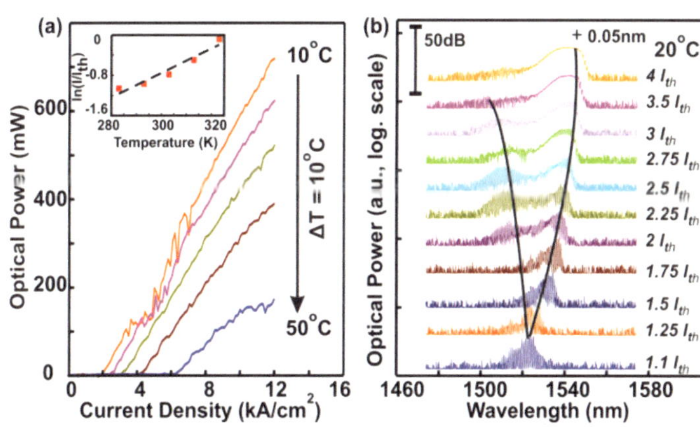

Fig. 7. (a) *L-I* characteristics of the 50 x 300 µm^2 broad area intermixed Qdash laser at different temperatures. Up to ~450 mW total output power (from both facets) has been measured at J = 4.0 x J_{th} at 20°C. (b) The progressive change of lasing spectra above threshold condition.

Compared to the laser with long cavity, the shorter cavity laser exhibits the progressive appearance of short wavelength emission line with an increase in injection level. The *L-I* curve of the short cavity laser shows kinks as compared to the long cavity laser. The jagged *L-I* curve below ~3 x J_{th} implies that the lasing actions from different confined energy levels are not stable due to the occurrence of energy exchange between short and long wavelength

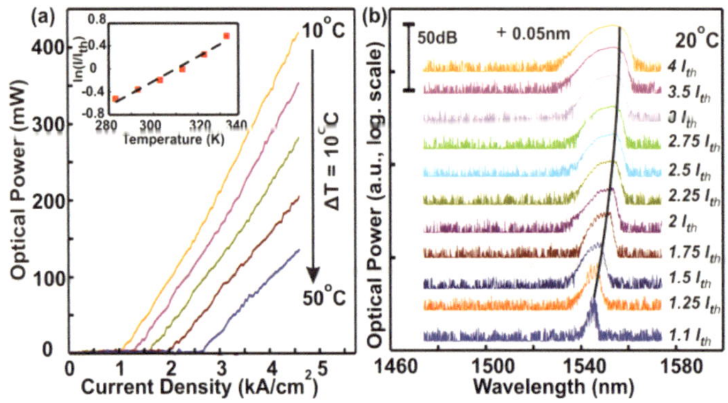

Fig. 8. (a) *L-I* characteristics of the 50 x 1000 μm² broad area intermixed Qdash laser at different temperatures. Up to ~340 mW total output power (from both facets) has been measured at $J = 4.0 \times J_{th}$ at 20°C. (b) The progressive change of lasing spectra above threshold condition.

lasing modes (Hadass et al., 2004), as can be seen in the lasing spectra of Fig. 7(b). In addition, the observation of kink in the *L-I* curve for device tested at low temperature might also be a result of mode competition in the gain-guided, broad area cavity devices. The calculated Fabry-Perot mode spacing of ~1.1 nm is well resolved in the measurement across the lasing wavelength span at low injection before a quasi-supercontinuum lasing is achieved, where the spectral ripple is less than 1 dB.

Subsequent injections contribute to the stimulated emission from longer wavelength or lower order subband energies while suppressing higher order subbands as shown in Fig. 7(b). This Qdash laser behavior is fundamentally different from the experimental observation from Qdot lasers with short cavity length, where the gain of lower subband is too small to compensate for the total loss, and lasing proceeds via the higher order subbands (Markus et al., 2003; Markus et al., 2006). In short-cavity Qdash laser, the initial lasing peak at shorter wavelength (~1525 nm) is dominantly emitted from different groups of smaller size Qdash ensembles instead of higher order subbands of Qdash. Hence, the significant difference of ~11 meV as compared to the dominant lasing peak of ~1546 nm at high injection will contribute to photon reabsorption by larger size Qdash ensembles and seize the lasing actions at shorter wavelength. Regardless, a smooth *L-I* curve at the injection above $3 \times J_{th}$ due to the only dominant lasing modes at long wavelength demonstrates the high modal gain of the Qdash active core (Lelarge et al., 2007). These observations indicate that carriers are easily overflows to higher order subbands (Tan, et al., 2009) because of the large cavity loss and the small optical gain (Shoji et al., 1997) at moderate injection. At high injection, carrier emission time becomes shorter, when equilibrium carrier distribution is reached and lasing from multiple Qdash ensembles is seized (Jiang & Singh, 1999).

On the other hand, a relatively smooth *L-I* curve above the threshold is observed from the long cavity intermixed Qdash laser regardless of the injection levels. The corresponding electroluminescence spectra show only one dominant lasing emission at long wavelengths, unlike, the short cavity Qdash lasers. This observation can be attributed to the effect of long cavity parameter that results in smaller modal loss as compared to short cavity Qdash devices. The progressive red-shift (~10 nm) of lasing peak with increasing injection up to $J = 4 \times J_{th}$ and the insignificant observation of band filling effect indicates that photon reabsorption occurs due to the photon-carrier coupling between different sizes of Qdash ensembles in addition to the high modal gain of the Qdash active core (Lelarge et al., 2007). Injection above $J = 4 \times J_{th}$ is expected to contribute to broader lasing span at long wavelength owing to the high modal gain characteristics (Tan et al., 2008) although the comparison scheme of the two devices with different cavity lengths may not be fair without applying threshold current density.

Distinctive lasing lines are observed from different cavity intermixed Qdash lasers at the near-threshold injection of $J = 1.1 \times J_{th}$. The similarity of lasing wavelength (inset of Fig. 9) from devices with different cavity lengths further shows promise that the Qdash structures have high modal gain characteristics (Lelarge et al., 2007). However, the Qdash laser with increasing cavity length shows progressive red-shift (total of ~20 nm up to $L = 1000$ μm) of peak emission. This may be ascribed to the wide distribution of energy levels because of highly inhomogeneous broadening and photon reabsorption among Qdash families. At the intermediate injection of $J = 2.25 \times J_{th}$, simultaneous two-state laser emission, which is attributed to two groups of Qdash ensembles as mentioned previously, is noticed from short cavity Qdash lasers. On the other hand, a broad linewidth laser emission from a single

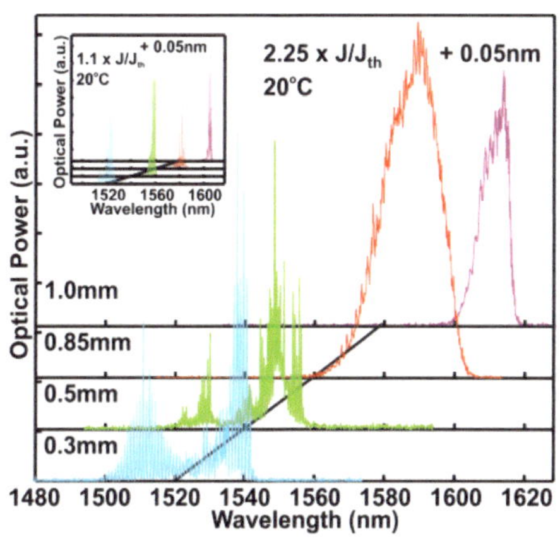

Fig. 9. The presence of different lasing Qdash ensembles with cavity length at the injection of $J = 2.25 \times J_{th}$. The inset shows the progressive red-shift of lasing peak emission with cavity length at the injection of $J = 1.1 \times J_{th}$.

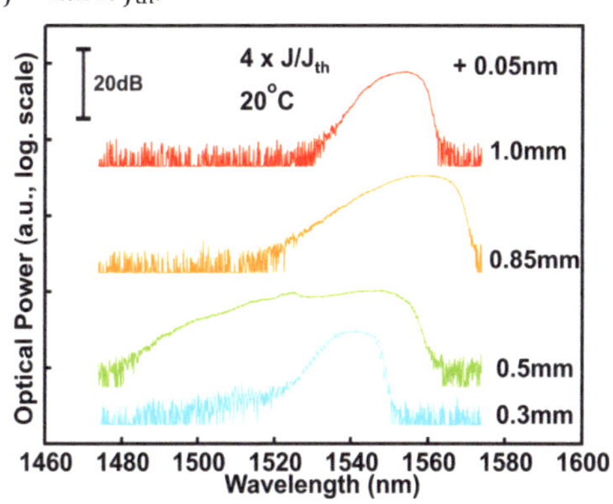

Fig. 10. The effect of cavity dependent on quasi-supercontinuum broadband emission from intermixed Qdash laser at an injection of $J = 4 \times J_{th}$.

dominant wavelength is shown in longer cavity Qdash lasers of 850 μm and 1000 μm, as depicted in Fig. 9. As a result, a quasi-supercontinuum broad laser emission could be achieved at high injection, as shown in Fig. 7. An ultrabroad quasi-supercontinuum lasing coverage from Qdash devices with L = 500μm (Tan et al., 2008) results from emission in different order of energy subbands and groups of ensemble, which will be discussed in the following section.

The broad lasing spectra from devices with different L suggest there is collective lasing from Qdashes with different geometries. However, the broad laser spectra of Qdash lasers obtained at room temperature are different from that of Qdot lasers which shows similar phenomenon but occur at low temperature below 100 K (Shoji et al., 1997; Jiang & Singh, 1999). In Qdot lasers, with increasing temperature, carriers can be thermally activated outside the dot into the well and/or barrier and then relax into a different dot (Tan et al., 2007). Carrier hopping between Qdot states can favor a drift of carriers towards the dots where the lasing action preferentially takes place, thus resulting in a narrowing of the laser mode distribution. However, in Qdash lasers, carriers will be more easily trapped in the dash ensembles due to the elongated dimension in addition to random height distribution in each ensemble. These profiles of energy potential will support more carriers, thus retarding the emission of carriers (Jiang & Singh, 1999) and resulting in a smaller homogeneous broadening at each transition energy level (Tan et al., 2007). Hence, the actual carrier distribution in Qdash nanostructures will be at high nonequilibrium and lead to broadband lasing even at room temperature.

4.3 Ultrabroadband lasers - as-grown and bandgap tuned devices

Fig. 11(a) shows the light-current (L-I) characteristics of the as-grown Qdash laser (L = 600 μm). The corresponding J_{th} and slope efficiency are 2.6 kA/cm^2 and 0.165 W/A. Up to 400 mW total output power has been measured at J = 4.5×J_{th} at 20°C, which is significantly higher than the SLED fabricated from the same wafer (Djie et al., 2006). From the dependence of J_{th} on temperature, the temperature characteristic T_0 of 43.6 K in the range of 10 to 70°C has been obtained. At J < 1.5×J_{th}, there is only ground state lasing E_0 with the wavelength coverage of ~10 nm [Fig. 11(b)]. The broad E_0 lasing spectrum suggests the collective lasing from Qdashes with different geometries. At J > 1.5×J_{th}, the bi-state lasing is noted. The simultaneous lasing from both E_0 and E_1 is attributed to the relatively slow carrier relaxation rate and population saturation in the ground state in low-dimensional quantum heterostructures (Zhukov et al., 1999). The transition from mono-state to bi-state lasing is marked with a slight kink in the L-I characteristics. The bi-state lasing spectrum is progressively broadened with increasing carrier injection up to a wavelength coverage of 54 nm at J = 4.5×J_{th}. The corresponding side-mode suppression ratio is over 25 dB and a ripple measured from the wavelength peak fluctuation within 10 nm span is less than 3 dB.

Bangap-tuned broad area lasers with optimum cavity length (L = 500 μm) that gives largest quasi-supercontinuum coverage of lasing emission, as presented in Fig. 10, are fabricated. The L-I curve of the Qdash laser yields an improved J_{th} and slope efficiency of 2.1 kA/cm^2 and 0.423 W/A, which is depicted in Fig. 12(a), as compared to that of as-grown laser with 2.6 kA/cm^2 and 0.165 W/A, respectively ([b]Djie et al., 2007). The L-I curve of the intermixed laser shows kinks, which is similar to that of short cavity L = 300 μm Qdash lasers. The energy-state-hopping instead of mode-hopping occurs due to the wide distribution of the energy levels across the highly inhomogeneous Qdash active medium, as derived from the

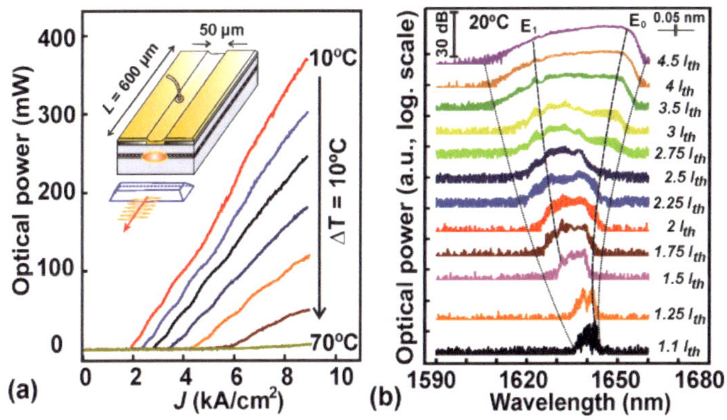

Fig. 11. (a) The L-I characteristics of the 50×600 µm² broad area Qdash laser at different temperatures. The inset shows the schematic illustration of oxide stripe lasers with [110] cavity orientated perpendicular to the dash direction. (b) The lasing spectrum above the threshold condition at 20°C (curves shifted vertically for clarity). The lines are as the guide to the eyes indicating the confined state lasing lines, E_0 and E_1 (dashed lines) and the wavelength coverage of laser emission (dotted lines). The spectra are acquired using an optical spectrum analyzer with wavelength resolution of 0.05 nm.

PL results. In spite of that, a smooth *L-I* curve above 6 kA/cm² yields a total high power of ~1 W per device at room temperature before any sign of thermal roll-over. This shows that injection above 6 kA/cm² provides enough carriers for population inversion in all the available or possible radiative recombination energy states and thus the energy-state-hopping is absent.

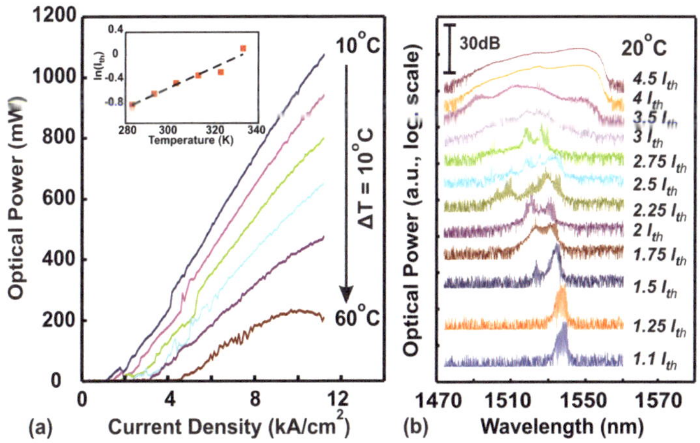

Fig. 12. (a) *L-I* characteristics of the 50 x 500 µm² broad area Qdash laser at different temperatures. Up to ~1 W total output power has been measured at $J = 5.5 \times J_{th}$ at 20°C before showing sign of thermal roll-off. (b) The lasing spectra above threshold condition that are acquired by an optical spectrum analyzer with wavelength resolution of 0.05 nm.

Measuring the temperature dependence J_{th} over a range of 10-60 °C reveals the improved T_o of 56.5 K as compared to the as-grown laser of 43.6 K ([b]Djie et al., 2007). This result is comparable to the T_o range (50-70 K) of the equivalent QW structure. In Fig. 12(b), only a distinctive ground state lasing with the wavelength coverage of ~15 nm is observed below injection of 1.5 x J_{th}. This broad lasing linewidth, again suggests collective lasing actions from Qdashes with different geometries. In addition, the quasi-supercontinuum lasing spectrum at high current injection (4 x J_{th}) without distinctive gain modulation (Harris et al., 1997) further validates the postulation of uniform distribution of dash electronic states in a highly inhomogeneous active medium. At J > 1.5 x J_{th}, the bistate lasing is evident. The simultaneous lasing from both transition states (Hadass et al., 2004) is attributed to the relatively slow carrier relaxation rate and population saturation in the ground state in low-dimensional quantum heterostructures. The bistate lasing spectrum is progressively broadened with increasing carrier injection up to a wavelength coverage of 85 nm at J = 4 x J_{th}, which is larger than that of the as-grown laser (~76 nm), as shown in Fig. 11 and Fig. 13. A center wavelength shift of 100 nm and an enhancement of the broadband linewidth, which is attributed to the different interdiffusion rates on the large height distribution of noninteracting Qdashes at an intermediate intermixing, are achieved after the intermixing. The inset of Fig. 13, showing the changes of FWHM of the broadband laser with injection depicts that energy-state-hopping and multi-state lasing emission from Qdashes with

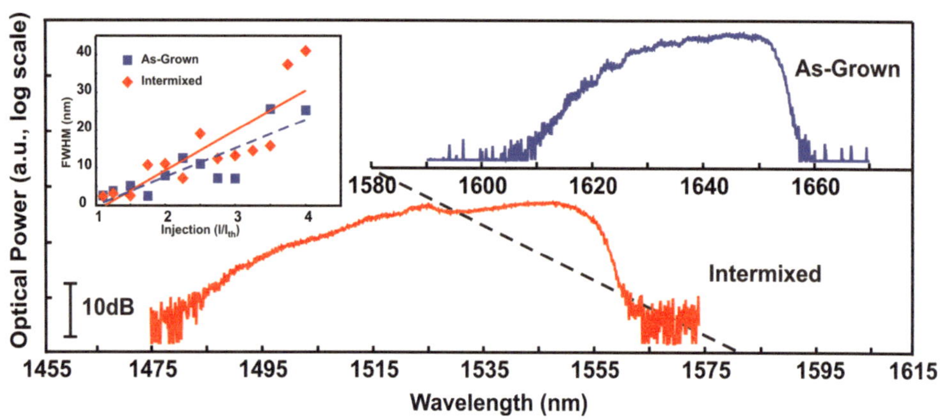

Fig. 13. The wavelength tune quasi-supercontinuum quantum dash laser from 1.64 μm to 1.54 μm center wavelength. The lasing coverage increases from 76 nm to 85 nm after intermixing process. The inset shows the FWHM of the broadband laser in accordance to injection above threshold up to J = 4 x J_{th}.

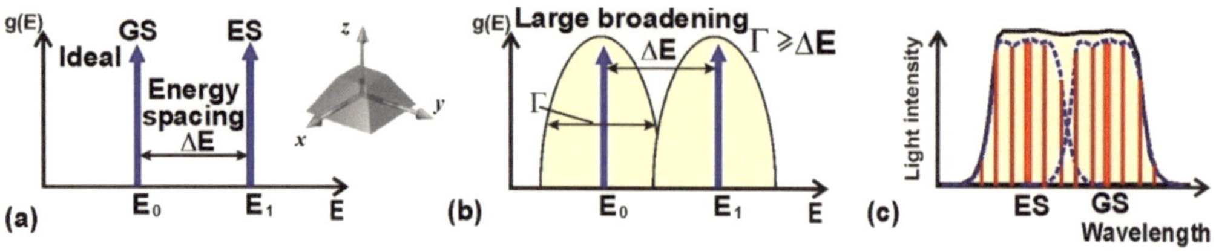

Fig. 14. (a) Spaced and quantized energy states from ideal Qdot samples. (b) Large broadening of each individual quantized energy state contributes to laser action across the resonantly activated large energy distribution. (c) Variation in each individual quantized energy state owing to inhomogeneous noninteracting quantum confined nanostructures in addition to self broadening effect demonstrate a broad and continuous emission spectrum.

different geometries occur before a quasi-supercontinuum broad lasing bandwidth with a ripple of wavelength peak fluctuation that is less than 1 dB is achieved. This idea can be illustrated clearly in Fig. 14, when a peculiarly broad and continuous spectrum is demonstrated from a conventional quantum confined heterostructures utilizing only interband optical transitions. The effect of variation in each individual quantized energy state owing to large ensembles of noninteracting nanostructures with different sizes and compositions, in addition to self inhomogeneity broadening within each Qdot/Qdash ensemble, will contribute to active recombination and thus quasi-supercontinuum emission.

5. Conclusion

In conclusion, the unprecedented broadband laser emission at room temperature up to 76 nm wavelength coverage has been demonstrated using the naturally occurring size dispersion in self-assembled Qdash structure. The unique DOS of quasi-zero dimensional behavior from Qdash with wide spread in dash length, that gives different quantization effect in the longitudinal direction and band-filling effect, are shown as an important role in broadened lasing spectrum as injection level increases. After an intermediate degree of postgrowth interdiffusion technique, laser emission from multiple groups of Qdash ensembles in addition to multiple orders of subband energy levels within a single Qdash ensemble has been experimentally demonstrated. The suppression of laser emission in short wavelength and the progressive red-shift of peak emission with injection from devices with short cavity length indicate the occurrence of photon reabsorption or energy exchange among different sizes of localized Qdash ensembles. These results lead to the fabrication of the wavelength tuned quasi-supercontinuum interband laser diodes via the process of IFVD to promote group-III intermixing in InAs/InAlGaAs quantum-dash structure. Our results show that monolithically integration of different gain sections with different bandgaps for ultra-broadband laser is feasible via the intermixing technique.

6. Acknowledgement

This work is supported by National Science Foundation (Grant No. 0725647), US Army Research Laboratory, Commonwealth of Pennsylvania, Department of Community and Economic Development. Authors also acknowledge IQE Inc. for the growth of Qdash material, and D.-N. Wang and J. C. M. Hwang for the TEM work.

7. References

Alfano, R. R. & Shapiro, S. L. (1970). Emission in the region 4000 to 7000 Å via four-photon coupling in glass. *Phys Rev. Lett.*, Vol. 24, No. 11, (March 1970) 584-587

Bimberg, D.; Kirstaedter, N.; Ledentsov, N. N.; Alferov, Zh. I.; Kop'ev, P. S. & Ustinov V. M. (1997). InGaAs-GaAs quantum-dot lasers. *IEEE J. Sel. Top. Quantum Electron.*, Vol. 3, No. 2, (April 1997) 196-205

Crank, J. (1975). *The Mathematics of Diffusion*, Oxford University Press, 0198534116, Clarendon

Dery, H.; Benisty, E.; Epstein, A.; Alizon, R.; Mikhelashvili, V.; Eisenstein, G.; Schwertberger, R.; Gold, D.; Reithmaier, J. P. & Forchel, A. (2004). On the nature of quantum dash structures. *J. Appl. Phys.*, Vol. 95, No. 11, (June 2004) 6103-6111

Djie, H. S.; Dimas, C. E. & Ooi, B. S. (2006). Wideband quantum-dash-in-well superluminescent diode at 1.6 μm. *IEEE Photon. Technol. Lett.*, Vol. 18, No. 16, (August 2006) 1747-1749

[a]Djie, H. S.; Ooi, B. S.; Fang, X. –M.; Wu, Y.; Fastenau, J. M.; Liu, W. K. & Hopkinson, M. (2007). Room-temperature broadband emission of an InGaAs/GaAs quantum dots laser. *Opt. Lett.*, Vol. 32, No. 1, (January 2007) 44-46

[b]Djie, H. S.; Tan, C. L. ; Ooi, B. S.; Hwang, J. C. M.; Fang, X. –M.; Wu, Y.; Fastenau, J. M.; Liu, W. K.; Dang, G. T. & Chang, W. H. (2007). Ultrabroad stimulated emission from quantum-dash laser. *Appl. Phys. Lett.*, Vol. 91, No. 111116, (September 2007) 111116 1-3

Djie, H. S.; Wang, Y.; Ding, Y. H.; Wang, D. –N.; Hwang, J. C. M.; Fang, X. –M.; Wu, Y.; Fastenau, J. M.; Liu, A. W. K.; Dang, G. T.; Chang, W. H. & Ooi, B. S. (2008). Quantum dash intermixing. *IEEE J. Sel. Top. Quantum Electron.*, Vol. 14, No. 4, (July/August 2008) 1239-1249

Garbuzov, D.; Kudryashov, I. & Dubinskii, M. (2005). 110 W (0.9 J) pulsed power from resonantly diode-laser-pumped 1.6-μm Er:YAG laser. *Appl. Phys. Lett.*, Vol. 87, No. 121101, (September 2005) 121101 1-3

Gmachl, C.; Sivco, D. L.; Colombelli, R.; Capasso, F. & Cho, A. Y. (2002). Ultra-broadband semiconductor laser. *Nature*, Vol. 415, No. 6874, (February 2002) 883-887

Gontijo, I.; Krauss, T.; Marsh, J. H. & De La Rue, R. M. (1994). Postgrowth control of GaAs/AlGaAs quantum well shapes by impurity-free vacancy diffusion. *IEEE J. Quantum Electron.*, Vol. 30, No. 5, (May 1994) 1189-1195

Hadass, D.; Alizon, R.; Dery, H.; Mikhelashvili, V.; Eisenstein, G.; Schwertberger, R.; Somers, A.; Reithmaier, J. P.; Forchel, A.; Calligaro, M.; Bansropun, S. & Krakowski, M. (2004). Spectrally resolved dynamics of inhomogeneously broadened gain in InAs/InP 1550 nm quantum-dash lasers. *Appl. Phys. Lett.*, Vol. 85, No. 23, (December 2004) 5505-5507

Harris, L.; Mowbray, D. J.; Skolnick, M. S.; Hopkinson, M. & Hill, G. (1998). Emission spectra and mode structure of InAs/GaAs self-organized quantum dot lasers. *Appl. Phys. Lett.*, Vol. 73, No. 7, (August 1998) 969-971

Jiang, H. & Singh, J. (1999). Nonequilibrium distribution in quantum dots lasers and influence on laser spectral output. *J. Appl. Phys.*, Vol. 85, No. 10, (May 1999) 7438-7442

Kovsh, A.; Krestnikov, I.; Livshits, D.; Mikhrin, S.; Weimert, J. & Zhukov, A. (2007). Quantum dot laser with 75 nm broad spectrum of emission. *Opt. Lett.*, Vol. 32, No. 7, (April 2007) 793-795

Lelarge, F.; Dagens, B.; Renaudier, J.; Brenot, R.; Accard, A.; Dijk, F. V.; Make, D.; Gouezigou, O. L.; Provost, J. G.; Poingt, F.; Landreau, J.; Drisse, O.; Derouin, E. ; Rousseau, B. ; Pommereau, F. & Duan, G. H. (2007). Recent advances on InAs/InP quantum dash based semiconductor lasers and optical amplifiers operating at 1.55 μm. *IEEE J. Sel. Top. Quantum Electron.*, Vol. 13, No. 1, (January/February 2007) 111-124

Lin, C. & Stolen, R. H. (1976). New nanosecond continuum for excited-state spectroscopy. *Appl. Phys. Lett.*, Vol. 28, No. 4, (February 1976) 216-218

Markus, A.; Chen, J. X.; Paranthoen, C.; Fiore, A.; Platz, C. & Gauthier-Lafaye, O. (2003). Simultaneous two-state lasing in quantum-dot lasers. *Appl. Phys. Lett.*, Vol. 82, No. 12, (March 2003) 1818-1820

Markus, A.; Rossetti, M.; Calligari, V.; Chek-Al-Kar, D.; Chen, J. X.; Fiore, A. & Scollo, R. (2006). Two-state switching and dynamics in quantum dot two-section lasers. *J. Appl. Phys.*, Vol. 100, No. 113104, (December 2006) 113104 1-5

Morioka, T.; Mori, K. & Saruwatari, M. (1993). More than 100-wavelength-channel picosecond optical pulse generation from single laser source using supercontinuum in optical fibres. *Electron. Lett.*, Vol. 29, No. 10, (May 1993) 862-864

Matthews, D. R.; Summers, H. D.; Smowton, P. M. & Hopkinson, M. (2002). Experimental investigation of the effect of wetting-layer states on the gain-current characteristics of quantum-dot lasers. *Appl. Phys. Lett.*, Vol. 81, No. 26, (December 2002) 4904-4906

Nisoli, M.; De Silvestri, S. & Svelto, O. (1996). Generation of high energy 10 fs pulses by a new pulse compression technique. *Appl. Phys. Lett.*, Vol. 68, No. 20, (May 1996) 2793-2795

Ooi, B. S.; Mcllvaney, K.; Street, M. W.; Helmy, A. S.; Ayling, S. G.; Bryce, A. C.; Marsh, J. H. & Roberts, J. S. (1997). Selective quantum-well intermixing in GaAs-AlGaAs structures using impurity-free vacancy diffusion. *IEEE J. Quantum Electron.*, Vol. 33, No. 10, (Oct 1997) 1784-1793

Ooi, B. S.; Djie, H. S.; Wang, Y.; Tan, C. L.; Hwang, J. C. M.; Fang, X. –M.; Fastenau, J. M.; Liu, A. W. K.; Dang, G. T. & Chang W. H. (2008). Quantum dashes on InP substrate for broadband emitter applications. *IEEE J. Sel. Top. Quantum Electron.*, Vol. 14, No. 4, (July/August 2008) 1230-1238

Popescu, D. P. & Malloy, K. J. (2006). Anisotropy of carrier transport in the active region of lasers with self-assembled InAs quantum dashes. *IEEE Photon. Technol. Lett.*, Vol. 18, No. 22, (November 2006) 2401-2403

Ranka, J. K.; Windeler, R. S. & Stentz, A. J. (2000). Visible continuum generation in air-silica microstructure optical fibers with anomalous dispersion at 800 nm. *Opt. Lett.*, Vol. 25, No. 1, (January 2000) 25-27

Sek, G.; Poloczek, P.; Podemski, P.; Kudrawiec, R.; Misiewicz, J.; Somers, A.; Hein, S.; Hofling, S. & Forchel, A. (2007). Experimental evidence on quantum well-quantum dash energy transfer in tunnel injection structures for 1.55 μm emission. *Appl. Phys. Lett.*, Vol. 90, No. 081915 (February 2007) 081915 1-3

Shoji, H.; Nakata, Y.; Mukai, K.; Sugiyama, Y.; Sugawara, M.; Yokoyama, N. & Ishikawa, H. (1997). Lasing characteristics of self-formed quantum-dot lasers with multistacked dot layer. *IEEE J. Sel. Top. Quantum Electron.*, Vol. 3, No. 2, (April 1997) 188-195

Tan, C. L.; Wang, Y.; Djie, H. S. & Ooi, B. S. (2007). Role of optical gain broadening in the broadband semiconductor quantum-dot laser. *Appl. Phys. Lett.*, Vol. 91, No. 061117, (August 2007) 061117 1-3

Tan, C. L.; Djie, H. S.; Wang, Y.; Dimas, C. E.; Hongpinyo, V.; Ding, Y. H. & Ooi, B. S. (2008). Wavelength tuning and emission width widening of ultrabroad quantum dash interband laser. *Appl. Phys. Lett.*, Vol. 93, No. 111101, (September 2008) 111101 1-3

Tan, C. L.; Djie, H. S.; Wang, Y.; Dimas, C. E.; Hongpinyo, V.; Ding, Y. H. & Ooi, B. S. (2009). The influence of nonequilibrium distribution on room-temperature lasing spectra in quantum-dash lasers. *IEEE Photon. Technol. Lett.*, Vol. 21, No. 1, (January 2009) 30-32

Van der Poel, M.; Mork, J.; Somers, A.; Forchel, A.; Reithmaier, J. P. & Eisenstein, G. (2006). Ultrafast gain and index dynamics of quantum dash structures emitting at 1.55 µm. *Appl. Phys. Lett.*, Vol. 89, No. 081102, (August 2006) 081102 1-3

Wang, R. H.; Stintz, A.; Varangis, P. M.; Newell, T. C.; Li, H.; Malloy, K. J. & Lester, L. F. (2001). Room-temperature operation of InAs quantum-dash lasers on InP (001). *IEEE Photon. Technol. Lett.*, Vol. 13, No. 8, (August 2001) 767-769

Wang, Y.; Djie, H. S. & Ooi, B. S. (2006). Group-III intermixing in InAs/InGaAlAs quantum dots-in-well. *Appl. Phys. Lett.*, Vol. 88, No. 111110, (March 2006) 111110 1-3

Wei, J. H. & Chan, K. S. (2005). A theoretical analysis of quantum dash structures. *J. Appl. Phys.*, Vol. 97, No. 123524, (June 2005) 123524 1-12

Xing, C. & Avrutin, E. A. (2005). Multimode spectra and active mode locking potential of quantum dot lasers. J. Appl. Phys., Vol. 97, No. 104301, (April 2005) 104301 1-9

Zheltikov, A. M. (2003). Supercontinuum generation: Special issue. *Appl. Phys. B*, Vol. 77, No. 2-3, (September 2003) 143-376

Zhukov, A. E.; Kovsh, A. R.; Ustinov, V. M.; Egorov, A. Y.; Ledentsov, N. N.; Tsatsulnikov, A. F.; Maximov, M. V.; Kopchatov, V. I.; Lunev, A. V.; Kopev, P. S.; Bimberg, D. & Alferov, Zh. I. (1999). Gain characteristics of quantum dot injection lasers. *Semicond. Sci.Technol.*, Vol. 14, No. 1, (January 1999) 118-123

2

Photonic Quantum Ring Laser of Whispering Cave Mode

O'Dae Kwon, M. H. Sheen and Y. C. Kim
Pohang University of Science & Technology
S. Korea

1. Introduction

In early 1990s, an AT&T Bell Laboratory group developed a microdisk laser of thumb-tack type based upon Lord Rayleigh's 'concave' whispering gallery mode (WGM) for the optoelectronic large-scale integration circuits (McCall et al., 1992). The above lasers were however two dimensional (2D) WGM which is troubled with the well-known WGM light spread problem. For the remedy of this problem, asymmetric WGM lasers of stadium type (Nockel & Stone, 1997) were then introduced to control the spreading light beam. Quite recently, a novel micro-cavity of ***limaçon*** shape has shown the capability of highly directional light emission with a divergence angle of around 40-50 degrees, which is a big improvement to the light spreading problem.(Wiersig & Hentschel, 2008)

On the other hand, when we employ a new micro-cavity of vertically reflecting distributed Bragg reflector (DBR) structures added below and above quantum well (QW) planes, say a few active 80Å (Al) GaAs QWs, a 3D toroidal cavity is formed giving rise to helix standing waves in 3D whispering cave modes (WCMs) as shown Fig. 1 (Ahn et al., 1999). The photonic quantum ring (PQR) laser of WCMs is thus born without any intentionally fabricated ring pattern structures, which will be elaborated later. The PQR's resonant light is radiating in 3D but in a surface-normal dominant fashion, avoiding the 2D WGM's in-plane light spread problem.

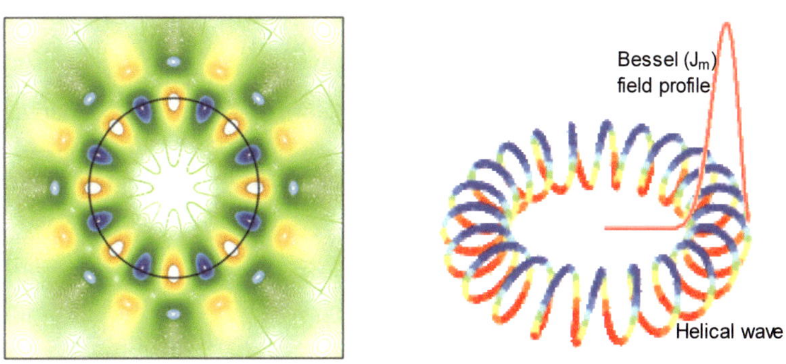

Fig. 1. Planar 2D Bessel function WGMs vs. toroidal 3D knot WCM (Park et al., 2002). The 3D WCM is a toroid with a circular helix symmetry not reducible to the simple 2D rotational symmetry

2. Basic properties of PQR lasers

The 3D WCM laser of PQR, whose simulation work will be shown later, behaves quite differently due to its quantum wire-like nature as follows: First of all, the PQR exhibit ultra-low threshold currents - for a mesa-type PQR device of 15 um diameter, the PQR at the peripheral Rayleigh band region lases with about one thousandth of the threshold current needed for the central vertical cavity surface emitting laser (VCSEL) of the same semiconductor mesa as illustrated in Fig. 2.

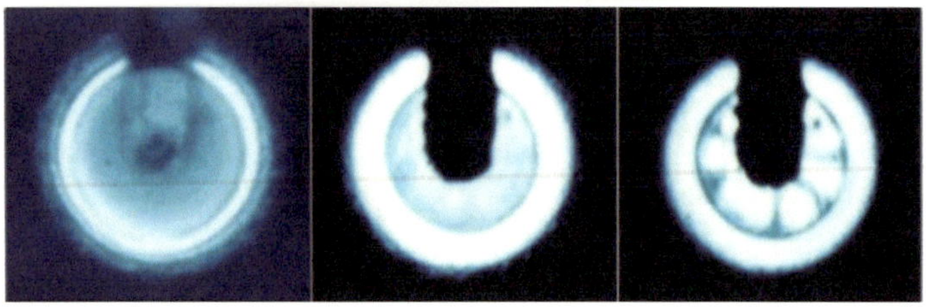

Fig. 2. CCD pictures of emisssions at 12 μA, near PQR threshold, at 11.5 mA, below VCSEL threshold, and at 12.2 mA, above VCSEL threshold, respectively.

We can however make theoretical formulae consistent with above concentric PQRs and do some calculations for comparing with the transparency and threshold current data observed. The PQR formulae can be derived by assuming that the pitch of concentric rings is 'photonic' kind of one half wavelength - optical λ/2 period: The transparency (I_{tr}: curve T) and threshold (I_{th}: curve A) current expressions for the case of PQRs occupying the annular Rayleigh region is now given by (1).

$$I_{th} = I_{tr} + I_i = N^{1D} \times W_{Rayleigh} / (\lambda / 2n_{eff}) \times \pi\phi \times (e / \eta\tau) + I_i \tag{1}$$

N^{1D} is the 1D transparency carrier density, τ the carrier lifetime, η the quantum efficiency, and I_i stands for internal loss (Ahn et al., 1999; Kwon et al., 2006). The PQR formulae are now compared with the actual data in Fig. 3, which show quite an impressive agreement except some random deviations due to device imperfections. For smaller diameters (ϕ) the active volume decreases below 0.1 μm^3, and with the cavity Q factor over 15,000. The corresponding spontaneous emission coefficient β will become appreciable enough for threshold-less lasing without a sharp turn-on threshold, which often occurs in the PQR light-current analyses. As listed in Fig. 3, the wide-spread data suggest a fuzzy ring trend growing as the device shrinks due to the growing leaky implantation boundary around the implant-isolated holes, and the hole PQR threshold data are actually approaching the curve B, whose formula is derived for the mesa by assuming that the Rayleigh region is now nothing but a piece of annular quantum well plane of random recombinant carriers instead:

$$I = N^{2D} \times W_{Rayleigh} \times \pi\theta \times (e / \eta\tau) \tag{2}$$

Figure 4 shows a collection of linewidth data being roughly inversely proportional to the device size as expected. The narrowest linewidth observed with an optical spectrum analyzer to date from a 10 um PQR is 0.55 Å at an injection current of 800 uA. We also note that with wet etching steps employed instead of dry etching, the Q factor reached up to 20,000 while the linewidth approached 0.4 Å (M. Kim et al., 2004). Although we did not attempt it for GaAs, a CALTECH group devised a laser baking process for achieving ultrahigh Q values of multi-millions involving a SiO2 microcavity. It is interesting to be a toroidal microcavity whose 3D WCM properties is unknown yet (Armani et al., 2003; Min et al., 2004).

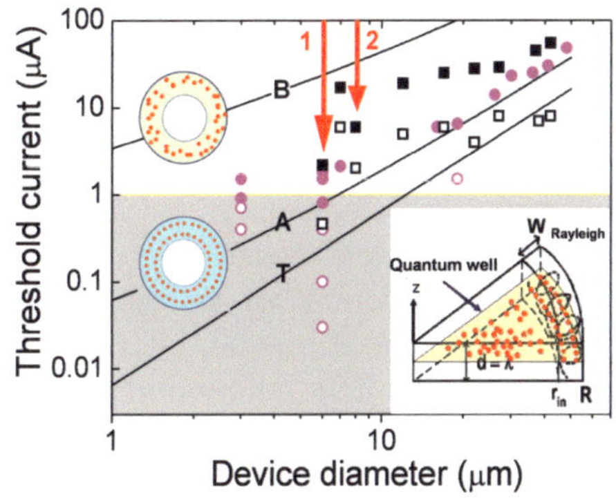

Fig. 3. Threshold curves A and B from PQR and quantum well formulae, respectively, with corresponding Rayleigh toroid schematics (defined by Rayleigh width between rin and R) and transparency curve T for the PQR case. Data for transparency (empty symbols) and threshold (solid symbols) currents: circles for PQRs and squares for PQR holes implant isolated. Data at 6 and 8 μm correspond to the case of 256×256 hole arrays without implantation (see the arrows 1 and 2).

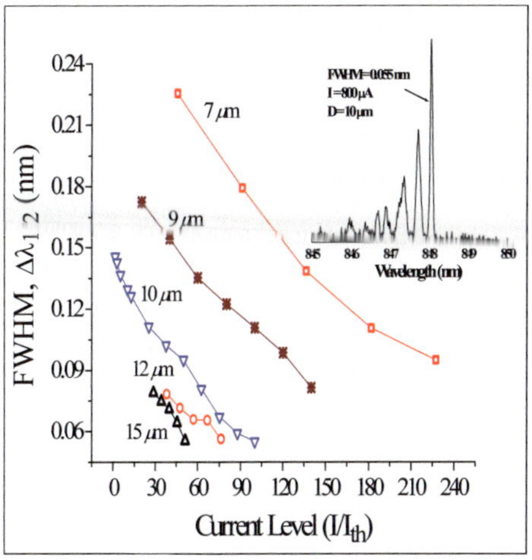

Fig. 4. Linewidth data vs. current s with various device sizes

Now we figure that the helical WCM standing wave manifold transiently induces concentric PQRs for imminently recombinant carriers present in the Rayleigh region $W_{Rayleigh}$ of the 2D quantum well. This in turn exhibits extremely small thresholds in the the μA-to-nA range with the given $\sqrt{T}$ -dependent thermal stabilities. It is attributed to a photonic (de Broglie) quantum corral effect, similar in character to the well-known electronic quantum corral image from room temperature scanning tunneling microscope studies of Au atomic island plane at a given bias.

The photonic (de Broglie) quantum corral effect imposes a $\lambda/2$ period transient ordering upon the imminently recombinant carriers, although the optical $\lambda/2$ period for GaAs semiconductor will be substantially larger than the electronic de Broglie spacing. We note that the Rayleigh region of quantum well planes is deeply buried beneath a few micron thick AlAs/GaAs Bragg reflectors not accessible for direct observation. However, recent experiments and modeling work on dynamic interactions between carriers and transient field in a quantum well plane is a close case in point (Gehrig & Hess, 2004). It thus appears that the transient quantum wire-like features considered here seem to persist within the relevant time scale through thermal fluctuations. For an ensemble of carriers randomly distributed in the regional quantum well plane of concentration 10^{12} cm^{-2} for instance, tens-of-nm scale local field-driven drifts of given carriers to a neighboring imminent PQR site should generate the proposed PQR ordering for an imminent recombination event of annihilating electron-hole pairs. For example, one can imagine a transient formation of the two separate Rayleigh rings instantly via light field-induced migration of random carriers within the $W_{Rayleigh}$ region as schematically shown for curve A in Fig. 3. We expect the standing waves in the Rayleigh region to give rise to a weak potential barrier for such a dynamic electron-hole pair process, perhaps an opposite case of extremely shallow quantum well excitons at room temperature where even the shallow barriers tend to assure at least one bound state according to square well quantum mechanics.

3. Spatio-temporal dynamic simulation of PQR standing waves and carriers

Although it is limited to 2D cases, recent spatiotemporal dynamic simulation work in a straight waveguide case (see Fig.5) faithfully reveals such a tangled but otherwise quantum-wire-like ordering of recombinant carriers undergoing some picosecond-long exciton process, consistent with the photonic quantum corral effect due to a strong carrier-photon coupling. The images of several standing light-wave-like carrier distribution patterns within a 1 micron wide quantum well stripe emerge, as a function of time from-5-to-8 psec after about 5 psec chaotic regime as indicated along the horizontal time axis of 10 psec full range, shown in Fig. 6 (Kwon et al., 2009). They are curiously reminiscent of the tangled web of the 2D electron gas due to impurity atom potentials studied by a Harvard group (Topinka et al., 2003).
The assumed concentric quantum ring pattern of carrier distribution within the Rayleigh region is not observable directly since they are buried below a few micron thick top DBR structures. Instead the CCD pictures are their distant images refracted and smeared out through the semiconductor medium.
As said before, the resonance of the PQR laser results in 3D WCM of helical standing waves, which is surface-normal dominant, in contrast to the in-plane 2D WG mode. The data taken with a home-built solid angle scanner setup, which will be discribed later, shows a tangential polarization dominance which supports strong carrier-photon couplings behaviors needed for the PQR formation (Kim et al., 2007)

4. 3D WCM mode analysis and single mode PQR laser

A 3D WCM mode analysis, based upon the helix mode of the PQR consisting of a bouncing wave between the two DBRs and a circulating wave of in-plane total reflection, gives an angular quantization rule for easy PQR mode analysis of 3D spectra taken with tapered single mode fiber probes as shown in Fig. 7 (Bae et al., 2003).

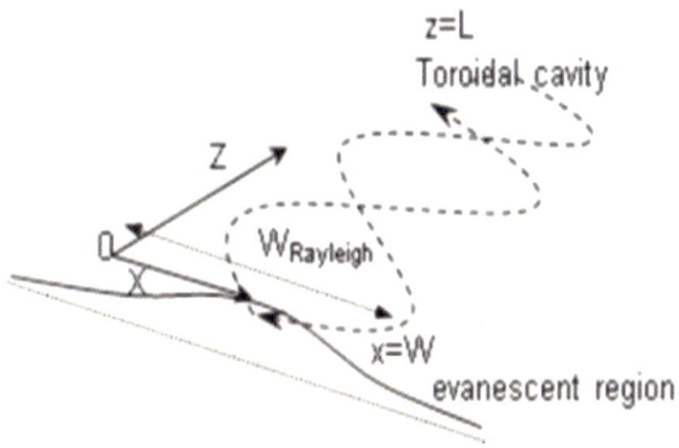

Fig. 5. Flattened top view of helix modes within a Rayleigh bandwidth

Fig. 6. Spatiotemporal 2D simulation results: top –standing waves are formed after a few picoseconds of chaotic regime in the case of flattened and straight rectangular wave guide version [x-axis span of 10 psec.]; bottom – carrier distribution dynamics shown for 10 picoseconds, where similar patterns emerge after a few psec. Y-axis indicates a 1 um wide central waveguide in the middle of 3 um boundary.

For single mode lasers we have made non-conventional PQRs of hyperboloid drum shape like Figs. 8 (a) and (b) (Kim et al., 2003) having a submicron active diameter with $\phi = 0.9\ \mu m$, where as its top region of a few micron diameter serves as metallic contact area for electro pumping. Figs. 8 (c) and (d) show the threshold data with a 0.46 $\AA$ linewidth exhibit the smallest threshold of about 300 nA, (Yoon et al., 2007) observed so far among the injection lasers of quantum well, wire, or dot type to the best of our, although the external quantum efficiency observed right after the threshold is poor suffering from the soft lasing turn-on behavior here.

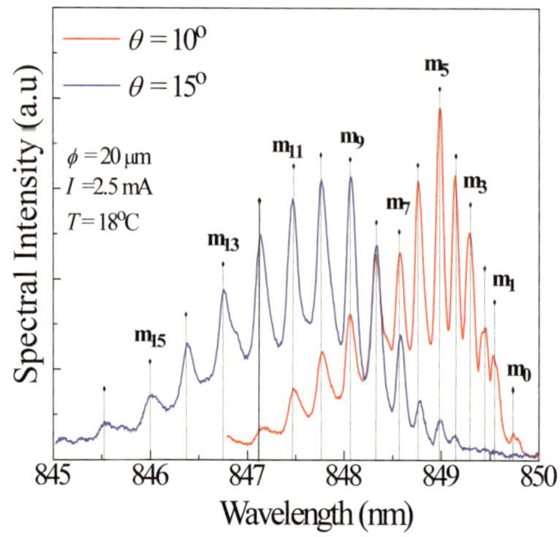

Fig. 7. Angular measurement set up for 3D WCM and some typical spectra

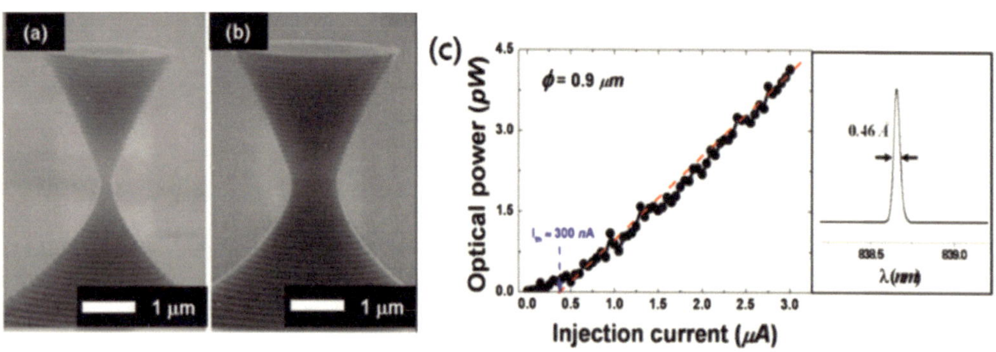

Fig. 8. Hyperboloid drum PQR: SEM micrograph, L-I curve, and single mode spectrum

5. Mega-pixel laser chips of photonics quantum ring holes

We have succeeded in fabricating the high density array chip of PQR hole lasers of one mega (M) integration. 1M PQR hole array chips has ultra low threshold current of 0.736 nA per single hole due to photonic crystal-like cooperative effect (Kwon et al., 2008) 1M PQR hole laser array chip is fabricated in tandem type with four 256K PQR hole arrays for uniformly injecting current on the device surface. The used epitaxial wafer structure of a p-*i*(MQW: multi quantum well)-n diode was grown on an n-type GaAs (001) substrate by metal-organic vapor-phase epitaxy. The structure consists of two distributed Bragg reflector (DBR) mirrors surrounding the *i*-region of a one-λ cavity active region (269.4 *nm* thick) including three GaAs/$Al_{0.3}Ga_{0.7}As$ quantum well structures, tuned to yield a resonance wavelength of 850 *nm*. The p- and n- type DBR mirrors consist of alternating 419.8 *Å* $Al_{0.15}Ga_{0.85}As$ and 488.2 *Å* $Al_{0.95}Ga_{0.05}As$ layers, 21.5 periods and 38 periods respectively. Figures 9(a) and (b) show scanning electron microscopy (SEM) images for top view and cross section of 1*M* PQR hole laser array, respectively, whose SEM pictures exhibit a bit rough cross section as compared with single device side walls in Figs. 9(c) and (d).

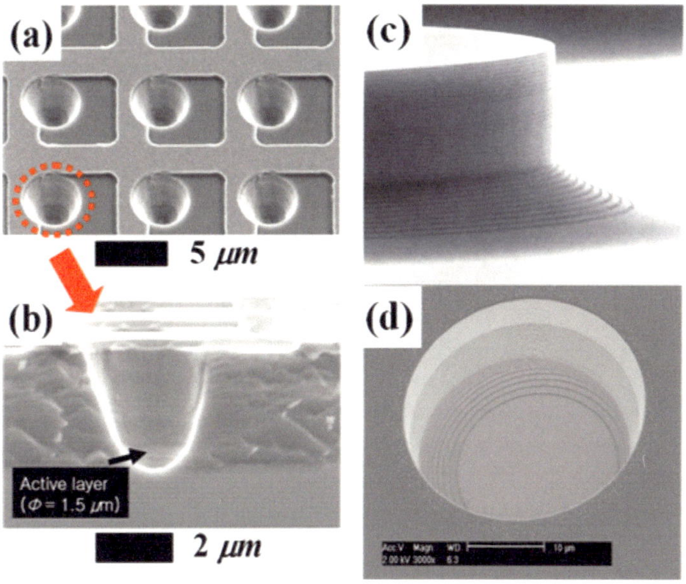

Fig. 9. (a) Top and (b) cross section SEM images of 1M PQR hole array (c) SEM micrographs of mesa and hole type PQR structures.

Figure 10(a) shows the CCD images of the illuminant 1M PQR hole array near the transparent current, 0.08 *A* (80 *nA/cell*) and near the threshold current, 0.7 *A* (700 *nA/cell*).
To measure the *L-I* curve for 1M PQR hole array, we used a conventional power meter (Adventest Mo.Q211) and measured directly 1M PQR hole array. For measurement of threshold current and angle-resolved spectra shown in Fig. 11, we used a piece of 1/32M PQR hole array, because the total size of 1M PQR hole array chip is 1 cm^2 which is larger than the aperture size (diameter = 0.8 *cm*) of the power meter.

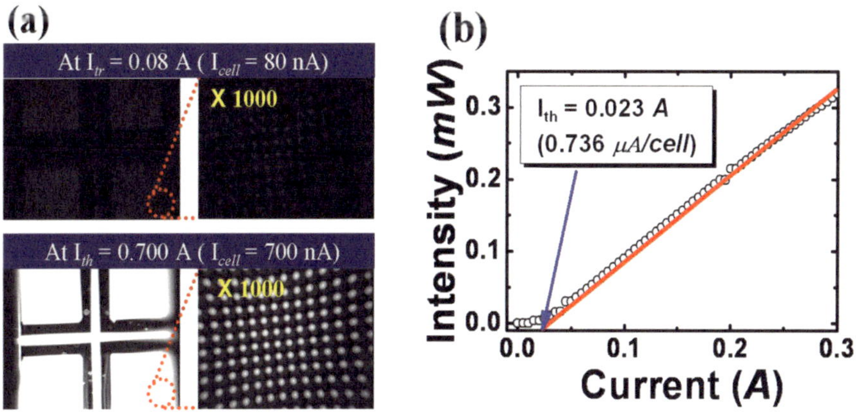

Fig. 10. (a) CCD (right) and 1000 times magnified (left) images of the illuminant 1M PQR hole array (4x250K arrays) at transparent and near threshold current. (b) *L-I* curve of 1/32M PQR hole array chip. As shown in Fig. 2(b), the threshold current is measured 0.736 *μA/hole* by using linear fitting.

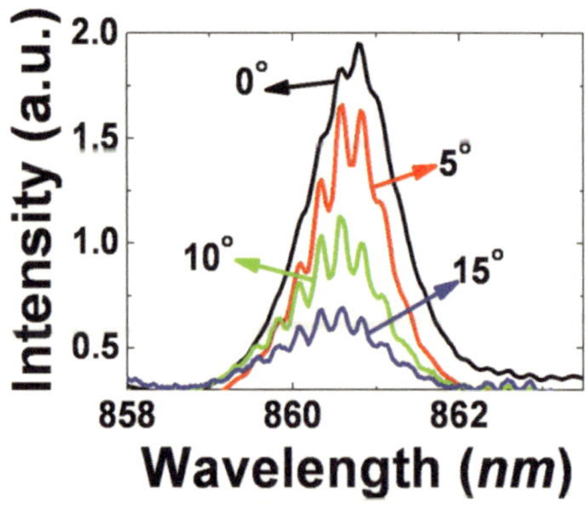

Fig. 11. (color online) Angle-resolved spectra of single hole among 1M PQR hole array at 32 *μA/hole.*

6. PQR light sources for display

We now discuss the properties of the PQR technology applicable for the next generation display. Light-emitting diodes (LEDs) display has become a multi-million dollar industry, and it is growing. LEDs are under intensive development worldwide for advanced display applications (Schubert, 2003)

However, high- power LEDs being bulk devices faces problems like the notorious LED extraction factor associated with internal heating problems, large concentration of impurity scatters, and low modulation frequencies less than MHz ranges. Although the LED performances are improving, lasers can be the alternative answer with the usual GHz range modulation capability. In particular, the PQR laser is an attractive candidate for next generation display, based upon the special PQR characteristics as explained in the preceding sections like extremely low threshold currents, thermally stable spectra, and high-density chip capabilities. The PQR of WCMs can have both concave and convex modes, which are the fundamental properties exploited for fabricating high power flower type PQR lasers as elaborated in the end for display applications.

The high power PQR laser properties will now be presented to compare with conventional LEDs, in terms of properties such as power-saving features, color purity, luminous efficiency, and beam shape properties:

The spectral data for a conventional LED has a linewidth of about 25 nm which may be reduced further down to several nm in the case of resonant cavity LEDs, while the linewidth of the PQR is usually around or below 0.1 nm, as illustrated in Fig. 12, the spectra for a PQR of ϕ = 7 μm. Namely, if the linewidths of the PQR and LED are about 0.1 and 25 nm, respectively, the electric power consumption of the PQR is about 1/250 of the LED power consumption. It means that the low threshold current and sharp discrete mode PQRs offer high brightness as LED with much less amount of electric current because the sum of each sharp peak can replace the broad peak of LED spectrum. The PQR's color purity is about 1 which means high color rendering ability.

Fig. 13(a) shows the emission image of the 16x16 mesa type red PQR laser array. A single red PQR emission reveals two different regions at a given injection current (I=24uA/cell). The PQR lasing occurs in the periphery of the active disk called the Rayleigh band and the

LED emission occurs in the middle part of the disk. Luminous efficiency of the 16x16 red PQR array is 7.20lm/w at the 670nm wavelength, which, if translated to 620nm with the color conversion factor multiplied, becomes two times better than the commercial 620nm LED products as shown in Fig. 13(b).

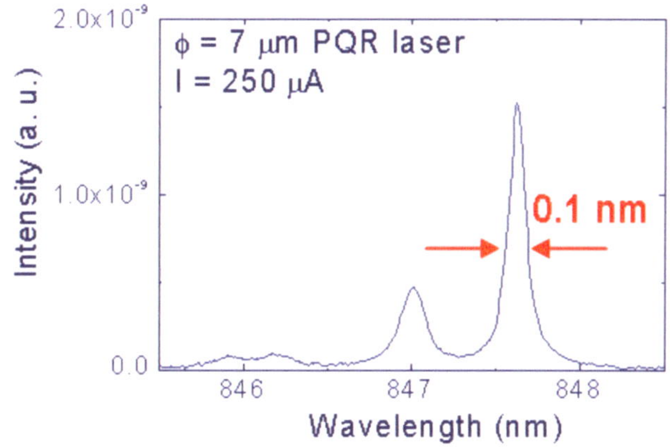

Fig. 12. Spectrum of ϕ = 7 μm PQR laser.

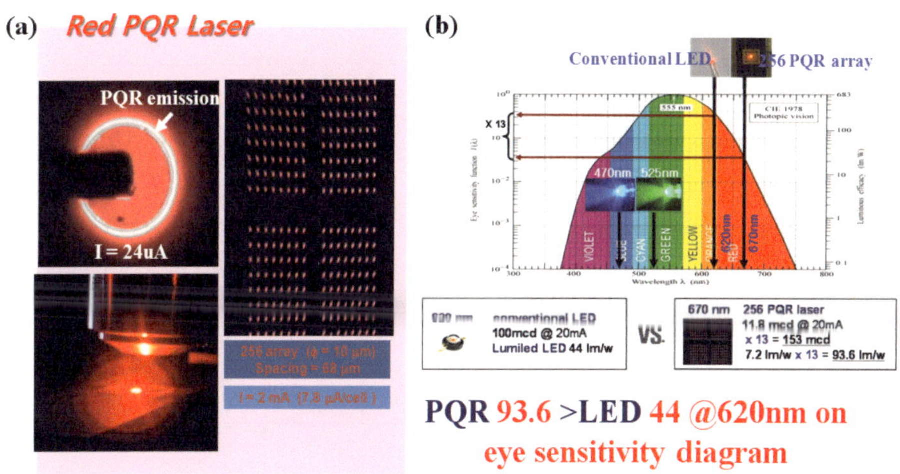

Fig. 13. Photometric characteristics (PQR vs LED) (a) Emission image of the 16x16 red PQR laser array (Φ= 7um, pitch = 68um). (b) Comparison of the photometric characteristics between 16x16 red PQR array and conventional high power LED.

Blue GaN surface-emitting lasers are notoriously difficult to fabricate and we give a couple of recent examples of GaN surface-emitting laser work: First, a photonic crystal based surface emitting laser was developed Japanese researchers where their photonic crystal structure consists of a 2 dimensional array of airholes. Their result is however far from practical applications. The threshold current obtained was rather large as 6.9A in pulsed mode operation (Yoshimoto et al., 2008)

A Taiwanese group also reported GaN hybrid VCSEL laser work where they used n type crack- free AlN/GaN DBR and Ta_2O_5/SiO_2 dielectric DBR. Still, the operation was at liquid

nitrogen temperature (77K) (Lu et al., 2008). Practical GaN VCSEL lasers thus seem very hard to achieve CW at room temperature.
On the other hand, we are making the blue PQR lasers which is CW operated at room temperature lasing in 3D but emitting dominantly in surface normal direction. Our blue PQR lasers with wavelengths between 420 and 470nm are fabricated using a GaN wafer with sapphire substrates removed via laser lift-off (LLO) procedures (Fig. 14).
The multi mode lasing spectra from the blue PQR as shown in Fig. 15 and this tentative result was reported in the reference (Kim et al., 2006).

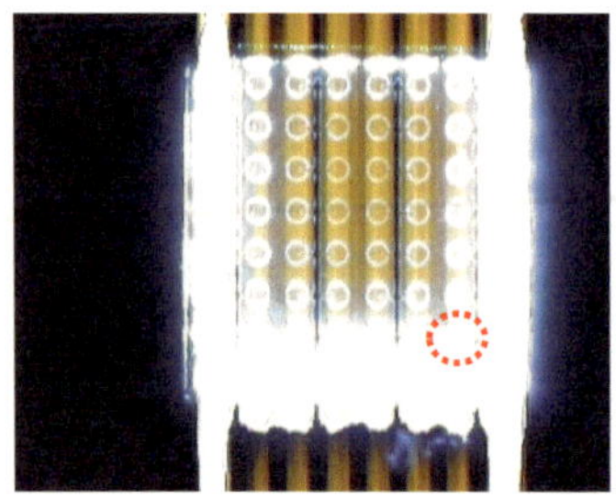

Fig. 14. Blue PQR array with the edge region affected by spontaneous background emission (in red circle).

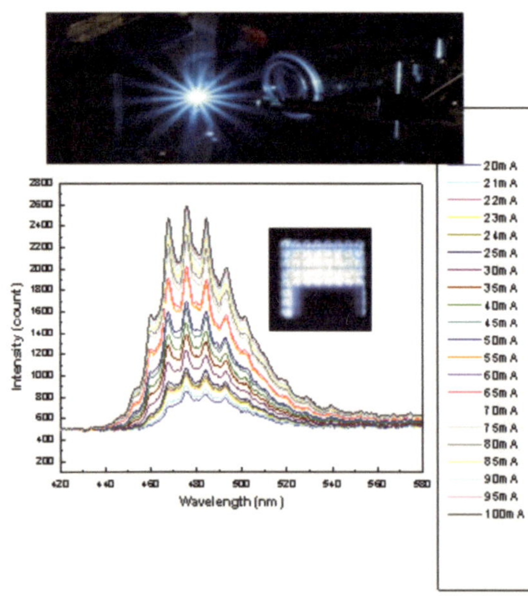

Fig. 15. Multimode spectra from a blue PQR (in red circle in Fig. 14) CW at room temperature with I = 60uA/cell to 1.63mA/cell.

7. PQR laser beam propagation characteristics

For 3D beam profile studies, we have used a home-built 2D/3D single photon scanning system for measuring the PQR beam profile and polarization with a resolution of 0.5μm/step.

As shown in the schematic Fig. 16, a tapered single mode fiber tip about 300nm in diameter was made by chemical etching for the photon collection, and a step motor generates relative motions of the tip against probed PQR laser device. The collected photon signal goes through single photon counting module, photon counter, and computer.

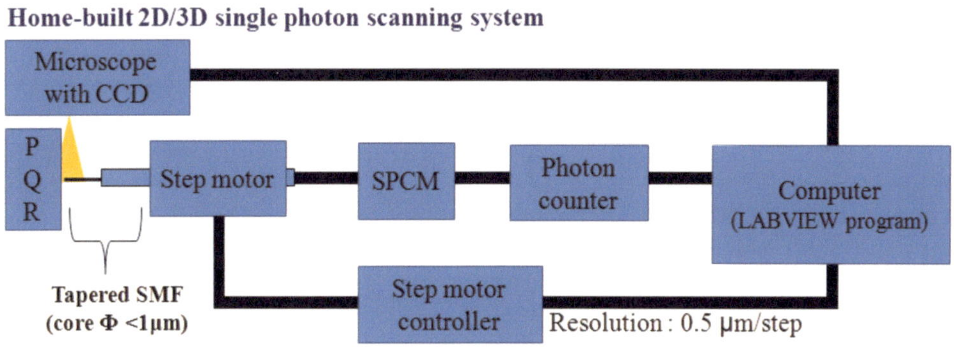

Fig. 16. A schematic diagram of home-built 2D/3D single photon scanning system.

Figure 17 shows some 2D scan results over a scan area of 60x60 um square, where, on the surface of the PQR, Fig. 17(a) exhibits that the emission pattern of the PQR beam is Laguerre Gaussian for the case of a mesa PQR, and Fig. 17(b) shows another Laguerre Gaussian pattern for the case of hole PQRs.

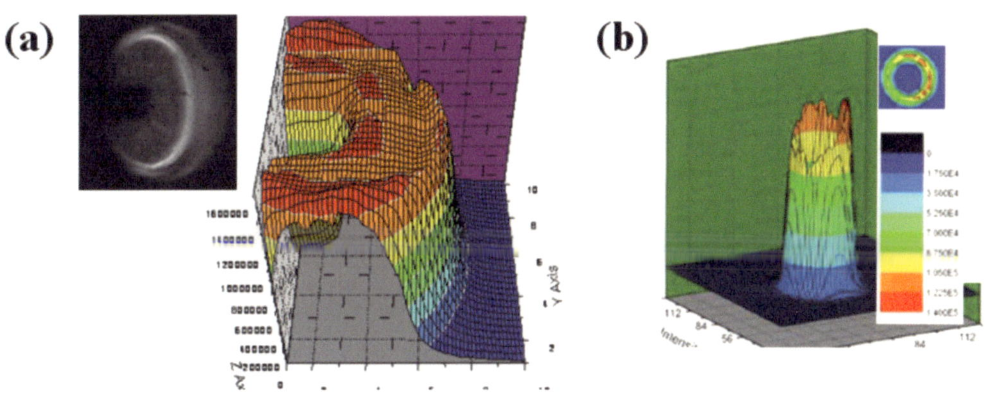

Fig. 17. (a) Lagurre Gaussian beam of the mesa PQR (b) Lagurre Gaussian beam of the hole PQR

8. Fabrication of micro collimators for PQR beam guiding

Laser printers with mechanically rotating polygon mirrors have been used widely in offices, whereas new LED printers, quiet and all-electronic drive circuitry with no moving parts, begin to replace them. However, the LED, being a spontaneous emission device with some disadvantages as stated earlier, can further be replaced by an efficient laser like the PQR laser diode with extremely low threshold currents and $\sqrt{T}$ -dependent thermally stable spectral properties which are good for fast, high density array applications. Moreover, typical LED printers use selfoc-lens arrays (SLAs) to concentrate and guide individual light, but the expensive SLA technology is complicated. We note that the PQR laser with the micro collimator (MC) for non-parallel to parallel beam guiding described previously may replace the LED + SLA technology. In order to find such a possibility, we will now describe several fundamental features of the PQR laser such as the beam shape and propagation behaviors, MC-guided PQR beams, beam divergence, and high power capabilities.

For beam divergence studies, Fig. 18(a) represents a PQR emission pattern observed from a device of a 48um diameter which is rather close to the Lambertian emission pattern of a conventional LED. However noting that the Gaussian beam is characterized by the spot size and divergence angle θ, recent 3D PQR beam profile studies of 15um PQR lasers also show possibilities of controlling the beam divergence to the narrower ranges, for example a divergence angle of θ = 2 x 6.3 degree as shown in Fig. 18(b). This analysis results from the 3D scans made at 30, 60 and 90 um heights respectively as shown in Figs. 18(c),(d) and (e), where divergence points are determined as half maximum intensity points. We find from the 3D scans that the initial beam profile of Laguerre Gaussian is evolving to Gaussian as a function of scan height. The beam shapes are nearly Gaussian at 30um height and perfectly Gaussian at 60um height, which gives rise to a cross-over from Laguerre Gaussian to Gaussian at around 40~50um height. In our divergence analysis we may regard the PQR ring as a Bessel beam formed at the rim of the PQR device surface from an imaginative point light at the origin located deep below the device surface.

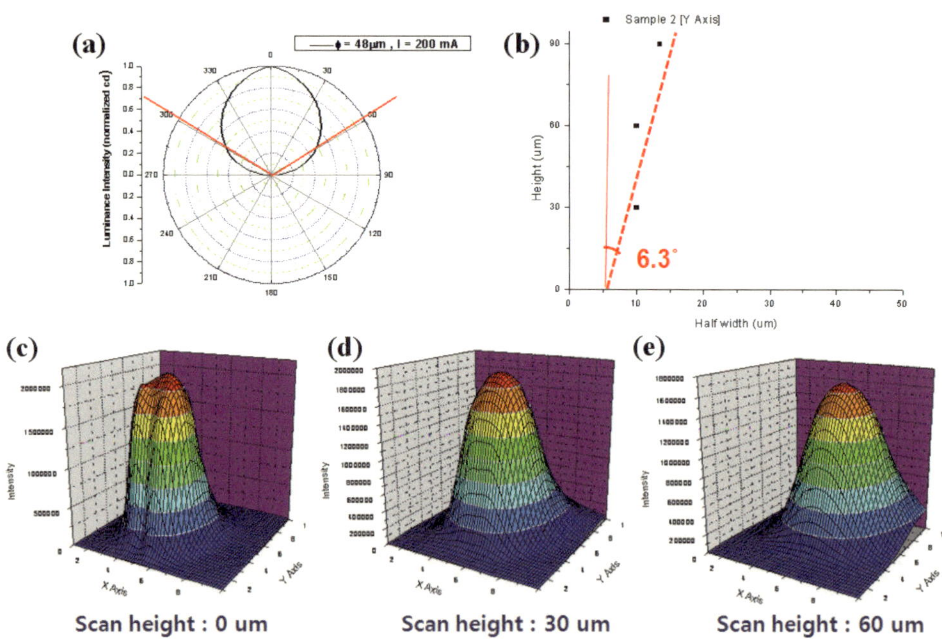

Fig. 18. (a) Lambertian profile of a PQR (b) Divergence angle of a PQR (c) (d) and (e) represent beam scans taken at different heights, 0, 30, and 60 um

For practical system applications of light sources one often has to find how to control, or focus and guide, the laser beam. Therefore we now turn to an active beam control method employing convex and concave MCs for focusing and guiding the PQR light through lens media and free space. The convex and concave lenses of the MC are designed and fabricated as shown in Figs. 19(a) and (b).

A master lens array is made by a photoresist (PR) reflow method, and the PR microlens array is transferred to a polydimethyl-siloxane (PDMS) master by a casting method. Finally, the PDMS is spin-coated again on the PDMS master, whose details are described (O'Neill & Sheridan, 2002). Fig.19(a) are SEM images of the final micro lens arrays fabricated to be 17um in diameter and 10um in height for the convex lenses (top) and 36um in diameter for the concave lenses (bottom).

Fig. 19(c) represents a series of CCD snap shots taken at various distances from the PQR laser surface where the microlens set on the fifth spot happens to be absent. The snap shots vividly shows that the propagating Gaussian beam is guided to the point of minimum spot at 160um distance and reconstructs the original PQR laser image at around 400um distance. The fact that the missing 5th spot is not affected by any possible neighbor's diffraction ghost means that the PQR beam behaves as a Bessel beam.

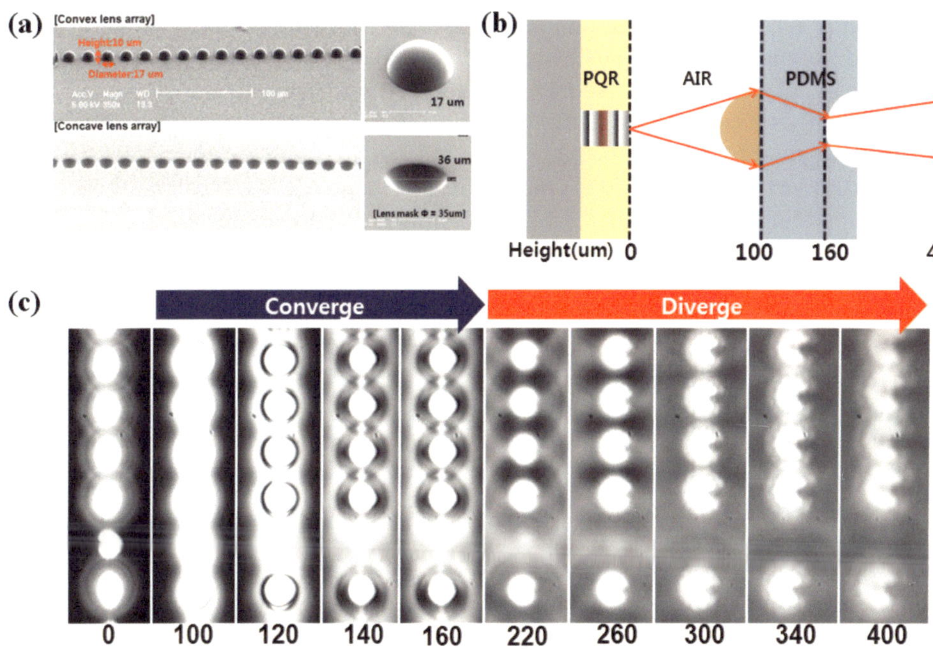

Fig. 19. (a) Convex and concave lens arrays. (b) Outline of beam guiding optics. (c) CCD snap shots taken at different distances.

9. Design and SEM images of flower PQR laser

We now describe the design and fabrication of the new flower PQR laser for output power enhanced about 5 times the power expected from regular circular PQR lasers of the same size, where 4, 8, and 12 -petal flower designs, combining concave and convex whispering cave modes, result in the increased overall quantum wire length of the emitting PQR within the same device area.

As shown earlier in Fig. 13(a), the PQR region emitted first and much brighter than the central LED emission region, which means a very high emission efficiency of the PQR laser. We however note that the emission region is occupied mostly by the central LED emission

in this case. That is the reason why we make use of the flower design in enhancing the PQR light output power since the increase of the PQR region by sacrificing the central LED area are achieved with more number of petals in a fixed diameter mesa. When the current density is the same, the more the number of petals, say the more the area of peripheral PQR region, the more the flower PQR laser intensity. We however note that the total length of peripheral PQR curves is to be smaller than the critical length for GaAs PQRs, corresponding to the device perimeter of a critical diameter (ϕ= ~50 μm), so that the quantum ring whispering cave mode begins to disappear (Kwon et al., 2006).
The photonic quantum ring (PQR) laser is an attractive candidate for high-density "laser" displays, given the unique operating characteristics attendant on its quantum-wire-like nature, such as extremely low threshold currents and thermally stable spectra in the typical operating-temperature range. When vertical mesa cavities are made of $\lambda/4$ Al0.92Ga0.08As/Al0.16Ga0.84As distributed Bragg reflector (DBR) structures added below and above an active region of multi-quantum wells (QWs) of 7nm thick GaAs each separated by 8nm thick barriers of Al0.3Ga0.7As, emitting at 850nm. Moreover, we have observed unusual convex WCMs from reverse-mesa (=hole)-type micro-resonators, whose WCMs we interpreted with respect to gain-guiding and photonic quantum corral effects. We now re-stress that the light output power observed enhances roughly in proportion to the number of petals of the flower PQR laser, up to the point where the total PQR perimeter reached a critical length corresponding to that of a circular PQR laser of about 50 μm diameter.
Circular and 4, 8, and 12-petal flower PQR lasers of the same overall diameter (Φ = 20 μm) for example are designed and fabricated. We can calculate the various multi-petal PQR perimeters' total lengths corresponding to the respective quantum wire lengths of Rayleigh band. The circular PQR of Φ = 20 μm has a peripheral PQR length of about 63 μm. When the number of petals, in the same overall diameter (Φ = 20 μm) of flower, is 8, the total peripheral PQR length is about 84 μm, and when the number is 12 then the total length is about 115 μm. The increased number of petals is more or less proportional to the growth of flower PQR output power, which is roughly proportional to the total peripheral PQR length. The SEM image of a 12-petal flower PQR laser is shown as an example in Fig. 20. Mesas 4.2

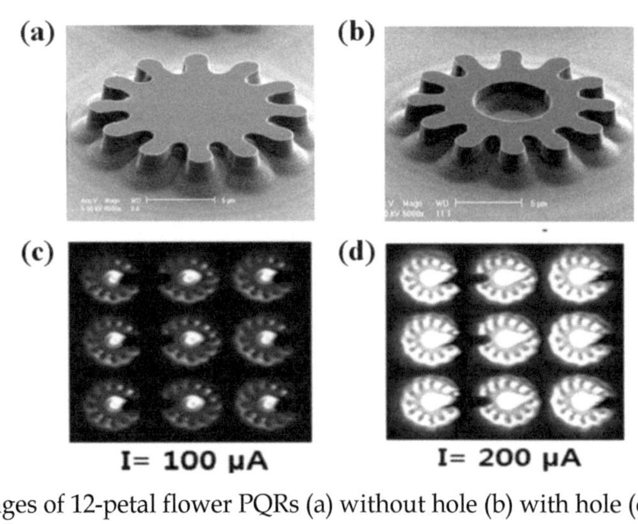

Fig. 20. SEM images of 12-petal flower PQRs (a) without hole (b) with hole (c) and (d) show illuminant PQRs at different injection levels.

μm high were etched by chemically assisted ion beam etching (CAIBE) with a photoresist mask. The smoothness of the side wall is an important factor in minimizing the spectral linewidth of PQR lasers. For side wall smoothness and highly anisotropic etching, we tilt and rotate the substrate in the CAIBE chamber during the etching process while adding BCl3 gas to facilitate Al2O3 removal in addition to an Ar/Cl2 gas mixture. Full details are given in a reference (Kim et al., 2004)

10. Fabrication of high power flower PQR laser

Fig. 21 shows emission images of various flower PQR lasers of Φ = 20 μm. For comparison, we simultaneously fabricated a circular mesa PQR laser of Φ = 18 μm. A tremendous intensity build-up occurred after increasing injection currents, so that appropriate neutral density filters had to be used for intensity attenuation. PQR lasing occurs along the perimeter of the active disk called the Rayleigh bandwidth, 0.63 μm width for Φ = 20 μm (Ahn et al., 1999), while LED emission occurs in the central bulk region of the PQR mesa. A threshold of 28 μm (= 11 A/cm2), observed through ring pattern schemes as shown in Fig. 21, is apparently smaller than the threshold range around 20 - 30 A/cm2 as estimated via usual extrapolation schemes, where the convex TIR effect of 'hole' PQR portions is involved in addition to the 'soft lasing turn-on' behavior (Kim et al., 2009).

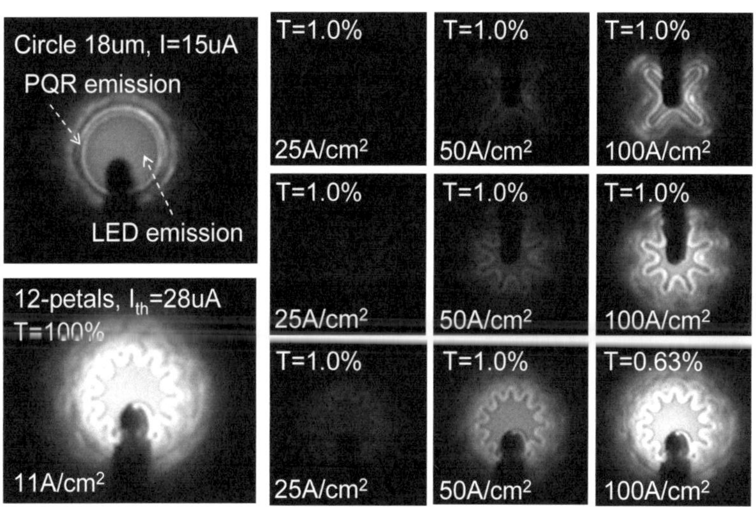

Fig. 21. Various emission patterns of 4-, 8-, 12- petal flower PQRs

As mentioned earlier, the flower design enhances the PQR light output power, thanks to the increase of the effective PQR region, while reducing the central LED area by means of a greater number of petals in a given diameter mesa. When the current density is the same, the greater the number of petals (the larger the area of the peripheral PQR region), the higher the flower PQR laser intensity. We can describe the light intensity as a function of the number of petals. For the devices with 20um width, the optical output power increased when the number of petals increased (Fig. 22). As the number of petals increased, the length of peripheral PQR region is larger so that the region occupied by the PQR emission in the whole emission region increased, leading to the final increase of the optical output power.

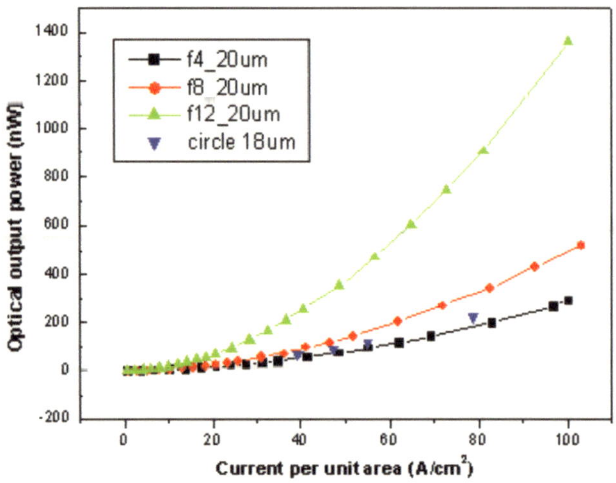

Fig. 22. Optical output power comparison (ϕ= 20 µm)

11. Panel-less TV display scheme with RGB PQR lasers

Today the market of display is dominated by LCD and PDP flat panel display (FPD) TVs, while expensive wider panels become too heavy to handle. The PQR laser is an attractive candidate for next generation display. We are currently developing a panel-less laser image chip for TV display using addressable PQR (photonic quantum ring) laser-pixels. For high brightness, wide-picture and full-color high definition TVs, we can design optimized projection systems involving RGB PQR laser array strategies, and lens optics for image magnification and projection similar to a light engine, where the RGB PQR display module will be the basic building block filling up the 2D/3D lattice of infinitely expansible TV display. Fig. 23 is a schematic diagram of a beam combination demonstrator for RGB color display.

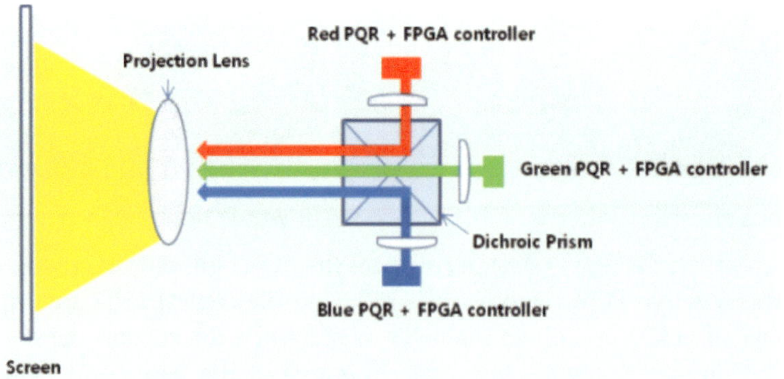

Fig. 23. A beam combination example

In the case of red beam, for example, the schematic may involve 1 or 2 lenses for beam guiding, resulting in an instantaneous frame of red beam scan implemented through an arrangement of optical components as shown Fig. 24. The blue and green beam combination structures are under development.

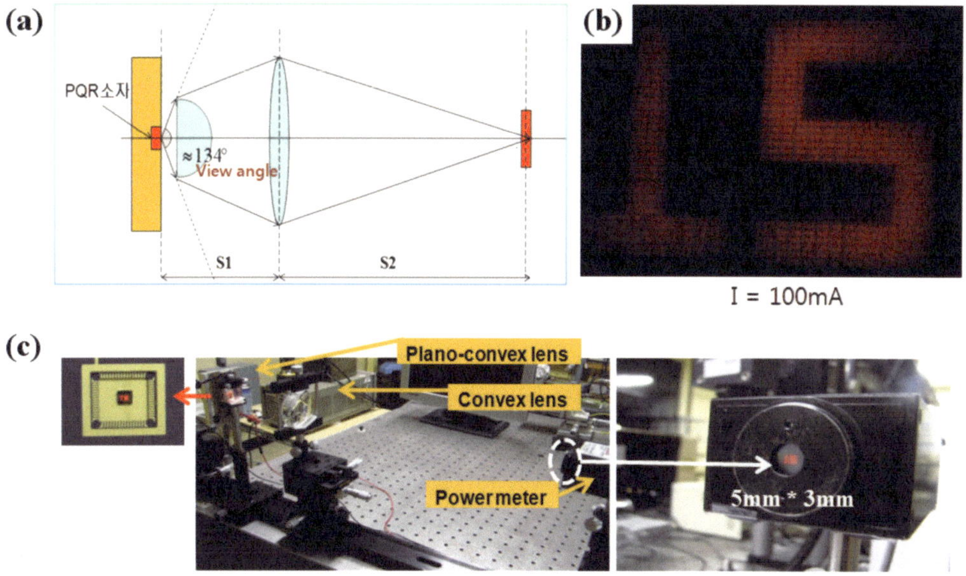

Fig. 24. (a) Red beam optics outlined. (b) Red letter image (c) Experimental set up of the optical components

12. Conclusions

We have presented studies of 3D WCM of PQRs. The 3D WCM laser is surface-normal dominant and has no in-plane resonance while the 2D WGM laser is in-plane dominant. Also the 3D WCM's major polarization state favors such a strong carrier-photon coupling that the powerful transient coupling generates PQRs, i.e., a photonic quantum corral effect. This gives rise to the low threshold currents and thermally stable spectra, important for easy optical mega-pixel ('Omega') chip fabrications which will be useful for next generation TV display. We have also presented Gaussian beam properties and guiding work of the PQR laser.

13. References

Ahn, J. C. et al., Photonic quantum ring, *Phys. Rev. Lett.* 82, No.3 pp 536-539 (1999).

Armani, D. K. et al., Optical microcavities, *Nature* 421, 925 (2003); Min, B. et al., Erbium-implanted high-Q silica toroidal microcavity laser on a silicon chip, *Phys. Rev.* A70, 033803 (2004).

Bae, J. et al., Spectrum of three-dimensional photonic quantum-ring microdisk cavities: comparison between theory and experiment, *Opt. Lett.* 26, 632 (2003).

Feidhlim, T. & O'Neill, J., Photoresist reflow method of microlens production Part I, *International Journal for Light and Electron Optics*, 113. 391 (2002)

Gehrig ,E. et al., Dynamic filamentation and beam quality of quantum-dot lasers, *Appl. Phys. Lett.* 84, 1650 (2004).

Ide, K. et al., LaGuerre–Gaussian Emission Properties of Photonic Quantum Ring Hole-Type Lasers, *IEEE Trans. Nano.* 7, 185 (2008).

Kim, D. & Kwon, O., Polarization characteristics of photonic quantum ring laser with three-dimensional whispering gallery resonances, *J. Appl. Phys.* 102, 053104(2007).

Kim, J. Y. et al., Fabrication of Photonic Quantum Ring Laser using Chemically Assisted Ion Beam Etching, *J. Vac. Sci. Technol.* B. 19, 1334 (2001).

Kim, J. Y. et al., Effect of surface treatment on leakage current of GaAs/AlGaAs laser microcavitys, *Appl. Phys. Lett.* 82, 4504 (2003).

Kim, M. et al., Wet etching fabrication of photonic quantum ring laser, *J. Appl. Phys.* 96, 4742 (2004).

Kim, Y. C. et al., PQR laser can outdo LED, *IEEE-NMDC 2006* 21-24 (2006), Gyeongju, Korea; Laser Focus World (March 2008).

Kwon, O. et al., Photonic quantum ring laser of 3D whispering cave mode, *Microelectronics Journal*, 40, 570 (2009)

Kwon, O. et al., Hole emitter of photonic quantum ring, *Appl. Phys. Lett*, Vol. 89, 11108 (2006)

McCall, S. L. et al., Whispering-gallery mode microdisk lasers, *Appl. Phys. Lett.*60, 289 (1992).

Noeckel, J. & Stone D., Ray and wave chaos in asymmetric resonant optical cavities, *Nature* 385, 45-47 (1997); Gmachl, C., High-power directional emission from microlasers with chaotic resonators, *Science* 280, 1556 (1998).

Park, B. H. et al., Chiral wave propagation manifold of the photonic quantum ring laser, *Appl. Phys. Lett.* 81, 580 (2002).

Topinka, M.A. et al., Imaging Coherent Electron Flow, *Physics Today* 56, 12 (2003).

Wiersig, J. & Hentschel, M., Combining Directional Light Output and Ultralow Loss in Deformed Microdisks, *Phys. Rev. Lett.* 100, 033901 (2008)

Yoon, J. H. et al., Single mode photonic quantum ring laser fabricated in hyperboloid drum shape, *J. Appl. Phys.* 103, 053103 (2008)

3

A Tunable Semiconductor Lased Based on Etched Slots Suitable for Monolithic Integration

D. C. Byrne, W. H. Guo, Q. Lu and J. F. Donegan
School of Physics, Trinity College Dublin
Ireland

1. Introduction

Widely tunable semiconductor lasers will play a critical part in future technologies. Tunable lasers are rapidly replacing fixed wavelength lasers in dense wavelength division multiplexing DWDM optical communications. The performance specifications of tunable lasers are the same as fixed wavelength specifications plus additional specifications that include: wavelength tuning range; wavelength switching speed; and minimum wavelength spacing. Tunable lasers diodes (TLD) have been used in optical networks for some time now starting with devices with small wavelength coverage and moving towards full band coverage.

Wavelength-agile networks are also simplified with tunable lasers. Reconfigurable optical add–drop multiplexers (ROADMs) and wavelength-based routing enable service providers to offer differentiated services, meet the ever-increasing demand for bandwidth and deliver all-optical networking. Tunable lasers are key to addressing this growing need to reconfigure networks remotely. The use of widely tunable lasers helps maximize existing network resources. The ability to dynamically provision bandwidth provides the ability to optimize the network configuration to meet demand. Widely tunable lasers move traffic from overcrowded channels to unused channels and are becoming essential for the network architecture.

Future DWDM networks will make more use of wavelength converters to increase network flexibility. Wavelength converters, such as, optical-electronic-optical (OEO) converters with the ability to detect a high data rate signal on any input wavelength channel and to convert to any output wavelength channel, will use tunable lasers. Future uses for tunable lasers will also include packet based selection of the wavelength on which the packet is to be transmitted. The tunable laser switching speed for these applications will be of the order of micro-seconds or longer. They will typically need to be widely tunable, i.e. tunable over a full C or L band and should be tunable to the 50 GHz channel spacing. In some UDWDM applications, channel spacing of 25 GHz and eventually as close as 12.5 GHz will be required.

Tunable lasers will also be used as a means to reduce costs as sparing lasers in wavelength division multiplexing (WDM) systems. New approaches to data transmission such as coherent WDM (CoWDM (Healy, Garcia Gunning et al. 2007)) require discrete tuning between particular wavelength channels on a grid. There is additionally an urgent need to integrate semiconductor lasers with other optical components such as amplifiers, modulators and detectors (Coldren 2000; Ward, Robbins et al. 2005; Welch, Kish et al. 2006; Raring & Coldren 2007) in order to reduce chip cost, system size and complexity. Tunable lasers are also needed in other important markets such as trace gas detection for environmental emission motoring (Phelan, Lynch et al. 2005).

Laser operation requires optical feedback which is conventionally obtained in a semiconductor Fabry-Pérot laser by cleaving the ends of the laser waveguide along either (011) or (01-1) crystallographic planes to form two semi-reflecting facets. However, due to the need for cleaving, it is difficult to integrate these lasers with other optical components on a single chip.

Distributed-Bragg-reflector (DBR) lasers and distributed feedback (DFB) lasers which employ a series of small refractive-index perturbations to provide feedback, do not rely on cleaved facets and therefore can be integrated with optical amplifiers and modulators.

However, complex processing with multiple epitaxial growth stages is required for fabricating these lasers. Another method to obtain feedback is to etch a facet. However, this approach is limited by difficulties in achieving the smoothness and verticality of the etched facet particularly for structures based on InP materials.

Previously it was shown that by introducing a shallow slot into the active ridge waveguide of a laser, the longitudinal modes of the Fabry-Perot (FP) cavity were perturbed according to the position of the slot with respect to the cleaved facets (Coldren & Koch 1984; Peters & Cassidy 1991; Corbett & McDonald 1995). By judicious placement of a sequence of low-loss slots with respect to the facets pre-selected FP modes could be significantly enhanced leading to robust single frequency lasing with wide temperature stability (John, Dewi et al. 2005; O'Brien & O'Reilly 2005) as well as tuning with fast switching characteristics (Phelan, Wei-Hua et al. 2008). More recently, we have characterized the properties of slots which are etched more deeply namely to the depth of, but not through, the core waveguide containing the quantum wells (Roycroft, Lambkin et al. 2007). In that case, the reflection of each slot is of the order of ~1% with transmission of ~80% and the slot will strongly perturb the mode spectrum of the FP cavity by creating sub-cavities. The loss introduced by the presence of the slot is compensated by gain in the laser. An array of such slots can provide the necessary reflectivity for the laser operation independent of a cleaved facet where the gain between the slots compensates for the slot loss producing an active slotted mirror region. Such a mirror has been used in conjunction with a cleaved facet permitting the integration of a photodetector with the laser. As the laser output facet is not cleaved this can provide a much easier integration platform on which complex devices such as Mach-Zehnder modulators (MZI) and semiconductor optical amplifiers (SOA) can be monolithically integrated with the laser to reduce chip cost and complexity significantly. In this chapter we demonstrate a tunable laser with an integrated SOA which is used to both increase and balance the output optical power of different channels.

2. Background on slot design

In this section a single slotted Fabry-Perot laser diode will be introduced which forms the basis for our tunable platform. The single slot laser is fabricated by etching into the waveguide of the FP laser diode as described in (DeChiaro 1991; McDonald & Corbett 1996; Fessant & Boucher 1998; Klehr, Beister et al. 2001; Lambkin, Percival et al. 2004; Engelstaedter, Roycroft et al. 2008). The slots act as reflection centres and produce a modulation of the reflection and transmission spectra dependent on the characteristics of the slot such as slot position, slot depth to which it is etched and slot width. Even though the slot is not etched into the active (waveguiding) regions it will still interact with the mode of the electric field (and magnetic field) of the waveguide as the mode profile is not fully confined to the active region and will expand into the surrounding cladding regions. The one-dimensional first order electric field mode profile modelled using the finite difference time domain (FDTD) technique for a simple laser structure with active region depth of 1 μm, upper cladding region of 1 μm and lower cladding of 1 μm with active region refractive index of 3.55 and cladding region refractive index 3.41, which are normal values for an InGaAsP active region sandwiched between InP cladding regions, are shown below in Fig. 1.

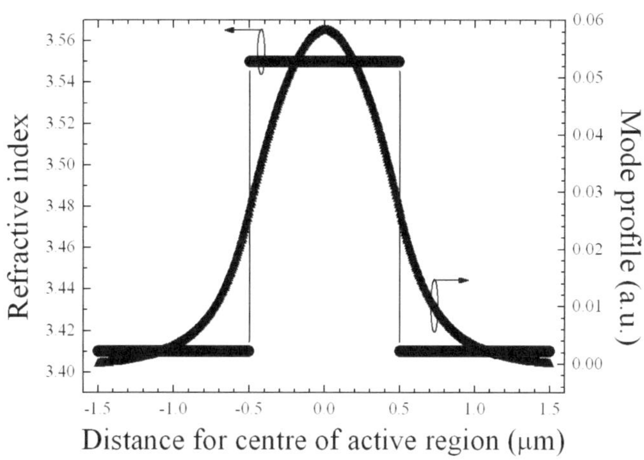

Fig. 1. Mode profile of the fundamental mode and refractive index profile through the laser structure.

From Fig. 1 the fundamental mode is seen to penetrate into the cladding region so any perturbation in this area will influence the mode profile of the laser diode.
The scattering matrix method (SMM) is a simple and accurate technique which can be used to determine the reflection and transmission from slots etched into the laser cavity. Numerous texts deal with the SMM of which (Buus, Amann et al. 2005) is a good introduction. Of particular importance in a laser structure is the ability to determine loss using the SMM method. This is an important advantage of the SMM over that transmission matrix method (TMM). A F-P laser with one etched slot can be described as three cavities with different interface reflections and transmissions as described below in Fig 2.

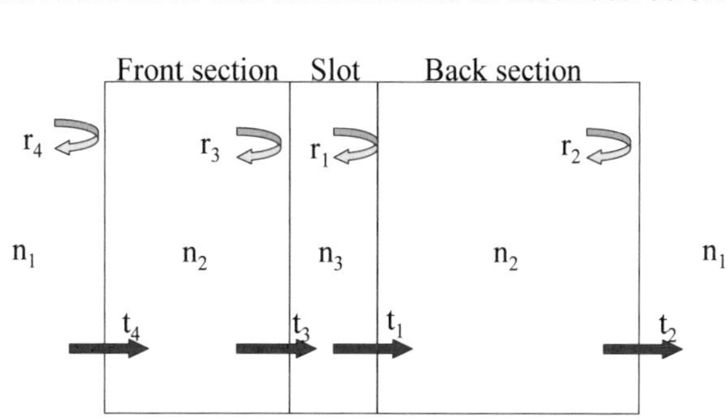

Fig. 2. Schematic description of single slot laser diode.

In fig. 2, n_i refers to the effective refractive index in these section of the laser structure, while r_i refers to the reflection from the interfaces as shown above. Each section can be described as a separated cavity and the total reflection and transmission is then found. The back section amplitude reflection from the left side and right side is described from the SMM as

$$r_{bl} = r_1 + \frac{t_1 r_2 (-t_1) \exp(-2i\tilde{\beta}_b L_b)}{1 - r_2(-r_1)\exp(-2i\tilde{\beta}_b L_b)} \quad (1)$$

and

$$r_{br} = -r_2 + \frac{t_2(-r_1)(-t_2)\exp(-2i\tilde{\beta}_b L_b)}{1 - r_2(-r_1)\exp(-2i\tilde{\beta}_b L_b)} \quad (2)$$

respectively where $\tilde{\beta}$ is the complex propagation constant ($\tilde{\beta} = \beta_{re} + i\beta_{im}$) and L_b is the back section cavity length. The back section amplitude transmission from the left side is described as

$$t_{bl} = \frac{t_1 t_2 \exp(-i\tilde{\beta}_b L_b)}{1 - r_2(-r_1)\exp(-2i\tilde{\beta}_b L_b)} \quad (3)$$

and

$$t_{br} = t_{bl}$$

giving a power reflection and transmission is $R_{bl} = (r_{bl})^2$ and $T_{br} = (t_{br})^2$ respectively. The reflection and transmission of the back section and slot region is found by including the back section reflection and transmission in the SMM calculation as follows

$$r_{bl+sl} = r_3 + \frac{t_3 r_{bl}(-t3)\exp(-2i\tilde{\beta}_s L_s)}{1 - r_{bl}(-r_3)\exp(-2i\tilde{\beta}_s L_s)} \quad (4)$$

and

$$t_{bl+sl} = \frac{t_3 t_{bl} \exp(-i\tilde{\beta}_b L_b)}{1 - r_{bl}(-r_3)\exp(-2i\tilde{\beta}_s L_s)} \quad (5)$$

again by a continuation of this method the reflection and transmission amplitudes for the full laser structure can be determined as

$$r_{totall} = r_4 + \frac{t_4 r_{bl+sl}(-t_4)\exp(-2i\tilde{\beta}_f L_f)}{1 - r_{bl+sl}(-r_4)\exp(-2i\tilde{\beta}_f L_f)} \quad (6)$$

and

$$t_{totall} = \frac{t_4 t_{bl+sl} \exp(-i\tilde{\beta}_f L_f)}{1 - r_{bl+sl}(-r_4)\exp(-2i\tilde{\beta}_f L_f)} \quad (7)$$

where the reflection and transmission from the right is found in a similar fashion to the total from the left. The calculated power reflection using an experimentally determined gain profile is shown in Fig. 3.

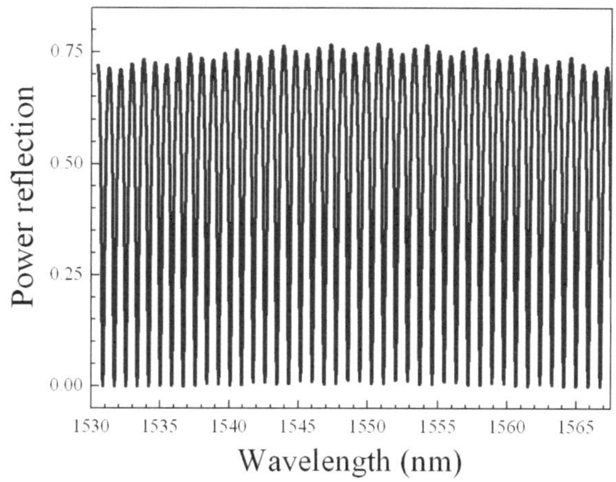

Fig. 3. Calculated reflection spectrum of a single slot laser diode versus wavelength operating near 1550 nm.

A modulation of the Fabry-Pérot modes is observed due to the slot. The strength and position of the modulation is dependent on the depth to which the slots are etched and the position of the slots in the laser cavity. An experimentally measured output spectrum recorded on an optical spectrum analyser (OSA) of a single slot laser diode with cleaved facets and a total cavity length of 350 μm and a slot position of 100 μm from the output facet is shown in Fig. 4. Coupled output power refers to the amount of output light collected by the measurement system.

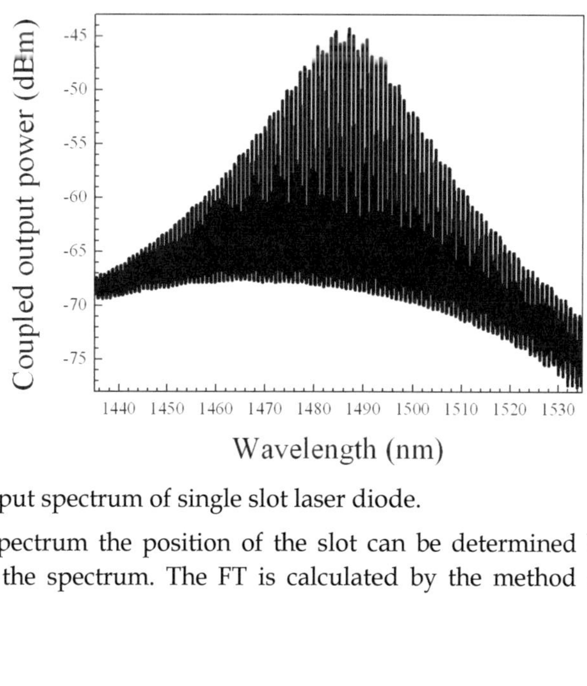

Fig. 4. Coupled output spectrum of single slot laser diode.

From the output spectrum the position of the slot can be determined by using a Fourier transform (FT) on the spectrum. The FT is calculated by the method described in (Guo,

Qiao-Yin et al. 2004) with the described deconvolution to remove the finite bandwidth resolution of the OSA. The FT spectrum is shown in Fig. 5 below.

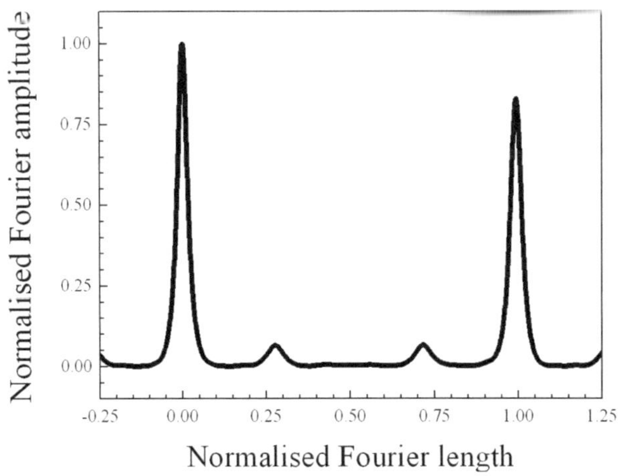

Fig. 5. Fourier transformed single slot spectrum, the large peaks at 0 and 1 are the facet reflection and harmonic while the two small peaks are the slot position and harmonic.

The single slot laser described here gives some mode selectivity however for single longitudinal mode operation more slots are needed and for a widely tunable laser multiple slot sections are included and the Vernier tuning method is utilised.

3. Electronic wavelength control

In order to have control of the output wavelength of a tunable laser diode we need to control the position of the gain peak wavelength of the cavity round trip gain (λ_p) and/or the longitudinal modes (λ_i). The gain peak wavelength (λ_p) is dependent on the injected carrier density however as the carrier density clamps above threshold widely tunable laser diodes cannot relay on this mechanism for large wavelength tuning.
Therefore in order to shift the output wavelength we need to change the positions of the longitudinal modes (λ_i) by changing the real part of the effective refractive as seen in the phase condition below (8),

$$\lambda_i = \frac{2n'_{eff}(\lambda_i)L}{m} \tag{8}$$

where n'_{eff} is the effective refractive index, L is the cavity length and m is the mode number. Therefore to control the output wavelength we need a waveguide with an electronically controllable effective refractive index where the amount of tuning is proportional to the product of the cavity length and the effective refractive index. With simple Fabry-Pérot laser diodes this provides little tuning of the output wavelength (a few nm) and so we look again at DBR and DFB type lasers. In DBR and DFB lasers the tuning of the cavity round-trip gain may be accomplished by tuning the Bragg reflector hence changing the position of the comb modes. Looking at equation (9)

$$\Lambda = \frac{\lambda_B}{2n'_{eff}} \tag{9}$$

we see that the only element that can be changed is the effective refractive index as the grating element is fixed during the fabrication of the Bragg mirror. As the effective refractive index is determined by the confinement factor and the refractive index of the layers of the laser diode any changes in the refractive index in any of these layers can change the effective refractive index and therefore the mirror loss $\alpha_m(\lambda)$. This type of tuning is employed commonly in many tunable laser diodes.

The extent of the lasers continuous tuning when the same cavity mode lases across the wavelength span can be determined easily form (10)

$$\frac{\Delta\lambda}{\lambda_0} = \frac{\left|\Delta n'_{eff}\right|}{n_{g,eff}} \tag{10}$$

where $\Delta\lambda$ is the wavelength tuning, λ_0 is the Bragg wavelength Δ'_{eff} is the change in the real part of the effective refractive index and $n_{g,eff}$ is the group effective refractive index. Allowing for mode hops between different longitudinal modes (discontinuous tuning) then the maximum tuning range is dependent on the spectral width of the gain envelope function. From this analysis we see that the electronic control of the wavelength of a tunable laser diode is dependent on our ability to control the effective refractive index of the laser. The effective refractive index may be controlled in practice by three different methods either carrier induced effects (free carrier plasma effect), by applying an electric across the device however this needs a reversed biased section(quantum confined Stark effect) or by varying the temperature of the device (thermal tuning).

DBR and DFB type laser diodes can be tuned by either changing the injected current or changing the temperature, however there is a limit to the tunability achieved by these methods (5-10 nm using refractive index changes) due to a limit in how much the refractive index or electronic bandgap can be changed. The gain bandwidth available to multiple quantum well lasers is ~ 100 nm and Erbium Doped Fibre Amplifiers (EDFA) have access to ~ 40 nm in the C or L band. Therefore there has been a large amount of research in extending the tunability of laser diodes beyond the refractive index limit (Jayaraman, Chuang et al. 1993; Rigole, Nilsson et al. 1995).

Increases in tunability can be obtained by using coupled cavities or array waveguide lasers which are discussed below however, these induce large complexities into the laser design and growth. In order to increase the tunability further we use a technique that allows us to use the change in the refractive index difference instead of the refractive index itself, therefore allowing a relative wavelength change to be used which can be much larger that the wavelength change due to a refractive index change. One method that exploits this relative refractive difference of semiconductor lasers is the Vernier effect. This effect requires two differing wavelength dependent mirror reflectivities to produce two spectrally different comb mode reflection spectra. The comb mode reflection spectra peaks will overlap at certain wavelengths which will produce lasing as at these wavelengths the gain will overcome loss since the round-trip loss is inversely proportional to the product of both mirror reflectivities. Both mirror reflection spectra can be directly and independently controlled by controlling the effective refractive index in these mirror regions. Any change in the effective refractive index in the mirror regions will produce a shift in the comb mode reflection spectra and allow different reflection peaks to overlap shifting the wavelength accordingly. This idea of Vernier tuning the output wavelength is shown in Fig. 6. This method can greatly increase the tuning range and lasers such as the Sample Grating Distributed Bragg Reflector (SGDBR) laser which have recorded tuning ranges of ever 60 nm (Oku, Kondo et al. 1998; Mason, Fish et al. 2000), however some limitations apply to this kind of tuning which are:

- If two modes are overlapped and on the gain curve some other means of suppressing the unwanted mode must be employed to preserve a high side mode suppression ratio (SMSR) which is a measure of the quality of the single mode of the laser.
- There must be large enough cavity gain to suppress other competing modes.
- To achieve continuous tuning there must be a phase control element meaning that the round trip phase must be controlled to be an integer multiple of 2π.

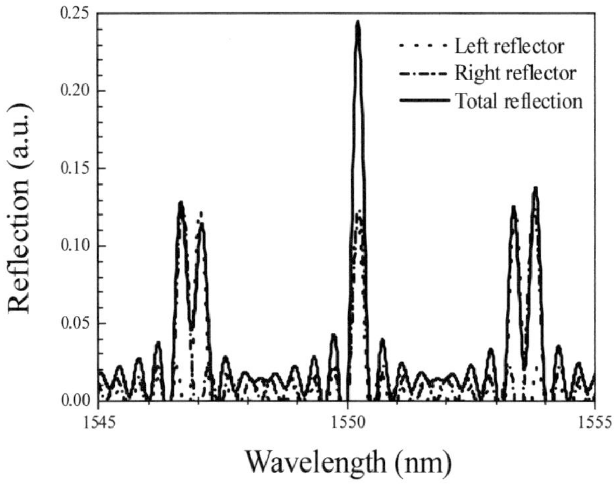

Fig. 6. Vernier tuning showing an overlap of reflection peaks at 1550 nm.

4. Multiple slot lasers

In this section a single section slotted laser with slots etched into the ridge of the waveguide down to the active region is introduced and characterised. This single section laser will form the basics for our three section tunable laser when combined with another similar section and a gain section which provides most of the optical gain. These slotted sections are termed "active slotted mirrors" as there is optical loss from each individual slot and therefore the reflection from a group of slots will saturate quickly if the slot loss is not effectively compensated. In the design presented here, the mirror regions are also actively pumped which provides the necessary gain under current injection to compensate for the loss introduced by the slots.

Firstly the scattering matrix method (SMM) as described earlier to model single slot lasers is used to determine the reflection, transmission, mirror loss and full width at half maximum (FWHM) of the reflection bandwidth profile with wavelength for a number of different slots at a particular slot spacing and gain conditions.

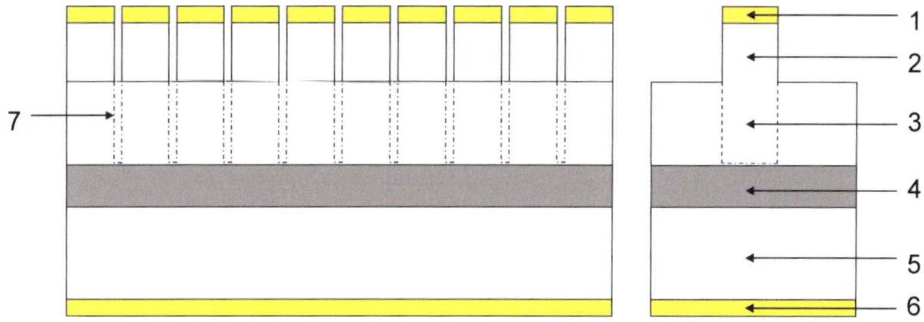

1 = Upper electrically conductive layer, 2 = Ridge of the waveguide, 3 = Upper cladding layer, 4 = Active region, 5 = Lower cladding layer, 6 = Lower electrically conductive layer, 7 = Etched slot.

Fig. 7. Single section slotted laser schematic showing nine etched slots.

A slot spacing of ~ 100 μm is chosen as this provides a reflection spectrum with super-mode peaks at ~ 400 GHz (3.2 nm) spacing. The free spectral range of the super-mode peaks is determined by the slot spacing through the following formula

$$FSR = \frac{\lambda^2}{2n_g L} \tag{11}$$

where n_g is the group effective index, L is the slot spacing and λ is the nominal wavelength. Starting from the SMM to describe a single slot with both slot interfaces straight etched into a medium that has gain and with anti-reflective (AR) coated facets, the reflection amplitude (r_{left}) and transmission amplitude (t_{total}) is determined simply by collecting SMM terms for slot sections and adding AR coated sections with the required slot spacing in between. The calculated power reflection for a single active section slotted laser with AR coated facets for differing numbers of slots is shown in Fig. 8.

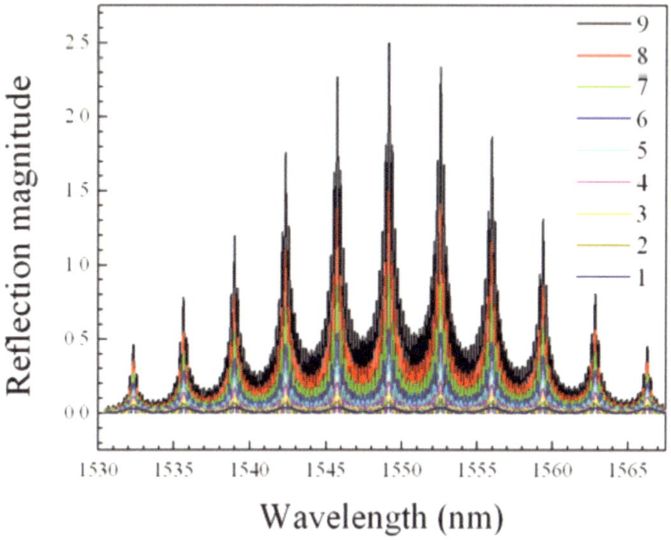

Fig. 8. Calculated power reflection spectrum for a single section active slotted laser with slot numbers form 1 to 9.

From Fig. 8 as the number of slots is increased the reflection at wavelengths determined by the slot spacing is also increased. A power reflection of over 1 is possible as there is considerable gain between the slots. As the mirror loss spectrum is inversely proportional to the reflection spectrum, the wavelength determined by these peaks will reach threshold first and lase at these positions provided the round trip loss is overcome by the gain between the slots. The FWHM of the main reflection peak also changes with slot number and is shown below in Fig. 9.

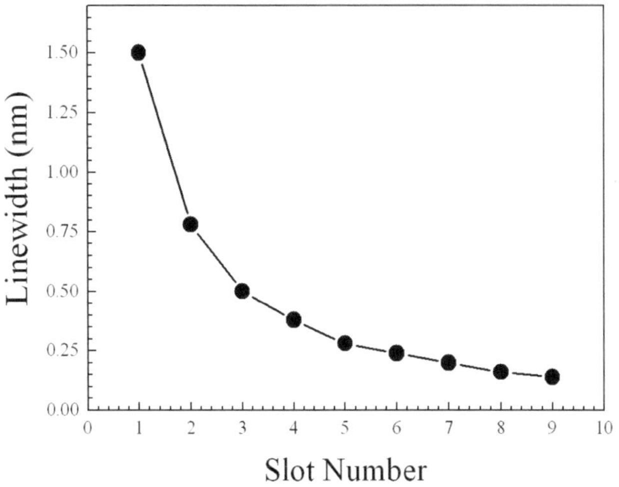

Fig. 9. Reflection linewidth versus slot number.

As the number of slots increases the linewidth of the reflection peaks decreases, therefore less cavity modes are covered by each super-mode meaning less cavity mode jumps are seen in the laser output spectrum. Increasing the number of slots also increases the length of the laser. Therefore a balance needs to be found between the reflection spectrum bandwidth and the laser length as it is better to keep the laser length as small as possible for integration on photonic chips.

As shown in (Lu, Guo et al. 2009) the slot can be described as a discontinuity as only the waveguide to slot interface provides meaningful reflection therefore using the SMM the total amplitude reflection can be approximated by the following formula

$$r = r_S \left[\frac{1 - \left(t_S^2 \exp(-2i\beta L) \right)^N}{1 - t_S^2 \exp(-2i\beta L)} \right] \tag{12}$$

where r_s is the slot reflection, t_s is the slot transmission and N is the slot number. The total amplitude transmission can also be approximated to

$$t = t_S \left[\frac{1 - t_S^{N-1} \exp(-(N-1)i\beta L)}{1 - t_S \exp(-i\beta L)} \right] \tag{13}$$

These formulae reduce the complexity of calculating the reflection and transmission of multiple slot laser diodes.

Including a gain section and another nine slot mirror section gives increased tunability using the Vernier effect as described above. A schematic of the laser structure is give in Fig. 10 below.

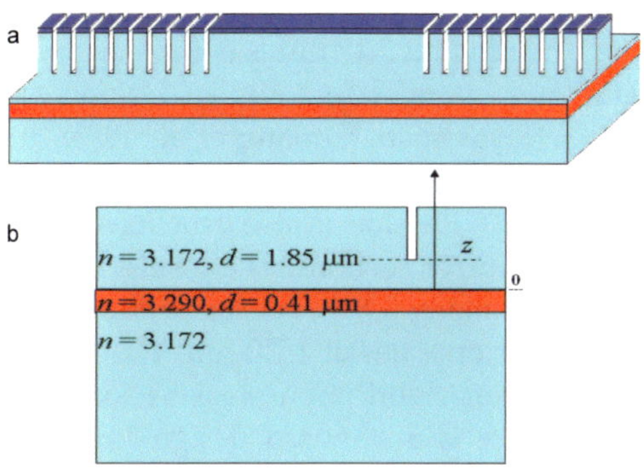

Fig. 10. a - Three section laser schematic showing the presence of etched slots, b - cross section of one slot area showing the slot depth.

Using the SMM to simulate the design shown above with the gain fixed in all sections and the refractive index changed by changing the carrier density in the mirror sections the output wavelength versus both mirror injection currents is found as shown below in Fig. 11.

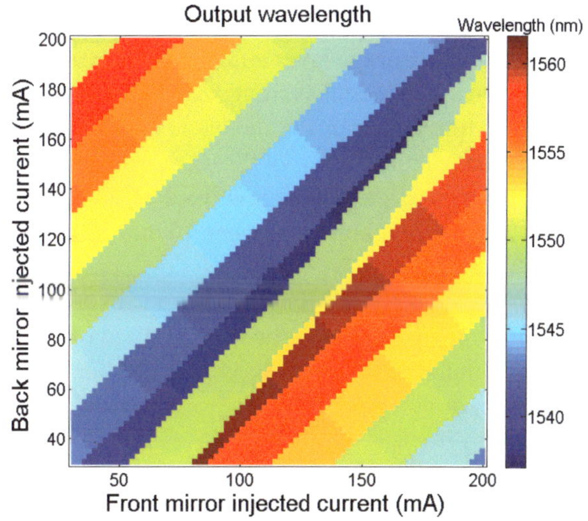

Fig. 11. Simulated three section tunable laser output wavelength versus mirror section injection currents (Gain section is held at a constant injection current).

5. Experimental laser diode characterization

The laser presented here consists of a central gain section of 500 µm and two mirror sections, the back mirror section has nine slots at a slot spacing of 108 µm giving a back mirror section length of 972 µm and the front mirror section which also has nine slots however the slot spacing is 97 µm giving a front mirror section length of 873 µm. The large gain section is chosen to provide sufficient gain as the mirrors have a high loss associated with them. The

total cavity length is ~ 2345 μm and each slot has a length in the propagation direction of ~ 1 μm. By having slightly differing slot spacing in the front and back mirror sections the Vernier tuning mechanism describe above can be used to extend the tuning range greatly. A discrete tuning of 400 GHz is achieved with this slot spacing as with a group index of 3.5 a free spectral range (FSR) of 3.53 nm and 3.17 nm is observed for both front and back mirror reflectors respectively. The laser operates in a similar fashion to a surface grating Bragg reflector laser as reported in (Jayaraman, Chuang et al. 1993), however due to the large distance between the slots (~100 μm) the mirror sections operates as a very high order grating. The large distance between the slots is also beneficial in directly injecting carriers to these regions to produce active mirror sections. The results from two lasers are given here one in which quantum well intermixing (QWI) is used to change the bandgap from 1550 nm to 1500 nm, while the other laser operates at 1550 nm. The tunable laser design described earlier was realized using the same fabrication steps as for a standard ridge waveguide laser. The laser epitaxial structure is a standard design employing an active region of 5 AlGaInAs quantum wells surrounded by InP n- and p-doped cladding regions. 2.5-μm-wide ridge waveguides were formed by inductively coupled plasma etching using Cl_2/N_2 gas. The slots are etched simultaneously with the ridge to a depth just into the waveguide core. The sidewalls are passivated with SiO_2 and an opening is made to the top of the ridge where a patterned Ti/Pt/Au electron-beam-evaporated ohmic contact is formed by lift-off lithography. The etched slot is sufficient to isolate the different longitudinal sections of the device allowing independent current injection. Following thinning of the substrate to 120 μm, an Au/Ge/Ni/Au contact is evaporated on to the n-type substrate. The devices are cleaved to the desired lengths and a single-layer antireflection coating applied to the facets.
To characterise the laser, three independent current sources are used to independently inject current into the gain and two mirror sections of the laser. The first device was mounted on a heat sink and held at a constant temperature of 20° C using a thermoelectric cooling unit. The current injected into the central gain section is fixed at 100 mA. The currents into the front and back mirror sections were scanned between 10 and 100 mA with a step of 1 mA. The wavelength and peak power of the laser emission spectrum and the side-mode-suppression-ratio (SMSR) were recorded using an optical spectrum analyzer with a resolution bandwidth of 0.1 nm. Fig. 12 shows the fibre coupled output power spectra under different current settings. A high SMSR (>30 dB) is required for single mode laser diodes used in optical communications.

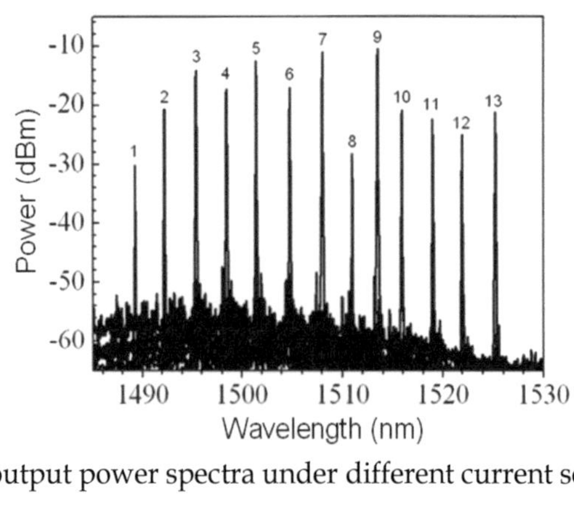

Fig. 12. Fiber coupled output power spectra under different current settings.

Relatively large power variations can be seen mainly because the front mirror current has been changed significantly in the scan in order to fully explore the tuning characteristics of the laser. Fig. 13 shows a diagram of the wavelength peaks and their corresponding SMSRs. A discrete tuning behaviour can be clearly seen over a tuning range of over 30 nm. With this experimental arrangement, a total of 13 discrete wavelengths can be accessed with a wavelength spacing around 3 nm as expected for the present design. 11 of the modes have a SMSR larger than 30 dB, except the 1st and 8th modes whose SMSR is around 20 dB.

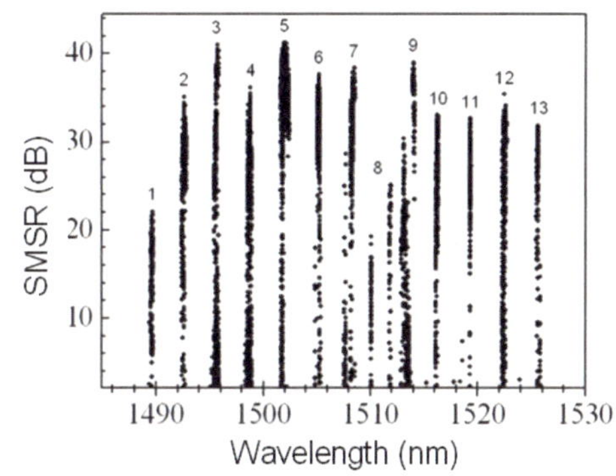

Fig. 13. Three section tunable laser SMSR versus wavelength for different mirror section injection currents.

The second laser described here is similar to the one described above however no QWI is used and therefore the wavelength is tuned around 1550 nm. Fig. 14 shows a wavelength tuning map versus both mirror section injection currents. Discrete mode hopping occurs at the boundaries of each different color section within this map. A total discontinuous tuning range of more than 40 nm is observed. The SMSR map versus both mirror currents is shown in Fig. 15. Clear islands of stable wavelength and high SMSR are observed in the maps.

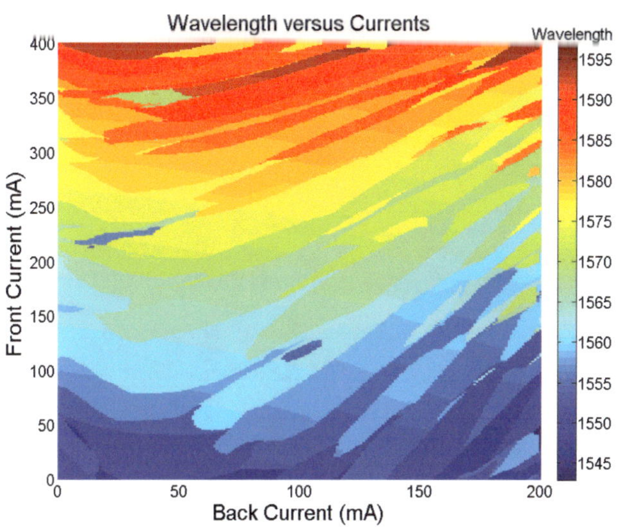

Fig. 14. Wavelength tuning map versus both mirror section injection currents.

The threshold current is difficult to determine accurately as the device has three sections but when both mirror section injection currents are set for a particular mode a threshold current of 56 mA in the gain section is observed. When all three sections are biased together a threshold current of 146 mA is observed.

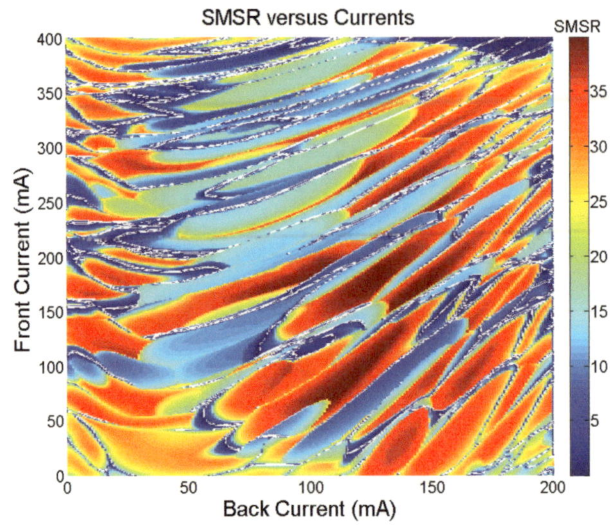

Fig. 15. SMSR tuning map versus both mirror section injection currents.

For comparison a four section sample grated distributed Bragg reflector (SG-DBR) laser wavelength map versus mirror section currents is shown in Fig. 16 below. The SG-DBR is a state of the art semiconductor tunable laser and is used extensively in optical communications and trace gas detection.

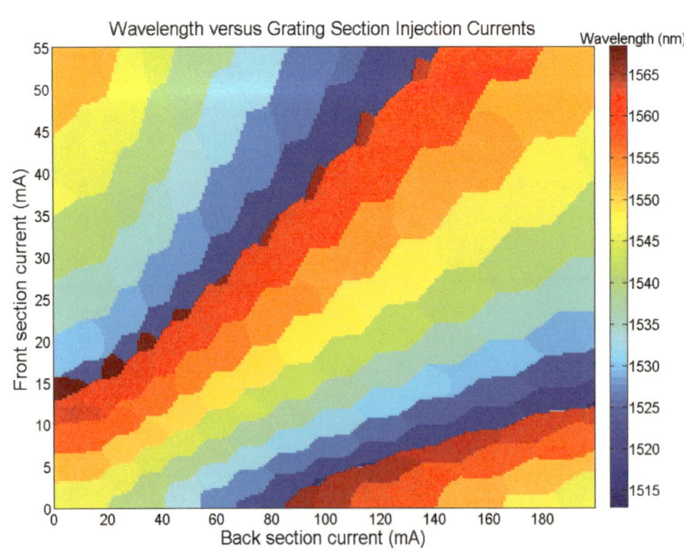

Fig. 16. Wavelength tuning map versus both mirror section injection currents for SG-DBR laser diode.

In order for accurate tuning to the ITU grid the super mode positions need to be fine tuned to a particular wavelength. To do this the laser needs to be continuously tunable over some wavelength range. The continuous tunability of one mode of the laser described above operating around 1550 nm is shown below in Fig. 17. This mode exhibit a continuous tuning range of 1.6 nm which allows for accurate setting of the laser to precise optical frequencies. The continuous tuning of this mode by current injection suggests that full carrier clamping does not take place in the mirror sections of this laser. In comparison, an SGDBR laser has a continuous tuning range of <0.4 nm for all discrete modes which is limited by the longitudinal mode spacing, although its quasi-continuous tuning range is much greater (Oku, Kondo et al. 1998; Mason, Fish et al. 2000).

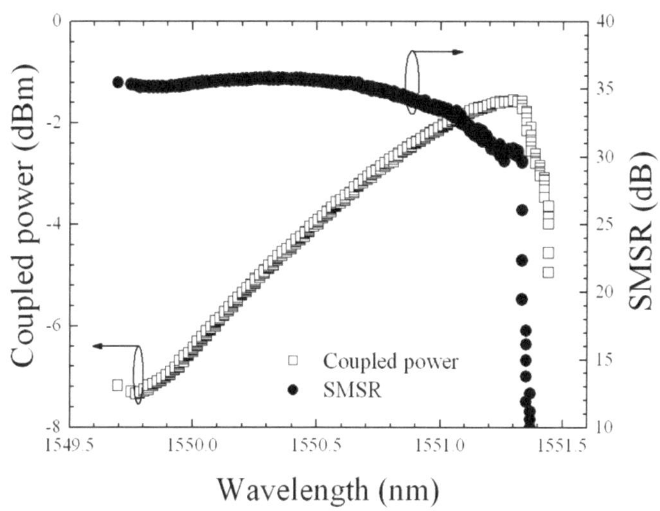

Fig. 17. Measured SMSR versus tuning wavelength due to a linear decrease in both mirror currents

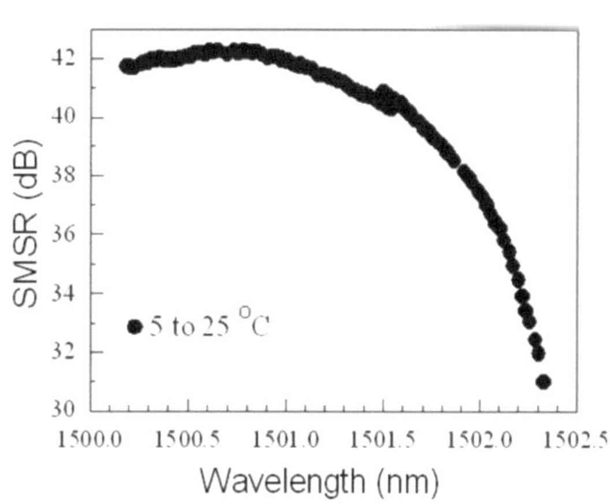

Fig. 18. SMSR versus wavelength for a discrete mode of the QWI laser with change in substrate temperature from 5 to 25° C. The temperature is increased linearly from left to right.

Fig. 18. shows the evolution of the wavelength and the associated SMSR due to thermal effects associated with a change of heat sink temperature from 5 to 25 °C, here the temperature is varied linearly over this range increasing from left to right in Fig. 17 below. A continuous tuning of over 2 nm while maintaining a SMSR of over 30 dB is measured. The change in wavelength with temperature is in line with the change in the index of InP which is $1.9x10^{-4}$ /K.

6. Integration of an optical amplifier

In order to demonstrate the compatibility with different photonic components, a semiconductor optical amplifier (SOA) was monolithically integrated with the tuneable laser source. The SOA consists of an 800µm long waveguide section on the output section. The SOA waveguide is curved to meets the output facet at a 5° angle reducing the requirement on the antireflection coating. This method reduces the back reflections to a negligible level. Figure 19 shows seven wavelength channels spaced 400 GHz apart which are accessible by the device. The optical output power is significantly increased by the SOA with channel powers ranging from 10 dBm to 14.2 dBm. All seven channels exhibit a SMSR greater than 30dB with a maximum SMSR of approximately 40dB. No deterioration of the maximum SMSR was observed compared to the laser without the SOA. Figure 20 shows the device output power as a function of the total laser drive current for three different SOA currents. The gain and tuning sections of the laser were connected together for this measurement. The device exhibits an optical output power in excess of 30 mW for a SOA current of 250 mA.

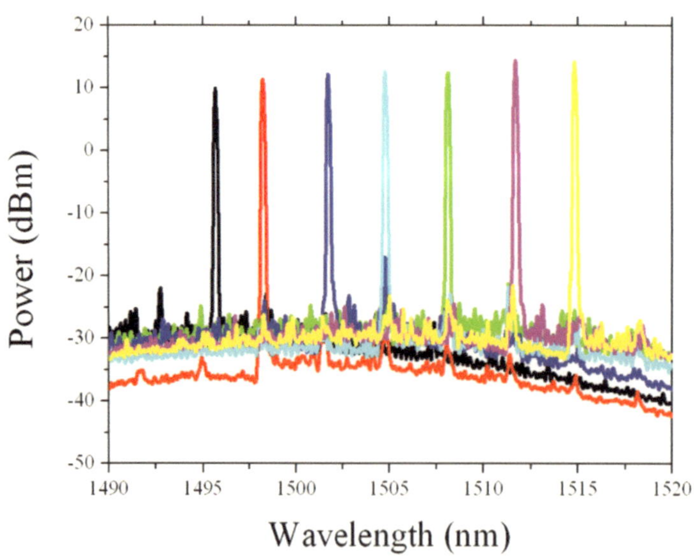

Fig. 19. Seven wavelength channels accessible by the laser integrated with an SOA showing maximum channel power of 14.2 dBm and a maximum SMSR of approx. 40 dB

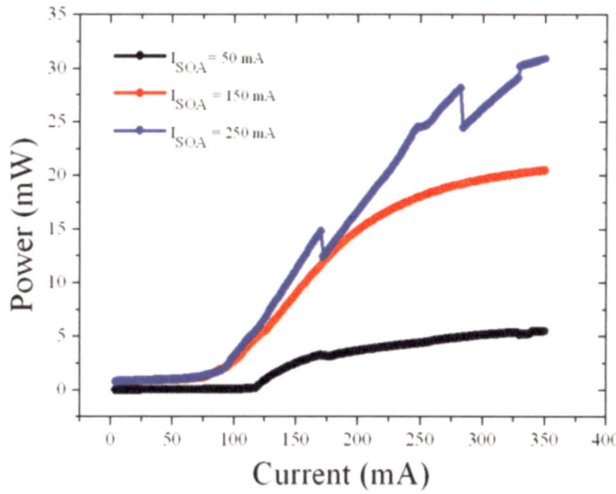

Fig. 20. Optical power as a function of laser drive current for three different SOA currents. The SOA increases the maximum device output power to 30 mW.

The graph shows how increased current injection into the amplifier increases the output power and delays the onset of gain saturation. As the SOA is located adjacent to the mirror section the high SOA drive currents can lead to significant heating of the reflector sections and current leakage into the mirror section. The resulting temperature and current changes cause a slight offset in the front reflector refractive index and a resultant change in the reflectivity spectrum. The blue graph in Fig 20 shows how the output power can change abruptly when the laser performs a mode jump due to thermal and current feedback from the SOA. In a tuneable laser with control over the individual sections, these effects can be offset by readjustment of the reflector currents.

7. Conclusion

The slotted tunable laser described here has many advantages over other state of the art semiconductor tunable laser diodes, however there are also some disadvantages with the slotted tunable laser design.

The key advantages of this laser are:

a. no re-growth step is required during manufacturing
b. no output facet necessary for operation so cleaving is not required
c. highly compatible with integration
d. insensitive to feed-back, therefore may not require optical isolator
e. high switching speed of the order of 1 ns
f. potentially very narrow line-width (of the order of MHz, unconfirmed)

The major advantages of the SFP tunable laser relate to the simpler manufacturing process enabled by the lack of any re-growth step being required. In addition no cleaving is required and this provides its compatibility with integration. This combination should provide an opportunity to obtain high yields with complex integrated devices, such as, a tunable laser, modulator and SOA.

The key disadvantages of this laser are:
a. current devices are significantly longer than competitive lasers, such as, sampled grating distributed Bragg reflector lasers (SG-DBR).
b. current designs have a large channel spacing, of the order of 400 GHz.

The fact that the slotted lasers are longer than competitive lasers reduces the yield advantage of the slotted tunable lasers. However, this should be proportionately less significant in highly integrated devices that include modulators, etc.

Direct comparison with the SG-DBR laser shows that this laser is easier and cheaper to fabricate however it cannot achieve full wavelength coverage of the C or L bands with high SMSR as the SG-DBR can.

One of the most important considerations for a tunable laser is the ability to tune to 50 GHz channel spacing in the C or L band for applications in DWDM applications. In order to address this 50 GHz issue, we are now investigating ways to incorporate a phase section that will allow more continuous tuning. The tunable laser described here also has a major advantage over most other tunable semiconductor lasers as it can be very easily integrated with other photonic components as describe above for integration with a SOA. More work is needed to integrate with Mach-Zehnder modulators and other such photonic devices.

8. Acknowledgements

The authors would like to acknowledge the help received from B. Corbett, J. P. Engelstaedter, B. Roycroft and F. Peters from Tyndall National Institute, Cork, Ireland.

The authors would like to acknowledge the funding received Science Foundation Ireland during the course of this work.

9. References

Buus, J., M.-C. Amann, et al., Eds. (2005). Tunable Laser Diodes and Related Optical Sources, Wiley-IEEE Press.

Coldren, L. and T. Koch (1984). "Analysis and design of coupled-cavity lasers--Part I: Threshold gain analysis and design guidelines." Quantum Electronics, IEEE Journal of 20(6): 659-670.

Coldren, L. A. (2000). "Monolithic tunable diode lasers." Selected Topics in Quantum Electronics, IEEE Journal of 6(6): 988-999.

Corbett, B. and D. McDonald (1995). "Single longitudinal mode ridge waveguide 1.3 micon Fabry-Perot laser by modal perturbation." Electronics Letters 31(25): 2181-2182.

DeChiaro, L. F. (1991). "Spectral width reduction in multilongitudinal mode lasers by spatial loss profiling." Lightwave Technology, Journal of 9(8): 975-986.

Engelstaedter, J. P., B. Roycroft, et al. (2008). "Laser and detector using integrated reflector for photonic integration." Electronics Letters 44(17): 1017-1019.

Fessant, T. and Y. Boucher (1998). "Additional modal selectivity induced by a localized defect in quarter-wave-shifted DFB lasers." Quantum Electronics, IEEE Journal of 34(4): 602-608.

Guo, W. H., L. Qiao-Yin, et al. (2004). "Fourier series expansion method for gain measurement from amplified spontaneous emission spectra of Fabry-Perot semiconductor lasers." Quantum Electronics, IEEE Journal of 40(2): 123-129.

Healy, T., F. C. Garcia Gunning, et al. (2007). "Multi-wavelength source using low drive-voltage amplitude modulators for optical communications." Opt. Express 15(6): 2981-2986.

Jayaraman, V., Z. M. Chuang, et al. (1993). "Theory, design, and performance of extended tuning range semiconductor lasers with sampled gratings." Quantum Electronics, IEEE Journal of 29(6): 1824-1834.

John, P., J. Dewi, et al. (2005). Specifying the wavelength and temperature tuning range of a Fabry-Perot laser containing refractive index perturbations (Invited Paper), SPIE.

Klehr, A., G. Beister, et al. (2001). "Defect recognition via longitudinal mode analysis of high power fundamental mode and broad area edge emitting laser diodes." Journal of Applied Physics 90(1): 43.

Lambkin, P., C. Percival, et al. (2004). "Reflectivity measurements of intracavity defects in laser diodes." Quantum Electronics, IEEE Journal of 40(1): 10-17.

Lu, Q. Y., W. H. Guo, et al. (2009). "Analysis of leaky modes in deep-ridge waveguides using the compact 2D FDTD method." Electronics Letters 45(13): 700-701.

Mason, B., G. A. Fish, et al. (2000). Characteristics of sampled grating DBR lasers with integrated semiconductor optical amplifiers. Optical Fiber Communication Conference, 2000.

McDonald, D. and B. Corbett (1996). "Performance characteristics of quasi-single longitudinal-mode Fabry-Perot lasers." Photonics Technology Letters, IEEE 8(9): 1127-1129.

O'Brien, S. and E. P. O'Reilly (2005). "Theory of improved spectral purity in index patterned Fabry-Perot lasers." Applied Physics Letters 86(20): N.PAG.

Oku, S., S. Kondo, et al. (1998). Surface-grating Bragg reflector lasers using deeply etched groove formed by reactive beam etching. Indium Phosphide and Related Materials, 1998 International Conference on.

Peters, F. H. and D. T. Cassidy (1991). "Model of the spectral output of gain-guided and index-guided semiconductor diode lasers." J. Opt. Soc. Am. B 8(1): 99-105.

Phelan, R., M. Lynch, et al. (2005). "Simultaneous multispecies gas sensing by use of a sampled grating distributed Bragg reflector and modulated grating Y laser diode." Appl. Opt. 44(27): 5824-5831.

Phelan, R., G. Wei-Hua, et al. (2008). "A Novel Two-Section Tunable Discrete Mode Fabry-Perot Laser Exhibiting Nanosecond Wavelength Switching." Quantum Electronics, IEEE Journal of 44(4): 331-337.

Raring, J. W. and L. A. Coldren (2007). "40-Gb/s Widely Tunable Transceivers." Selected Topics in Quantum Electronics, IEEE Journal of 13(1): 3-14.

Rigole, P. J., S. Nilsson, et al. (1995). "114-nm wavelength tuning range of a vertical grating assisted codirectional coupler laser with a super structure grating distributed Bragg reflector." Photonics Technology Letters, IEEE 7(7): 697-699.

Roycroft, B., P. Lambkin, et al. (2007). "Transition From Perturbed to Coupled-Cavity Behavior With Asymmetric Spectral Emission in Ridge Lasers Emitting at 1.55 μm." Photonics Technology Letters, IEEE 19(2): 58-60.

Ward, A. J., D. J. Robbins, et al. (2005). "Widely tunable DS-DBR laser with monolithically integrated SOA: design and performance." Selected Topics in Quantum Electronics, IEEE Journal of 11(1): 149-156.

Welch, D. F., F. A. Kish, et al. (2006). "The Realization of Large-Scale Photonic Integrated Circuits and the Associated Impact on Fiber-Optic Communication Systems." J. Lightwave Technol. 24(12): 4674-4683.

4

Monolithic Integration of Semiconductor Waveguide Optical Isolators with Distributed Feedback Laser Diodes

Hiromasa SHIMIZU
Tokyo University of Agriculture and Technology
Japan

1. Introduction

Monolithically InP-based photonic integrated circuits, where more than two semiconductor optoelectronic devices are integrated in a single InP substrate, have long history of research and development. Representatives of these InP-based photonic integrated circuits are, electroabsorption modulator integrated distributed feedback laser diodes (DFB LDs) (Kawamura et al., 1987, H. Soda et al., 1990) and arrayed waveguide grating (AWG) integrated optical transmitters and receivers (Staring et al., 1996, Amersfoort et al., 1994). Recently, dense wavelength division multiplexing (DWDM) optical transmitters and receivers have been reported with large-scale photonic integrated circuits having more than 50 components in a single chip (Nagarajan et al., 2005).

However optical isolators have been one of the most highly desired components in photonic integrated circuits in spite of their important roles to prevent the backward reflected light and ensure the stable operation of LDs. Although commercially available "free space" optical isolators are small in size and high optical isolation (>50dB) with low insertion loss (<0.1dB) is already realized, they are composed of Faraday rotators and linear polarizers, which are not compatible with InP based semiconductor LDs. Especially, Faraday rotators are based on magneto-optic materials such as rare earth iron garnets, and they are quite incompatible with InP based materials. Monolithically integrable semiconductor waveguide optical isolators are awaited for reducing overall system size and the number of the assembly procedure of the optical components. Also, such nonreciprocal semiconductor waveguide devices could enable flexible design and robust operation of photonic integrated circuits.

To overcome these challenges, we have demonstrated monolithically integrable transverse electric (TE) and transverse magnetic (TM) mode semiconductor active waveguide optical isolators based on the nonreciprocal loss (Shimizu & Nakano, 2004, Amemiya et al., 2006), and reported 14.7dB/mm optical isolation at λ=1550nm (Shimizu & Nakano, 2006). In this chapter, we report monolithic integration of a semiconductor active waveguide optical isolator with distributed feedback laser diode (DFB LDs).

2. Fabrication of the integrated devices

The semiconductor active waveguide optical isolators in the integrated devices are based on the nonreciprocal loss. In our TE mode semiconductor active waveguide optical isolators, ferromagnetic metal (Fe) at one of the waveguide sidewalls provides the TE mode nonreciprocal loss, that is, larger propagation loss for backward traveling light than forward traveling light. The gain of the semiconductor optical amplifier (SOA) compensates the forward propagation loss by the ferromagnetic metal (Shimizu & Nakano, 2004 & 2006). Fig. 1 shows the cross sectional image of the TE mode semiconductor active waveguide optical isolator taken by a scanning electron microscope. Since our waveguide optical isolators are not based on Faraday rotation, polarizers are not necessary for optical isolator operation. This is great advantage for monolithic integration of waveguide optical isolators with DFB LDs. The principle of the semiconductor active waveguide optical isolators is schematically shown in Fig. 2 (Takenaka & Nakano, 1999, Zaets & Ando, 1999). Discrete TE mode semiconductor active waveguide optical isolators have been reported in previous papers [Shimizu & Nakano, 2004 & 2006]. In TE mode semiconductor active waveguide optical isolators of Fig. 1, the waveguide width (*w*) determines the optical isolation and propagation loss characteristics. In narrow waveguides (w = 1.6μm), the optical confinement factor in the Fe thin film at one of the waveguide sidewalls is 0.16%, and the optical confinement factor of 0.16% brings the optical isolation of 14.7dB/mm (Shimizu & Nakano, 2006). Here, the optical isolation and propagation loss are almost proportional to the optical

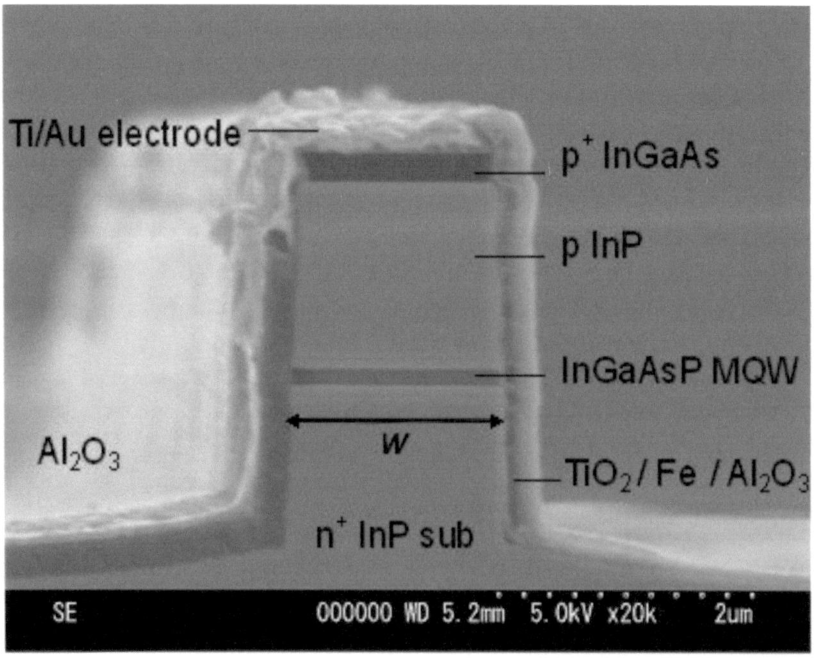

Fig. 1. A cross sectional scanning electron microscope image of a TE mode semiconductor active waveguide optical isolator having Fe layer at one of the waveguide sidewalls. *w* denotes the waveguide stripe width.

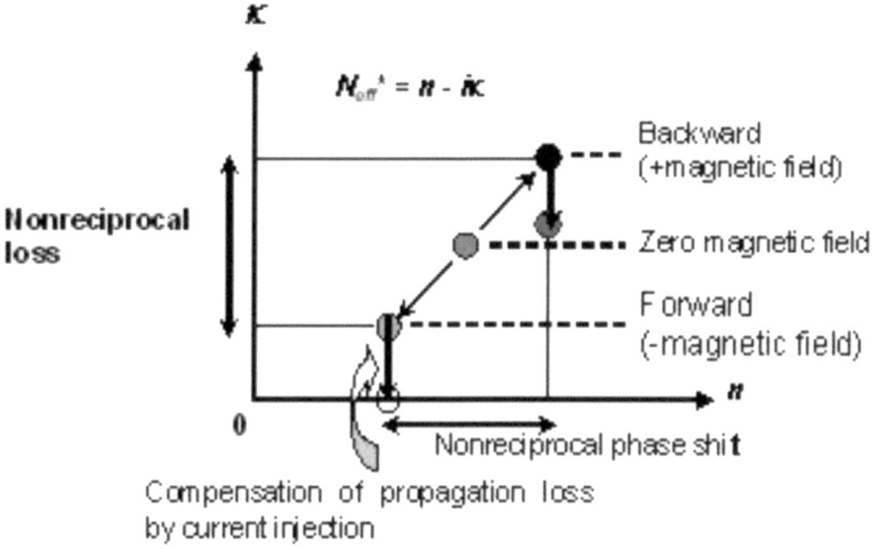

Fig. 2. Schematic operation principle of the semiconductor active waveguide optical isolators based on the nonreciprocal loss.

confinement factor in the Fe layer. As a result, the narrow waveguides work as optical isolators. On the other hand, in wide waveguides (w = 3μm), the optical confinement factor in the Fe thin film at one of the waveguide sidewalls is 0.02%, the propagating light receives small magneto-optic effect and absorption loss from the Fe layer. Hence, the wide waveguides work as LD. Higher optical transverse modes are absorbed by the Fe layer. Fig. 3 shows light output - current characteristics of TE mode semiconductor active waveguide optical isolators with the waveguide width w of 1.7 - 4.5 μm. TE mode semiconductor active waveguide optical isolators of w > 2.2 μm show lasing. On the other hand, TE mode semiconductor active waveguide optical isolators of w < 2.1 μm do not show lasing. This is because the Fe layer at the sidewall provides propagation loss, and non-radiative surface recombination at the etched sidewall reduces the internal quantum efficiency and gain of the MQW active layer. The reduced internal quantum efficiency is one of the problems of TE mode semiconductor active waveguide optical isolators. Thus, we have fabricated the monolithically integrated devices of DFB LDs and semiconductor active waveguide optical isolators in a simple fabrication process (Shimizu & Nakano, 2006).
The monolithically integrated devices are composed of 0.25mm-long index-coupled DFB LD and 0.75mm-long TE mode semiconductor active waveguide optical isolator sections on single InP chip. The DFB LD/semiconductor active waveguide optical isolator layer structures were grown by two steps of metal-organic vapor phase epitaxy (MOVPE) process. The active layer and grating layer were grown by the first step MOVPE. The DFB LD and the optical isolator section have the same InGaAsP compressively strained multiple quantum well (MQW) active layers. The MQW is composed of 14 compressively strained (+0.7%) quantum wells and 15 tensile strained (-0.4%) InGaAsP barriers. The MQW active layer is sandwiched by 50nm-thick InGaAsP separated confinement heterostructure (SCH)

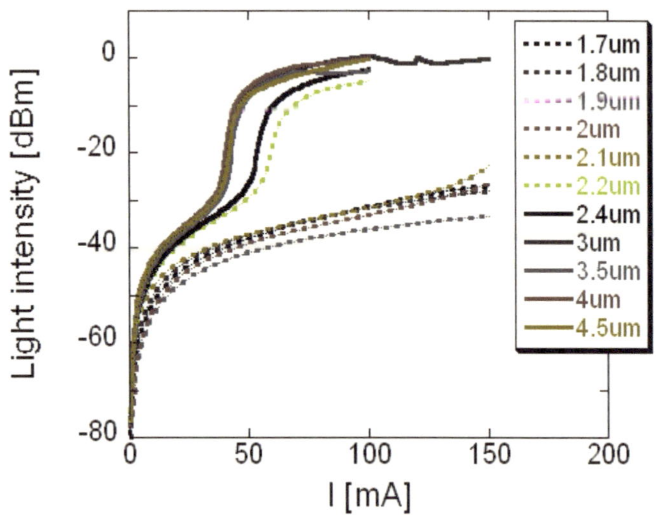

Fig. 3. Light output - current charactristics of TE mode semiconductor active waveguide optical isolators with waveguide width *w* of 1.7-4.5μm. Measurement temperature is 15°C.

layers. The photoluminescence peak wavelength of the MQW active layer was set at 1540nm. The InGaAsP index-coupled grating layer thickness is 20nm. The p-InP spacer layer thickness between the upper InGaAsP SCH layer and the grating layer is 50nm. A grating is defined by electron-beam lithography in DFB LD section. After the InGaAsP grating formation by wet chemical etching, 1μm-thick p-InP upper cladding layer and p^+InGaAs contact layer were grown by the second step MOVPE. The deep-etched waveguides were fabricated by Cl_2/Ar reactive ion etching, as shown in Fig. 1. The waveguide widths were

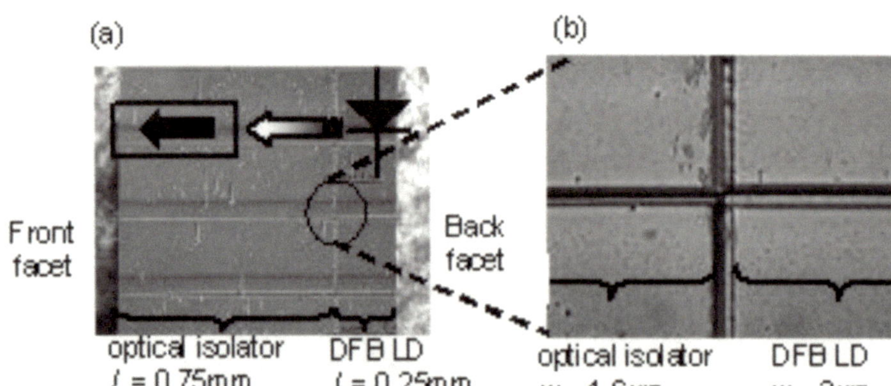

Fig. 4. Top views of the fabricated device bar with three integrated devices of waveguide optical isolators and DFB LDs by an optical microscope. (a) is the whole image and (b) is the magnified image of the optical isolator / DFB LD junction. Three horizontal waveguide stripes in (a) are corresponding to three integrated devices. The vertical line is a 5μm-width electrode separation region. *L* and *w* denote the device length and waveguide stripe width. The distance between the adjacent waveguide stripes is 250μm.

3μm for DFB LDs and 1.6μm for waveguide optical isolators. The tapered waveguide region where the waveguide width w gradually changes, is 10μm-long. Fig. 4 shows the top views of the integrated devices taken by an optical microscope. The basic fabrication process including the waveguide stripe formation, and the ferromagnetic / electrode metal deposition, is the same as that of previous discrete TE mode semiconductor active waveguide optical isolators (Shimizu & Nakano, 2004, 2006). The Ti/Au top electrodes and p^+InGaAs contact layers of the DFB LD / optical isolator sections are separated by each other, as shown in Fig. 4(b). The electrical isolation resistance between the two top electrodes is 1-5kΩ. It should be stressed that unlike conventional free space optical isolators, no polarizers are needed between the DFB LD and the optical isolator section. The device facets are as cleaved for the characterizations in this paper.

3. Characterizations

We measured the emission spectra of the integrated devices from the front and back facets under permanent magnetic fields of +/-0.1T and 0T. The front and back facets correspond to the optical isolator and the DFB LD sides, respectively (Fig. 4). The front facet emission is from the DFB LD with propagating through the waveguide optical isolator. The back facet emission is the direct emission from the DFB LD without propagating through the waveguide optical isolator. Fig. 5 shows the emission spectra by an optical spectrum analyzer from the (a) front and (b) back facets of the integrated devices under permanent magnetic fields of +/-0.1T and 0T. The emitted light was coupled by lensed optical fibers. The bias currents are 90 and 150mA for the DFB LD and active waveguide optical isolator, respectively. The threshold current of the DFB LD is larger than 40mA. The fabricated chips were kept at 15°C. The DFB LDs showed single mode emissions with λ = 1543.8nm. A 4dB emission intensity change was observed for waveguide-optical-isolator-propagated DFB LD light under magnetic field of +/-0.1T as shown in Fig. 5(a). On the other hand, such intensity change was much smaller (0.4dB) for the back facet emission, as shown in Fig. 5(b). These results show that the waveguide-optical isolator- propagated DFB LD light received the nonreciprocal loss. Therefore, this is the first demonstration of monolithic integration of the semiconductor active waveguide optical isolators with DFB LDs. Although the output light intensity of the waveguide optical isolator is weak (-56dBm), an anti-reflection (AR) coating at the front facet, and a high-reflection (HR) coating at the back facet could enhance the output intensity. Also, the optical reflection at the tapered waveguide region brings the internal reflections along the DFB LD section, which leads to weak output intensity. By solving these issues, the output intensity can be improved and the optical isolation can be enhanced with the Fe layer closer to the active layer. At this stage, maximum optical isolation is 14.7dB/mm for discrete TE mode semiconductor active waveguide optical isolators (Shimizu & Nakano, 2006).

4. Conclusion

We have demonstrated monolithic integration of the semiconductor active waveguide optical isolators with DFB LDs. By controlling the waveguide width of the TE mode semiconductor active waveguide optical isolators, we established simple monolithic

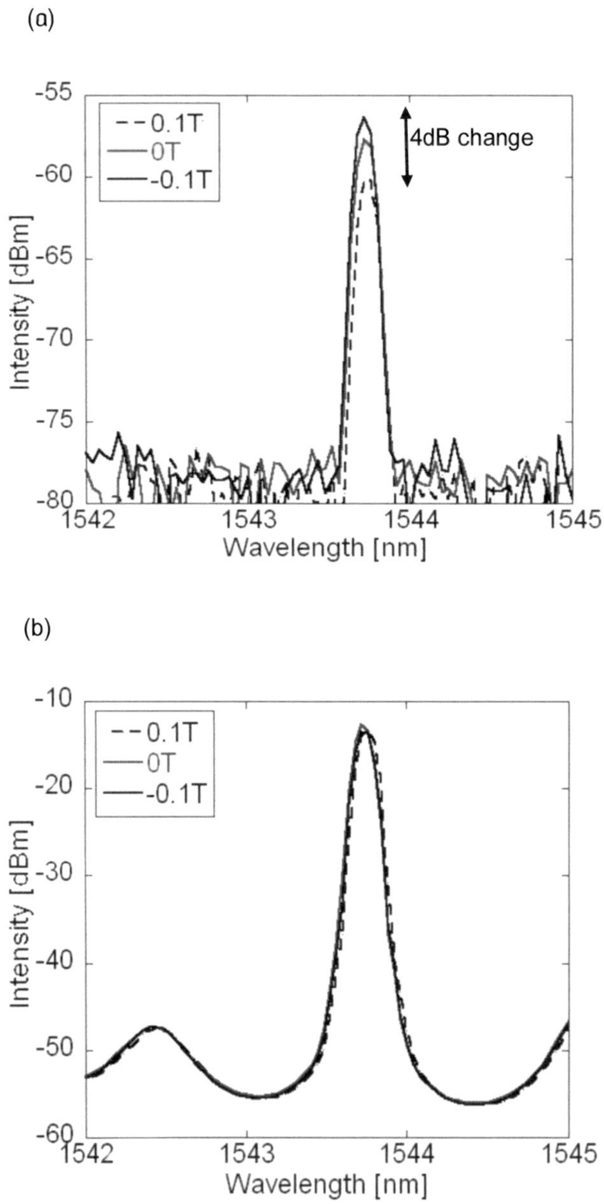

Fig. 5. Emission spectra of the integrated device from the (a) front and (b) back side facets under the permanent magnetic field of +/-0.1T and 0T. Note that the three curves in (b) are almost overlapped.

integration process of the waveguide optical isolators with DFB LDs. The integrated devices showed a single mode emission at λ = 1543.8nm and 4dB optical isolation. Although the optical isolation is smaller than commercially available "free space" optical isolators at this stage, this is the first step towards monolithically integrated isolator-DFB LD devices.

5. References

Kawamura, Y.; Wakita, K; Yoshikuni, Y; Itaya, & Asahi, H. (1987). *IEEE. J. Quantum Electron.* Vol. QE-23, No. 6, (Jun. 1987) 915-918.

Soda, H.; Furutsu, M.; Sato, K.; Okazaki, N.; Yamazaki, Y.; Nishimoto, H.; & Ishikawa, H. (1990). *Electron. Lett.*, Vol. 26, (1990) 9-10.

Staring, A. M.; Spiekman, L. H.; Binsma, J. J. M.; Jansen, E. J.; van Dongen, T.; Thijs, P. J. A.; Smit, M. K.; & Verbeek, B. H. (1996). *IEEE. Photon. Technol. Lett., Vol. 8,* No. 9, (Sep. 1996) 1139-1141.

Amersfoort, M. R.; de Boer, C. R.; Verbeek, B. H.; Demeester, P.; Looyen, A.; & van der Tol, J. J. G. M. (1994). *IEEE. Photon. Tech. Lett.*, Vol. 6, No. 1, (Jan. 1994) 62-64.

Nagarajan, R.; Joyner, C. H.; Schneider, R. P.; Bostak, J. S.; Butrie, T.; Dentai, A. G.; Dominic, V. G.; Evans, P. W.; Kato, M.; Kauffman, M.; Lambert, D. J. H.; Mathis, S. K.; Mathur, A.; Miles, R. H.; Mitchell, M. L.; Missey, M. J.; Murthy, S.; Nilsson, A. C.; Peters, F. H.; Pennypacker, S. C.; Pleumeekers, J. L.; Salvatore, R. A.; Schlenker, R. K.; Taylor, R. B.; Tsai, H. S.; Van Leeuwen, M. F.; Webjorn, J.; Ziari, M.; Perkins, D.; Singh, J.; Grubb, S. G.; Reffle, M. S.; Mehuys, D. G.; Kish, F. A.; & Welch, D. F.(2005). *IEEE. J. Select. Topics Quantum. Electron.* Vol. 11, No. 1, (Jan. 2005) 50-65.

Shimizu, H.; & Nakano, Y.(2004). *Jpn. J. Appl. Phys.* Vol. 43, (Dec. 2004) L1561-1563.

Shimizu, H.; & Nakano, Y.(2006). *IEEE J. Lightwave Technol.* Vol. 24, No. 1, (Jan. 2006) 38-43.

Amemiya, T.; Shimizu, H.; Nakano, Y.; Hai, P. N.; Yokoyama, M.; & Tanaka, M. (2006) *Appl. Phys. Lett.* Vol. 89, (2006) 021104.

Amemiya, T.; Shimizu, H.; Yokoyama, M.; Hai, P. N; Tanaka, M; & Nakano, Y. (2007) *Appl. Opt.* Vol. 46, No. 23, (Aug. 2007) 5784-5791.

Amemiya, T.; Ogawa, Y; Shimizu, H.; Munekata, H; & Nakano, Y. (2008) *Appl. Phys. Expr.* Vol. 1, No. 2, (Jan. 2008) 022002.

Takenaka, M.; & Nakano. (1999), Y. Proceeding of 11th International Conference on Indium Phosphide and related materials, 289.

Zaets, W.; & Ando, K. (1999) IEEE., Photon. Tech. Lett. Vol. 11, (Aug. 1999) 1012-1014.

Vanwolleghem, M.; Van Parys, W.; Van Thourhout, D.; Baets, R.; Lelarge, F.; Gauthier-Lafaye, O.; Thedrez, B.; Wirix-Speetjens, R.; & Lagae, L.(2004) *Appl. Phys. Lett.*, Vol. 85, (Nov. 2004) 3980.

Van. Parys, W.; Moeyersoon, B.; Van. Thourhout, D.; Baets, R.; Vanwolleghem, M.; Dagens, B.; Decobert, J.; Gouezigou, O. L.; Make, D.; Vanheertum, R.; & Lagae, L. (2006) *Appl. Phys. Lett.*, Vol. 88, (Feb. 2006) 071115.

Shimizu, H.; & Nakano, Y. (2006) Proceeding of 2006 International Semiconductor Laser Conference, (Sep. 2006) TuA6.

Shimizu, H.; & Nakano, Y. (2007) *IEEE. Photon. Tech. Lett.*, Vol. 19, No. 24, (Dec. 2007) 1973-1975.

5

Optical Injection-Locking of VCSELs

Ahmad Hayat, Alexandre Bacou,
Angélique Rissons and Jean-Claude Mollier
Institut Supérieur de l'Aéronautique et de l'Espace (ISAE),
Toulouse
France

1. Introduction

Since the telecommunication revolution in the early 90s, that saw massive deployment of optical fibre for high bit rate communications, coherent optical sources have made tremendous technological advances. The technological improvement has been multi dimensional; component sizes have been reduced, conversion efficiencies increased, power consumptions decreased and integrability into compact optoelectronic sub-modules improved. Semiconductor lasers, emitting in the 1.1-1.6 µm range, have been the most prominent beneficiaries of these technological advances. This progress is a result of research efforts that consistently came up with innovative solutions and components, to meet the market demand. This in-phase, demand and supply, problem and solution and consumer need and innovation cycle, has ushered us in to the present information technology era, where stable high speed data links make the backbone of almost every aspect of life, from economy to entertainment and from health sector to defence production.

By the start of twenty-first century, a new, low cost, low power consumption and miniaturized generation of lasers had started to capture its own market share. These lasers, named Vertical-Cavity Surface-Emitting Lasers (VCSELs) due to the presence of an optical cavity which is normal to the fabrication plane , have established themselves as premier optical sources in short-haul communications such as Gigabit Ethernet, in optical computing architectures and in optical sensors. While shorter wavelength VCSEL (< 1µm) fabrication technology was readily mastered, due to the ease in manipulation of AlGaAs-based materials, long wavelength VCSELs especially VCSELs emitting in the 1.3-1.5 µ range have encountered several technical challenges. There importance as low-cost coherent optical sources for the telecommunication systems is primordial, since they are compatible with the existing infrastructure.

VCSEL utilization in low-cost systems imply the application of direct modulation for high bit rate data transmission which engenders the problems of frequency chirping which increases laser linewidth and severely limits the system performance. Furthermore, relatively lower VCSEL intrinsic cut-off frequencies translated in to impossibility of achieving high bit rates. Optical injection-locking is proposed as a solution to these problems. It enhances the intrinsic component bandwidth and reduces frequency chirp considerably.

2. Emergence of Vertical-Cavity Lasers

2.1 Historical background and motivation

It must be noted that the Vertical-Cavity Surface-Emitting Lasers (VCSELs) or simply SELs (Surface-Emitting Lasers, as they were referred to as at that time) were not proposed to overcome the bottlenecks that had hindered the progress of FTTX systems. The lasers usually used for long-haul telecommunications have cleaved structures with edge emission. Consequently they are referred to as Edge Emitting Lasers (EELs). This structure does pose some problems, e.g. the initial probe testing of these devices is impossible before there separation into individual chips. Their monolithic integration is also limited due to finite cavity length. The cavity length implies generation of undesirable longitudinal modes and the non-monolithic fabrication process implies the impossibility of fabricating laser arrays and matrices. It was specifically in order to overcome these problems that, K. Iga, a professor at that time at Tokyo University, proposed a vertical-cavity laser in 1977.

These surface-emitting lasers provided following advantages:

- Probe-testing during the manufacturing process.
- Fabrication of a large number of devices by fully monolithic processes yielding a very low-cost chip-production.
- Very small cavity length guaranteeing longitudinal single mode operation.
- Possibility of production as arrays and matrices.
- Very low threshold currents due to ultra small cavity volume.
- Monolithic integration compatibility with other devices.
- Circular far-field pattern as compared to elliptical pattern for EELs.

A pulsed operation at 77K with a threshold current of 900mA was demonstrated in 1979 with a GaInAsP-InP vertical-cavity laser emitting at 1.3μm (Soda et al., 1979). However, more pressing issues regarding the delivery of higher bit rates using the conventional EELs meant that the research into vertical-cavity lasers progressed very slowly. Consequently VCSEL research and development stagnated through out the decade that followed its first demonstration.

Continuous Wave (CW) operation of a VCSEL was presented in 1989, by Jewell et. al, for a device emitting at 850nm (Jewell et al., 1991). This VCSEL presented two unique features as compared to the previous generation of components. It had a QW-based active region and the semiconductor DBR mirrors were grown by means of Molecular Beam Epitaxy (MBE) which replaced the dielectric mirrors previously being used. The VCSEL technology then progressed steadily over the next ten years. A 2mA threshold quantum-well device was presented in 1989 (Lee et al., 1989). In 1993 Continuous Wave (CW) operation for a VCSEL emitting at 1.3μm was demonstrated (Baba et al., 1993). A high power VCSEL emitting at 960nm and with an output of 20mW CW output was reported in 1996 (Grabherr et al., 1996).

Despite these advances and maturity in fabrication technology, the VCSELs could not replace the EELs as optical sources for long-haul telecommunications and were hence confined to other applications such as optical computing, sensors, barcode scanners and data storage etc.

The reason for this shortcoming lies in the VCSEL physical structure that gives priority to:

- Monolithic integration favouring vertical emission
- Low threshold current
- On chip testing

These priorities impose a set of design guidelines for VCSEL fabrication which, when implemented, induce certain unwanted and unforeseen traits in the device behaviour. These undesirable characteristics rendered the VCSEL unsuitable for utilization in prevalent telecommunication systems.

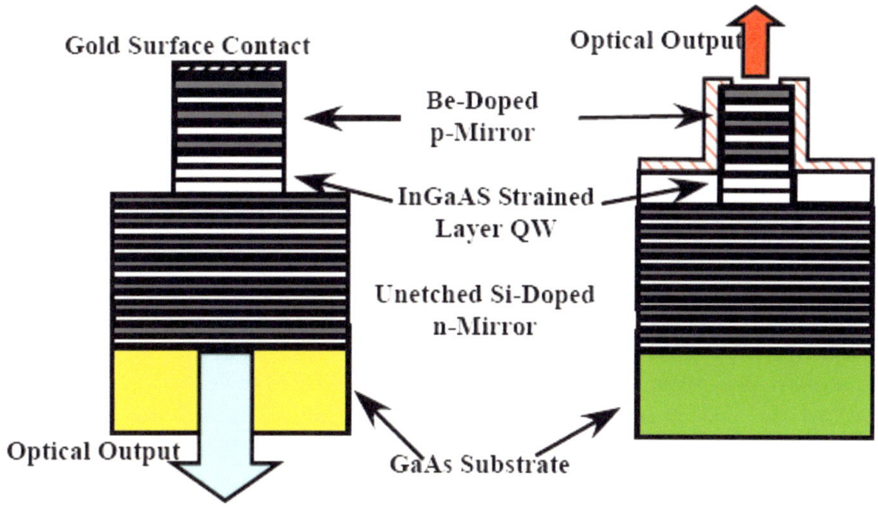

Fig. 1. An early design schematic for top-emitting and bottom-emitting VCSELs presented by Jewell et. al. in 1989.

Following is a concise analysis of these shortcomings. We would present the basic VCSEL structure that would try to achieve the above given objectives. Following this discussion we would present the drawbacks in the device performance related to the realization of design objectives. Certain remedies and improvements would then be presented in order to render the device more performing and efficient.

2.2 VCSEL structure

A VCSEL is essentially a gain medium based active region vertically stacked between two Distributed Bragg Reflectors (DBRs). In order to achieve a single mode operation it is proposed that the length of the active region be very small: Effectively of the order of the desired lasing wavelength. A short cavity eliminates the generation of longitudinal modes associated to Fabry-Pérot cavities. This however imposes a severe restriction on VCSEL DBR design.

The threshold gains for the surface-emitting and edge-emitting devices must be comparable regardless of the cavity length. The threshold gain of an EEL is approximately $100cm^{-1}$. For a VCSEL of active layer thickness of 0.1 μm, this value corresponds to a single-pass gain of about 1%. Thus for a VCSEL to lase with a threshold current density comparable to that of an EEL, the mirror reflectivities must be greater than 99% in order to ensure that the available gain exceeds the cavity losses during a single-pass.

Achieving a reflectivity of 99% with DBRs is a formidable task and thus central to the conception of low threshold VCSELs is the capacity to fabricate high reflectivity mirrors. Let's consider the example of a VCSEL operating at 850nm. The active region would consist of several ultra thin layers composed alternately of GaAs and AlGaAs materials. The difference between the refractive index of layers of a pair determines the number of pairs required to achieve a reflectivity of 99% or more. In the case of AlAs-$Al_{0.1}Ga_{0.9}As$ the refractive index difference between two alternate layers is 0.6 as is shown in fig. 2 (Adachi, 1985). Consequently only 12 pairs are needed to achieve a reflectivity of 99% or more. As far as AlAs and $Al_xGa_{1-x}As$ alloys go, the situation is conducive, even desirable, for the fabrication of VCSELs using these materials. The band gap energy of AlAs–$Al_xGa_{1-x}As$ alloys is about 1.5eV which eventually corresponds to a wavelength in the 800-900nm region.

Fabrication technology for VCSELs emitting in this wavelength band therefore has perfectly been mastered since monolithic growth of 12-15 DBR pairs does not pose serious fabrication challenges. Furthermore AlAs-GaAs alloy DBRs have an excellent thermal conductivity which allows the dissipation of heat fairly rapidly and avoids device heating which eventually could have been responsible for VCSEL underperformance.

2.3 Performance drawbacks

As far as the fabrication of near infrared VCSELs was concerned, the existing technologies and fabrication processes proved to be quite adequate. However, applying a similar methodology to telecommunication wavelength VCSELs proved to be much more challenging. Long wavelength VCSELs operating in the 1.1µm- 1.6µm range are of considerable interest for optical fibre telecommunications since the hydroxyl absorption and pulse dispersion nulls for silicon optical fibres are found at 1.5µm and 1.3µm respectively.

Although several material systems were considered, the combination InGaAsP-InP turned out to be the most suitable in view of the near perfect lattice match. The active layer is composed of the $In_{1-x}Ga_xAs_yP_{1-y}$ quaternary alloy. By varying mole fractions x and y, almost any wavelength within the 1.1–1.6µm can be selected.

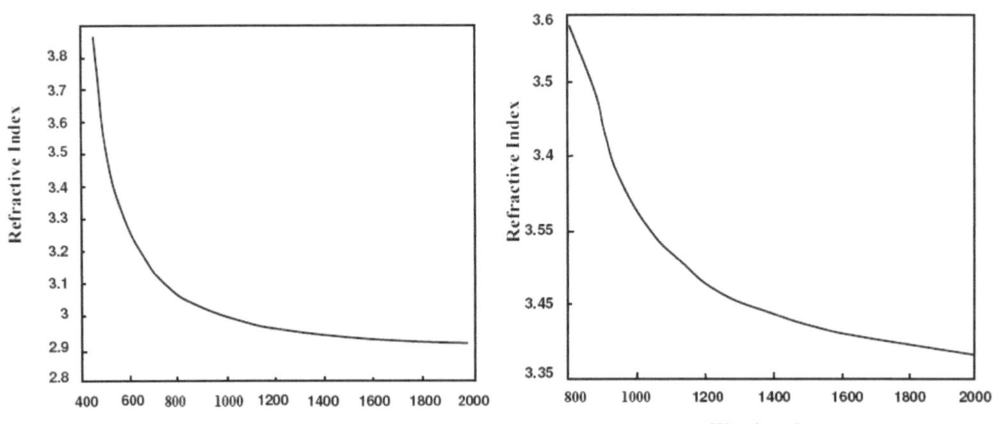

(a) Refractive Index of AlAs (b) Refractive Index of $Al_{0.1}Ga_{0.9}As$

Fig. 2. Refractive indices of AlAs and $Al_{0.1}Ga_{0.9}As$ as a function operating wavelengths.

2.4 DBR growth

Only 12–15 AlAs–$Al_xGa_{1-x}As$ pairs are needed to fabricate a DBR with a 99% reflectivity. By contrast, the refractive index difference between an InP- InGaAsP pair is only 0.3 and hence more than 40 pairs would be needed to achieve a reflectivity of 99%. The problem consequently encountered concerns thermal properties of InP−based materials that intervene to affect the process in following ways (Shau et al., 2004), (Piprek, 2003):

- For the fabrication of long wavelength VCSELs, there are mainly $In_{1-x}Ga_xAs_yP_{1-y}$ alloys available which have to be grown on InP substrates. Due to the effects of non negligible Auger's recombination effects and intra-valence band absorption, these materials suffer from temperature-dependent losses.
- The thermal conductivity is greatly reduced due to alloy disorders which causes phonon scattering. This reduction in thermal conductivity is particularly adverse for effective heat sinking through the VCSELs' DBRs usually having a thickness of several μms.
- AlAs-$Al_xGa_{1-x}As$ DBRs have a good thermal conductivity and could be thinner but due to lattice mismatch could not be grown on the InP substrate.

DBR growth has been one of the fundamental problems regarding the fabrication of long wavelength VCSELs that has hampered the entry of VCSELs in high-speed data, command and telecommunications domain.

2.5 Optical and electrical confinement

Growing stacks of DBRs was not the only problem encountered by VCSEL manufacturers. One of the primary objectives of VCSEL design was to fabricate short cavity single mode devices. The short cavity did eliminate the undesirable longitudinal modes but it gave birth to another unforeseen problem. Initial VCSEL designs suggested that the carriers and the photons share a common path traversing the DBRs. This led to the heating of certain zones of the DBRs due to carrier flow and resulted in a variable refractive index distribution inside the VCSEL optical cavity. This phenomenon is known as "Thermal Lensing". Instead of being concentrated in the centre in the form of a single transverse mode, the optical energy is repartitioned azimuthally inside the optical cavity. This particular optical energy distribution is observed in the form of transverse modes. Higher bias currents therefore imply high optical power and in consequence a higher number of transverse modes.

An oxide-aperture is employed, principally in shorter wavelength emission VCSELs, in order to block the unwanted transverse modes. The oxide-aperture diameter then determines the multimode or single mode character of a VCSEL. VCSELs having oxide aperture diameter greater than 5μm exhibit multimode behaviour. It can also be inferred from the above discussion that for the type of VCSELs employing the oxide-aperture technology for optical confinement, single mode VCSELs almost always have emission powers less than those of multimode VCSELs.

The problems of optical and electrical confinement are hence interrelated. It is evident that in order to attain single mode emission the thermal lens effect must be avoided. This can only be achieved by segregating the carrier and photon paths. Although challenging technically, it can be achieved using a tunnel junction. The concept and functioning of a tunnel junction is explained in the following sub-section.

2.6 The tunnel junction

The "Tunnel Junction" was discovered by L. Esaki in 1951 (Esaki, 1974) and the tunnel junction diodes used to be labeled "Esaki Diodes" for quite some time after this discovery (Batdorf et al., 1960), (Burrus, 1962). Esaki observed the tunnel junction functioning while working on Ge layers but soon after his discovery, tunnel junction diodes were presented by

other researchers on other semiconductor materials such as GaAs, InSb, Si and InP. The tunnel junction is formed by joining two highly doped (degenerate) "p" and "n" layers. It has a particular current-voltage characteristic curve. A negative differential resistance region (− dI/dV) over part of the forward characteristics can be observed.
In the case of a VCSEL the tunnel junction serves a "Hole Generator". Under the tunnel effect, the electrons move from valence band (doped p++) to conduction band (doped n++), leaving holes in their place. Fig.1.12 shows the schematic diagram of a tunnel diode in reverse bias conditions. The existence of a tunnel junction in a VCSEL presents following advantages:

- It reduces the intra valence band absorption due to P doping.
- It serves to reduce the threshold current, by improving the carrier mobility.
- It is used for electrical as well as optical confinement.

Due to these properties, the tunnel junction has become an integral part of long wavelength VCSELs.

2.7 Technological breakthroughs and advances in long wavelength VCSEL fabrication

Although by the start of the 21st century serial production and delivery of VCSELs was in full flow for diverse applications, they had failed to fulfil the two following essential criteria for utilization in optical networks.

- They did not emit in the 1.3µm and 1.5µm range: The so-called "Telecoms Wavelengths". This meant not only definition and standardization of new standards at 850nm wavelength but also the deployment and manufacturing of a host of optical components such as optical fibres, couplers, multiplexers and photodiodes compatible with the 850nm emission range.
- As has been explained above, transverse-mode operation starts to manifest itself from a few milli-amperes above the threshold current rendering the VCSELs multimode in character. This multimodality is disconcerting in two ways:
  - It reduces the effective channel bandwidth hence reducing the maximum deliverable bit rate.
  - It requires the utilization of multimode optical fibre which although being less expensive than the single mode fibre, affects the VCSEL operation in another way. When high optical powers are injected in a multimode fibre, several undesired fibre modes are excited thus reducing the effective bandwidth.

It is clear from the above discussion that a suitable substitute for EELs, for applications in short to medium distance optical fibre networks, must possess the following properties:

- It must emit at either 1.3µm or at 1.5µm wavelength so that the existing standards, infrastructure, optoelectronic components and devices could be utilized.
- It must have a single mode emission spectrum so as to profit from the high bandwidths offered by the employment of single mode optical fibres.

As late as 2000, there were no serial production and mass deployment of VCSELs that fulfilled these two essential criteria. As has been discussed above, this was due to the technical challenges posed by a combination of several different factors which rendered the fabrication of long wavelength VCSEL devices very difficult.

2.8 Emergence of long wavelength VCSELs

Regarding the manufacturing of long wavelength VCSELs, several different research groups kept trying to realize long wavelength emission devices. In 1993, Iga et al. demonstrated the

CW operation of a 1.3μm InGaAs-InP based VCSEL at 77K (Soda, 1979). The upper DBR consisted of 8.5 pairs of p-doped MgO-Si material with Au-Ni- Au layers at the top while the bottom DBR consisted of 6 pairs of n-doped SiO-Si material (Dielectric Mirror). In 1997, Salet et.al demonstrated the pulsed room-temperature operation of a single mode InGaAs-InP VCSEL emitting at 1277nm. The bottom mirror consisted of n-doped InGaAsP-InP material grown epitaxially to form a 50 pair DBR mirror with a 99.5% reflectivity (Salet et al., 1997).

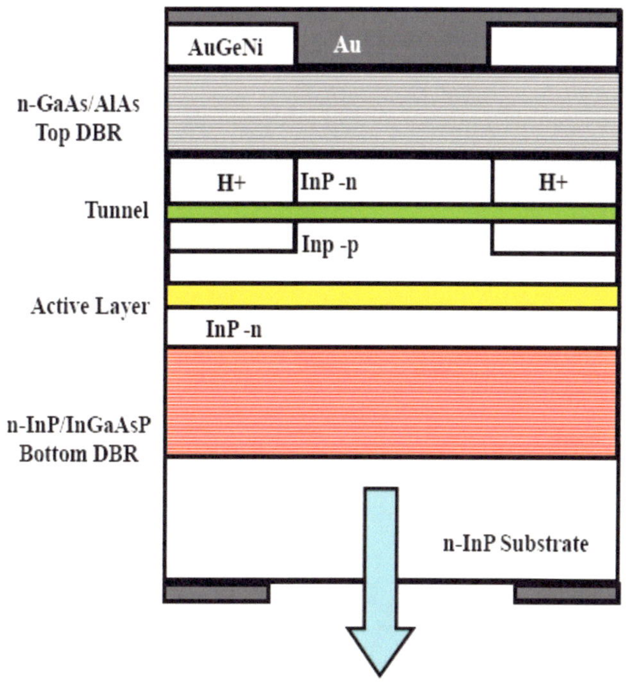

Fig. 3. A long wavelength VCSEL with a tunnel junction emitting at 1.55μm presented by Boucart et. al in 1999.

The device threshold current at 300K was 500mA. The top mirror was realized using p-doped SiO_2-Si reflectors. A year later, in 1998, Dias et al. reported the growth of InGaAsP-InP, AlGaInAs-AlInAs and AlGaAsSb-AlAsSb based DBRs on InP substrates to achieve reflectivities up to 99.5% (Dias et al., 1998). Soon afterward, in 1999, Boucart et al extended their previous work to demonstrate the room temperature CW operation of a 1.55μm VCSEL. In this case the top DBRs consist of 26.5 n-doped GaAs-AlAs pairs which were grown directly on an n-InP substrate (Metamorphic mirrors). A tunnel junction was fabricated to localize the current injection. The bottom mirror consisted of 50 pairs of n-doped InGaAsP-InP layers having a reflectivity of 99.7%. The device had a threshold current of only 11mA and had been fabricated using gas-based Molecular Beam Epitaxy (MBE) (Boucart et al., 1999).

The tunnel junction proved benificial in two ways:

- It enabled the utilization of two n-doped DBRs;
- Once the conductive properties of the tunnel junction were neutralized using H+ ion implantation, it served to localize the current injection without having to etch a mesa.

The resulting device was therefore coplanar in structure. It can be ascertained from Table.1.1 that several different materials such as InGaAsP, InGaAsAl, InGaAsSb and InGaAsN were chosen to fabricate the active layer. The material choice for DBRs and the fabrication processes were equally diverse. Although most of the research groups chose "Monolithic Integration Techniques" for the fabrication of VCSELs, "Wafer Fusion", and "Fusion Bonding" were also applied.
Meanwhile, in 1998, the Institute of Electrical and Electronics Engineers (IEEE) defined the "1000BASEX-Gbps Ethernet over Fibre-Optic at 1Gbit/s" standard. This standard for the transmission of "Ethernet Frames" at a rate of at least one Gbps was defined using light sources emitting at 850nm. The definition of Gigabit Ethernet standards using 850nm optical sources boosted the research and development of near infrared emission VCSELs. By the year 2000, 850nm VCSELs had firmly established themselves as standard optical sources for short-haul communication applications. This development was a setback for ongoing research in long wavelength VCSELs and as a result many research groups shifted their focus from long wavelength VCSEL development to other emerging fields. Furthermore, the research focus, even in the long wavelength VCSEL development field, shifted toward a new dimension. Long wavelength VCSELs were no longer being developed solely as telecommunication sources, an emerging field of spectroscopy was beginning to play an increasingly important part in eventual long wavelength VCSEL applications.

2.9 Vertilas VCSELs

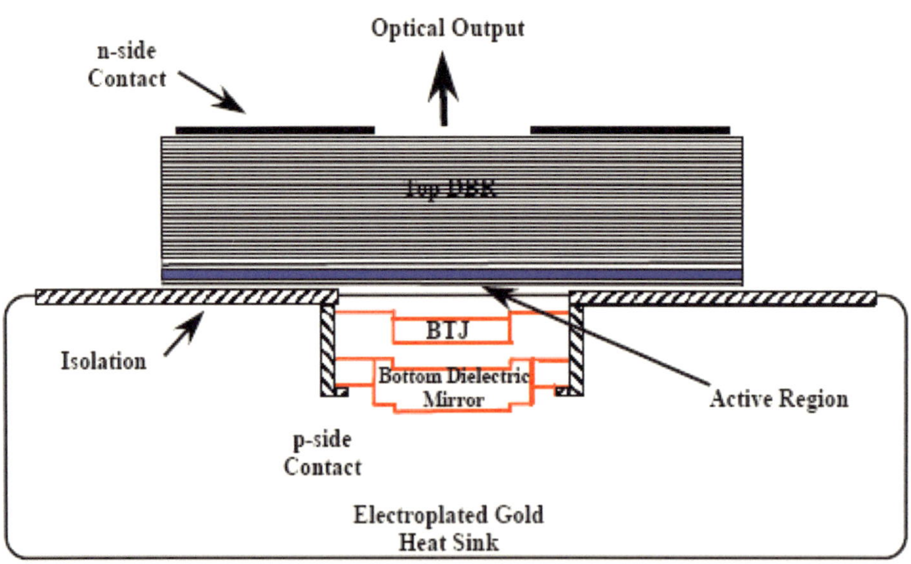

Fig. 4. A Vertilas BTJ structure with an emission wavelength of 1.55μm [28].

Although long wavelength VCSEL operation using a tunnel junction device was already demonstrated by Boucart et al. in 1999, Ortsiefer et al. presented a variation to this concept. Soon the single mode room temperature operation of an InP-based VCSEL operating at 1.5μm was demonstrated by the same research group (Ortsiefer et al., 1999), (Ortsiefer et al., 2000). The top DBR is composed of 34.5 InGaAlAs-InAlAs pairs. The bottom mirror is comprised of 2.5 pairs of CaF2-Si with Au-coating. The gold coating, apart from serving as a high reflectivity mirror (99.75%), serves as an integrated heat sink (Shau et al., 2004). The successful incorporation of tunnel junction in the long wavelength VCSEL design proved to be the technical breakthrough that would present VCSELs as standard devices for short to medium distance optical fibre communications. By 2002 Vertilas was delivering 1.55µm single mode VCSELs for 10Gbps operation.

2.10 BeamExpress VCSELs

The manufacturing of a long wavelength VCSEL requires the growth of an InP-InGaAsP alloy active region on an InP substrate. These alloys however are difficult to grow as DBR stacks above and below the active region since the restrictions imposed by the material thermal conductivity render proper device functioning impossible. On the other hand, AlAs-$Al_xGa_{1-x}As$ DBRs have a good thermal conductivity but they can not be monolithically grown on InP-based substrates due to lattice mismatch. The solution to the matching of disparate materials to optimize VCSEL performance was developed at the University of California Santa Barbara (UCSB) in 1996 by Margalit et al. (Margalit et al., 1996). The technique utilized is known as "Wafer Fusion" or "Wafer Bonding" and consists of establishing chemical bonds directly between two materials at their hetero-interface in the absence of an intermediate layer (Black et al., 1997). The first demonstration constituted of fabrication of a 1.55µm VCSEL. The device was fabricated by wafer fusion of MOVPE-grown InGaAsP quantum well active region to two MBEgrown AlGaAs-GaAs DBR reflectors (Margalit et al., 1996).

By applying a variant of the "Wafer Fusion" technique in 2004, Kapon et. al demonstrated that it was possible to grow separate components of a VCSEL cavity on separate host substrates (Syrbu et. al, 2004), (Syrbu et. al, 2005). These separate components were then bonded (fused) together to construct the complete VCSEL optical cavity. This process was developed at the Ecole Polytechnique Fédérale de Lausanne (EPFL) and patented as "Localized Wafer Fusion". Fig. 5 presents the structure of a BeamExpress VCSEL with an emission wavelength of 1.55µm. This is a double intracavity contact single-mode VCSEL with coplanar access. The InP-based optical cavity consists of five InAlGaAs quantum wells. The top and bottom DBRs comprise of 21 and 35 pairs respectively and are grown by Metal-Organic Chemical Vapor Deposition (MOCVD) epitaxy method. Using the technique of localized wafer fusion, the top and the bottom AlGaAs-GaAs DBRs are then bonded to the active cavity wafer and the tunnel junction mesa structures. Using VCSELs with double intracavity contacts has its own advantages. These contacts are much nearer to the active region than the classical contacts. Their utilization combined with the presence of tunnel junction allows having lower series resistance as compared to oxidized-aperture VCSELs. Due to this proximity of the contacts to the active region these VCSELs tend to have high quantum efficiency. Their location near the active region results in no current passage through DBRs.

The process used for the fabrication of Beam Express VCSELs is not monolithic. The bottom AlGaAs-GaAs DBR is grown on the GaAs substrate. The InP-based cavity is then bonded to this DBR. After the growth of an isolation layer on the active region, the epitaxially grown AlGaAs-GaAs top DBR is fused to complete the optical cavity. This double fusion increases the complexity of the fabrication process but it presents certain advantages. Waferfusion allows replacing the InAlGaAs DBRs by GaAs DBRs. Not only the GaAs DBRs have a better thermal conductivity, they are much cheaper than InAlGaAs DBRs which allows increasing the performance and decreasing the cost of the component at the same time. The biggest advantage of "Wafer Fusion" is the possibility of serial production of VCSELs which further serves to reduce the component cost.

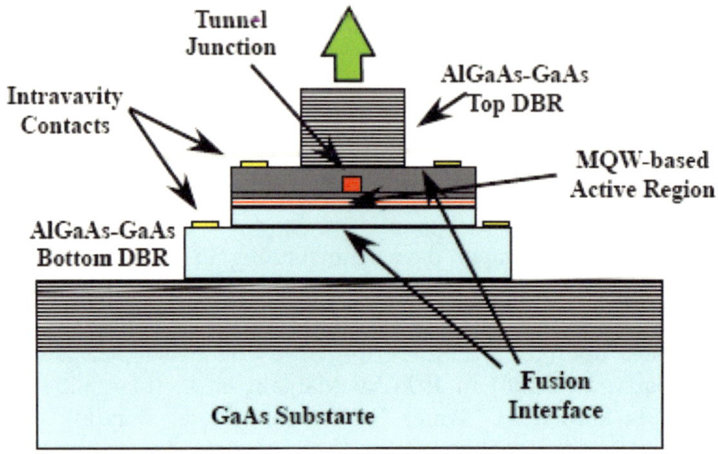

Fig. 5. Schematic diagram of a wafer-fused Beam-Express VCSEL with an emission wavelength of 1.5μm.

2.11 RayCan VCSELs

Starting as a spin-off company from the Korean government funded Electronics and Telecommunications Research Institute (ETRI) in 2002, RayCan launched an ambitious project for manufacturing of long wavelength VCSELs. Instead of using the above described specialized technologies for long wavelength VCSEL manufacturing, RayCan decided to embark upon a different course. They decided to monolithically grow InAlGaAs DBRs and an InGaAs-based quantum well active region on an InP substrate. As has been discussed above, this technique was previously not considered because in order to achieve 99% reflectivity using InAlGaAsbased DBRs, a growth of more than 40 pairs is needed. RayCan employed Metal-Organic Chemical Vapour Deposition (MOCVD) technique to fabricate a long wavelength VCSEL.

For 1.55μm VCSELs, the top and bottom DBRs were grown as 28 and 38 pairs of un-doped InAlGaAs-InAlAs schemes. The top and bottom DBRs consisted of 33 and 50 layers respectively for 1.3μm emission VCSELs. The 0.5λ thick active region consists of seven pairs of strain-compensated (SC) InAlGaAs quantum wells (Park et al., 2006). The lower number of top DBRs in both the VCSELs was compensated by using an InAlGaAs phasematching layer and Au metal layer. Fig. 6 presents the structure of a RayCan VCSEL emitting at 1.5μm. RayCan has been shipping 1.3μm and 1.5μm VCSELs since 2004. In November 2005 RayCan shipped its first 10GBit/s long wavelength CWDM VCSEL module.

2.12 Long wavelength VCSEL direct modulation

Up to this point we have discussed the prospects of long wavelength VCSELs in the context of high bit rate data delivery over medium and short distance links. It would not be an exaggeration to state that consumer demand for multimedia and interactive applications and therefore bandwidth has increased to an unprecedented level. Current electrical-electrical infrastructures can not support this demand. The major obstacle in switching from electrical/ hertzian systems to optical/fibred systems is the cost of the coherent optical source compatible with existing infrastructure. Recent advances in the fabrication, development and serial production of VCSELs emitting at 1.3μm and 1.5μm have paved the way for future FTTX systems.

Having been able to solve the problem at component level, by developing reliable long wavelength VCSELs, the next logical approach is the development of new systems incorporating these components. Conventionally the EELs used in the long-haul fibre links are externally modulated i.e. the photon generation process inside the cavity is independent of the modulation mechanism. While being extremely effective, this method necessitates the utilization of an external modulator which increases the system cost. Such a scheme is inherently unfeasible for FTTX systems due to the cost of the external modulators. The elimination of external modulators as a component of choice for FTTX systems decrees the employment of direct modulation techniques. In this technique the laser diode bias current is varied to achieve the optical output intensity variation. Apparently the scheme is simple and easy to implement, but when put into practice, it presents two major problems which are detailed in the following two sub-sections.

2.13 Phase-amplitude coupling

Semiconductor lasers, whether EELs or VCSELs, are different from other lasers in one respect. The refractive index of a semiconductor laser depends on the carrier concentration inside the cavity. The carrier concentration variation affects the refractive index of the cavity which eventually changes the emission wavelength of the component. The consequences of this uniqueness manifest themselves during the process of direct modulation. A variation in bias currents varies the optical output power as well as the optical frequency of the cavity. These variations are proportional to the variation in carrier concentration and therefore the bias current.

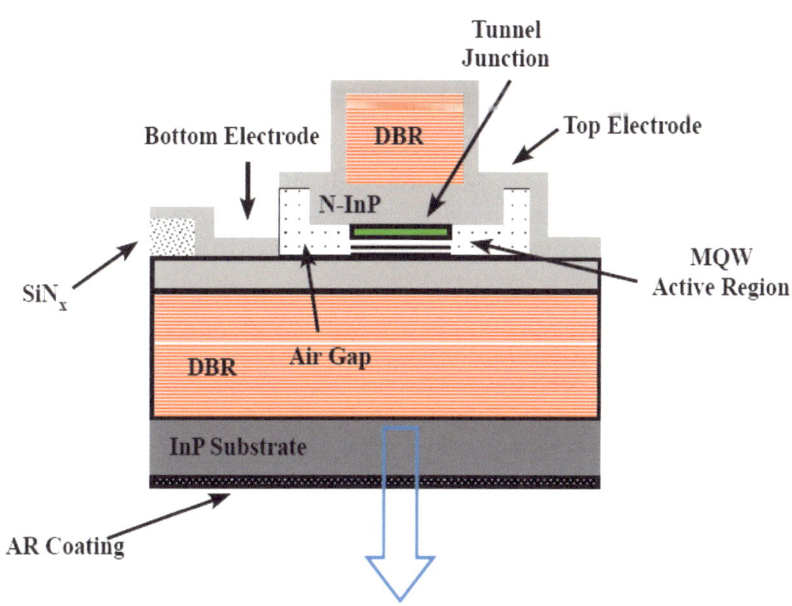

Fig. 6. MOVCD Grown monolithic structure of a 1.5μm RayCan VCSEL.

The device is modulated in amplitude and frequency at the same time. This phenomenon of "Phase-Amplitude Coupling" or the dynamic shift of the lasing frequency during modulation is known as "Frequency Chirping" or simply "Chirping".

Chirping broadens the linewidth of a laser. The extent to which a pulse broadens depends upon the amplitude of the modulating signal. Larger modulation amplitudes result in linewidths of the order of GHz 1. This spectral broadening at the time of modulation becomes more pronounced during the passage of the modulated pulse through an optical channel and the effective channel bandwidth is reduced. Direct modulation while being costeffective proves to be inefficient, in terms of deliverable bit rates, when compared to external modulation.

2.14 Intrinsic modulation limits

A semiconductor optical cavity, in essence, is a resonator. Like every resonator, or electrical circuit for that matter, its frequency response depends on its intrinsic parameters. In case of semiconductor lasers these parameters might be cavity volume, photon and electron populations, group velocity, gain compression factor etc. When directly modulated, a laser can not better the modulation frequency response already defined by these intrinsic parameters. On the other hand, the utilization of an external modulator provides a means to bypass the laser intrinsic parameters. The modulation response (or the deliverable bit rate) of the system is then defined by the external modulator and not the laser.

2.15 Long wavelength VCSEL optical injection-locking

It is clear from the description of the two above given problems that a viable optical system must minimize the effects of "Amplitude-Phase Coupling" and "Intrinsic Modulation Limits" in order to be efficient and acceptable. Once injection-locked, the master laser holds the frequency of the follower laser and makes it immune to carrier variations. This isolation from carrier variations appears as the reduction of chirp during direct modulation. In 1984, Lin et al. demonstrated the reduction of frequency chirping in a directly modulated semiconductor laser by the application of injection-locking technique (Lin et al., 1984).

Henry presented an approximate formula for the calculation of resonance frequency of optically injection-locked semiconductor lasers (Henry et al., 1985) but its significance was not appreciated at that time until Simpson and Meng demonstrated bandwidth and resonance frequency enhancements in late 90's (Simpson et al., 1996), (Meng et al., 1998). In 2002, a research group in University of California Berkley (UCB), led by Connie J. Chang-Hasnain reported the first optical injection-locking of a long wavelength VCSEL for 2.5Gbps transmission (Chang et al., 2002).

In 2003 long wavelength VCSEL chirp reduction and bandwidth enhancement were presented by the same research group (Chang et al., 2003) but there was a marked technical difference from their first publication. Whereas the first time optical injection-locking of a long wavelength VCSEL was carried-out using an identical VCSEL, the second demonstration used a Distributed FeedBack (DFB) laser to injection-lock a long wavelength VCSEL. The group has extensively published on the subject of the optical injection-locking of long wavelength VCSELs, but this pattern of locking a VCSEL with a DFB has remained unchanged since.

Several optical injection-locking studies regarding semiconductor lasers have reported frequency-chirp reduction (Lin et al., 1984), (Sung et al., 2004) increased RF link gain

(Chrostowski et al. 2003), (Chrostowski et al. 2007), improved relative intensity noise (Yabre et al., 2000) and diminished non-linear distortion (Chrostowski et al. 2007). Although the utilization of a DFB laser to injection-lock a VCSEL is excellent for demonstration of phenomena related to optical injection-locking, its practical application presents two major drawbacks. Without immediately entering into the details of these drawbacks, it can be logically inferred that both these drawbacks are related to the utilization of the DFB laser.
First of all the physical symmetry of the two lasers used is not the same. The VCSELs are a vertical emission device while the DFB lasers emit in the horizontal direction. This asymmetry renders the integration of an optical injection-locking system consisting of a DFB laser and a VCSEL very difficult. The second reason, of course, is the cost. One of the reasons of employing VCSELs in optical networks for high-speed data communication is their cost-effectiveness. Utilization of a DFB laser to improve the transmission and the component characteristics compromises this very objective. Due to these reasons despite all these advances regarding this very potent combination of semiconductor lasers and optical injection-locking, the phenomenon and its practical applications have not got any commercial breakthrough as yet.
With the arrival of Vertical-Cavity Surface-Emitting Lasers (VCSELs) on the commercial scene as low-cost, integrable sources, the efforts to revive the optical injection-locking phenomena were once again undertaken and follower VCSEL resonance frequencies ranging from 27 Ghz to 107 GHz have been reported in recent years (Chrostowski et al. 2007). The problem of non-integrability however is still unresolved due to the utilization of a distributed feedback (DFB) laser as master optical source to injection-lock a follower VCSEL. The DFB lasers have horizontal optical cavities. This physical asymmetry renders the monolithic integration very complicated. On the other hand the utilization of a powerful DFB laser compromises the economy of the setup by increasing the cost dramatically and fails the purpose of using a VCSEL in the first place. Clearly the solution to afore-mentioned problems would be to try a VCSEL-by-VCSEL optical injection-locking approach.

3. VCSEL rate equations

The previous chapter introduced the overall historical background of the subject and the motivation for undertaking this research work. In this chapter we will present a complete theoretical analysis of the optical injection-locking phenomenon in semiconductor lasers. A semiconductor laser cavity is essentially a resonator and its input (electrons) and output (photons) can be demonstrated to be interrelated to each other via cavity parameters. Like any other resonator cavity, the quality factor "Q" and the resonance frequency of this cavity can be controlled by manipulating its physical dimensions or intrinsic parameters.
Ordinarily, the only externally manipulable variable is the electron concentration that can be varied by changing the bias current. During the optical injection-locking process the internal parameters of the cavity are changed by varying the photon concentration inside the cavity. Since the locking effect is the result of interaction between two optical fields, the phase difference between the master and follower VCSELs can also be varied to achieve the desired effect.
Ordinarily, the only externally manipulable variable is the electron concentration that can be varied by changing the bias current. During the optical injection-locking process the internal parameters of the cavity are changed by varying the photon concentration inside the cavity. Since the locking effect is the result of interaction between two optical fields, the phase

difference between the master and follower VCSELs can also be varied to achieve the desired effect.

$$\frac{dN(t)}{dt} = \frac{\eta_i I}{qV_{act}} - (A + BN(t) + CN(t)^2)N(t) - v_g GS(t) \tag{1}$$

$$\frac{dS(t)}{dt} = \Gamma\beta BN(t)^2 + \Gamma v_g GS(t) - \frac{S(t)}{\tau_P} \tag{2}$$

Where N(t) and S(t) are the electron and photon densities, η_i the internal quantum efficiency, q the electron charge, V_{act} the active region volume, v_g the group velocity, β the spontaneous emission coefficient, Γ the confinement factor and τ_P the photon lifetime.
The spontaneous emission rate, R_{sp} is defined in terms of the constants A, B and C where A represents the Shockly-Read-Hall non-radiative recombination coefficient, B the bimolecular recombination coefficient and C the Auger non-radiative recombination coefficient. The gain G can be expressed as

$$G = a_0 \frac{N(t) - N_{tr}}{1 + \epsilon S(t)} \tag{3}$$

Where N_{tr} is the transparency carrier density, a_0 the differential gain coefficient and ε the gain compression factor.
A third equation describing the phase behaviour of the device can be introduced as follows:

$$\frac{d\phi(t)}{dt} = \frac{\alpha_H \Gamma v_g a_0}{2}(N(t) - N_{tr}) \tag{4}$$

α_H is the "Phase-Amplitude" coupling factor and is referred to as "Henry's Factor". It might be important to note here that equation 2.4 is not a coupled equation i.e. the term does not appear in equations 2.1 and 2.2. Lang proposed the utilization of three equations, instead of two, to model an optically injection-locked system (Lang, 1982). Lang's equations coupled the electric field variations in the cavity directly to carrier and phase variations and as such rendered the physical interpretation of the phenomenon somewhat cumbersome. In 1985, P. Gallion et al. presented the optical injection-locking rate equations that replaced cavity electrical field by photon number (Gallion & Debarge, 1985), (Gallion et al., 1985). Following the injection of optical power in the optical cavity, the dynamics of the follower laser change. This change can be mathematically presented by modifying the VCSEL rate equations to compensate for optical injection.

$$\frac{dN(t)}{dt} = \frac{\eta I}{qV_{act}} - (A + BN(t) + CN(t)^2)N(t) - v_g GS(t) \tag{5}$$

$$\frac{dS(t)}{dt} = \Gamma v_g GS(t) - \frac{S(t)}{\tau_P} + \frac{v_g}{L}\sqrt{S(t)S_{inj}}\cos(\theta) + \Gamma B\beta N(t)^2 \tag{6}$$

$$\frac{d\phi(t)}{dt} = \frac{\alpha_H \Gamma v_g a_0}{2}(N(t) - N_{tr}) - \Delta\omega - \frac{v_g}{2L}\sqrt{\frac{S_{inj}}{S(t)}}\sin(\theta) \tag{7}$$

It must be remarked that while the equation concerning the carrier density remains unchanged, the equations regarding the phase and the photon density are modified to accommodate for the effects of external light injection.

Two very important parameters of note, S_{inj} and θ are added to equations 2.6 and 2.7. S_{inj} represents the photon density injected inside the follower VCSEL optical cavity while θ denotes the phase difference between the master and follower optical fields so that:

$$\theta = \phi_{inj} - \phi(t) \tag{8}$$

$$\Delta\omega = \omega_{Master} - \omega_{Follower} \tag{9}$$

Apart from frequency detuning, phase difference and injected optical power, the fourth parameter which characterizes an optically injection-locked system is the "coupling coefficient" of a laser. It is defined as k_c and can be expressed mathematically as

$$k_c = \frac{v_g}{2L} \tag{10}$$

This coefficient describes the rate at which the injected electric field adds to the follower cavity electric field as a function of the VCSEL optical cavity length. 'L' is the length of the VCSEL optical cavity.

2.2 Locking Range Calculations

Solving equations (5) and (6) in the steady-state regime which renders $\frac{dS}{sT}$ and $\frac{dN}{sT}$ equal to zero gives the very important parametric equation:

$$\Delta\omega = k_c\sqrt{\frac{S_{inj}}{S}}\left[sin(\theta) - \alpha_H cos(\theta)\right] \tag{11}$$

The dependence of equation (11) on α_H can be elaborated by using the linear combination property for sinuses and cosines. Using this property we can write that:

$$\Delta\omega = k_c\sqrt{\frac{S_{inj}}{S}}\left[\sqrt{1+\alpha_H^2}\right] sin(\theta - \tan^{-1}\alpha_H) \tag{12}$$

This relation is important because it helps the calculation of effective locking bandwidth of an injection-locked system. Moreover it can be deduced that due to the presence of the sine function, the inequality is limited to the range of:

$$|\Delta\omega| \leq k_c\sqrt{\frac{S_{inj}}{S}}\left[\sqrt{1+\alpha_H^2}\right] \tag{13}$$

On the other hand, it appears that the oscillation limit for θ is between $-\pi/2$ and $\pi/2$. $\Delta\omega$ is then bounded by:

$$-k_c\sqrt{\frac{S_{inj}}{S}}\left[\sqrt{1+\alpha_H^2}\right] \leq \Delta\omega \leq k_c\sqrt{\frac{S_{inj}}{S}} \tag{14}$$

The asymmetry of the locking range can be explained both mathematically and physically. Mathematically speaking, if we observe (14), we can see that due to the multiplication with the term α_H on the left hand side, this relation becomes asymmetric with respect to α_H.

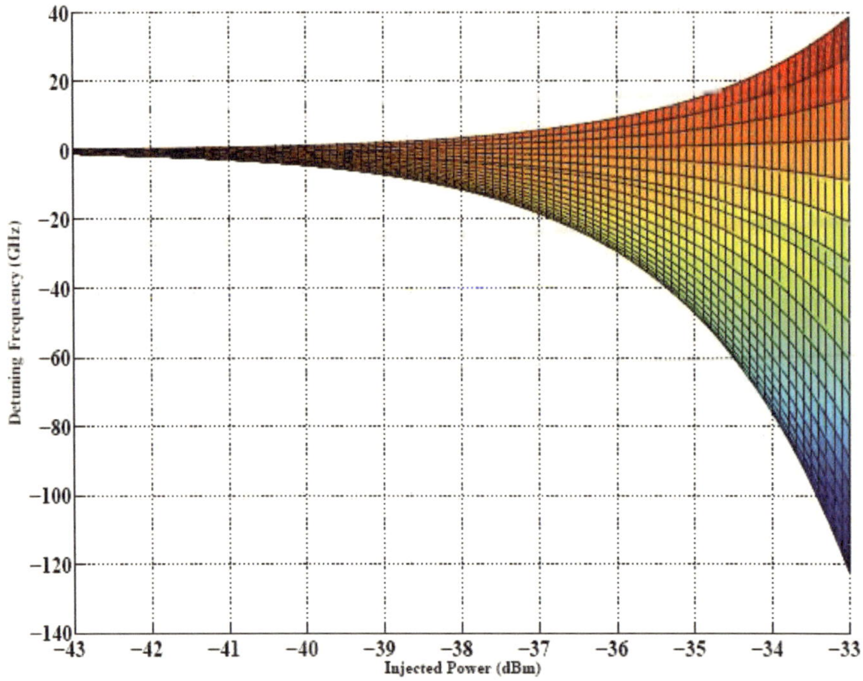

Fig. 7. 2D presentation of calculated locking range of a long wavelength VCSEL with $\alpha_H = 3$ showing the locking-range dependence on injected optical power.

Physically speaking, during the injection-locking of a semiconductor laser the increased photon population changes the refractive index and leads to a cavity wavelength shift in the longer wavelength direction and finally an asymmetric locking range. Calculated locking-range for $\alpha_H = 3$ is presented in fig. 7. It can be observed from equation (14) that a higher value of α_H leads to higher locking-range: A higher value of α_H favours locking in the negative frequency detuning range. In terms of locking-range characteristics, VCSELs are different from EELs. Locking range determines the extent of frequency enhancement of an optically injection-locked laser. Equation (14) shows that the locking-range depends on injected power and coupling coefficient k_c. Therefore mathematically it can be stated that the locking-range follows the variation of the term $k_c \sqrt{S_{inj}/S}$.

Since a VCSEL cavity is much shorter than an EEL cavity, VCSELs have typically very high values of k_c (10) as compared to those of conventional lasers. This implies that VCSEL locking-ranges are higher compared to EEL locking-ranges and can potentially lead to much higher resonance frequencies.

3.1 Small signal analysis

We begin by presenting once again the "Modified VCSEL Rate Equations". The small signal analysis is performed to derive the S_{21} response of an injection-locked VCSEL. Consider that a sinusoidal signal modulates a laser biased at current I. The resulting expression for current I then becomes:

$$I(t) = \bar{I} + \Delta I e^{j\omega t} \tag{15}$$

Similarly, the carrier, photon and phase variations can be described as follows:

$$N(t) = \overline{N} + \Delta N e^{j\omega t} \tag{16}$$

$$S(t) = \overline{S} + \Delta S e^{j\omega t} \tag{17}$$

$$\phi(t) = \overline{\phi} + \Delta\phi e^{j\omega t} \tag{18}$$

By putting

$$\dot{N} = \frac{dN}{dt} \tag{19}$$

$$\dot{S} = \frac{dS}{dt} \tag{20}$$

$$\dot{\phi} = \frac{d\phi}{dt} \tag{21}$$

We have:

$$\Delta\dot{N}(I, N, S) = \frac{\partial\dot{N}}{\partial I}\cdot\Delta I + \frac{\partial\dot{N}}{\partial N}\cdot\Delta N + \frac{\partial\dot{N}}{\partial S}\cdot\Delta S \tag{22}$$

$$\Delta\dot{S}(N, S, \phi) = \frac{\partial\dot{S}}{\partial N}\cdot\Delta N + \frac{\partial\dot{S}}{\partial S}\cdot\Delta S + \frac{\partial\dot{N}}{\partial\phi}\cdot\Delta\phi \tag{23}$$

$$\Delta\dot{\phi}(N, S, \phi) = \frac{\partial\dot{\phi}}{\partial N}\cdot\Delta N + \frac{\partial\dot{\phi}}{\partial S}\cdot\Delta S + \frac{\partial\dot{\phi}}{\partial\phi}\cdot\Delta\phi \tag{24}$$

The gain, as defined in (3), contains both the carrier and the photon terms. Partial differentiation of (3), with respect to the carrier and photon densities N and S, yields two new variables G_N and G_S, where G_N and G_S are defined as:

$$G_N = \frac{\partial G}{\partial N} = \frac{a_0}{1+\epsilon S} \tag{25}$$

$$G_S = -\frac{\partial G}{\partial S} = \frac{a_0\epsilon(N-N_{tr})}{(1+\epsilon S)^2} \tag{26}$$

Differentiating equation (5) with respect to N, S and ϕ therefore results in the following set of three equations:

$$\frac{\partial\dot{N}}{\partial N}\cdot\Delta N = (-(A+2BN+3CN^2) - v_g G_N S)\Delta N \tag{27}$$

$$\frac{\partial\dot{N}}{\partial S}\cdot\Delta S = (-v_g G + v_g G_S S)\Delta S \tag{28}$$

$$\frac{\partial\dot{N}}{\partial I}\cdot\Delta I = \frac{\eta_i}{qV_{act}}\Delta I \tag{29}$$

Similarly if we define a new variable ρ as:

$$\rho - \frac{v_g}{2L}\sqrt{\frac{S_{ini}}{S}} \tag{30}$$

And differentiate equation (6) with respect to N, S and ϕ we have the following set of equations:

$$\frac{\partial \dot{S}}{\partial N} \cdot \Delta N = (\Gamma v_g G_N S - 2\beta BN)\Delta N \tag{31}$$

$$\frac{\partial \dot{S}}{\partial S} \cdot \Delta S = (-\Gamma v_g G_S S + \rho cos(\theta))\Delta S \tag{32}$$

$$\frac{\partial \dot{S}}{\partial \phi} \cdot \Delta \phi = (-2 \cdot \rho S sin(\theta))\Delta \phi \tag{33}$$

The partial differentiation of the phase equation (7) with respect to N, S and ϕ results in the following set of equations:

$$\frac{\partial \dot{\phi}}{\partial N} \cdot \Delta N = \frac{\alpha_H \Gamma v_g a_0}{2}\Delta N \tag{34}$$

$$\frac{\partial \dot{\phi}}{\partial S} \cdot \Delta S = \frac{\rho sin(\theta)}{2S}\Delta S \tag{35}$$

$$\frac{\partial \dot{\phi}}{\partial \phi} \cdot \Delta \phi = -\rho cos(\theta)\Delta \phi \tag{36}$$

Linearised rate equations can then be expressed as:

$$\Delta \dot{N} = \frac{\eta_i}{qV_{act}}\Delta I - (A + 2BN + 3CN^2 + v_g G_N S)\Delta N - (v_g G - v_g G_S S)\Delta S \tag{37}$$

$$\Delta \dot{S} = (\Gamma v_g G_N S - 2\beta BN)\Delta N - (\Gamma v_g GS - \rho cos(\theta))\Delta S - (2 \cdot \rho S sin(\theta))\Delta \phi \tag{38}$$

$$\Delta \dot{\phi} = (\frac{\alpha_H \Gamma v_g a_0}{2})\Delta N + (\frac{\rho sin(\theta)}{2S})\Delta S - (\rho cos(\theta))\Delta \phi \tag{39}$$

Replacing the partial derivatives by intermediate variables gives

$$\Delta \dot{N} = \frac{\eta_i}{qV_{act}}\Delta I - \gamma_{NN}\Delta N - \gamma_{NS}\Delta S \tag{40}$$

$$\Delta \dot{S} = \gamma_{SN}\Delta N - \gamma_{SS}\Delta S - \gamma_{S\Phi}\Delta \phi \tag{41}$$

$$\Delta \dot{\phi} = \gamma_{\Phi N}\Delta N + \gamma_{\Phi S}\Delta S - \gamma_{\Phi\Phi}\Delta \phi \tag{42}$$

This can be readily arranged into a three equation matrix system as follows:

$$\frac{d}{dt}\begin{bmatrix}\Delta S\\ \Delta N\\ \Delta\Phi\end{bmatrix}=\begin{bmatrix}-\gamma_{SS} & \gamma_{SN} & -\gamma_{S\Phi}\\ -\gamma_{NS} & -\gamma_{NN} & \gamma_{N\Phi}\\ \gamma_{\Phi N} & \gamma_{\Phi S} & -\gamma_{\Phi\Phi}\end{bmatrix}\cdot\begin{bmatrix}\Delta S\\ \Delta N\\ \Delta\Phi\end{bmatrix}+\frac{\eta_i}{qV_{act}}\cdot\begin{bmatrix}0\\ \Delta I\\ 0\end{bmatrix} \tag{43}$$

Taking the Laplace transform of the equation set in order to pass from time-domain to frequency-domain, and arranging, yields:

$$\begin{bmatrix}\gamma_{SS}+j\omega & -\gamma_{SN} & \gamma_{S\Phi}\\ \gamma_{NS} & \gamma_{NN}+j\omega & 0\\ -\gamma_{\Phi S} & -\gamma_{\Phi N} & \gamma_{\Phi\Phi}+j\omega\end{bmatrix}\cdot\begin{bmatrix}\Delta\widetilde{S}\\ \Delta\widetilde{N}\\ \Delta\widetilde{\Phi}\end{bmatrix}=\frac{\eta_i\widetilde{I}}{V_{act}}\cdot\begin{bmatrix}0\\ 1\\ 0\end{bmatrix} \tag{44}$$

In order to solve this three-equation matrix system we have to calculate the determinant of the intermediate variable matrix:

$$\Lambda=\begin{vmatrix}\gamma_{SS}+j\omega & -\gamma_{SN} & \gamma_{S\Phi}\\ \gamma_{NS} & \gamma_{NN}+j\omega & 0\\ -\gamma_{\Phi S} & -\gamma_{\Phi N} & \gamma_{\Phi\Phi}+j\omega\end{vmatrix} \tag{45}$$

Where

$$\begin{aligned}\Lambda = & -j\omega^3\\ & -\omega^2\left(\gamma_{NN}+\gamma_{SS}+\gamma_{\Phi\Phi}\right)\\ & +j\omega\left(\gamma_{S\Phi}\gamma_{\Phi S}+\gamma_{SS}\gamma_{NN}+\gamma_{NS}\gamma_{SN}+\gamma_{NN}\gamma_{\Phi\Phi}+\gamma_{SS}\gamma_{\Phi\Phi}\right)\\ & +\left(\gamma_{S\Phi}\gamma_{\Phi S}\gamma_{NN}-\gamma_{S\Phi}\gamma_{NS}\gamma_{\Phi N}+\gamma_{SS}\gamma_{NN}\gamma_{\Phi\Phi}+\gamma_{SN}\gamma_{NS}\gamma_{\Phi\Phi}\right)\end{aligned} \tag{46}$$

Using the Kramer's rule, the photon density variation can be expressed as:

$$\Delta\widetilde{S}=\frac{\eta_i\widetilde{I}}{qV_{act}}\cdot\frac{\begin{vmatrix}0 & -\gamma_{SN} & \gamma_{S\Phi}\\ 1 & \gamma_{NN}+j\omega & 0\\ 0 & -\gamma_{\Phi N} & \gamma_{\Phi\Phi}+j\omega\end{vmatrix}}{\Lambda} \tag{47}$$

Simplifying equation 1.55 leads to:

$$\Delta\widetilde{S}=\frac{\eta_i\widetilde{I}}{qV_{act}}\cdot\frac{\left(\gamma_{SN}\gamma_{\Phi\Phi}-\gamma_{\Phi N}\gamma_{N\Phi}\right)+j\omega\gamma_{SN}}{\Lambda} \tag{48}$$

(6) and (7) can alternatively be solved to obtain a relation in terms of the phase difference between two lasers and is presented below:

$$\theta \;=\; \sin^{-1}\left[\frac{\Delta\omega}{k_c\sqrt{\frac{S_{inj}}{S}}\sqrt{1+\alpha_H^2}}+\tan^{-1}\alpha_H\right] \tag{49}$$

3.2 Numerical simulations

The mathematical model proposed above is implemented in MATLAB in order to observe the small-signal response of an injection-locked system. Table 1 summarises the VCSEL intrinsic parameters used to calculate the S_{21} response of an injection locked system (Bacou, 2008).

3.3 Simulation results

Recently the most significant application of optical injection-locking has been in the domain of resonance frequency enhancement. The enhanced resonance frequency can lead to an extended bandwidth many times the original device bandwidth. The modulation response of an injection-locked laser can be characterized as one of the following three:

- High Resonance Frequency, Low Bandwidth
- High Resonance Frequency, High Bandwidth
- Low Resonance Frequency, Low Bandwidth

Although the resonance frequency of an optically injection-locked laser increases with increasing injected power levels, the frequency detuning between the two lasers plays a very important role in determining the eventual characteristics of the S_{21} curve and finally the effective bandwidth. The above presented three different kinds of modulation responses depend on different locking conditions and parameters and are described in the following section.

3.4 High resonance frequency, low bandwidth

The high resonance frequency, low bandwidth operation regime can be attributed to a positive frequency detuning. Since the resonance frequency of an injection-locked system is the difference between the master laser frequency and the down shifted follower cavity frequency, positive frequency detuning results in very high resonance frequencies.

Parameter	Units	Value
η_i	-	0.8
τ_e	ns	0.61
N_{th}	cm^{-3}	5.33x10^{18}
N_{tr}	cm^{-3}	3.24x10^{18}
A	s^{-1}	1.1x10^{8}
B	cm^3/s	1x10^{-10}
C	cm^6/s	3.57x10^{-29}
a_0	cm^2	4.8x10^{-16}
v_g	cm/s	7.7x10^{9}
ϵ	cm^3	2.2x10^{-17}
τ_P	ps	6.4
S	cm^{-3}	2.5x10^{15}

Table 1. Long wavelength VCSEL intrinsic parameters used to simulate the small-signal injection-locking behaviour (Bacou, 2008).

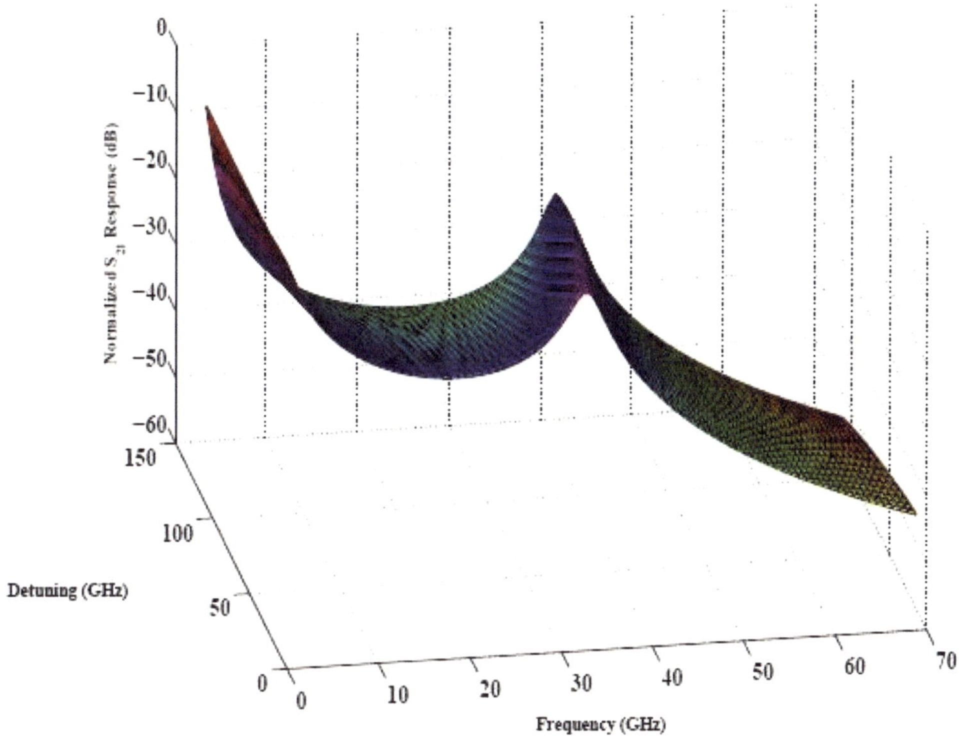

Fig. 8. Calculated S_{21} response of an optically injection-locked VCSEL with constant injected power and variable positive frequency detuning. The detuning is varied from 10 GHz to 110 GHz.

On the other hand, optically injection-locked systems can be mathematically defined as third-order systems and suffer from low-frequency dips due to the presence of a parasitic pole. Fig. 8 presents the simulated S_{21} response of an optically injection-locked VCSEL operating in the positive frequency detuning regime. The injected optical power is maintained constant for this set of curves in order to study the effects of variation in positive frequency detuning. The resonance frequency increases with increasing difference between the master and follower VCSEL frequencies.

Although from a telecommunication point of view, enhancement in resonance frequency is desired but the low frequency dip of an optically injection-locked system operating in the positive frequency detuning regime limits the effective bandwidth of the system and renders the system inefficient. This configuration therefore is not desired for operation in Datacom and telecommunication environments.

Very high resonance frequencies however can be beneficial for another very important application i.e. the generation of millimetre-wave signals. Since the proposal of the 60GHz band for the radio link frequency in broad-band cellular systems, the utilization of optical fibre for signal distribution has attracted much interest. This is due to low-loss nature of the optical fibres that are capable of transmitting data at very high bit rates. The main obstacle in the implementation of this scheme is the conception of a high frequency oscillator. Goldberg et. al had already demonstrated the generation of microwave signals using injection-locked laser diodes in 1983 [16], but the enthusiasm in the implementation of this scheme faded away due to the incipient nature of semiconductor lasers at that time.

3.5 High resonance frequency, high bandwidth

Positive frequency detuning can be employed to achieve very high resonance frequencies that could be useful for certain applications such as microwave and millimetre-wave signal generation but such high resonance frequencies imply very low cut-off frequencies due to low-frequency dip associated to positive frequency detuning. This situation can be improved by operating the laser at close to zero detuning. In such a configuration, the cut-off frequency increases with increase in injected power but due to very low frequency detuning value there is no loss at low frequency values. Frequency detuning has little or no effect on the resonance frequency of such a system and the bandwidth increase is dependent only on optical injected power. This configuration can be employed for broadband digital communications that require the transmission of very high bit rates. The third operation regime is defined by negative frequency detuning. Fig. 2.10 presents a set of simulated S_{21} curves with increasing negative frequency detuning. It is clear from Fig. 2.10 that for positive frequency detuning values, the follower VCSEL S_{21} response is un-damped with high resonance frequencies. However when the detuning between the two VCSELs is varied in the negative detuning operation regime, the S_{21} response curves start to become highly damped.

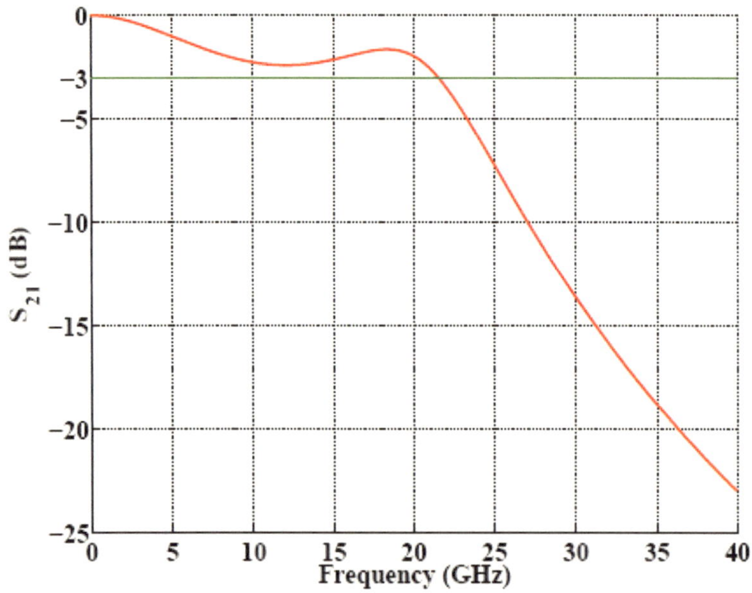

Fig. 9. Simulated S_{21} response of an optically injection-locked follower VCSEL showing cut-off frequency enhancement.

3.6 Low resonance frequency, low bandwidth

At the same time, the low frequency dip, exhibited due to positive frequency detuning operation regime starts to disappear. Finally at relatively high values of negative frequency detuning the S_{21} curves become over-damped and gradually the resonance peak vanishes.

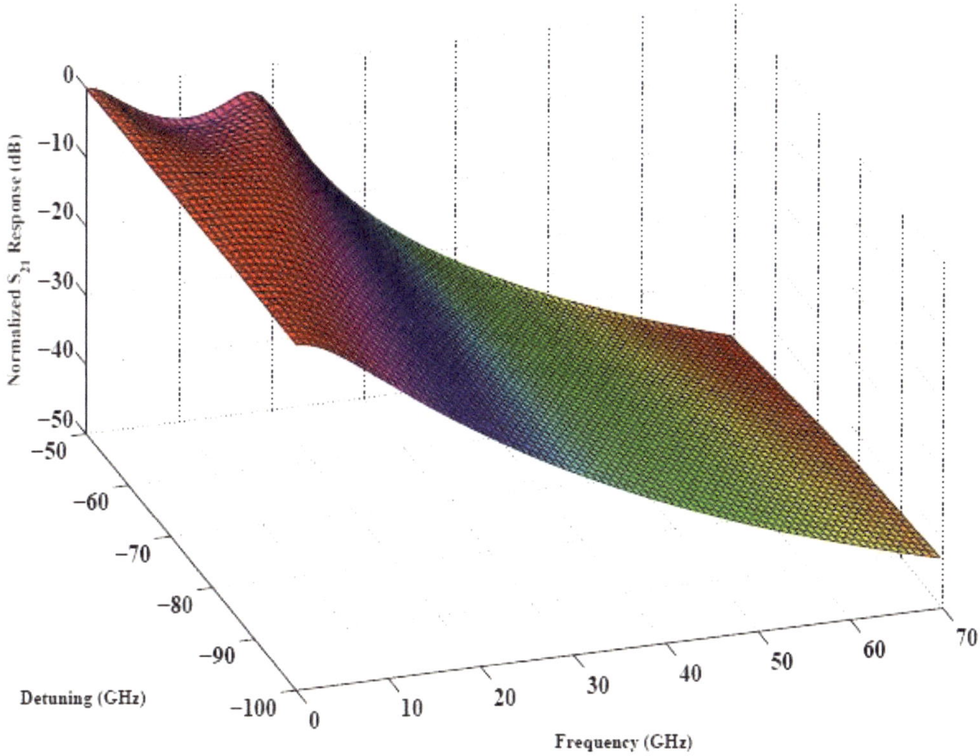

Fig. 10. Calculated S_{21} response of an optically injection-locked VCSEL with constant injected power and variable negative detuning. The detuning is varied from 10 GHz to -190 GHz.

The negative frequency detuning can hence be used to generate high low frequency gain S_{21} curves. This is particularly important for directly modulated optical fibre links. The losses in such links, apart from coupling and connector losses, are due to Electrical- Optical (E/O) and Optical-Electrical (O/E) conversion. Sung et al. have demonstrated that by injection locking a laser in negative frequency detuning regime the RF link gain can be improved by up to 10 dB [21].

3.7 Comparison between free-running and injection-locked VCSEL models

Fig. 11 presents a comparison between the free-running and injection-locked S_{21} response of a VCSEL. The frequency responses are plotted on a logarithmic scale in order to highlight the difference between the respective slopes of the two systems. The injection-locked system has a slope of -18dB/octave as compared to a slope of - 12dB/octave for a free-running VCSEL.

Another important difference of note is the low frequency dip in the optically injection-locked VCSEL S_{21} response which is due to the extra pole in the transfer function denominator. By putting S_{inj} and $\Delta\omega$ to zero the modified VCSEL rate equations are reduced to classical VCSEL rate equations.

The simulations, under different operating conditions, of optically injection-locked VCSELs presented in this chapter reveal certain interesting patterns. First of all, it must be noted that due to the very highly selective nature of the DBR mirrors used in the VCSEL manufacturing, a very small amount of light enters in the cavity. This is clear from the locking-range calculations presented in Fig. 7. It is therefore not the injected optical power intensity that is mainly responsible for injection-locked VCSELs' S_{21} curves variations. It is in fact the coupling factor k_c whose numerical value is responsible for high locking ranges, facility of injection-locking and high resonance frequencies.

Another important point is the S_{21} curve shape dependence on the frequency detuning value between the two VCSELs. The frequency detuning is the dominant factor in determining the shape of the S_{21} curve and whether it would be high resonance frequency under-damped response or a low resonance frequency high bandwidth flat response. This phenomenon can be explained by understanding the beat-frequency generation effect produced inside the follower VCSEL optical cavity.

Finally, due to optical coupling with the master laser, the dynamic response characteristics of the follower VCSEL change. Usually a two-equation mathematical model is utilized in VCSEL dynamic response simulations. This model gives way to a three-equation system which incorporates the effect of external light injection. Due to this third equation, the presence of a 3rd pole is observed in the transfer function of the optically injection-locked VCSEL. At positive detuning frequency values, this pole becomes dominant at low frequencies and causes the S_{21} response to suffer dips of several dBs which in turn severely limits the effective bandwidth of the system.

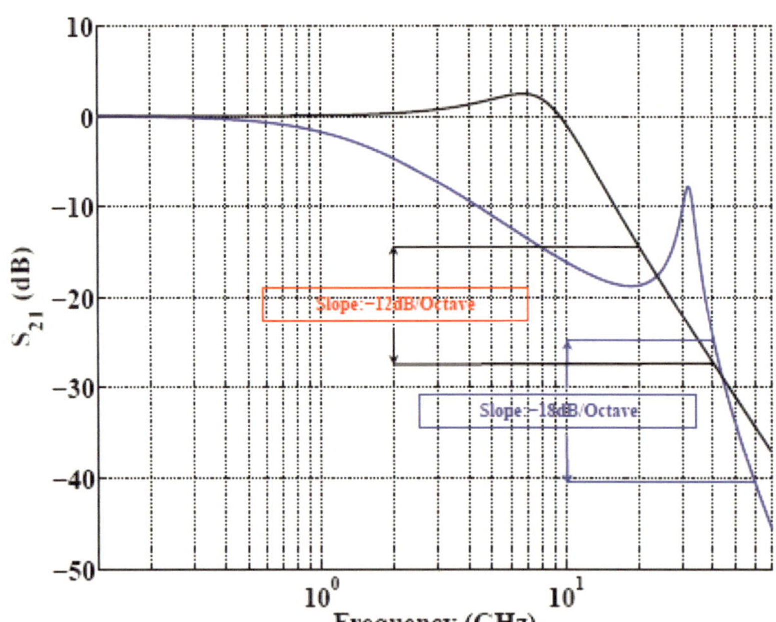

Fig. 11. Comparison between the free-running and injection-locked transfer functions of a VCSEL.

The injection-locking experiments carried-out during the course of this work evolved progressively in their complexity. The objective was to demonstrate and understand the VCSEL-by-VCSEL optical injection-locking phenomena under different operating constraints. Our focus was the study of variations in S_{21} response of injection-locked VCSELs under different injection powers and varying detuning frequencies.

4. Experiments using multimode lasers

4.1 Multimode Edge Emitting Lasers (EELs)

Optically injection-locked lasers are known to overcome many fundamental limitations of free-running systems. One of the very important improvements proposed by the employment of the optical injection-locking technique is the side-mode suppression of a multimode laser (Iwashita and Nakagawa, 1982). Fig. 12 presents the superimposed optical spectra of a free-running and an injection-locked laser diode. The Fabry-Pérot modes, visible in the free-running regime, undergo approximately 35 dB suppression when injection-locked using a DFB laser diode.

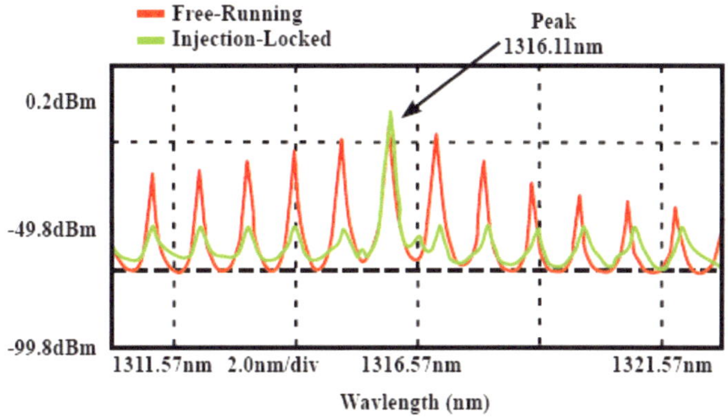

Fig. 12. The super-imposed spectra of a free running and an injection locked Fabry-Pérot EEL. Mode suppression can be observed in the injection locked spectrum.

In the stable locking regime the follower laser frequency is locked to the master laser lasing frequency. The injection-locked Fabry-Pérot mode therefore becomes dominant and the unlocked modes are suppressed. Iwashita et. al demonstrated the utilization of this method for the suppression of mode-partition noise [1]. The employment of optical-injection locking for side-mode suppression in VCSELs however is not very effective. This is due to the difference in the side-mode generation mechanism between the EELs and the VCSELs. A detailed analysis of side-mode generation is presented in the following section.

Single-mode operation of the follower laser however is highly desirable due to another very important reason. As presented in figure 3.2, the locking-range of an injection-locked laser, in the "stable operation region", is dependent on the injected optical power. This effective locking-range is exploitable only if the follower laser is single-mode. If the follower laser is multimode, the achievable detuning frequency is limited by the Free Spectral Range (FSR) of the follower laser. At large detuning frequencies, the master laser might come closer to an adjacent longitudinal mode and in that case, it will lock the adjacent longitudinal mode instead of sweeping the entire locking range with previously locked mode. This mode-hopping reduces the effective "locking" and hence "operation range" of an injection-locked system.

4.2 Multimode VCSELs

Fig. 13 presents the optical spectrum of a multimode VCSEL. The VCSEL in question is manufactured by Vertilas with a threshold current of 6 mA and peak output optical power of 20 mW. The VCSEL chip was powered-up using a probe-station. The master laser is single-mode Vertilas VCSEL emitting in the 1.55μm range. A comparison with Fig. 14 shows that optical injection-locking fails to produce an effect similar to that demonstrated previously on multimode EELs. Although nominal side-mode suppression is observed in the injection-locked follower VCSEL spectrum, the emission spectrum rests multimode.

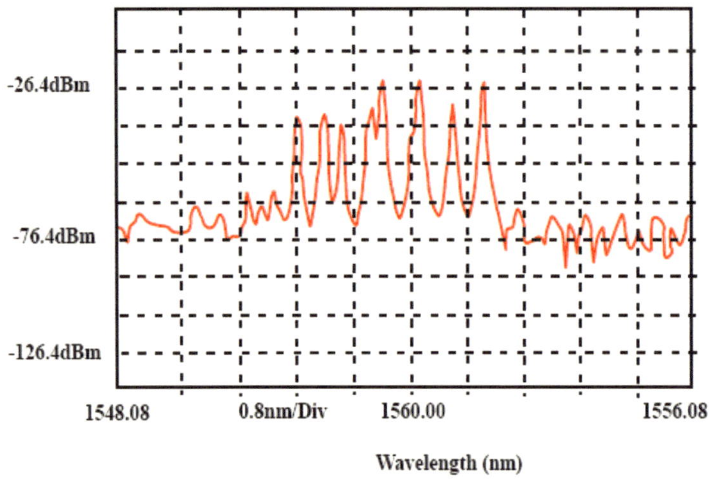

Fig. 13. Optical spectrum of an Vertilas multimode "Power" VCSEL. The VCSEL threshold current is about 6 mA.

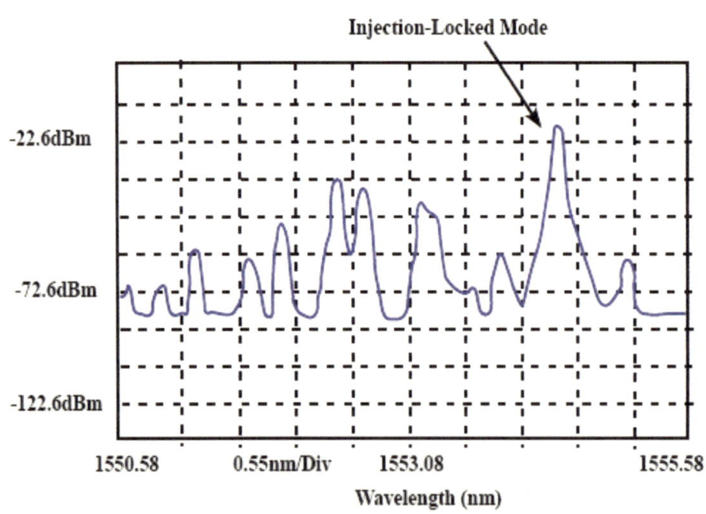

Fig. 14. Spectrum of an optically injection-locked multimode Vertilas VCSEL. The threshold current is about 6 mA. Very feeble side-mode suppression is observed due to injection-locking.

4.3 Experiments using single-mode VCSELs

This can be explained by developing an understanding of the side-mode generation phenomena in VCSELs. The active region of a VCSEL is very short as compared to that of an EEL, essentially of the order of the emission wavelength. Consequently, only one Fabry-Pérot mode exists in the VCSELs, since the physical dimensions of the cavity eliminate the

possibility of longitudinal multi-mode lasing action. Therefore VCSELs are fundamentally single-mode emission devices. However, the confinement and guiding of the optical field thus generated is made very difficult due to a very peculiar VCSEL structural characteristic.
VCSEL design suggests the sharing of a common path for photons and carriers, moving through the DBRs. This leads to the heating of the DBRs due to carrier flow and results in a variable refractive index distribution inside the VCSEL optical cavity. The creation of non-uniform refractive index zones inside the optical cavity leads to different optical paths and has an overall dispersive effect. This phenomenon is known as "Thermal Lensing".
The electrons passing through the DBRs tend to concentrate on the edge of the active zone due to the oxide aperture-based carrier guiding. A higher carrier concentration at the fringes of the active zone translates into higher photon generation at the edges of the active zone. Instead of being concentrated in the centre of the optical cavity, in the form of a single transverse mode, the optical energy is repartitioned azimuthally inside the optical cavity. The creation of non-uniform refractive index zones within the VCSEL optical cavity, changes the effective optical path inside the cavity which manifests itself in the form of undesired side-modes. Since the VCSEL sidemodes are a consequence of spatial energy distribution, they are referred to as "Spatial" or "Transverse Modes". Higher bias currents therefore imply high optical power and in consequence a higher number of transverse modes. An oxide-aperture is employed in order to achieve optimal current confinement and to block unwanted transverse modes. The oxide-aperture diameter determines the multimode or single mode character of a VCSEL. VCSELs having oxide aperture diameters greater than 5µm exhibit a multimode behaviour.
It can also be inferred from the above discussion that for the type of VCSELs employing the oxide-aperture technology for optical confinement, single mode VCSELs almost always have emission powers less than those of multimode VCSELs. Since the Vertilas VCSEL used here is a high power device, it has a Buried Tunnel Junction (BTJ) diameter of 20µm and is therefore distinctly multimode. Since optical injection-locking favours single-mode operation by eliminating longitudinal modes and since the modes generated in VCSELs are not longitudinal, the employment of optical injection-locking for single-mode VCSEL operation is not very effective.

4.4 Experiments using vertilas VCSELs

A logical step, after trying optical injection-locking of multimode VCSELs, was to attempt the injection-locking of single-mode VCSELs. The VCSELs used for initial injection-locking experiments were manufactured by Vertilas GmbH. These are single-mode, TO-46 packaged, pigtailed, Buried Tunnel Junction (BTJ) devices with an emission wavelength of 1.55µm. The L-I curve of the follower VCSEL is presented in figure 3.5 (a). The mode suppression ratio between the fundamental and the side-mode is approximately 40 dBs. The injection-locking experiments using Vertilas VCSELs were simple to carry-out due to the pigtailed nature of the components that made the optical power-injection inside the follower VCSEL cavity relatively easy. The well known phenomenon of sidemode suppression (as demonstrated with EELs and presented in figure 12) was observed. When the VCSEL satellite mode is optically injection-locked, the fundamental mode undergoes a rapid diminution and the VCSEL output optical power shifts to the side-mode wavelength. However, other than being a proof of concept demonstration, this exercise proved to be of little significance. The real price of this ease of manipulation was paid in terms of a degraded frequency response.

The TO-46 package cut-off frequency was about 5 Ghz which was well below the component cut-off frequency (11 GHz). The observation of injection-locked VCSELs' S_{21} response under various injection conditions was therefore not possible.

4.5 Experiments using RayCan VCSELs

The optically injection-locked follower VCSEL S_{21} responses presented above provide very interesting results. Especially the availability of on-chip components allows the observation of parasitics-free free-running and injection-locked S_{21} responses. It was noticed however that the Master VCSEL is not modulated for these injection-locking experiments and hence needs not be on-chip.

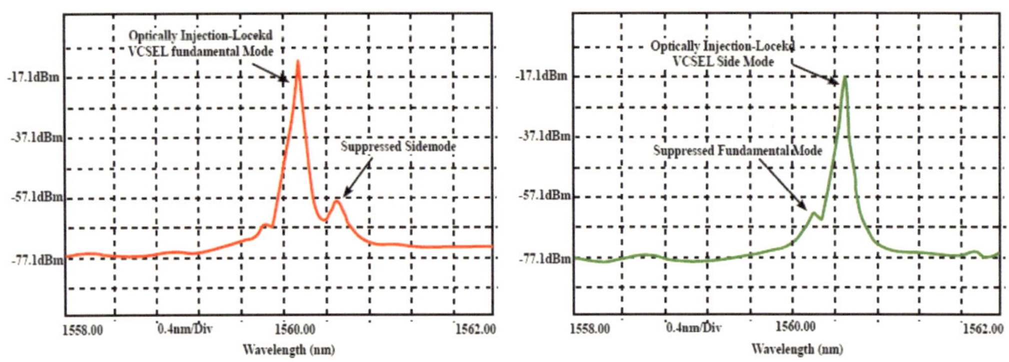

(a) (b)

Fig. 15. (a) Optical spectrum of an optically injection-locked Vertilas VCSEL. The locking of fundamental mode further suppresses the side-mode. (b) Optical spectrum of an optically injection-locked Vertilas VCSEL. The locking of side mode has suppressed the fundamental lasing mode. Notice the position of the suppressed modes in the two different cases.

The employment of a fibred master VCSEL will facilitate the injection-locking experiments in the following ways:

- This will allow the utilization of only one probe-station instead of two thus reducing the test-bench size and minimizing its complexity.
- This will increase the magnitude of available optical power since the coupling losses on the master VCSEL side would be eliminated.

Also, injection-locking experiments in the static domain such as linewidth, polarization and RIN measurements could be carried out using fibred follower VCSEL without suffering from packaging parasitics performance penalties. It was then decided to carry-out injection-locking experiments using commercially available RayCan VCSELs.

4.6 RayCan VCSELs structure

The structure of a 1.3µm RayCan VCSEL is presented in Fig. 6. RayCan VCSELs are bottom-emitting type, as has been explained above. As far as the incorporation of a bottom-emitting VCSEL in an optical sub-assembly is concerned, the application of normal integration techniques such as wire-bonding or flip-chip designs is easily applicable. However, probe-station testing of bottom-emitting components poses some challenging problems. Bottom-emission implies the existence of electrodes on the reverse side of the VCSEL chip, as shown in figure 3.20. This means that in order to power-up the VCSEL, using coplanar probes, the chip has to be inverted.

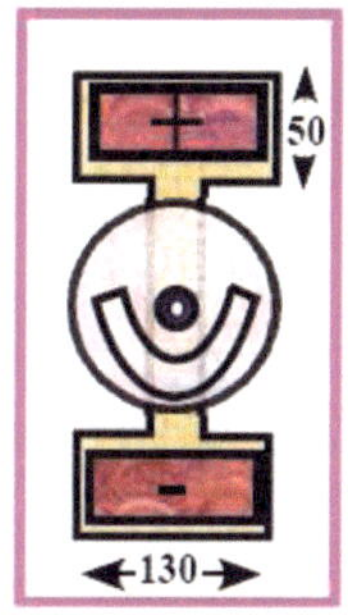

Fig. 16. Bottom-emitting on-chip RayCan VCSEL with 1.3μm operation wavelength.

The chip-inversion, in turn, implies the impossibility of optical power collection with a single-mode or multimode fibre. On the other hand, if the chip is used in the top-emitting configuration, it becomes impossible to power-up the chip using probes.

Another problem was the distance between the two electrodes. The probes used for VCSEL testing have a pitch of 125 μm. However the distance between the two RayCan VCSEL electrodes is about 300 μm. Without using 300 μm pitch probes, it would have been impossible to power-up the VCSELs anyway. These two problems were solved by getting the VCSEL chip integrated to a sub-mount. The sub-mount was prepared by RayCan for VCSEL integration with a monitoring photodiode, inside a TO-46 package. As per our demand, the VCSEL chips were integrated to the sub-mounts and delivered to us unpackaged. Furthermore, the intent of optical injection-locking experiments was observation of the enhanced S_{21} response. This objective was compromised by the employment of the sub-mount, as the S_{21} response was limited by the parasitic transmission line frequency.

The presence of air-gaps in the VCSEL structure implies lower intrinsic cut-off frequencies. The inevitable utilization of the sub-mount assembly, combined with the above-mentioned structural deficiency, renders these VCSELs relatively low frequency operation devices. It is perhaps due to this reason that the 10 Gbps modules supplied by RayCan employ four VCSELs in parallel configuration to achieve 10Gbps bit rate, as opposed to Vertilas 10Gbps modules that are composed of only one VCSEL.

4.7 Injection locking experiments

The availability of fibred components however simplified the test-bench considerably. In stead of using two probe-stations for master and follower VCSELs respectively, only one probe-station was used since only the follower VCSEL was used in the on-chip configuration.

The utilization of a pigtailed master VCSEL also increased the available optical power and allowed the elimination of the OSA from the injection-locking setup. Fig. 17 presents the optical injection-locking test-bench used for RayCan VCSEL experiments schematically. The utilization of a pigtailed master VCSEL made the testbench considerably compact and increased the available optical power but despite these advantages, the follower VCSEL injection-locked S_{21} spectra do not exhibit very large resonance frequencies. Fig. 18 presents the S_{21} response of an optically injection-locked RayCan follower VCSEL, in the positive frequency detuning regime. Compared to the free-running responses presented, it is clear that an increased resonance frequency is observed. Also, due to operation in the positive frequency detuning regime, the S_{21} is un-damped and therefore the resonance peak is very pronounced.

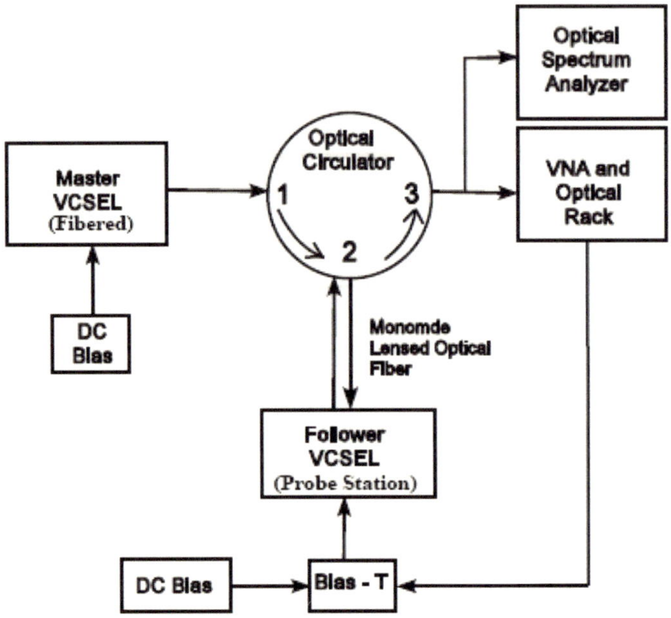

Fig. 17. Schematic representation of the test-bench employed for injection-locking experiments using RayCan VCSELs emitting at 1.3μm.

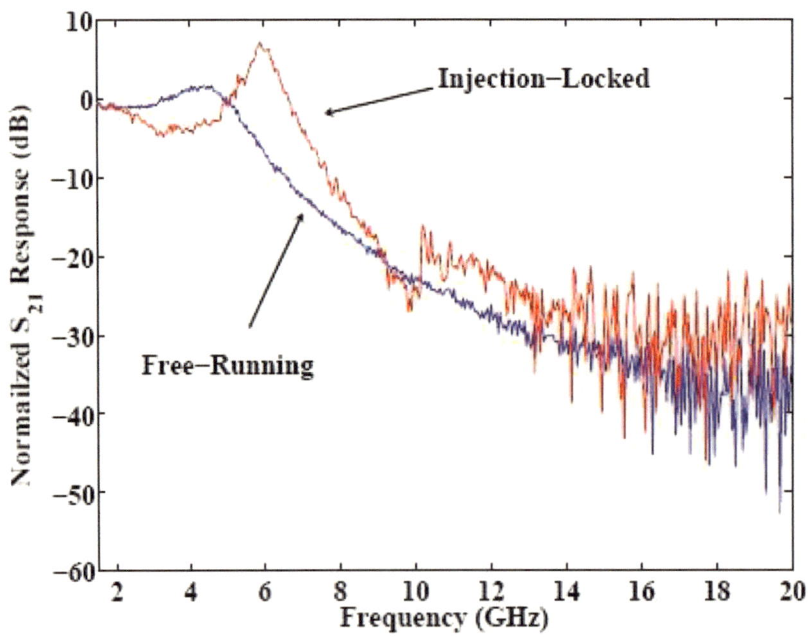

Fig. 18. S_{21} response of an optically injection-locked RayCan VCSEL emitting at 1.3μm operating in the positive frequency detuning regime.

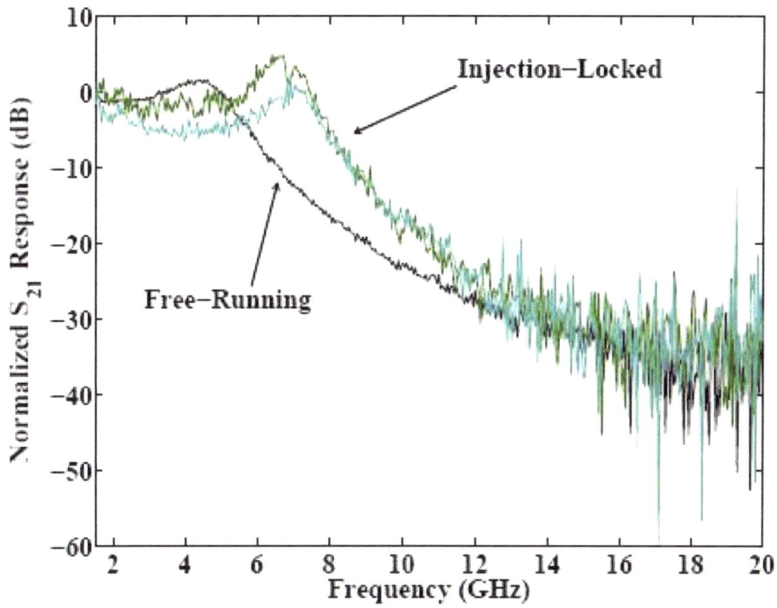

Fig. 19. S_{21} response of an optically injection-locked RayCan VCSEL emitting at 1.3µm operating in the positive frequency detuning regime.

5. Conclusion and discussion

Experimental studies of VCSEL-by-VCSEL optical injection-locking phenomena were presented in this chapter. It was demonstrated that optical injection-locking suppresses only the Fabry-Pérot modes of an optical cavity. The transverse modes commonly found in VCSELs remain largely unaffected by optical injection-locking. VCSEL-by- VCSEL optical injection-locking was presented using fibred single-mode VCSELs and fundamental and sidemode suppression phenomena were demonstrated.

Optical injection-locking of on-chip VCSELs was suggested, in order to observe the parasitics free S_{21} response. Three different operation regimes were explored using VCSEL-by- VCSEL optical injection-locking. Resonance frequencies as high as 7 GHz were presented for follower VCSELs operating in positive frequency detuning regimes. It was however observed that positive frequency detuning increases the resonance frequency but limits the effective bandwidth of the injection-locking system which is not desirable for VCSEL employment in high bit rate telecommunication system.

The zero or slightly negative detuning regime proposes flat, highly damped S_{21} curves. An increase in injected optical power, while remaining keeping the VCSELs in negative detuning configuration, results in the increase of effective bandwidth. Effective bandwidths as high as 10 GHz, using optical injection-locking, have been demonstrated. It must be noted that the free-running cut-off frequency of the VCSELs used is about 5 GHz. In order to simplify the optical injection-locking setup, the utilization of a fibred master VCSEL has been proposed. Such a configuration also increases the effective available optical power. Optically injection-locked follower VCSEL S_{21} response has been presented in different operating conditions. Experimental results and numerical calculations using the mathematical model have been compared.

6. References

T. Baba, Y. Yogo, K. Suzuki, F. Koyama, and K. Iga, "Near room temperature continuous wave lasing characteristics of GaInAsP/InP surface emitting laser," *Electronics Letters*, vol. 29, no. 10, pp. 913–914, May 1993.

A. Bacou, A. Rissons, and J.-C. Mollier, "Spectral behavior of long wavelength VCSELs," *SPIE Vertical-Cavity Surface-Emitting Lasers* XII, vol. 6908, no. 1, 2008.

A. Bacou, "Caractérisation et Modélisation optoélectronique de VCSEL à grande longueur d'onde pour sousensembles optiques intégrés," *Ph.D. dissertation*, Institut Supérieur de l'Aéronautique et de l'Espace, 2008.

R. L. Batdorf, G. C. Dacey, R. L. Wallace, and D. J. Walsh, "Esaki Diode in InSb," *Journal of Applied Physics*, vol. 31, no. 3, pp. 613–614, 1960.

A. C. Bordonalli, C. Walton, and A. J. Seeds, "High-Performance Phase Locking of Wide Linewidth Semiconductor Lasers by Combined Use of Optical Injection Locking and Optical Phase-Lock Loop," *J. Lightwave Technol.*, vol. 17, no. 2, p. 328, 1999.

J. Boucart, C. Starck, F. Gaborit, A. Plais, N. Bouche, E. Derouin, L. Goldstein, C. Fortin, D. Carpentier, P. Salet, F. Brillouet, and J. Jacquet, "1-mW CW-RT monolithic VCSEL at 1.55 µm," *IEEE Photonics Technology Letters*, vol. 11, no. 6, pp. 629–631, Jun 1999.

C. A. Burrus, "Indium Phosphide Esaki Diodes," *Solid-State Electronics*, vol. 5, p. 357, 1962.

C.-H. Chang, L. Chrostowski, C. Chang-Hasnain, and W. Chow, "Study of long wavelength VCSEL-VCSEL injection locking for 2.5-Gb/s transmission," *IEEE Photonics Technology Letters*, vol. 14, no. 11, pp. 1635–1637, Nov 2002.

C.-H. Chang, L. Chrostowski, and C. Chang-Hasnain, "Injection locking of VCSELs," *IEEE Journal of Selected Topics in Quantum Electronics*, vol. 9, no. 5, pp. 1386–1393, Sept.-Oct. 2003.

L. Chrostowski, C.-H. Chang, and C. Chang-Hasnain, "Reduction of Relative Intensity Noise and Improvement of Spur-Free Dynamic Range of an Injection-Locked VCSEL," *The 16th Annual Meeting of the IEEE Lasers and Electro-Optics Society*. LEOS 2003, vol. 2, pp. 706–707 vol.2, Oct. 2003.

L. Chrostowski, X. Zhao, and C. Chang-Hasnain, "Microwave Performance of Optically Injection-locked VCSELs," *IEEE Transactions on Microwave Theory and Techniques*, vol. 54, no. 2, pp. 788–796, Feb. 2006.

L. Chrostowski, F. B., W. Hoffman, M.-C. Amann, S. Wieczorek, and W. Chow, "40 GHz Bandwidth and 64 GHz Resonance Frequency in Injection-Locked 1.55 µm VCSELs," *IEEE Journal of Selected Topics in Quantum Electronics*, Vol. 13, No. 5, 2007.

A. G. Chynoweth, W. L. Feldman, and R. A. Logan, "Excess Tunnel Current in Silicon Esaki Junctions," Physics Review, vol. 121, p. 684, 1961. C. A. Burrus, "Indium Phosphide Esaki Diodes," *Solid-State Electronics*, vol. 5, p. 357, 1962.

I. Dias, B. Nabet, A. Kohl, J. Benchimol, and J. Harmand, "Electrical and optical characteristics of n-type-doped distributed Bragg mirrors on InP," *IEEE Photonics Technology Letters*, vol. 10, no. 6, pp. 763–765, Jun 1998.

L. Esaki, "Long journey into tunneling," *Proceedings of the IEEE*, vol. 62, no. 6, pp. 825–831, June 1974.

M. Espana-Boquera and A. Puerta-Notario, "Noise Effects in Injection Locked Laser Simulation: Phase Jumps and Associated Spectral Components," *Electronics Letters*, vol. 32, no. 9, pp. 818–819, Apr 1996.

N. M. Froberg and K. Y. Lau, "Ultrahigh-efficiency Microwave Signal Transmission Using Tandem-Contact Single Quantum Well GaAlAs Lasers," in *Optoelectronic Signal Processing for Phased-Array Antennas* II, vol. 1217, no. 1, 1990, pp. 2–5.

P. Gallion and G. Debarge, "Influence of amplitude-phase coupling on the injection locking bandwidth of a semiconductor laser," *Electronics Letters*, vol. 21, no. 7, pp. 264–266, 28 1985.

P. Gallion, H. Nakajima, G. Debarge, and C. Chabran, "Contribution of spontaneous emission to the linewidth of an injection-locked semiconductor laser," *Electronics Letters*, vol. 21, no. 14, pp. 626–628, 4 1985.

M. Grabherr, B. Weigl, G. Reiner, and K. Ebeling, "Comparison of Proton Implanted and Selectively Oxidized Vertical-Cavity Surface-Emitting Lasers," in *Conference on Lasers and Electro-optics*, 1996. CLEO/Europe., Sep 1996, pp. 165–165.

K. Hasebe, Y. Onishi, and F. Koyama, "Novel polarization controller based on injection-locked vertical-cavity surface-emitting laser," *IEICE Electronics Express*, vol. 2, no. 8, pp. 274–279, 2005.

C. Henry, N. Olsson, and N. Dutta, "Locking Range and Stability of Injection Locked 1.54μm InGaAsp Semiconductor Lasers," *IEEE Journal of Quantum Electronics*, vol. 21, no. 8, pp. 1152–1156, Aug 1985.

N. Holonyak and I. Lesk, "Gallium-Arsenide Tunnel Diodes," *Proceedings of the IRE*, vol. 48, no. 8, pp. 1405–1409, Aug. 1960.

J. Jewell, J. Harbison, A. Scherer, Y. Lee, and L. Florez, "Vertical-cavity surfaceemitting lasers: Design, growth, fabrication, characterization," *IEEE Journal of Quantum Electronics*, vol. 27, no. 6, pp. 1332–1346, Jun 1991.

J. Kahn, "1 Gbit/s PSK homodyne transmission system using phase-locked semiconductor lasers," *IEEE Photonics Technology Letters*, vol. 1, no. 10, pp. 340–342, Oct 1989.

H. Kawaguchi, Y. Yamayoshi, and K. Tamura, "All-Optical Format Conversion Using an Ultrafast Polarization Bistable Vertical-Cavity Surface-Emitting Laser," *Lasers and Electro-Optics*, 2000. (CLEO 2000). Conference on, pp. 379–380, 2000.

S. Kobayashi and T. Kimura, "Coherence of Injection Phase-Locked AlGaAs Semiconductor Laser," *Electronics Letters*, vol. 16, no. 17, pp. 668–670, 14 1980.

R. Koda, "All-Epitaxial, Long-Wavelength, Vertical-Cavity Surface-Emitting Lasers using Bipolar Cascaded Active Region for High Differential Quantum Efficiency," *Ph.D. dissertation*, UNIVERSITY OF CALIFORNIA Santa Barbara, 2005.

R. Lang, "Injection Locking Properties of a Semiconductor Laser," *IEEE Journal of Quantum Electronics*, vol. 18, no. 6, pp. 976–983, Jun 1982.

Y. Lee, J. Jewell, A. Scherer, S. McCall, J. Harbison, and L. Florez, "Room-temperature continuous-wave verticalcavity single-quantum-well microlaser diodes," *Electronics Letters*, vol. 25, no. 20, pp. 1377–1378, Sept. 1989.

C. Lin and F. Mengel, "Reduction of Frequency Chirping and Dynamic Linewidth in High-Speed Directly Modulated Semiconductor Lasers by Injection Locking," *Electronics Letters*, vol. 20, no. 25, pp. 1073–1075, 1984.

N. Margalit, D. Babic, K. Streubel, R. Mirin, R. Naone, J. Bowers, and E. Hu, "Submilliamp long wavelength vertical cavity lasers," *Electronics Letters*, vol. 32, no. 18, pp. 1675–, Aug 1996.

M. Ortsiefer, M. Lohner, R. Shau, G. Böhm, and M.-C. Amann, "Low-resistance InGaAs tunnel junctions on InP for long-wavelength VCSELs," in *Semiconductor and Integrated Optoelectronics Conference* (SIOE), Cardiff, 1999.

M. Ortsiefer, R. Shau, G. Bohm, F. Kohler, and M.-C. Amann, "Room-temperature operation of index-guided 1.55 μm InP-based vertical-cavity surface-emitting laser," *Electronics Letters*, vol. 36, no. 5, pp. 437–439, Mar 2000.

M.-R. Park, O.-K. Kwon, W.-S. Han, K.-H. Lee, S.-J. Park, and B.-S. Yoo, "All-epitaxial InAlGaAs-InP VCSELs in the 1.3-1.6-μm Wavelength Range for CWDM Band

Applications, " *IEEE Photonics Technology Letters*, vol. 18, no. 16, pp. 1717-1719, Aug. 2006.

J. Piprek, *Semiconductor Optoelectronic Devices. Introduction to Physics and Simulation.* Academic Press, 2003.

Y. Qian, Z. Zhu, Y. Lo, H. Hou, M. Wang, and W. Lin, "1.3-µm Vertical-cavity surface-emitting lasers with double-bonded GaAs-AlAs Bragg mirrors," *IEEE Photonics Technology Letters*, vol. 9, no. 1, pp. 8–10, Jan. 1997.

P. Salet, F. Gaborit, P. Pagnod-Rossiaux, A. Plais, E. Derouin, J. Pasquier, and J. Jacquet, "Room temperature pulsed operating of 1.3 µm vertical cavity lasers including bottom InGaAsP/InP multilayer bragg mirrors," *Electronics Letters*, vol. Vol. 33 No. 24, pp. 2048–2049, 1997.

N. Schunk and K. Petermann, "Noise Analysis of Injection-Locked Semiconductor Injection Lasers," *IEEE Journal of Quantum Electronics*, vol. 22, no. 5, pp. 642–650, May 1986.

R. Shau, M. Ortsiefer, J. Rosskopf, G. Boehm, C. Lauer, M. Maute, and M.-C. Amann, "Long-wavelength InPbased VCSELs with buried tunnel junction: properties and applications," *SPIE*, Vertical-Cavity Surface- Emitting Lasers VIII, vol. 5364, no. 1, pp. 1–15, 2004.

T. Simpson, J. Liu, and A. Gavrielides, "Small-signal analysis of modulation characteristics in a semiconductor laser subject to strong optical injection," *IEEE Journal of Quantum Electronics*, vol. 32, no. 8, pp. 1456–1468, Aug 1996.

H. Soda; K. Iga, C. Kitahara, and Y. Suematsu, "GaInAsP/InP Surface Emitting Injection Lasers," *Japanese Journal of Applied Physics*, vol. 18, no. 12, pp. 2329–2330, 1979.

H.-K. Sung, T. Jung, D. Tishinin, K. Liou, W. Tsang, and M. Wu, "Optical Injection- Locked Gain-Lever Distributed Bragg Reflector Lasers With Enhanced RF Performance," *IEEE International Topical Meeting on Microwave Photonics*. MWP'2004., pp. 225–228, Oct. 2004.

A. Syrbu, V. Iakovlev, G. Suruceanu, A. Caliman, A. Mereuta, A. Mircea, C.-A. Berseth, E. Diechsel, J. Boucart, A. Rudra, and E. Kapon, "VCSELs Emitting in the 1310- nm Waveband for Novel Optical Communication Applications," *SPIE Vertical-Cavity Surface-Emitting Lasers* IX, vol. 5737, no. 1, pp. 167–173, 2005.

A. Syrbu, V. Iakovlev, G. Suruceanu, A. Caliman, A. Rudra, A. Mircea, A. Mereuta, S. Tadeoni, C.-A. Berseth, M. Achtenhagen, J. Boucart, and E. Kapon, "1.55-µm optically pumped wafer-fused tunable VCSELs with 32-nm tuning range," *IEEE Photonics Technology Letters*, vol. 16, no. 9, pp. 1991–1993, Sept. 2004.

H. Toba, Y. Kobayashi, K. Yanagimoto, H. Nagai, and M. Nakahara, "Injection-locking Technique Applied to a 170 km Transmission Experiment at 445.8 Mbit/s," *Electronics Letters*, vol. 20, no. 9, pp. 370–371, 26 1984.

K. J. Vahala, M. A. Newkirk, and T. R. Chen, "The Optical Gain Lever: A Novel Gain Mechanism in the Direct Modulation of Quantum Well Semiconductor Lasers," *Applied Physics Letters*, vol. 54, no. 25, pp. 2506–2508, 1989.

G. Yabre, H. De Waardt, H. van den Boom, and G.-D. Khoe, "Noise Characteristics of Single-Mode Semiconductor Lasers Under External Light Injection," *IEEE Journal of Quantum Electronics*, vol. 36, no. 3, pp. 385–393, Mar 2000.

X. Zhao, M. Moewe, L. Chrostowski, C.-H. Chang, R. Shau, M. Ortsiefer, M.-C. Amann, and C. Chang-Hasnain, "28 GHz Optical Injection-Locked 1.55µm VCSELs," *IEE Electronic Letters*, vol. Vol. 40 No. 8, 15 April 2004.

X. Zhao, E. K. Lau, D. Parekh, H.-K. Sung, W. Hofmann, M. C. Amann, M. C. Wu, and C. J. Chang-Hasnain, "107- GHz Resonance Frequency of 1.55µm VCSELs Under Ultra-high Optical Injection Locking," in *OSA/CLEO*, 2008.

6

Tunable, Narrow Linewidth, High Repetition Frequency Ce:LiCAF Lasers Pumped by the Fourth Harmonic of a Diode-Pumped Nd:YLF Laser for Ozone DIAL Measurements

Viktor A. Fromzel, Coorg R. Prasad, Karina B. Petrosyan, Yishinn Liaw, Mikhail A. Yakshin, Wenhui Shi, and Russell DeYoung[1]
Science and Engineering Services, Inc.
[1]NASA Langley Research Center
USA

1. Introduction

Ozone plays a crucially important role in all aspects of human life, although it is only a trace gas present in the middle and low atmosphere. Variations in ozone concentration in the stratosphere have an affect on the protection of the earth's biosphere from the harmful portion of the Sun's ultraviolet rays. Tropospheric ozone initiates the formation of photochemical smog and in high concentrations is harmful to human health and vegetation. Also ozone has a significant influence on the earth radiation budget. Human activities have produced adverse effects on atmospheric ozone distribution, which it left unchecked could lead to catastrophic changes to the biosphere . Hence the continuous measurement of ozone with good spatial resolution over large regions of the globe is an important scientific goal. A remote sensing technique for the monitoring of ozone concentration based on differential absorption lidar (DIAL) has been established as a method providing rapid and precise time and spatial resolutions [Browell, 1989, Richter, 1997]. Ozone absorbs strongly in the UV over the 240 - 340 nm region and also in the IR at near 9.6 μm. A two-wavelength differential absorption technique in the UV is commonly used for ozone measurement. After obtaining the lidar signals at two neighboring wavelengths (on- and off-line), the differential absorption due to ozone is obtained by taking the ratio of the two signals to eliminate the contribution to extinction from scattering commen to both signales. Since the ozone absorption in UV exhibits a smooth band structure, the separation between the on- and off-line wavelengths is required to be a few nanometers.

A number of ground-based [Profitt & Langford, 1997] and aircraft-based DIAL [Richter et al.,1997] systems for monitoring ozone concentrations in the planetary boundary layer, the free troposphere and the stratosphere have been developed by research groups all over the world [McGee et al, 1995, Mc Dermit et al,1995, Carswell et al,1991, Sunesson, et al,1994]. Most of the ground-based ozone DIAL instruments utilize large excimer gas lasers and Raman wavelength shifters, or flashlamp pumped frequency tripled and quadrupled Nd:YAG lasers and dye lasers, which are large complex systems requiring considerable maintenance. Many different approaches have been used to improve the efficiency and reduce the size and complexity of the UV lasers required, for example, for airborne ozone DIAL systems. These systems consist of multi-stage solid-state laser systems involving Nd:YAG pump lasers, and some combinations of optical parametric oscillators, or Ti:Sapphire lasers and frequency mixers and solid-state Raman frequency shifters [Richter, 1997, Profitt & Langford, 1977]. However, all of them are still large, and/or complex and they present enormous challenges for adapting them to autonomous operation.

In conventional lidar systems, high energy laser pulses (~100 mJ) are utilized to obtain a sufficiently large lidar signal to achieve adequate signal to noise ratio (SNR). A different approach can be used, wherein a much smaller laser energy (~ 1 mJ) is sufficient to achieve good lidar performance. This calls for a much smaller all-solid-state laser system that makes it possible to conform to the playload bay constraints of a small aircraft or other small movable platform. By operating the laser at much higher pulse repetition rate (PRR = 1 kHz), the average transmitted power (1 W) is maintained at the same level as that of the bigger laser (100 mJ, 10 Hz, 1 W), despite the much smaller laser energy output (1 mJ) per pulse. The smaller resulting signal is effectively measured by a low noise photon counting PMT detection system, whose dark noise counts are in the 10 to 100 Hz range, making the detector noise negligible. By averaging the signal over a few seconds it is possible to achieve adequate SNR by reducing the contribution of the signal shot noise to SNR. Overall system size and complexity are reduced by this approach making the system rugged, compact and easy to maintain. The recent advances in compact diode-pumped solid state lasers provide an attractive option for the development of compact and effective laser transmitter for ozone lidar. While the DPSS lasers are suited for providing only moderate pulse energies, they can operate at high pulse repetition rates of several kHz to produce reasonably high average power. It is possible to generate tunable UV output starting with the UV DPSS laser, by two different techniques both of which are now commercially available. The first method involves pumping an OPO with a frequency tripled Nd:YAG (355 nm) to generate continuously tunable output spanning 560 to 630 nm and then frequency doubling it to obtain the required range of 280 to 315 nm. But the efficiency of this system is very low in view of the multiple non-linear conversion steps. The second method is simpler and more efficient, and involves a Ce:LiCAF laser [Stamm, et al, 1997, Govorkov, et al, 1998, Fromzel & Prasad, 2003] pumped by an appropriate commercially available frequency quadrupled diode-pumped Nd laser to provide direct UV tunability.

In this chapter, a new development of all-solid-state Ce:LiCAF tunable UV laser (280nm - 315nm), which utilizes a single step conversion of the pump wavelength in Ce:LiCAF crystal, when pumped by frequency quadrupled diode-pumped Nd:YLF laser is described. This laser is the central component of a very compact ozone DIAL system. With moderate (~1mJ) pulse output but high pulse repetition rate (1 kHz) this laser system has a good performance capability. This laser is a further development of a previously reported Ce:LiCAF laser producing ~ 0.5 mJ pulse output at 1 kHz with a 46% conversion efficiency [Fromzel & Prasad, 2003].

2. Requirements to laser transmitter for DIAL ozone measurement

Specific character of ozone absorption line and its distribution in atmosphere as well as necessary accuracy of the ozone DIAL measurements determine requirements to parameters of the laser transmitter (energy, PRF, pulse duration, integration time). To establish this relationship, we will consider basic factors which have influence on this accuracy. As it was mentioned above, in DIAL measurements the differential resonant absorption $K(\lambda_n)$ - $K(\lambda_f)$ is obtained by taking the ratio of two atmospheric backscattered signals received by the lidar at the on- and off- wavelengths λ_n and λ_f. from range R. Ozone concentration is then calculated from the mean differential absorption coefficient K for the range cell layer of thickness ΔR by using the known ozone differential absorption cross section $\Delta\sigma = (\sigma_n - \sigma_f)$ where σ_n and σ_f are absorption cross sections at the on- and off-line wavelengths. A number of papers have analyzed the sensitivity and accuracy of the DIAL technique [Ismail & Browell, 1989, Korb et al, 1995]. The accuracy of the ozone concentration n_{O3} measurement is calculated by using the relation [Grant, et al, 1991]:

$$\frac{\Delta n_{O3}}{n_{O3}} = \frac{1}{n_{O3}\ 2\,\Delta\sigma\ \Delta R\ N_s^{1/2}\ (SNR)} \tag{1}$$

here N_s is the number of laser shots, and SNR is the signal to noise ratio of the DIAL measurement which includes the SNR of both the on-line and off line signals. The accuracy of the measurement is thus improved by: averaging over larger number shots, increasing the range cell size, increasing the differential absorption and increasing the signal to noise ratio of the measurement. The parameters which determine the range are: the ozone differential absorption cross section; the distribution of ozone along the path at the time of the measurement; other sources of extinction, such as aerosol loading, fog, etc; the choice of the on- and off-line wavelengths for ozone. From equation (1), it is seen that the accuracy and the range resolution can be improved by choosing the wavelengths so as to provide a large differential absorption cross section (i.e., a large $\Delta\sigma$). However this also makes the differential scattering cross section: $\Delta\alpha = \alpha(\lambda_n) - \alpha(\lambda_f)$ large. Correcting for this requires knowledge of the molecular and aerosol distributions also. Furthermore, the signal strength depends on the atmospheric extinction. Hence the choice of optimal wavelength depends on a number of parameters, which include: the required range, range resolution, temporal resolution (i.e., measurement time), measurement accuracy, and the expected spatial distribution of ozone in the atmosphere.

Figure 1 shows the ozone absorption spectrum between 240 and 340 nm. Below 300 nm, absorption is dominated by the Hartley continuum superimposed by weak Hartley bands. Band structures seen at wavelengths longer than 300 nm are the Huggins bands. While the strongest absorption occurs at 260 nm these wavelengths will be completely attenuated after traveling a short distance and are therefore unsuitable for achieving significant range. Conversely, wavelengths longer than 300 nm are able to penetrate into the high ozone concentrations that are characteristic of the stratosphere, but give small differential absorption signals at the typical tropospheric ozone concentrations. Further, since the absorption cross sections in the Huggins bands also vary significantly with temperature, this region of the spectrum is not very useful for tropospheric measurements where the temperature is highly variable.

Thus, the optimal wavelength range for tunable ozone laser transmitter depends on atmospheric region of interest setting in the UV spectrum between 280 and 300 nm. Comparison of calculated ozone lidar performance for two types of UV lasers operating in the required wavelength region with different characteristics: laser with low energy but high PRF (1 mJ/pulse, 1kHz) and photon counting for detection and laser with high energy but low PRF (100 mJ, 10 Hz) and conventional analog detection shows that the low energy

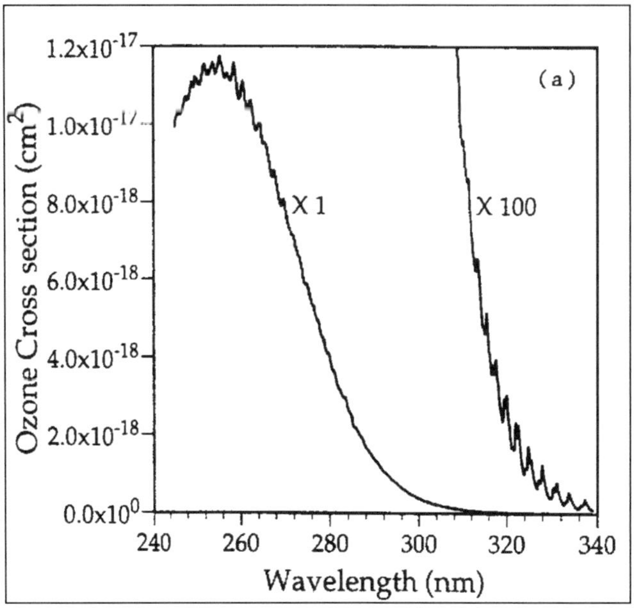

Fig. 1. Ozone absorption spectrum in UV.

laser gives a much higher SNR for all cases of lidar operation. The feasibility of such approach for ozone DIAL - using a low energy, high PRF laser along with photon counting detection have been also demonstrated experimentally [Prasad, et al, 1999]. The Ce:LiCAF laser, which is the best suited for modest energy outputs in the range of 1 to 10 mJ/pulse, presents a very effective direct method of generating the required wavelengths. The principal reasons for this are:

1. Laser linewidths of the order of 0.2 nm are adequate for ozone DIAL. Hence a fairly simple Ce:LiCAF laser system can be designed with a single intra-cavity prism for generating tunable wavelength with the necessary linewidth, with no need for highly selective dispersive elements.
2. The spectral bandwidth of the pump laser does not have to be narrow, because of the broad absorption spectrum of Ce:LiCAF material.
3. Directly tunable laser allows rapid change of wavelength, as it required in hopping from on- to off-wavelengths.

3. Spectroscopic and thermo-mechanical characteristics of Ce:LiCAF crystals

Cerium doped crystals Ce:$LiCaAlF_6$ and Ce:$LiSrAlF_6$ (Ce:LiCAF and Ce:LiSAF) are well established as efficient laser media, which can operate directly in the UV region. Both Ce:LiCAF and Ce:LiSAF crystals demonstrated good conversion efficiency (up to 46%) when pumped by the fourth harmonic of Nd:YAG or Nd:YLF laser (266 or 262 nm). Figure 2 shows the spectral absorption and fluorescence of Ce3+ in LiCAF and LiSAF. Their strong absorption at 266 nm (~ 7.5 x 10^{-18} cm^2 for π-polarization), broad emission spectrum (280 – 325 nm), high emission cross-section (~ 6.8 x 10^{-18} cm^2 for Ce:LiCAF at 290 nm for π-polarization), and broad tunability (280 -328 nm) make them well suited for ozone DIAL

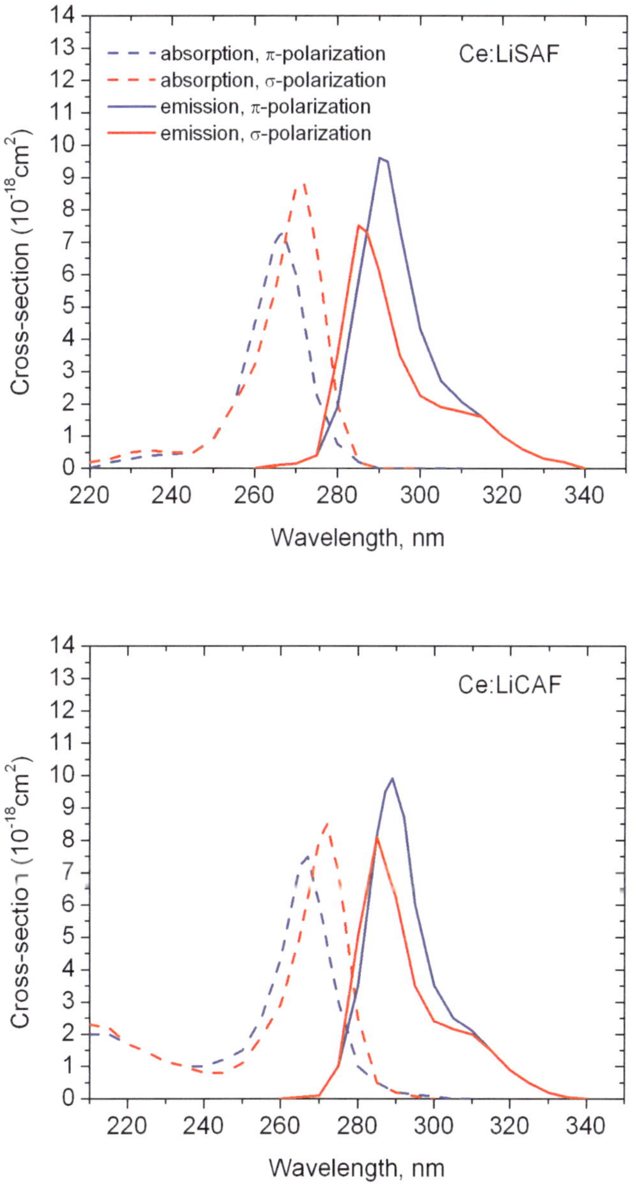

Fig. 2. Polarized absorption and emission spectra of Ce:LiSAF and Ce:LiCAF (π-parallel and σ-perpendicular to the optical axis)

application. Since the cross-section for absorption at 266 and 262 nm are fairly high, the Ce3+ dopant concentration of a few percent (1 - 4%) is enough for complete absorption of the pump. Of the two, Ce:LiCAF is better suited for the high PRF operation, because its spectroscopic properties are slightly better, it is more mechanically robust, and has much better solarization properties for withstanding high power pumping at 266 or 262 nm than

that of Ce:LiSAF. The fluorescence lifetime both Ce:LiCAF and Ce:LiSAF crystals are short (27 and 25 ns, respectively). This implies that the nanosecond pulse durations are required for pumping of Ce:LiCAF (or Ce:LiSAF) lasers and the laser output is gain-switched by the pump laser pulse. Also it means that a short resonator is preferred for the Ce:LiCAF laser.
The thermal conductivity of both LiCAF and LiSAF are low and anisotropic in nature (5.14 - 4.58 and 3.09 - 2.9 W/m°C, respectively). Thus even for the low thermal loading (~ 1 W), a noticeable temperature gradient is set up within the crystal. Considering a 3.5% Ce:LiCAF crystal of 8 x 3 x 10 mm (thickness 3 mm), the calculated temperature rise in the crystal will be as $\Delta T \sim 17^{0}C$.

4. Diode-pumped frequency quadrupled Nd:YLF laser

From our previous experience with designing of a tunable Ce:LiCAF laser producing 0.5 mJ pulse energy at 1 kHz PRF, it was estimated that in order to obtain ~1 mJ/pulse UV tunable output from Ce:LiCAF laser the pump Nd:YAG or Nd:YLF laser has to provide pulse energy in excess of ~ 11 - 12 mJ in a TEM_{00} beam profile at the second harmonic (532 or 527 nm) that will allow to have ~ 2.8 -3.0 mJ/pulse at the forth harmonic (266 or 263 nm). Such TEM_{00}-mode green laser was developed by Positive Light company on the base of the commercial multomode Nd:YLF Evolution 30 laser and supplemented by us with the fourth harmonic module (263 nm).
The optical layout of the Nd:YLF laser with the intracavity frequency doubling (Evolution - TEM_{00}) is shown in Figure 3. It consists of a Nd:YLF laser rod that is side-pumped by laser diode arrays. Two high reflective end mirrors M1 and M2 (HR @ 1053nm) form the Nd:YLF laser resonator. The resonator includes a reflective telescope (mirrors TM1 and TM2) that serves to increase the beam size incident on the Nd:YLF crystal. The laser beam is then intracavity frequency doubled by a non-critically phase matched LBO crystal and delivers an output green beam (527nm) through the harmonics separating mirror, which is highly transparent at 527 nm and highly reflecting at 1053nm. An acousto-optical Q-switch performs Q-switched laser operation at 1 kHz repetition rate. The LBO doubling crystal is placed in a temperature regulated oven (154°C) to achieve the non-critical phase matching conditions. Figure 4 shows the 527 nm output performance for the frequency doubled diode-pumped Nd:YLF laser, while Figures 5 shows a temporal pulse profile of the Evolution TEM_{00} laser. It may be noted that the pulse duration is very long and with a half width (FWHM) slightly smaller than 100 ns, when diode pump current is 24 Amp (close to

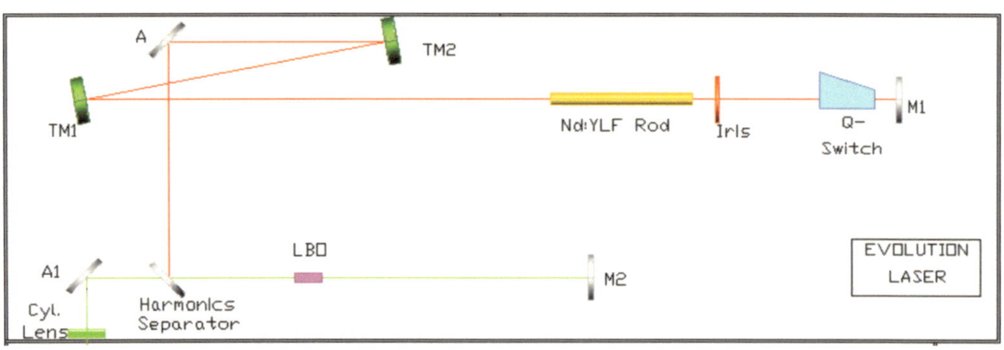

Fig. 3. Optical schematic of the intra-cavity frequency doubled diode pumped Nd:YLF TEM_{00} pump laser.

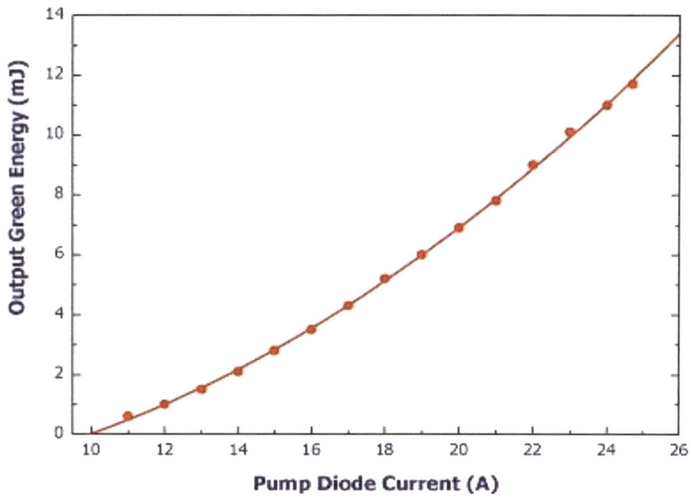

Fig. 4. Intracavity doubled green (527 nm) output of diode pumped Nd:YLF laser.

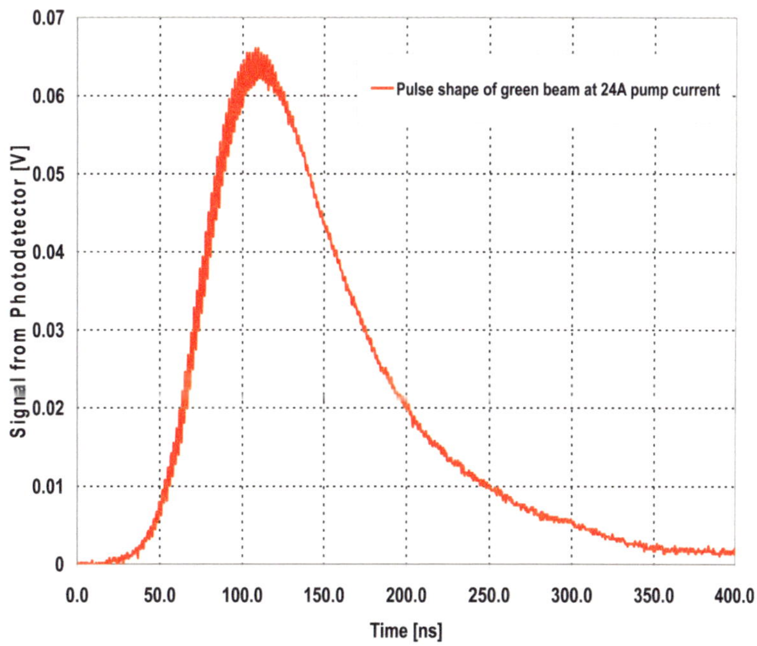

Fig. 5. Temporal profile of the green (527 nm) output pulse of the pump Nd:YLF laser

the maximum of pump current). This long pulse duration is caused by a large length of the laser resonator (~2 m) and also by the fact that the output coupling of resonator is not optimal, because the only load for the resonator is the second harmonic generation. Measurement of spatial profile of the green beam showed that the output beam being very close to the TEM_{00} mode ($M^2 \sim 1.5$) at the same time exhibited a significant amount of astigmatism, with the beam divergence being about 1 x 1.5 mrad in the X and Y directions, respectively. The second harmonic output beam (527 nm) measured at the output window of the laser was slightly elliptic with a diameter of about 0.9 x 1.1 mm for an output of 12 mJ.

The second step in building of appropriate UV pump laser for Ce:LiCAF was development of an efficient CLBO fourth harmonic generator for a 1 kHz, diode-pumped Nd:YLF laser. CLBO is well established as the nonlinear material of choice [Mori, et al, 1995] for efficient fourth harmonic conversion of diode-pumped solid-state neodymium lasers with moderate pulse energies but high average powers and high PRF. CLBO has a high nonlinear coefficient and a large temperature and angular acceptance. However, CLBO is highly hygroscopic [Taguchi, et al, 1997] in nature and thus any exposure of the crystal to humid (>20%) atmospheric conditions causes rapid degradation of the crystal surface, which can lead to a reduced performance and/or optical damage. A simple technique to avoid the problems associated with CLBO crystal is to maintain the crystal at >150°C, so that atmospheric humidity does not degrade the crystal. To avoid the crystal degradation, a special crystal ceramic oven for maintaining the crystal temperature at temperature of ~152°C has been constructed and was heated all the time being supplied from a battery backed UPS power source.t should be noted that the Evolution TEM_{00} laser output was no optimum for obtaining the best fourth harmonic conversion efficiencies because its pulse duration was fairly long (~ 100 ns) and the beam was not true TEM_{00}-mode showing some astigmatism: different beam divergences in the x- and y-directions. In spite of the non-optimal 527 nm beam, a fairly high fourth harmonic conversion efficiency (~ 25%) have been achieved in the 15 mm long uncoated CLBO crystal by using mode matching optics. At this output, the mean incident energy density on the CLBO crystal was ~ 25% lower than the damage threshold and the CLBO crystal was operated in a safe damage free regime. Figure 6 shows the fourth harmonic energy output as a function of the diode pump current for the Nd:YLF laser. At maximum diode current of the Nd:YLF laser of 25 A, the fourth harmonic output was as high as 2.85 mJ/pulse.

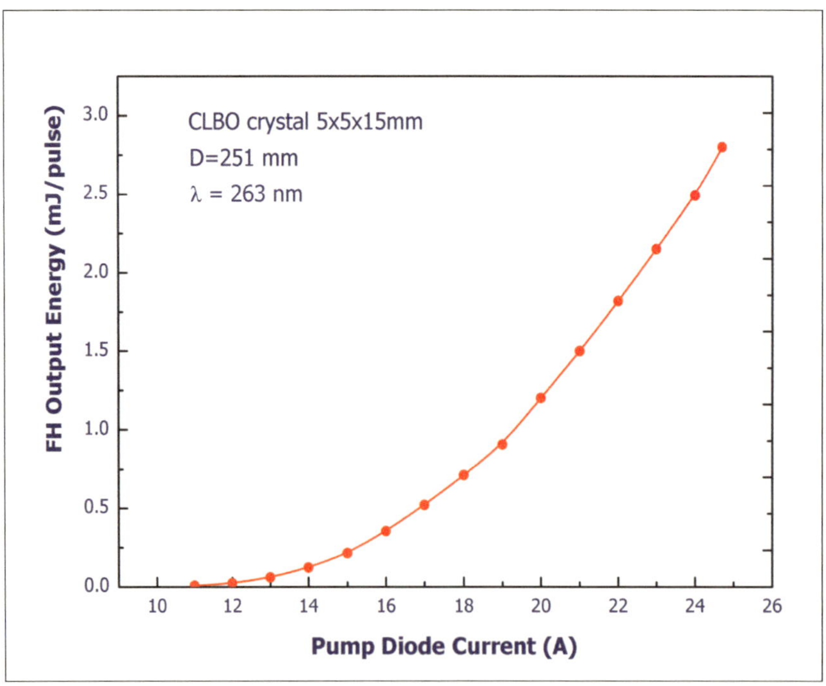

Fig. 6. Output from the optimized CLBO fourth harmonic generator shown as a function of the diode current for pump green laser.

5. Ce:LiCAF tunable UV laser

The optical schematic of the Ce:LiCAF laser is shown in Figure 7. A pair of CaF_2 rectangular prisms was used for separation of the second and fourth harmonic pump beams and for beam folding. After that the incoming 263 nm UV pump beam was split by a fused silica beam splitter (40% and 60%) into two parts and directed to the Ce:LiCAF crystal faces by four 100% reflecting folding mirrors. The pumped spot size on the Ce:LiCAF crystal has an elliptical shape with dimensions of ~ 0.4 x 0.65 mm. Pump spot sizes on the Ce:LiCAF crystal were chosen carefully to avoid optical damage of the crystal, to obtain good conversion efficiency and to provide TEM_{00} operation of Ce:LiCAF laser. A Brewster cut 3.5% doped Ce:LiCAF crystal with dimensions of 2 mm (thickness) x 8 mm (width) x 10 mm (length) is pumped from both faces. The measured absorption of this crystal at 263 nm (π-polarization) was found to be k_{263} = 4.47 cm^{-1}, and ~ 98% of the incident pump power is absorbed in the crystal. The Ce:LiCAF crystal is mounted on a copper crystal holder heat sink which is maintained at about 20°C. The Ce:LiCAF laser resonator consists of a flat mirror (HR @ 280-320 nm) and a curved output coupler (R_{out} = 0.6 @ 280-320 nm, RoC = 1 m) with an intra cavity fused silica (suprasil) prism as a wavelength selector which results in a linewidth of 0.15 - 0.2 nm. The pumping beams are focused into the Ce:LiCAF crystal by means of two fused silica lenses (200 mm focal length). The tilt angle between the pump beams and the Ce:LiCAF laser beam is ~ 2.5°. Wavelength tuning of the laser is performed by rotation of the flat HR mirror of the resonator in horizontal plane. Because direction of the beam between output coupler and Ce:LiCAF crystal stays unchangeable, such tunable laser resonator design provides output beam pointing stability and collinearity better than +/- 0.05 mrad whereas wavelength of the laser is tuned. The length of the resonator is ~ 12 cm. A slightly off-axis pumping scheme is used here. This configuration provides a significant advantage by spatially separating the pump and laser beams so that the pump beam does not have to pass through the laser mirrors or other optical components of the laser thus avoiding a common problem of optical damage caused by the pump beam.

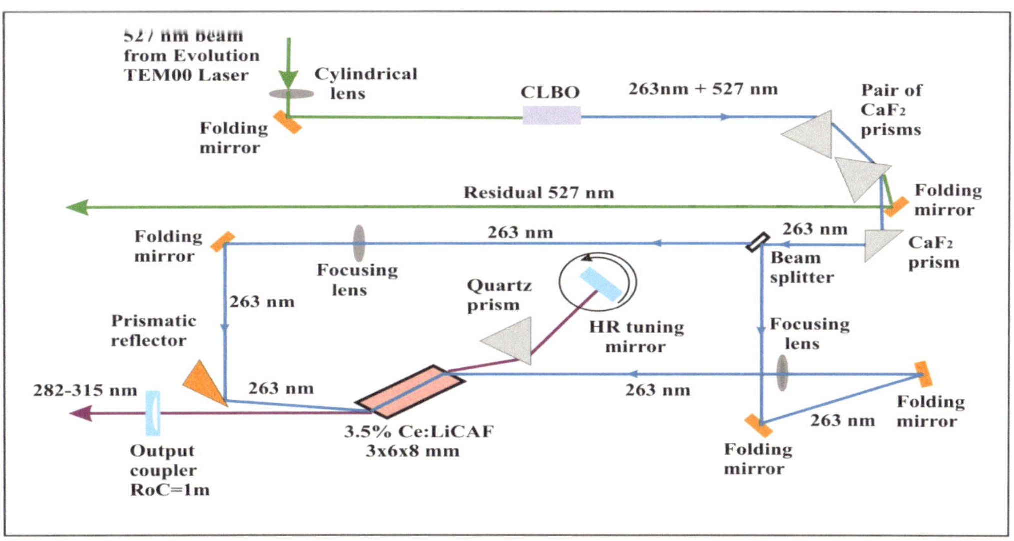

Fig. 8. Optical layout for the double-side pumped Ce:LiCAF laser.

6. Ce:LiCAF laser performance

Figure 9 shows the input - output performance of the Ce:LiCAF laser at wavelength of 290 nm with narrow wavelength bandwidth of ~ 0.15 nm when it is pumped by the FH laser beam at 263 nm with a pulse repetition rate of 1 kHz. Output pulse energy 1 mJ/pulse was obtained from the Ce:LiCAF laser when the total incident pump pulse energy on both faces of the laser crystal was 2.86 mJ/pulse. In our experiments, the slope efficiency is ~ 45%, which was found to be about 90% of the theoretical maximum value for the laser [Fromzel & Prasad, 2003]. This result shows that there is nearly full utilization of the pump energy.

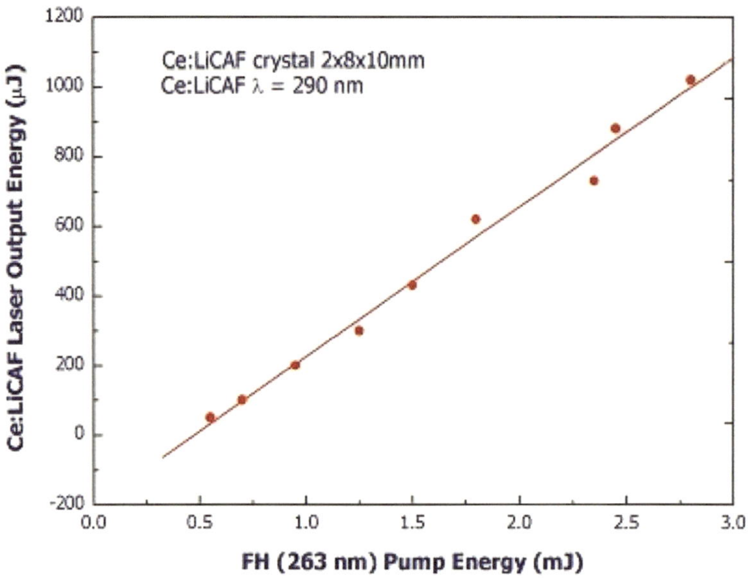

Fig. 9. Input-output performance of the Ce:LiCAF laser pumped from two sides by the fourth harmonic of Nd:YLF laser.

Figure 10 shows the typical temporal shape of the Ce:LiCAF laser pulse. The upper trace is the shape of the pump pulse at 263 nm and the lower trace is the corresponding Ce:LiCAF output pulse. It is noted that in spite of a short pulse duration, typical for Q-switched lasers, the Ce:LiCAF laser operates to the point at free running (gain-switch) regime. It can be clear seen from the fact that the Ce:LiCAF laser output pulse exhibits typical for free running laser operation transient behavior (relaxation oscillations). Thus the pulse length of Ce:LiCAF depends on the pump pulse length. Also shown is the pump laser pulse, and by comparing the two, the build up time for the Ce:LiCAF pulse is seen to be about 48 ns.
The transverse beam shape of the Ce:LiCAF laser output was measured with a beam profiler. It was found that the output laser beam has a true TEM_{00}-mode distribution (M^2 ~ 1.1) and the profiles are smooth without any hot spots.
Ability of Ce:LiCAF laser to be directly wavelength tuning is one of the advantages of this UV laser, which allows rapid change of wavelength, as it required in hopping from on- to off-line wavelengths, or for sensing ozone at different altitudes. The output wavelength of Ce:LiCAF laser was tuned by rotating the HR tuning mirror which was mounted on a rotary mirror mount. The laser wavelength and linewidth were determined by the intra-cavity dispersing prism. Figure 11 shows a sample laser tuning curve, which was obtained by

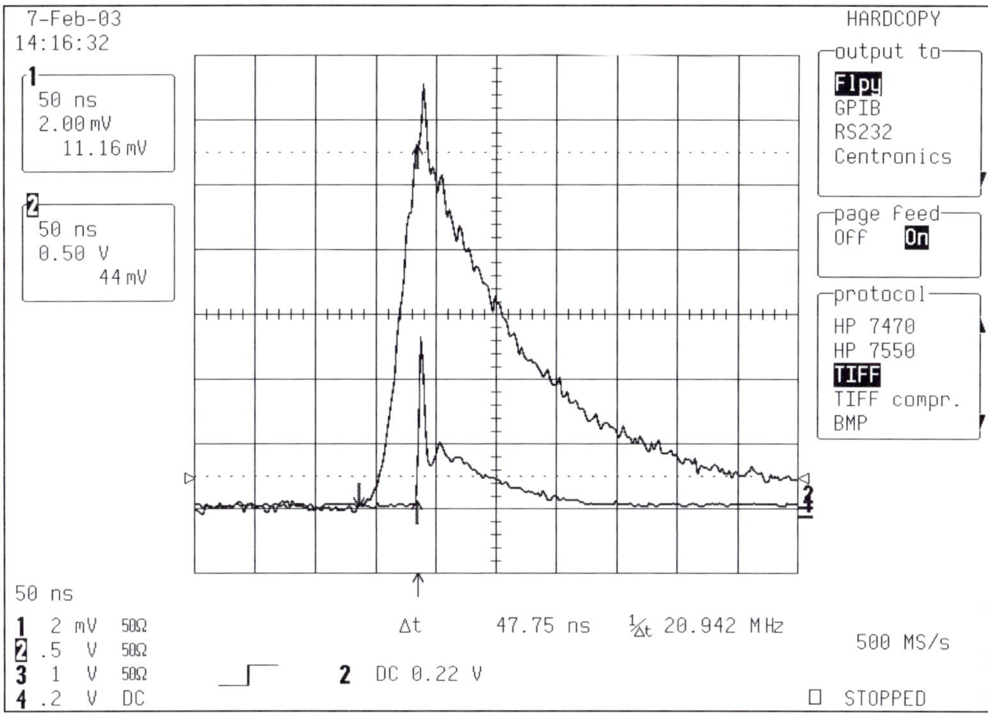

Fig. 10. Shape of the pump UV pulse (upper trace) and the Ce:LiCAF laser output pulse (lower trace).

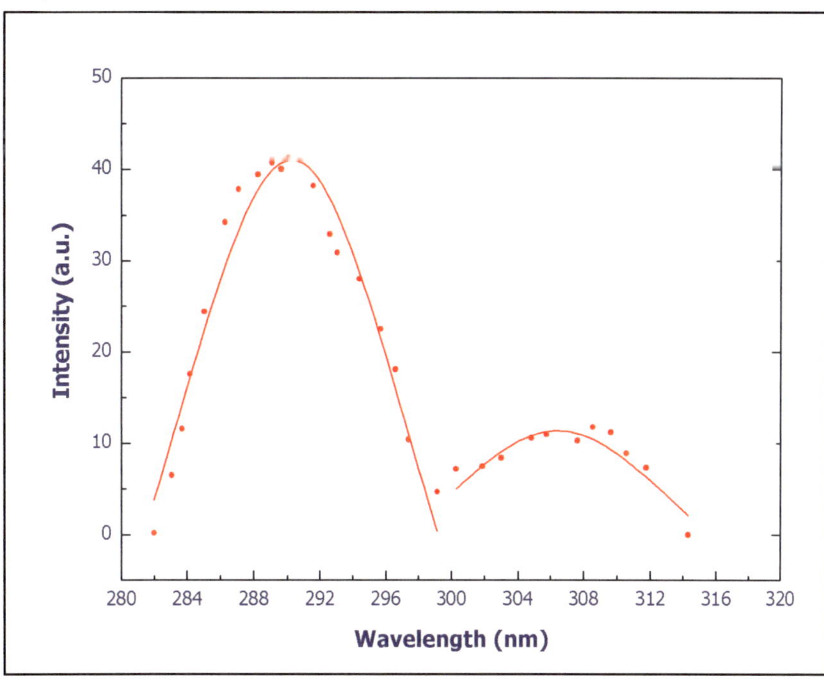

Fig. 11. Tuning curve of the Ce:LiCAF laser.

tuning over a broad spectral region from 281 nm to 316 nm while the 263 nm pump energy was ~ 2 mJ/pulse. Laser linewidth as measured with the Ocean Optics grating spectrometer with a resolution of 0.065 nm was approximately 0.15 – 0.2 nm, when a fused silica dispersing prism was used. The maximum laser output occurred at a wavelength of ~ 290 nm, the second much more weak maximum of laser output corresponded to ~ 308 nm. By using a different prism material with a larger dispersion, such as, sapphire the linewidth can be reduced. With a sapphire dispersion prism the linewidth was reduced to ~0.1 nm. The angular motion required for tuning over 10 nm is approximately 0.3□ for fused quartz and about 0.6□ for sapphire.

It can be concluded from Figure 11 that output pulse energy of the Ce:LiCAF laser reduces approximately four times regarding the maximum of 1 mJ/pulse output at ~ 290 nm, when laser wavelength is tuned to ~ 284 nm or to ~ 297 nm on the short- and long-wavelength edge of the tuning curve, respectively.

7. High speed wavelength tuning of Ce:LiCAF laser

Ce:LiCAF laser using as a transmitter for lidar has to supply lidar with both the on- and off-line wavelengths. As it was shown above, wavelength tuning of Ce:LiCAF laser was achieved by changing the angle of the rear mirror. A rapid tuning of the laser output wavelength from shot to shot at pulse repetition frequency of 1 kHz was achieved by mounting the HR mirror on a servo-controlled high speed galvanometric deflector. The tuner control system has been designed to provide pairs of pre-selected "on" and "off-line" wavelengths λ1 and λ2 at 1 kHz operation chosen for ozone differential absorption measurements. It is essential for the "on" and "off-line" wavelengths to be stable both in the short term (i.e., from shot-to-shot) and in the long term (over a period of a few hours). Mechanical backlash, hysterisis, thermal drift and other instabilities affect the short and long term wavelength stability. By utilizing sinusoidal small angle rotations we have eliminated the potential problems of hysterisis and backlash. Long-term drifts are corrected by the feedback control loop embedded into the servo motor drive circuit.The servo controlled mirror is continuously oscillated at 500 Hz to generate harmonic angular deflection as shown in Figure 12. By taking different time delay between the pump laser pulse and the clock, which generates the harmonic drive signal for varying the mirror position, different output wavelengths are produced on every pulse. Then by firing the pump laser with

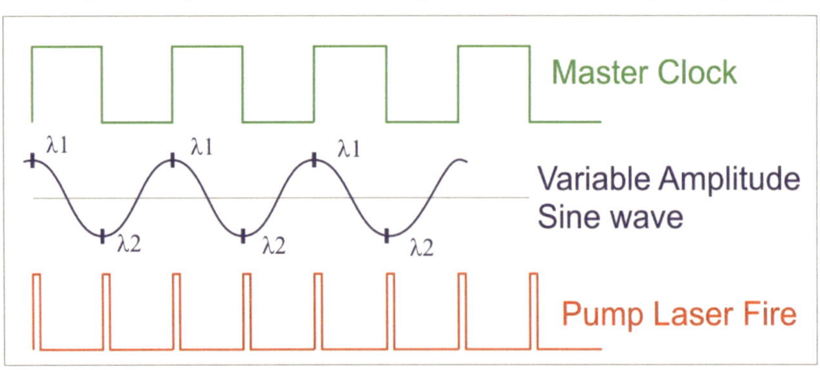

Fig. 12. Basic principle of the high speed tuner for generating pairs of "on-" and "off-"line adjustable wavelengths at 1 kHz PRF.

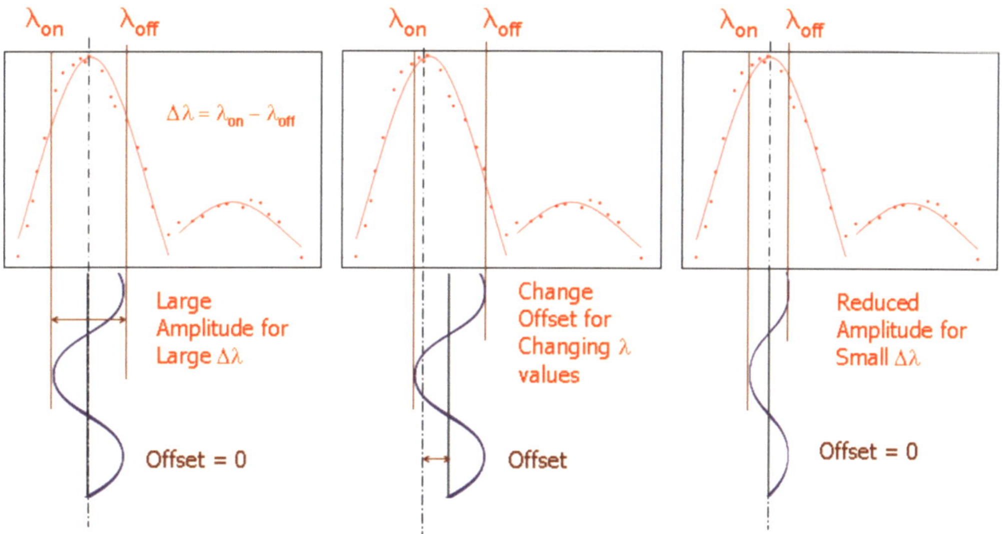

Fig. 13. Amplitude and offset of the harmonic motion of tuning mirror used to select "on" and "off" wavelength

proper time delays on both halves of the harmonic wave, pairs of wavelengths, i.e. an "on-line" and "off-line" wavelength on alternate pulses, are generated when the laser is operated at 1 kHz (see Figure 12). Since a stable master clock is used to synchronize the sine wave, the delays and the laser fire, the pulse variability and jitter are very small (<5ns). Variation of sine wave amplitude determined separation between pre-selected pair of on - and off-line wavelengths of Ce:LiCAF laser, while the offset change moved the pair along the tuning curve of the Ce:LiCAF laser, as it is shown in Figure 13.

8. Assembled Ce:LiCAF laser module

All the optical components required for generating the tunable UV output, including the FH generator are housed in a single modular assembly. A mono-block laser head machined from an annealed aluminum block provides a sealed enclosure. Many of the optical mounts were custom designed to achieve adequate rigidity and robustness needed for an airborne laser system. All the optical mounts and fixtures are also constructed out of aluminum to maintain an athermal optical alignment over a wide range of temperatures. Figure 14 shows the component layout of the tunable laser. The Ce:LiCAF crystal is mounted on water cooled heat sink, and the entire assembly sits on a motorized translation stage allowing for repositioning the Ce:LiCAF crystal. If any degradation of the crystal is observed due to solarization, the crystal can be remotely moved by the translation stage to utilize a fresh region of the crystal. It may be noted that the plumbing used for the water cooled heat sink is hard soldered to prevent the possibility of any water leak within the laser head.
The fourth harmonic beam from the CLBO crystal was passed through a pair of CaF_2 prisms, which deflect the beam by 90° and also separate the unused second harmonic beam from the UV beam. The UV beam is deflected by an additional 90° in a second set of prisms, to fold the beam. A 2X beam expander is used to expand the beam to avoid damage in the downstream optical components, which include the beam divider and 100% mirrors.

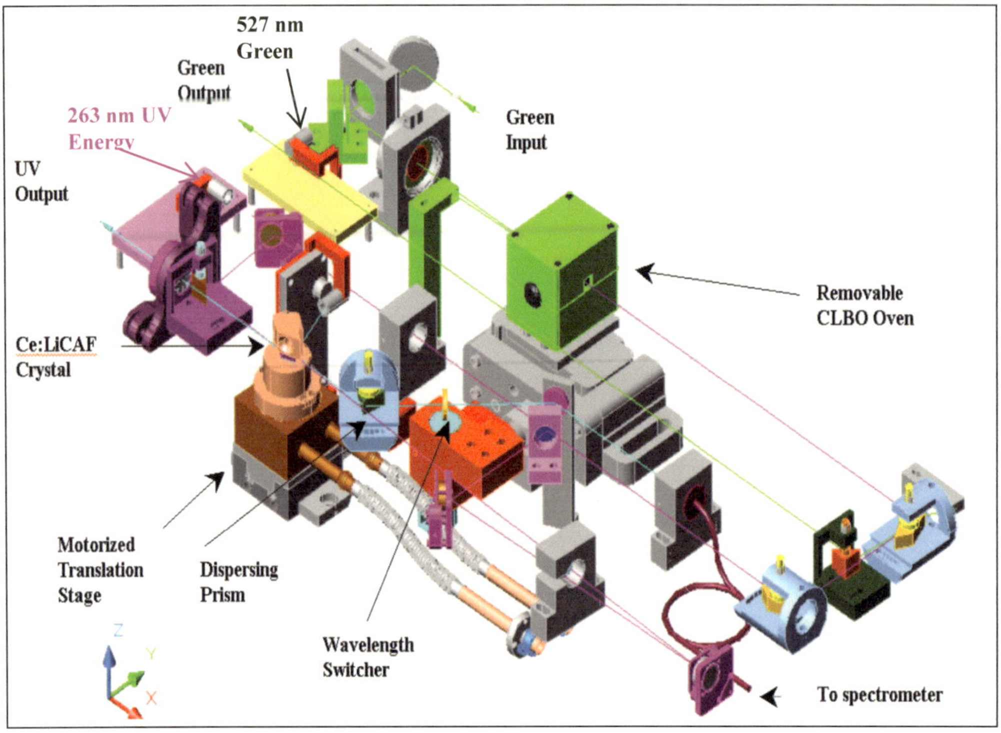

Fig. 14. Component layout of the Ce:LiCAF laser. CLBO crystal is in oven heated to ~ 150°C. CLBO and Ce:LiCAF crystals are placed on motorized mounts.

Figure 15 shows the complete view of tunable laser assembly together with the pump laser. The optical bench is designed such that the water hoses and cables are conveniently accessed from the back of the laser head and there are no cooling lines are inside the head. The main attributes of the laser system are simplicity and ruggedness.

The operation of the Evolution TEM_{00} laser was performed through a computer controlled operator interface residing on PC computer. The same computer was used to operate and control the tunable UV laser.

The central component in the control function is the General Control Unit (GCU) which generates the timing sequence and all the trigger signals required for the wavelength tuning and for the laser operation. It has been implemented using a master clock and a CPLD (128 macro-cell complex programmable logic device). The wavelength controller unit generated the variable amplitude harmonic modulation to dither the galvanometric rotary actuator. A simple interactive computer interface was provided for the operator to choose the values of the required pair of the output wavelengths. During operation the laser output cycles through the chosen wavelengths □1 and □2, sequentially at a 1 kHz pulse repetition frequency.

MS Windows based operator control interfaces have been designed to separately control the Evolution TEM_{00} pump laser, fourth harmonic generator (CLBO crystal) phase match control, and Ce:LiCAF crystal position motors. The pulse energies and temperature at several points in the laser head were monitored with an eight-channel ADC card. The motorized FHG crystal mount stage allowed remote adjustment of the CLBO crystal. Provision was made for feedback control of the stage to achieve optimal phase matched operation.

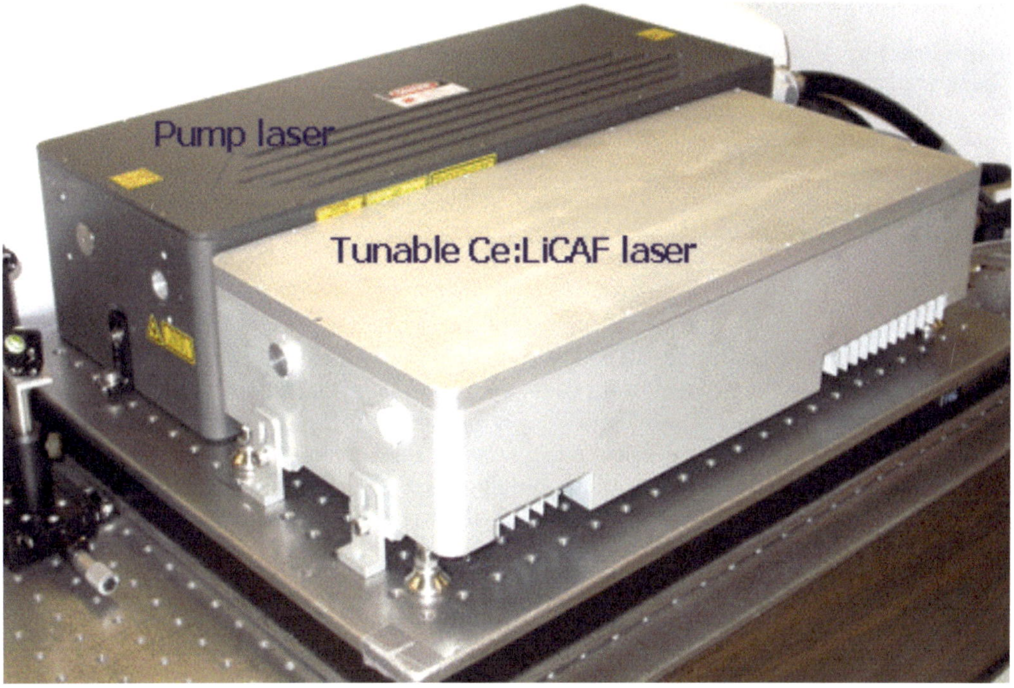

Fig. 15. Complete assembly of tunable UV laser source. The pump laser is the Evolution TEM_{00} (PositiveLight) intra-cavity doubled diode pumped Nd:YLF laser.

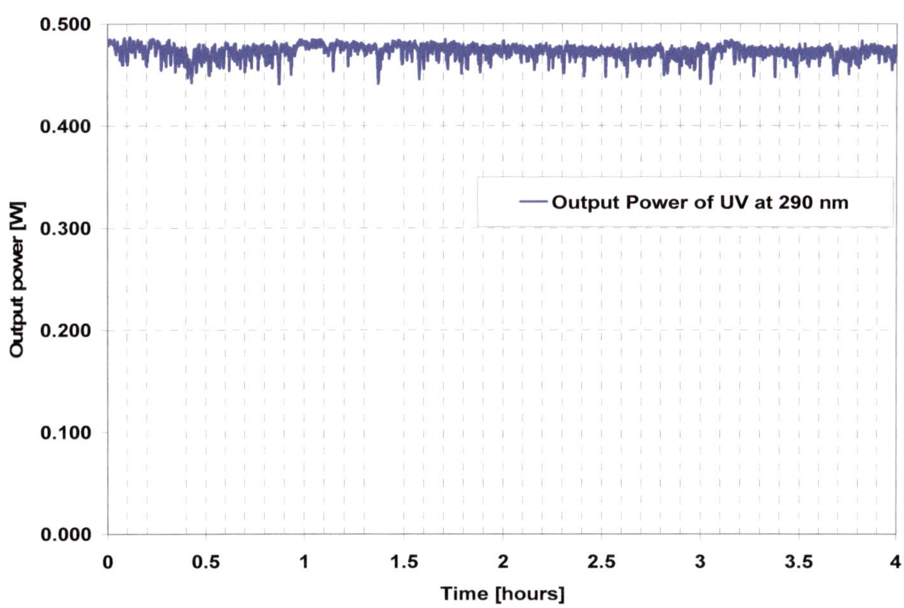

Fig. 16. Daily Long Term stability output power at 290 nm

9. Prolonged performance of the Ce:LiCAF laser

In this test, the Ce:LiCAF laser was operating continuously for 4 hours daily during 20 days. The operating conditions were maintained constant over the duration of the test. The output power at the pump wavelengths (527 nm, 262 nm) and of the Ce:LiCaF laser output at 290 nm was continuously monitored. The drift in the phase matching in the CLBO crystal has been periodically revised and eliminated.
The observed variations in Ce:LiCAF output (290 nm) follows those of green pump beam (527nm) and do not exceed 8 %. as showed at the Fig.16.

10. Conclusion

A highly efficient, compact and rugged 1 kHz tunable UV Ce:LiCAF laser pumped by the fourth harmonic of a diode-pumped commercial Nd:YLF laser for ozone DIAL measurements has been developed and the performance of this laser was investigated. The Ce:LiCAF laser delivered 1 mJ pulse energy at 290 nm wavelength and was able to be wavelength tuned from 281 to 316 nm that was achieved with a single fused silica dispersion prism in the laser cavity. Fast shot-to-shot wavelength switching was obtained by the harmonic motion of tuning mirror mounted on a servo-controlled high speed galvanometric deflector.

11. References

Browell, E.V., (1991). *Differential Adsorption Lidar Sensing of Ozone,* Proc. *IEEE,* 77, pp. 419-432, Carswell.

Fromzel, V.A., and Prasad C.R., (2003). A Tunable Narrow Linewidth 1kHz Ce:LiCAF Laser with 46% Efficiency, *OSA TOPS,* Vol.83, Advanced Solid-State Photonics, pp. 203-209.

Govorkov, S.V.; Weissner, A.O., Schroder, Th., Stamm, U., Zschoke,W., and Basting, D., 1998. "Efficient high average power and narrow spectral linewidth operation of Ce:LiCAF laser at 1 kHz repetition rate," *Advanced Solid State Lasers,* OSA TOPS 19, pp. 2-5.

McGee, T.J.; Gross, M.R., Butler, J.J., and Kimvilakani, P.E., (1995). "Improved stratospheric ozozne lidar", *Optical Engineering,* Vol. 34, pp. 1421-1430.

McDermit, S.; Walsh, T.D., Deslis, A., and White, M.L., (1995). "Optical system design for a stratospheric lidar system," *Applied Optics,* Vol. 34, pp. 6201-6210.

Mori,Y.; Kuroda, I., Nakajima, S., Sasaki, T., and Nakai, S., (1995). "New nonlinear optical crystal: cesium lithium borate," *Appl.Phys. Lett.,* 67, p.1818.

Profitt, M.H., and Langford, A.O., (1977). *Applied Optics,* 36, No.12, pp. 2568-2585,

Richter, D.A., Browell, E.V., Butler,C.F., and Noah,S.H., (1997). *"Advanced airborne UV DIAL system for stratospheric and tropospheric ozone and aerosol measurements",* Advances in Atmospheric Remote Sensing with Lidar, pp. 317-320, Springer, Berlin.

Stamm, U.; Zschocke, W., Schroder, T., Deutsch, N., and Basting, D., (1997). "High efficiency UV-conversion of a 1 kHz diode-pumped Nd:YAG laser system,"in *Advanced Solid State Lasers,* C.R.Pollock and W.R.Bosenberg, OSA TOPS vol.10, p. 7.

Sunersson, J.A.; Apituley, A., and Swart, D.P.J., (1994). "Differential absorption lidar system for routine monitoring of troposperic ozone," *Applied Optics,* Vol. 33, pp. 7046-705.

Taguchi, A.; Miyamoto, A.,Mori, Y., Haramura, S., Inoue, T., Nishijima, K., Kagebayashi, Y., Sakai, H., Yap, Y.K., and Sasaki, T., (1997). "Effects of moisture on CLBO,"in *Advanced Solid State Lasers,* C.R.Pollock and W.R.Bosenberg, OSA TOPS vol.10, p.19.

Optical and Photonic Devices

7

Single Mode Operation of 1.5-μm Waveguide Optical Isolators Based on the Nonreciprocal-loss Phenomenon

T. Amemiya[1] and Y. Nakano[2]
[1]*Quantum Nanoelectronics Research Center, Tokyo Institute of Technology,*
[2]*Research Center for Advanced Science and Technology, University of Tokyo,*
Japan

1. Introduction

The explosive growth of Internet traffic requires the development of advanced optical telecommunication networks that can enable the high-speed processing of this exponentially growing data traffic. Such advanced network systems will need an enormous number of optical devices, so photonic integrated circuits (PICs) are indispensable for constructing the system at low cost, reduced space, and high reliability. To date, monolithic integration on an indium phosphide (InP) substrate is the most promising way of making PICs because it has the capability to integrate both active and passive optical functions required in optical transport systems for the 1.3-um or 1.55-um telecom window. To develop large-scale, InP-based monolithic PICs, various planar optical devices such as lasers, modulators, detectors, multiplexers/demultiplexers, and optical amplifiers have been developed [1-4].

This paper provides an overview of the present state of research on waveguide optical isolators for InP-based monolithic PICs. Optical isolators are indispensable elements of PICs used to interconnect different optical devices while avoiding the problems caused by undesired reflections of light in the circuit. They must have the form of a planar waveguide because they must be monolithically combined with other semiconductor-waveguide-based optical devices such as lasers, amplifiers, and modulators. Conventional isolators cannot meet this requirement because they use Faraday rotators and polarizers, which are difficult to integrate with waveguide-based semiconductor optical devices. For this reason, many efforts have been expended in developing waveguide isolators [5-11]. Although the research on waveguide isolators is still in the experimental stage, it will probably reach a level of producing practical devices in the near future.

In Section 2, we first give a short sketch of conventional optical isolators. The conventional isolator is a mature device made with established technology and has sufficient performance (low insertion loss and large isolation ratio) for use in optical transport systems. However, it uses bulky components, a Faraday rotator and polarizers, and therefore cannot be used in PICs. We then turn to waveguide optical isolators and, in Section 3, outline two promising methods of making waveguide isolators on InP substrates. All of the methods use semiconductor optical waveguides combined with magnetic materials. One of them is based on the polarization conversion of light caused by the Faraday effect; another is based on a nonreciprocal phase shift in a waveguide interferometer; the third is based on nonreciprocal propagation loss in a magneto-optic waveguide. In the succeeding sections, we focus on the nonreciprocal-loss waveguide isolator and make a detailed explanation of the isolator. In Section 4, we explain the principle and theory of the nonreciprocal-loss phenomenon. Actual devices based on this phenomenon have been developed. In Sections 5, we report the experimental results for the devices consisting of semiconductor optical waveguides combined with manganese arsenide (MnAs), which are ferromagnetic material compatible with semiconductor manufacturing process. We hope that this paper will be helpful to readers who are aiming to develop photonic integrated circuits.

2. Conventional optical isolator

Optical isolators are one of the most important passive components in optical communication systems. The function of an optical isolator is to let a light beam pass through in one direction, that is, the forward direction only, like a one-way traffic. Optical isolators are used to prevent destabilizing feedback of light that causes undesirable effects such as frequency instability in laser sources and parasitic oscillation in optical amplifiers.

Ordinary optical isolators available commercially make use of the Faraday effect to produce nonreciprocity. The Faraday effect is a magneto-optic phenomenon in which the polarization plane of light passing through a transparent substance is rotated in the presence of a magnetic field parallel to the direction of light propagation. The Faraday effect occurs in many solids, liquids, and gases. The magnitude of the rotation depends on the strength of the magnetic field and the nature of the transmitting substance. Unlike in the optical activity (or natural activity), the direction of the rotation changes its sign for light propagating in reverse. For example, if a ray traverses the same path twice in opposite directions, the total rotation is double the rotation for a single passage. The Faraday effect is thus non-reciprocal.

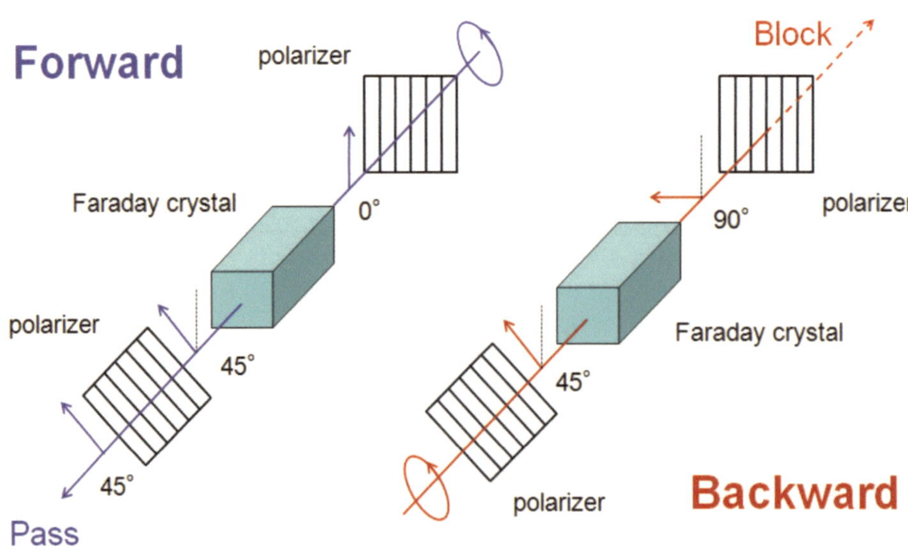

Fig. 1. Schematic structure of ordinary optical isolator.

Figure 1 shows the schematic structure of an ordinary optical isolator. The isolator consists of three components, i.e., a Faraday rotator, an input polarizer, and an output polarizer. The Faraday rotator consists of a magnetic garnet crystal such as yttrium iron garnet and terbium gallium garnet placed in a cylindrical permanent magnet and rotates the polarization of passing light by 45°. As illustrated in Fig. 1, light traveling in the forward direction (from A to B) will pass through the input polarizer and become polarized in the vertical plane (indicated by Pi). On passing through the Faraday rotator, the plane of polarization will be rotated 45° on axis. The output polarizer, which is aligned 45° relative to the input polarizer, will then let the light pass through. In contrast, light traveling in the reverse direction (from B to A) will pass through the output polarizer and become polarized by 45° (indicated by Pr). The light will then pass through the Faraday rotator and experience additional 45° of non-reciprocal rotation. The light is now polarized in the horizontal plane and will be rejected by the input polarizer, which allows light polarized in the vertical plane to pass through.
The ordinary optical isolator is bulky (therefore called a bulk isolator) and incompatible with waveguide-based optical devices, so it cannot be used in PICs. It has, however, superior optical characteristics (low forward loss and high backward loss) as shown in Fig. 2 [12]. Such good performance is a target in developing waveguide optical isolators.

Optical Component	Free Space Optical Isolatoı		unit
Structure	Single	Double	
Center Wavelength	1310, 1480, and 1550		[nm]
Insertion Loss	≦0.3	≦0.6	[dB]
Backward Loss	≧30	≧50	[dB]
Outer Diameter	ϕ3		[mm]
Length	1.5	NA	[mm]
Tilt	6		[deg.]
Clear Aperture	ϕ0.8	ϕ0.75	[mm]

Fig. 2. Optical characteristics of ordinary isolators available commercially [12]

3. Recent progress in waveguide optical isolators

3.1 How to make waveguide optical isolators

There are several strategies to develop waveguide optical isolators that can be integrated monolithically with waveguide-based semiconductor optical devices on an InP substrate. The strategies can be classified into two types. One is to use the Faraday effect as in conventional bulk isolators. Transferring the principle of bulk isolators to a planer waveguide geometry raises a number of inherent difficulties such as the discoherence of polarization rotation induced by structural birefringence. Therefore new idea is needed to use the Faraday effect in waveguide structure. Sophisticated examples are the Cotton-Mouton isolator [13, 14] and the quasi-phase-matching (QPM) Faraday rotation isolator [15, 16]. The latter in particular have attracted attention in recent years because of its compact techniques for producing the device. The other strategy to make waveguide isolators is to use asymmetric magneto-optic effects that occur in semiconductor waveguides combined with magnetic material. Leading examples are the nonreciprocal-phase-shift isolator [17-20] and the nonreciprocal-loss isolator [21-26]. The nonreciprocal-loss isolator uses no rare-earth garnet, so it is very compatible with standard semiconductor manufacturing processes. In the following sections, we give the outline of the QPM Faraday rotation isolator and the nonreciprocal-phase-shift isolator. The nonreciprocal-loss isolator, which has been developed in our laboratory, is explained in detail in Section 4.

3.2 Quasi-phase-matching faraday rotation isolator

Figure 3 shows a schematic of the QPM Faraday rotation isolator. The device consists of a Faraday rotator (non-reciprocal) section and a polarization rotator (reciprocal) section integrated with a semiconductor laser diode that provides an TE-polarized output. The Faraday rotator section consists of an AlGaAs/GaAs waveguide combined with a sputter-coated film of magnetic rare-earth garnet $CeY_2Fe_5O_x$. To obtain an appropriate polarization rotation, this device uses the QPM Faraday effect in an upper-cladding that periodically alternates between magneto-optic (MO) and non-MO media. Incident light of TE mode traveling in the forward direction will first pass through the Faraday rotator section to be rotated by +45°. The light then passes through the reciprocal polarization rotator section and is rotated by -45°. Consequently, the light keeps its TE mode and passes through the output edge. In contrast, backward traveling light of TE mode from the output filter is first rotated by +45° in the reciprocal polarization rotator and then nonreciprocally rotated by +45° in the Faraday rotator section. Consequently, backward light is transformed into a TM mode and therefore has no influence on the stability of the laser because the TE-mode laser diode is insensitive to TM-polarised light. The point of this device is TE-TM mode conversion in the waveguide. At the present time, efficient mode conversion cannot be achieved, so practical devices have yet to be developed.

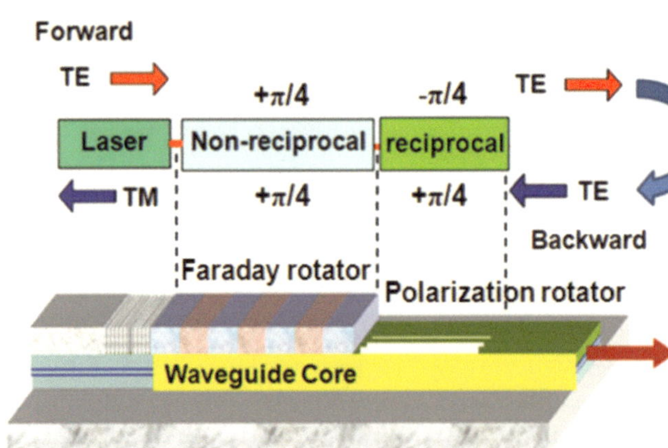

Fig. 3. Schematic of QPM Faraday rotation isolator.

Using magneto-optical waveguides made of $Cd_{1-x}Mn_xTe$ is effective to achieve efficient mode conversion [27, 28]. Diluted magnetic semiconductor $Cd_{1-x}Mn_xTe$ has the zincblende crystal structure, the same as that of ordinary electro-optical semiconductors such as GaAs and InP. Therefore, a single crystalline $Cd_{1-x}Mn_xTe$ film can be grown epitaxially on GaAs and InP substrates. In addition, $Cd_{1-x}Mn_xTe$ exhibits a large Faraday effect near its absorption edge because of the anomalously strong exchange interaction between the *sp*-band electrons and localized *d* electrons of Mn2+. Almost complete TE-TM mode conversion (98%+/-2% conversion) was observed in a $Cd_{1-x}Mn_xTe$ waveguide layer on a GaAs substrate [27, 28].

3.3 Nonreciprocal phase-shift isolator

The nonreciprocal-phase-shift isolator uses a modified Mach-Zehnder interferometer that is designed so that light waves traveling in two arms will be in-phase for forward propagation and out-of-phase for backward propagation. Figure 4 shows the structure of the isolator combined with a laser. The InGaAsP Mach-Zehnder interferometer consists of a pair of three-guide tapered couplers, and an ordinary reciprocal 90° shifter on one of the arms. Reciprocal phase shifting is achieved simply by setting a difference in dimensions or a refractive index between the optical paths along two arms. A magnetic rare-earth garnet YIG:Ce layer is placed on the arms to form a nonreciprocal 90° phase shifter on each arm. The garnet layer was pasted on the interferometer by means of a direct-bonding technique. Two external magnetic fields are applied to the magnetic layer on the two arms in an anti-parallel direction, as shown in Fig. 4; this produces a nonreciprocal phase shift in the interferometer in a push-pull manner. The isolator operates as follows. A forward-traveling light wave from the laser enters the central waveguide of the input coupler and divided between the two arms. During the light wave traveling in the arms, a -90° nonreciprocal phase difference is produced, but it is canceled by a +90° reciprocal phase difference. The divided two waves recouple at the output coupler, and output light will appear in the central waveguide. In contrast, for a backward-traveling wave from the output coupler, the nonreciprocal phase difference changes its sign to +90°, and it is added to the reciprocal phase difference to produce a total difference of 180°. Consequently, output light will appear in the two waveguides on both sides of the input coupler and not appear in the central waveguide.

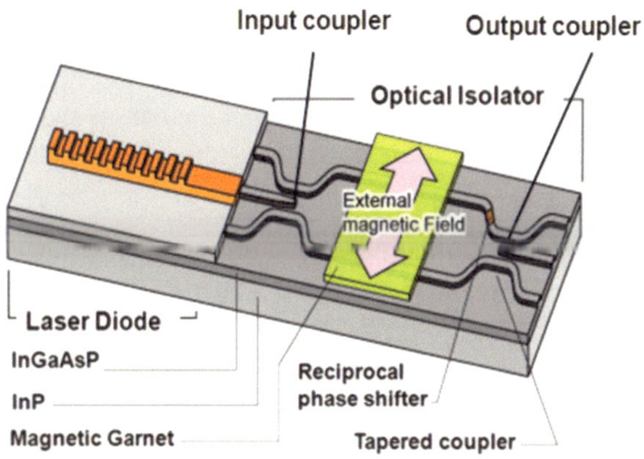

Fig. 4. Nonreciprocal-phase-shift isolator uses modified Mach-Zehnder interferometer.

4. Nonreciprocal loss phenomenon in magneto-optic waveguides

4.1 What is nonreciprocal loss phenomenon

One of the promising ways of creating waveguide optical isolators is by making use of the phenomenon of nonreciprocal loss. This phenomenon is a nonreciprocal magneto-optic phenomenon where——in an optical waveguide with a magnetized metal layer——the propagation loss of light is larger in backward than in forward propagation. Using this phenomenon can provide new waveguide isolators that use neither Faraday rotator nor polarizer and, therefore, are suitable for monolithic integration with other optical devices on an InP substrate. The theory of the nonreciprocal loss phenomenon was first proposed by Takenaka, Zaets, and others in 1999 [29, 30]. After that, Ghent University-IMEC and Alcatel reported leading experimental results in 2004; they made an isolator consisting of an InGaAlAs/InP semiconductor waveguide combined with a ferromagnetic CoFe layer for use at 1.3-μm wavelength [21, 22]. Inspired by this result, aiming to create polarization-insensitive waveguide isolators for 1.5-μm-band optical communication systems, we have been developing both TE-mode and TM-mode isolators based on this phenomenon. We built prototype devices and obtained a nonreciprocity of 14.7 dB/mm for TE-mode devices and 12.0 dB/mm for TM-mode devices——to our knowledge, the largest values ever reported for 1.5-μm-band waveguide isolators. The TE-mode device consisted of an InGaAsP/InP waveguide with a ferromagnetic Fe layer attached on a side of the waveguide [24]. For the TM-mode device, instead of ordinary ferromagnetic metals, we used ferromagnetic intermetallic compounds MnAs and MnSb, which are very compatible with semiconductor manufacturing processes. The following sections provide the details on this TM-mode isolator.

4.2 Structure of the TM-mode waveguide isolator

Figure 5 illustrates our TM-mode waveguide isolators with a cross section perpendicular to the direction of light propagation. Two kinds of structure are shown. The device consists of a magneto-optical planar waveguide that is composed of a TM-mode semiconductor optical-amplifying waveguide (SOA waveguide) on an InP substrate and a ferromagnetic layer attached on a top of the waveguide. To operate the SOA, a metal electrode is put on the surface of the ferromagnetic layer (a driving current for the SOA flows from the electrode to the substrate). Incident light passes through the SOA waveguide perpendicular to the figure (z-direction). To operate the device, an external magnetic field is applied in the x-direction so that the ferromagnetic layer is magnetized perpendicular to the propagation of light. Light traveling along the waveguide interacts with the ferromagnetic layer.

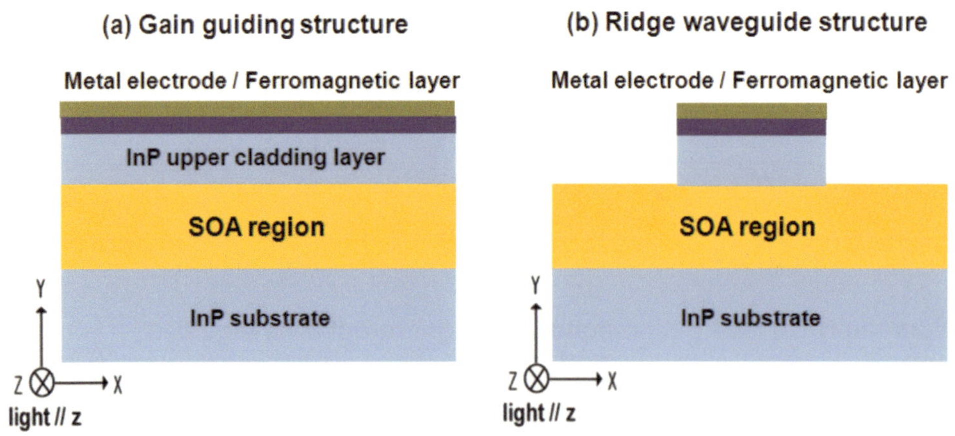

Fig. 5. Typical TM-mode nonreciprocal-loss waveguide isolators.

The nonreciprocal propagation loss is caused by the magneto-optic transverse Kerr effect in the magneto-optical planar waveguide. To put it plainly for TM-mode light, the nonreciprocity is produced when light is reflected at the interface between the magnetized ferromagnetic layer and the SOA waveguide. The light reduces its intensity when reflected from the ferromagnetic layer, which absorbs light strongly, and the reduction is larger for backward propagating light than forward propagating light because of the transverse Kerr effect. As a result, the propagation loss is larger for backward propagation (-z-direction) than for forward propagation (z-direction). Figure 6 illustrates the operation of the isolator on the propagation constant plane of the waveguide. The backward light is attenuated more strongly than forward light. Since forward light is also attenuated, the SOA is used to compensate for the forward loss; the SOA is operated so that the net loss for forward propagation will be zero. Under these conditions, the waveguide can act as an optical isolator.

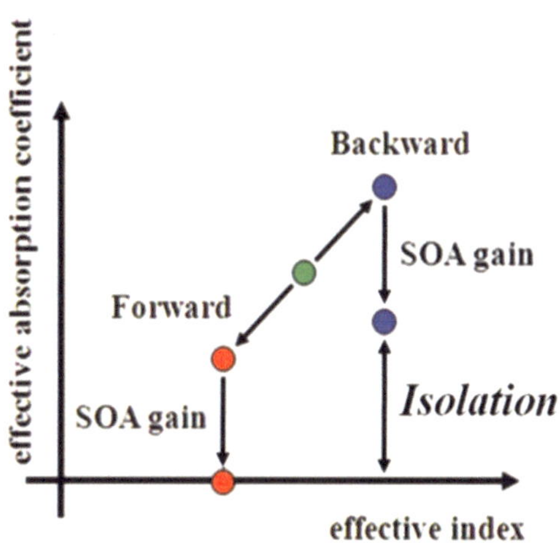

Fig. 6. Principle of nonreciprocal-loss waveguide isolator.

4.3 Theory of nonreciprocal loss in the waveguide isolator

Let us calculate the nonreciprocal loss in the magneto optic waveguide and design optimized structure for the isolator device, using electromagnetic simulation. In the TM-mode isolator, light traveling along the SOA waveguide extends through the cladding layer into the ferromagnetic layer to a certain penetration depth and interacts with magnetization vector in the ferromagnetic layer (see Fig. 5). Therefore, the thicknesses of the cladding layer and the ferromagnetic layer greatly affect the performance—the isolation ratio and forward loss (insertion loss) —of the isolator as follows:

i. A large isolation ratio can be obtained at small cladding-layer thickness because a thin cladding layer easily lets light through into the ferromagnetic layer to produce a large magneto-optic interaction. Therefore, the cladding layer has to be thin as long as the amplifying gain of the SOA can compensate for the absorption loss of light in the ferromagnetic layer.
ii. The ferromagnetic layer has to be thicker than its penetration depth of light. If it is not, light leaks out of the upper part of the ferromagnetic layer and is needlessly absorbed by the metal electrode. This reduces the isolation ratio because part of the propagating light in the device cannot interact with the ferromagnetic layer.

To determine the optimum thicknesses of the cladding and ferromagnetic layers, we calculated the isolation ratio and the insertion loss of the device as a function of the thicknesses by means of two-dimensional electromagnetic simulation based on the finite difference method (FDM).

In this device, the structure of the SOA has an influence on the device performance as well. However, the SOA structure cannot be changed greatly under the condition that the SOA should amplify TM-mode light at 1.5-μm-band wavelength. Therefore, we focus only on the thicknesses of the cladding and ferromagnetic layers to optimize the device performance.

The nonreciprocity of the device is caused by the off-diagonal elements in the dielectric tensor of the ferromagnetic layer. The dielectric tensor of each layer in the device is given by

$$\tilde{\varepsilon}_n = \begin{pmatrix} \varepsilon_n & 0 & 0 \\ 0 & \varepsilon_n & j\alpha \\ 0 & -j\alpha & \varepsilon_n \end{pmatrix}, \quad (3\text{-}1)$$

where ε_n is the diagonal element of the tensor in *n*th layer. The off-diagonal element α is 0 except in the ferromagnetic layer. Using these tensors, we write the Maxwell's equations in an isotropic charge-free medium as

$$\begin{aligned} \nabla \times \mathbf{H} &= j\omega\varepsilon_0\tilde{\varepsilon}_n\mathbf{E} \\ \nabla \times \mathbf{E} &= -j\omega\mu_0\mathbf{H} \;. \\ \nabla \cdot (\tilde{\varepsilon}_n\mathbf{E}) &= 0 \end{aligned} \quad (3\text{-}2)$$

Taking the rot of the second equation and using the first equation, we obtain the equation,

$$\nabla(\nabla \cdot \mathbf{E}) - \nabla^2\mathbf{E} = k_0^{\,2}\tilde{\varepsilon}_n\mathbf{E}, \quad (3\text{-}3)$$

where we used $\nabla \times (\nabla \times \mathbf{E}) = \nabla(\nabla \cdot \mathbf{E}) - \nabla^2\mathbf{E}$, and $k_0 = \omega\sqrt{\mu_0\varepsilon_0} = 2\pi/\lambda$ is the free-space propagation constant. Using the second and third equations in (3-2) and $\partial_z = j\beta$, the z component of eq. (3-3) can be written as

$$\partial_x^{\,2}E_z + \partial_y^{\,2}E_z + (k_0^{\,2}\varepsilon_n - \beta^2)E_z = j\alpha k_0^{\,2}E_y - \frac{\alpha\omega\mu_0}{\varepsilon_n}\partial_z H_x, \quad (3\text{-}4)$$

where β is the propagation constant in the device along *z* direction, E_t and H_t (t = x, y, z) are electric field (parallel to *t* axis) and magnetic field (parallel to *t* axis) of the light.

The y and z components of the first equation in (3-2) can be given by the equations for TM-mode light ($E_x = H_y = H_z = 0$),

$$\begin{aligned} \partial_z H_x &= -\alpha\omega\varepsilon_0 E_z + j\omega\varepsilon_0\varepsilon_n E_y \\ -\partial_y H_x &= \alpha\omega\varepsilon_0 E_y + j\omega\varepsilon_0\varepsilon_n E_z \end{aligned}. \quad (3\text{-}5)$$

Substituting the first equation of (3-5) into eq. (3-4) gives the equation for E_z,

$$\partial_x^{\,2}E_z + \partial_y^{\,2}E_z + \left(k_0^{\,2}\varepsilon_n - \beta^2 - \frac{k_0^{\,2}\alpha^2}{\varepsilon_n}\right)E_z = 0\,. \quad (3\text{-}6)$$

Using eqs. (3-5), we can express E_z with H_x as

$$E_z = \frac{j\varepsilon_n}{\omega\varepsilon_0(\varepsilon_n{}^2 - \alpha^2)}\left(\partial_y H_x - \frac{\alpha\beta}{\varepsilon_n} H_x\right). \tag{3-7}$$

From eqs. (3-6) and (3-7), we can obtain the scalar wave equation for magnetic field H_x of TM waves in each layer. The wave equation in non-magnetic layers (α=0) is given by

$$\partial_x{}^2 H_x + \partial_y{}^2 H_x + \phi H_x = 0 \quad (\phi = k_0{}^2\varepsilon_n - \beta^2)\,. \tag{3-8}$$

For the ferromagnetic layer, the wave equation has first-order and third-order derivative terms because of the nonzero off-diagonal element α in the dielectric tensor. For ordinary values of α in ferromagnetic materials, third-order terms of $\partial_x{}^2\partial_y H_x$ and $\partial_y{}^3 H_x$ are small and can be ignored. In consequence, the wave equation in the ferromagnetic layer is given by

$$\partial_x{}^2 H_x + \partial_y{}^2 H_x - \frac{\varepsilon_n}{\alpha\beta}\varphi\cdot\partial_y H_x + \varphi H_x = 0 \quad (\varphi = k_0{}^2\varepsilon_n - \beta^2 - \frac{k_0{}^2\alpha^2}{\varepsilon_n})\,. \tag{3-9}$$

Because of the nonzero off-diagonal elements in the dielectric tensor, the equation involves a linear term in the propagation constant β; this leads to a nonreciprocal solution to the propagation direction. The nonreciprocal solution gives a difference in absorption coefficient between forward (z-direction) and backward (-z-direction) TM waves and, therefore, gives the isolation ratio (or the difference between forward absorption and backward absorption) in the device.

To solve the wave equation numerically, we partition the domain in space using a mesh x_0, x_1,…x_p,… in x direction and mesh y_0, y_1,…y_q,… in y direction with a mesh width (the difference between two adjacent space points) of m in x direction and n in y direction. We represent the magnetic field on each mesh point (x_p, y_q) by $H_{p,q}$. Using a second-order central difference for the space derivative at position (x_p, y_q), we obtain the recurrence equation

$$\frac{1}{m^2}H_{p-1,q} + \frac{1}{m^2}H_{p+1,q} + \frac{1}{n^2}H_{p,q-1} + \frac{1}{n^2}H_{p,q+1} + (\phi - \frac{2}{m^2} - \frac{2}{n^2})H_{p,q} = 0 \tag{3-10}$$

for eq. (3-8), and recurrence equation

$$\frac{1}{m^2}H_{p-1,q} + \frac{1}{m^2}H_{p+1,q} + \frac{1}{n^2}H_{p,q-1} + \frac{\varepsilon_n}{2n\alpha\beta}\varphi H_{p,q-1} + \frac{1}{n^2}H_{p,q+1} - \frac{\varepsilon_n}{2n\alpha\beta}\varphi H_{p,q+1} + (\varphi - \frac{2}{m^2} - \frac{2}{n^2})H_{p,q} = 0 \tag{3-11}$$

for eq. (3-9). Solving eqs. (3-10) and (3-11) numerically, we can calculate the forward and backward propagation loss and the isolation ratio, as a function of the thicknesses of the cladding layer and the ferromagnetic layer, where the SOA is not operated. (In actual operation, the SOA is operated so that it compensates for the forward propagation loss.)

Before calculating the optimum thicknesses of the cladding and ferromagnetic layers, we must design the appropriate structure of the SOA region to amplify 1.5-μm TM-mode light. The structural parameters we used for the SOA was as follows. The substrate is a highly doped n-type InP (refractive index n = 3.16). The constituent layers of the SOA are: (i) lower guiding layer: 100-nm thick InGaAlAs (bandgap wavelength λ_g = 1.1 μm, n = 3.4), (ii) MQW: five InGaAs quantum wells (-0.4% tensile-strained, 15-nm-thick well, n_{MQW} = 3.53) with six InGaAlAs barriers (+0.6% compressively strained, 12-nm-thick barrier, λ_g = 1.2 μm), and (iii) upper guiding layer: 100-nm-thick InGaAlAs (λ_g = 1.1 μm, n = 3.4).

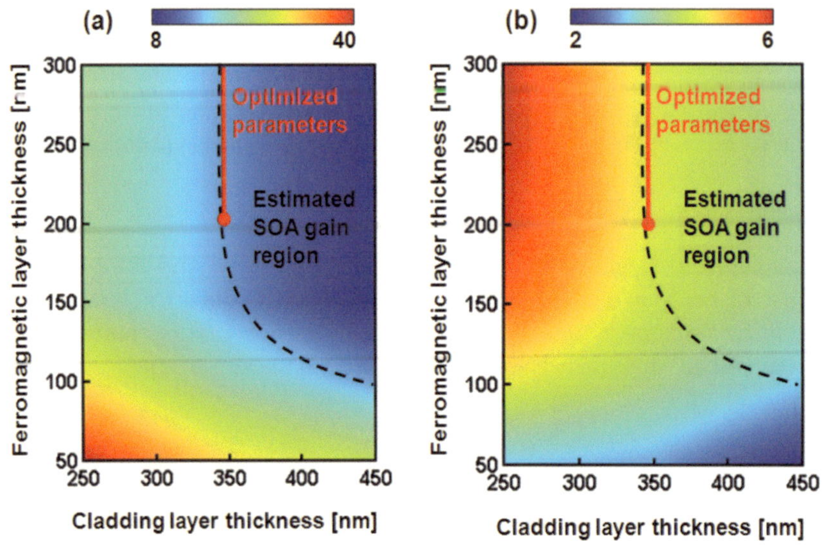

Fig. 7. (a) Forward absorption loss (propagation loss) and (b) isolation ratio (nonreciprocity) in the device as a function of Ferromagnetic-layer thickness and cladding layer thickness, calculated for 1.55-μm TM mode.

For the isolator with this SOA region, we calculated the propagation loss and the isolation ratio, using the method described above. Figure 7 shows an example of the results, i.e., (a) the absorption loss for forward propagation and (b) the isolation ratio as a function of the InP-cladding and ferromagnetic layer thicknesses. In this simulation, we assumed a device consisting of a ridge-shaped optical amplifying waveguide (see Fig. 5(b)) covered with a ferromagnetic MnAs layer and an Au-Ti metal electrode. The reason we used manganese pnictides as the ferromagnetic layer will be explained in Section 5. The parameters we used in the simulation are given in Table 1. The forward absorption loss in the device is large and the isolation ratio is small at small MnAs thickness because part of the propagating light in the device leaks out of the MnAs layer and is needlessly absorbed by the Au-Ti electrode. As MnAs layer thickness increases, forward absorption loss decreases and isolation ratio increases, both approaching a constant in MnAs layers thicker than 200 nm. This means that light penetrates to a depth of about 200 nm in the MnAs layer. Therefore, more than 200 nm can be considered a necessary and sufficient thickness for the MnAs layer when fabricating devices.

Figure 7 also shows that both the isolation ratio and the absorption loss increase as the thickness of the InP-cladding layer decreases. This is so because a thinner cladding layer lets a higher percentage of light through into the MnAs layer, producing a larger interaction. A thin cladding layer is preferable for obtaining a large isolation ratio as long as the forward absorption loss can be compensated for by the amplifying gain of the SOA. We expected an SOA gain of 16 dB/mm, and therefore decided that the optimum thickness of the cladding layer was 350 nm.

Figure 8 illustrates the distribution profile of light traveling in the isolator for forward and backward propagation, with the results calculated for a device with a 350-nm InP-cladding layer and a 200-nm MnAs layer. Figures 8(a-1) and 8(b-1) show the contour lines for TM magnetic field vector intensity—large magnetic fields in the central part—on the cross

Parameters	Values	Layers
$\varepsilon_1, \varepsilon_5$	$(3.16)^2$	InP substrate, InP cladding layer
$\varepsilon_2, \varepsilon_4$	$(3.4)^2$	InGaAsP down, upper guiding layer[a]
ε_3	$(3.53)^2$	MQW layer[b]
ε_6	$(2.8+4i)^2$	Ferromagnetic MnAs layer
ε_7	$(0.6+10i)^2$	Au/Ti metal electrode
α	$-1.62+0.27i$	Ferromagnetic MnAs layer
m	200×10^{-9}	Mesh width of x axis
n	50×10^{-9}	Mesh width of y axis

[a] The thickness of each guiding layer is fixed to 100 nm.
[b] The thickness of the MQW layer is fixed to 150 nm.

Table 1. Example of parameters used for calculating device characteristics.

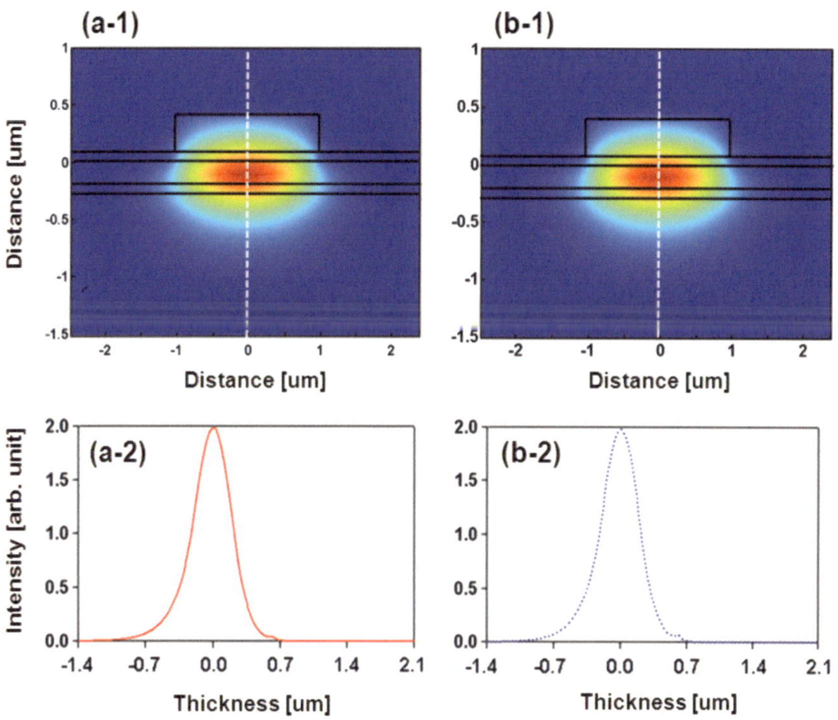

Fig. 8. Distribution profile of light traveling in isolator, calculated for 1.55 μm TM mode, with a 350-nm cladding layer and a 200-nm MnAs layer: cross section of distribution for (a-1) forward and (b-1) backward propagating light; distribution along vertical center line (dashed lines in (a-1) and (b-1)) of device for (a-2) forward and (b-2) backward propagating light.

section (x-y plane) of the device, where Fig. 8(a-1) is for forward propagating light and Fig. 8(b-1) is for backward light. Figures 8(a-2) and 8(b-2) depict the magnetic field vector intensity along the vertical center line (dashed lines in Figs. 8(a-1) and 8(b 1)) of the device, where Fig. 8(a-2) is for forward light and Fig. 8(a-2) is for backward light. Unlike forward propagating light, backward propagating light shifts its distribution tail to the MnAs layer and, therefore, suffers a larger absorption loss in the MnAs layer. Therefore, the propagation loss of light is larger in backward than in forward propagation.

5. Prototype device with ferromagnetic MnAs

5.1 Using manganese pnictides as a ferromagnetic material

The point of our device is its use of manganese arsenide (MnAs) as a ferromagnetic material, instead of ordinary ferromagnetic metals such as Fe and Co. In our device structure-- which is necessary for TM-mode operation--the ferromagnetic layer used to produce the nonreciprocity is also used as a contact to supply a driving current to the SOA. This means that the ferromagnetic layer has to meet a dual requirement of (i) producing a large Kerr effect at the wavelength of 1.5 μm and of (ii) providing a low-barrier contact for p-type III-V semiconductors. Ordinary ferromagnetic metals are not suited for this purpose because they produce a Schottky barrier on III-V semiconductors, thereby producing a high-resistance contact on the contact layer. In addition, during contact annealing, they produce undesirable nonferromagnetic compounds such as FeAs and CoAs at the contact interface and simultaneously degrade the microscopic flatness of the interface; this reduces optical nonreciprocity in the device. To solve these problems, we used manganese arsenide, MnAs, for the ferromagnetic layer. MnAs are ferromagnetic, intermetallic compounds with a NiAs-type hexagonal structure (see Fig. 9). They can be grown epitaxially on GaAs, InP, and related semiconductors by means of molecular beam epitaxy (MBE), without producing a solid-phase reaction at the interface [31-34]. MnAs is suitable ferromagnetic materials for our device because they have enough Kerr effect at 1.5-μm wavelength to produce practical nonreciprocity and, at the same time, can make a low-resistance contact on III-V semiconductors. The Currie temperature is 40°C for MnAs.

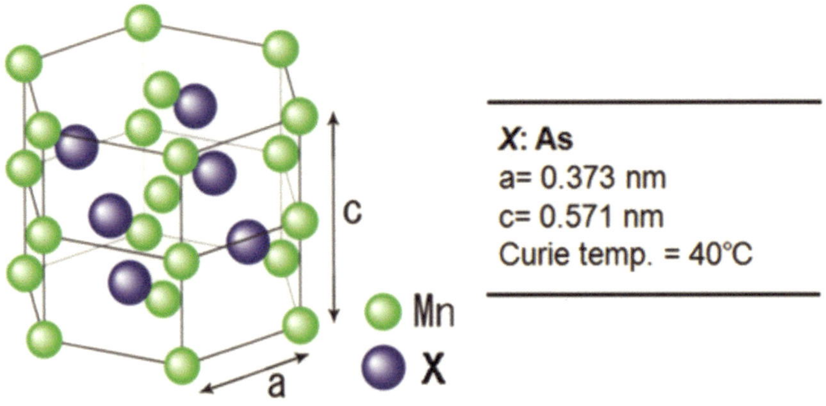

Fig. 9. Structure of manganese pnictides.

To take the first step, we made a device with a MnAs layer because the epitaxial growth technology of MnAs layers on III-V semiconductors was well established [31-33]. To reduce the propagation loss of light and obtain a single-mode operation, we used the ridge waveguide structure with a large lateral-confinement factor (see Fig. 5(b)). In the following sections, we provide details of the fabrication process and operation characteristics of the device that uses ferromagnetic MnAs.

5.2 Constructing the device

Figure 10(a) is a cross-sectional diagram of our TM-mode waveguide isolator with a ferromagnetic MnAs layer. The MnAs layer covers the SOA surface, and two interface layers (a highly doped p-type InGaAs contact layer and a p-type InP cladding layer) are inserted between the two. The InGaAs contact layer has to be thin so that 1.5-μm light traveling in the SOA will extend into the MnAs layer (the absorption edge of the contact layer is about 1550 nm). An Au/Ti double metal layer covers the MnAs layer, forming an electrode for current injection into the SOA. Light passes through the SOA waveguide in a direction perpendicular to the figure (z direction). An Al_2O_3 insulating layer separates the SOA surface from the Au-Ti electrode except on the contact region. Incident light passes through the SOA waveguide perpendicular to the figure (z direction).

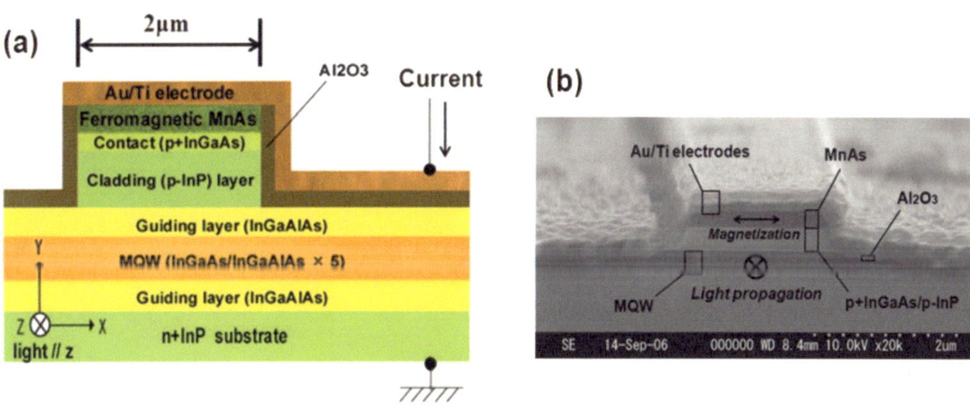

Fig. 10. (a) Schematic cross section of our waveguide isolator for 1.5-μm TM mode, consisting of a ridge-shaped optical amplifying waveguide covered with a MnAs layer magnetized in *x*-direction. Light propagates along *z*-direction. (b) SEM cross section of device.

On the basis of the simulation results mentioned in Section 4.3, we fabricated a device as follows. The substrate was a highly doped, [100]-oriented n-type wafer of InP. The SOA was formed on the substrate by metalorganic vapor-phase epitaxy (MOVPE). The MQW showed a photoluminescence peak at 1.54 μm— —this means that the SOA had a gain peak at 1.54 μm. The thicknesses of the p-InP cladding and p^+InGaAs contact layers were set to be 350 nm and 10nm. After the formation of the SOA, a 200 nm MnAs layer was grown on the surface of the p^+InGaAs contact layer by MBE. The wafer was first heat treated at about 550

°C under As_2 flux in the MBE chamber to remove a native oxide layer on the contact layer. The wafer temperature was then lowered to 200°C, and the As_2 flux was kept supplying to form an As template on the surface. This As template on the surface is important to grow high quality MnAs, as in the growth of MnAs layers on GaAs [31, 32] and InP [33]. The surface of the InGaAs contact layer with the As template showed spotty refraction high energy electron diffraction (RHEED) pattern. After that, Mn and As_2 fluxes were supplied on the surface to grow a 200 nm MnAs thin film. During the growth process, we confirmed (1×2) reconstruction in RHEED, indicating that the MnAs structural properties were improved. An X-ray diffraction pattern showed strong MnAs peaks in [1-100] directions.

After the growth of MnAs, the ridge waveguide structure was formed as follows. First, a photoresist mask in the form of a 2-μm-wide waveguide pattern was made on the surface of the MnAs layer. Then, the MnAs layer, InP cladding layer, and InGaAs contact layers were selectively etched in this order to fabricate a ridge waveguide—the MnAs layer was etched by reactive ion etching with Ar, and the cladding and the contact layers were wet-etched with a Br_2-HBr-H2O solution. An Al_2O_3 layer was deposited on this ridge waveguide using electron-beam (EB) evaporation. Then, the Al_2O_3 on the contact layer was removed using a lift-off process. Finally, a Ti layer and an Au layer were deposited to make a top electrode, using EB evaporation. This was the process we used to fabricate the structure depicted in Fig. 10(a). Finally, both ends of the device were cleaved, and the cleaved surfaces were left uncoated. Figure 10(b) is a cross section of the device as observed with scanning electron microscopy (SEM).

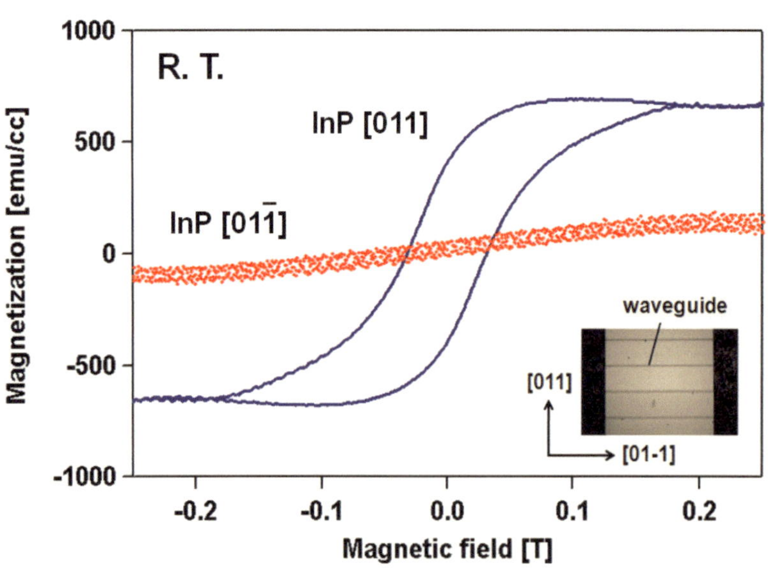

Fig. 11. Magnetization curve for MnAs layer, measured with a AGFM. MnAs layer can be easily magnetized along [011] direction of InP substrate. In contrast, magnetization is difficult along [01-1] direction.

MnAs thin films grown on the InGaAs contact layer showed strong magnetocrystalline anisotropy— —an intrinsic property of a ferrimagnet, independent of grain size and shape; the MnAs thin films were easily magnetized along the [011] direction of the InP substrates. Based on the fact, we formed the waveguide stripe parallel to the [0-11] direction of the InP substrate, and applied an external magnetic field to the [011] direction (x-direction in Fig. 10). However, in addition to the magnetocrystalline anisotropy, the shape anisotropy of the MnAs layer must be taken into consideration for the fabricated device because our device (or the MnAs layer) had the form of the 2-μm-wide waveguide structure. Therefore, we confirmed a magnetization curve of the MnAs layer in our device before measuring device characteristics. Figure 11 shows a plot of the magnetization curve, measured by alternating gradient force magnetometry (AGFM). Along the [011] direction of the InP substrate, the MnAs layer showed a soft hysteresis curve and was easily magnetized with a small coercive field of 0.07 T. In contrast, the magnetization was not easy along the [01-1] direction and was insufficient even in a magnetic field of 0.5 T. This means that the magnetocrystalline anisotropy is larger than the shape anisotropy in our device, and the device was expected to work with an external magnetic field of 0.07-0.1 T (initial magnetizing requires 0.15-0.2 T).

5.3 Device operation

We confirmed that the device functioned successfully as an optical isolator with nonreciprocal loss for TM-polarized, 1.5-μm light. Figure 12 shows our experimental setup for the measurement. It consisted of a wavelength-tunable laser, two polarization controllers, two circulators, two optical switches, an output coupler, an optical power meter, and an optical spectrum analyzer (OSA). Light from a tunable laser was transmitted to the device through a polarization controller and a circulator. The light was transferred into and out of the device using lensed-fiber couplers. A magnetic field was applied using a permanent magnet along the [011] direction of the device, i.e., parallel to the surface of the device and perpendicular to the direction of light propagation. Light propagation in the device was switched between forward direction (switch node 1-upper circulator-device-lower isolator- switch node 4 in Fig. 12) and backward direction (node 2-lower circulator-

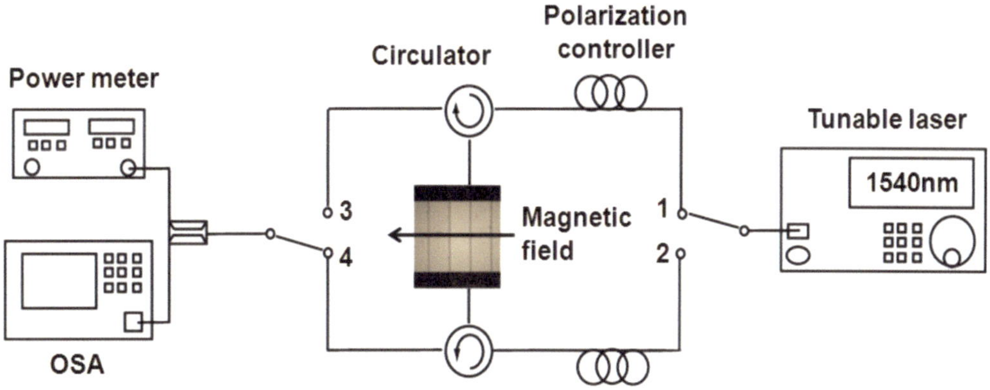

Fig. 12. Experimental setup for measuring isolation ratio and propagation loss of light in device.

device–upper isolator–node 3) by controlling the optical switches. The intensity of light transmitted in the device (or the output light from the device) was measured using the optical spectrum analyzer and the power meter. The output of the tunable laser was set to 5 dBm, and the magnetic field for the device was set to 0.1 T. During measurement, the device was kept at 20°C and operated with a SOA driving current of 100 mA. The MnAs layer successfully provided a low-resistance contact for the InGaAs contact layer. The voltage drop across the device 0.65 mm in length was only 1.7 V (SOA diode drop 0.9 V plus ohmic contact drop 0.8 V), whereas the drop across a control device with Fe-Ni layers instead of MnAs was 3.0 V (SOA diode drop 0.9 V plus ohmic contact drop 2.1 V) [35].

Figure 13 shows the transmission spectra of the device with a length of 0.65 mm. The intensity of the output light from the device is plotted as a function of wavelength for forward (dashed line) and backward (solid line) propagation of (a) TM-polarized and (b) TE-polarized light. The wavelength of incident light was fixed at 1.54 μm, which was the gain peak wavelength of the SOA. For TM-mode light, the output intensity changed by 4.7 dB by switching the direction of light propagation. The device operated efficiently as a TM-mode isolator with an isolation ratio of 7.2 dB/mm (= 4.7 dB/0.65 mm). In contrast, the output intensity for TE-mode light was not dependent on the direction of the light propagation. Small periodic ripples in amplified spontaneous emission spectra are shown in Fig. 13. They are caused by Fabry-Perot interference due to reflection from cleaved facets; the period was consistent with the value predicted from the length and effective refractive index of the device. The inset in Fig 13(a) is the near-field pattern of the TM-mode forward propagating light and shows that the device operated successfully in a single mode.

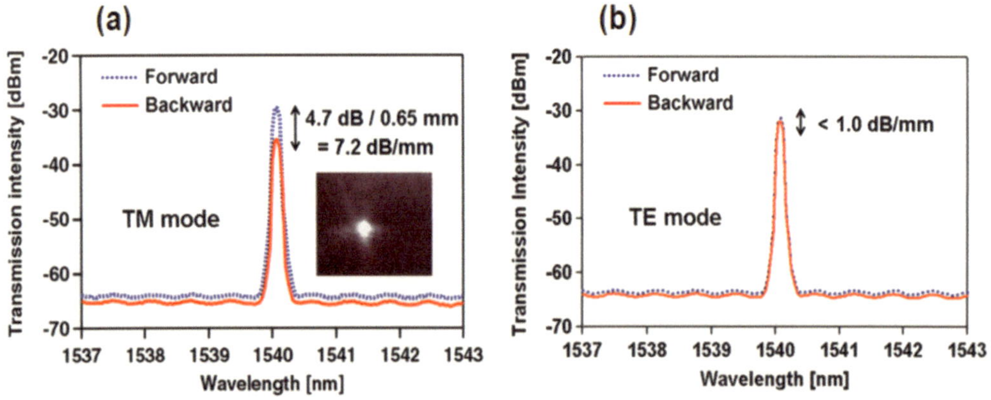

Fig. 13. Transmission spectra of device for forward transmission (dashed line) and backward transmission (solid line), measured for (a) TM-mode and (b) TE-mode, at 1.54-μm wavelength, 100-mA driving current, and 0.1-T magnetic field. Device is 0.65 mm long. Data on transmission intensity include loss caused by measurement system. Inset is near-field pattern of TM-mode forward propagating light.

The data of transmission intensity in Fig. 13 include the loss caused by the measurement system. To examine the intrinsic transmission loss of the device, we measured the transmission intensity for devices with different lengths. Figure 14 shows the results, i.e., the output intensity for forward and backward transmission as a function of device length (isolation ratio is also plotted). The slope of the forward line gives the intrinsic transmission loss (or absorption loss) per unit length. We estimated that forward loss in the device was 10.6 dB/mm—still large for practical use. This is so because the gain of the SOA was lower than we had expected, and therefore, insufficient to compensate for the intrinsic transmission loss in the device. The loss caused by the measurement system can also be calculated using the vertical-axis intercept of the forward line and the output intensity of the tunable laser. It was estimated to be 28 dB—output coupler loss 3 dB plus lensed-fiber coupling loss 12.5 dB/facet × 2 between the measurement system and the device.

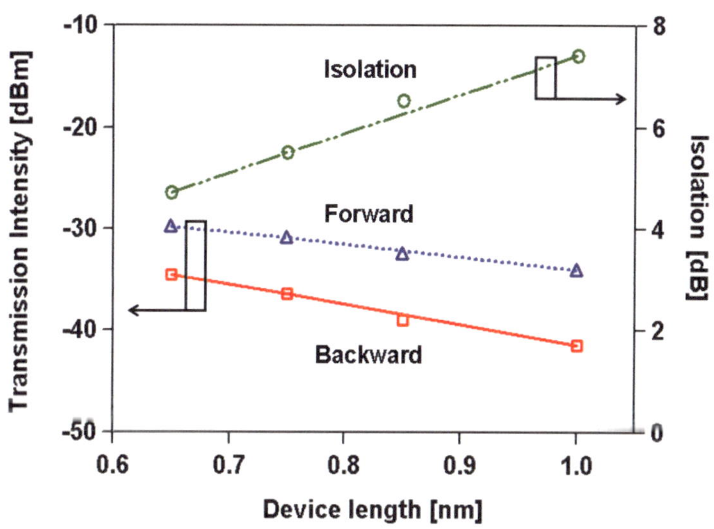

Fig. 14. Transmission intensity as a function of device length, measured for 1.54 μm TM mode, with 100-mA driving current and 0.1-T magnetic field. Isolation ratio is also plotted.

Figure 15 is a plot of the isolation ratio, as a function of wavelength from 1.53 to 1.55 μm. The device was 0.65-mm long. The output intensities for forward and backward propagations are also plotted (including the measurement system loss). In this range of wavelength, the isolation ratio was almost constant. The isolation ratio 7.2 dB/mm of this waveguide isolator was still small for practical use. In addition, the device was unable to operate at temperatures higher than room temperature because the Currie temperature of MnAs is only 40°C. To improve the device performance, we have to seek other superior ferromagnetic materials. In the next section, we present a device that uses MnSb instead of MnAs.

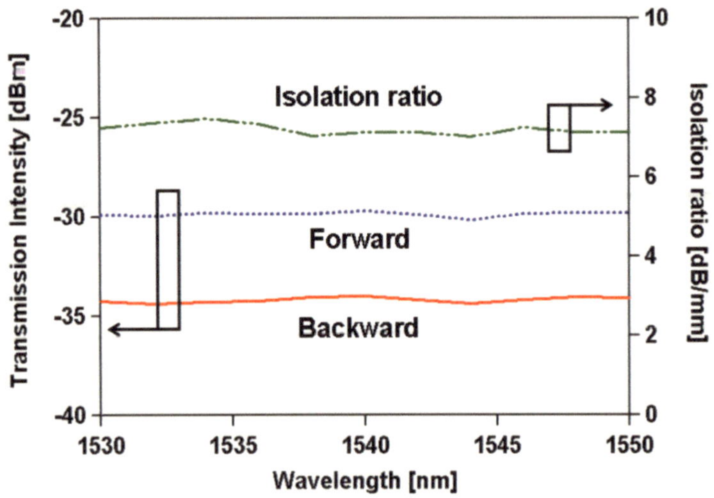

Fig. 15. Isolation ratio as a function of a wavelength from 1.53 to 1.55 μm for a 0.65-mm long device. Transmission intensity is also plotted for forward and backward propagation (including measurement system loss).

6. Conclusion

An important element for developing photonic integrated circuits is waveguide optical isolators that can be monolithically combined with other waveguide-based devices such as lasers. One promising way of creating such waveguide isolators is by using the phenomenon of nonreciprocal loss in magneto-optical waveguides. Making use of this phenomenon, we have been developing TE- and TM-mode waveguide isolators operating at 1.5-μm telecommunication band. As a fromagnetic material for the magneto-optical waveguide isolator, manganese pnictides such as MnAs are more superior than ordinary ferromagnetic metals because they can be formed on GaAs, InP, and related materials using semiconductors manufacturing process. Although MnAs is not common material at present for integrated optics, it will soon bring technical innovation in functional magneto-optic devices for large-scale photonic integrated circuits.

7. References

[1] O. Wada; T. Sakurai; T. Nakagami IEEE J. Quantum Electron. 1986, 22, 805-821.
[2] T. L. Koch; U. Koren IEEE J. Quantum Electron. 1991, 27, 641-653.
[3] A. A. M. Staring; L. H. Spiekman; J. J. M. Binsma; E. J. Jansen; T. van Dongen; P. J. A. Thijs; M. K. Smit; B. H. Verbeek IEEE Photon. Technol. Lett. 1996, 8, 1139-1141.
[4] Nagarajan, R.; Joyner, C.H.; Schneider, R.P., Jr.; Bostak, J.S.; Butrie, T.; Dentai, A.G.; Dominic, V.G.; Evans, P.W.; Kato, M.; Kauffman, M.; Lambert, D.J.H.; Mathis, S.K.; Mathur, A.; Miles, R.H.; Mitchell, M.L.; Missey, M.J.; Murthy, S.; Nilsson, A.C.;

Peters, F.H.; Pennypacker, S.C.; Pleumeekers, J.L.; Salvatore, R.A.; Schlenker, R.K.; Taylor, R.B.; Huan-Shang Tsai; Van Leeuwen, M.F.; Webjorn, J.; Ziari, M.; Perkins, D.; Singh, J.; Grubb, S.G.; Reffle, M.S.; Mehuys, D.G.; Kish, F.A.; Welch, D.F. IEEE J. Select. Topics Quantum Electron. 2005, 11, 50-65.

[5] N. Sugimoto; H. Terui; A. Tate; Y. Katoh; Y. Yamada; A. Sugita; A. Shibukawa; Y. Inoue J. Lightwave Technol. 1996, 14, 2537-2546.

[6] M. Levy; R. M. Osgood; H. Hegde; F. J. Cadieu; R. Wolfe; V. J. Fratello IEEE Photonics Technol. Lett. 1996, 8, 903-905.

[7] M. Fehndrich; A. Josef; L. Wilkens; J. Kleine-Börger; N. Bahlmann; M. Lohmeyer; P. Hertel; H. Dötsch Appl. Phys. Lett.1999, 74, 2918-2920.

[8] J. Fujita; M. Levy; R. M. Osgood; L. Wilkens; H. Dotsch Appl. Phys. Lett. 2000, 76, 2158-2160.

[9] M. Levy IEEE J. Sel. Top. Quantum Electron. 2002, 8, 1300-1306.

[10] J. S. Yang; J. W. Roh; S. H. Ok; D. H. Woo; Y. T. Byun; W. Y. Lee; T. Mizumoto; S. Lee IEEE Trans. Magn. 2005, 41, 3520-3522.

[11] T. R. Zaman; X. Guo; R. J. Ram IEEE Photonics Technol. Lett. 2006, 18, 1359-1361.

[12] http://www.namiki.co.jp/

[13] K. Ando; N. Takeda; N. Koshizuka; T. Okuda J. Appl. Phys. 1985, 57, 1277-1281.

[14] H. Dammann; E. Pross; G. Rabe; W. Tolksdorf; M. Zinke Appl. Phys. Lett. 1986, 49, 1755-1757.

[15] B. M. Holmes; D. C. Hutchings Appl. Phys. Lett. 2006, 88, 061116.

[16] B. M. Holmes; D. C. Hutchings Proc. of IEEE Lasers and Electro-Optics Society 2006, 897-898.

[17] H. Yokoi; T. Mizumoto; N. Shinjo; N. Futakuchi; Y. Nakano Appl. Optics 2000, 39, 6158-6164.

[18] H. Yokoi; T. Mizumoto; Y. Shoji Appl. Optics 2003, 42, 6605-6612.

[19] K. Sakurai; H. Yokoi; T. Mizumoto; D. Miyashita; Y. Nakano Jpn. J. Appl. Phys. 2004, 43, 1388-1392.

[20] Y. Shoji; T. Mizumoto Appl. Optics 2006, 45, 7144-7150.

[21] M. Vanwolleghem; W. Van Parys; D. Van Thourhout; R. Baets; F. Lelarge; O. Gauthier-Lafaye; B. Thedrez; R. Wirix-Speetjens; and L. Lagae Appl. Phys. Lett. 2004, 85, 3980-3982.

[22] W. Van. Parys; B. Moeyersoon; D. Van. Thourhout; R. Baets; M. Vanwolleghem; B. Dagens; J. Decobert; O. L. Gouezigou; D. Make; R. Vanheertum; L. Lagae Appl. Phys. Lett. 2006, 88, 071115.

[23] H. Shimizu; Y. Nakano Jpn. J. Appl. Phys. 2004, 43, L1561-L1563.

[24] H. Shimizu; Y. Nakano IEEE J. Lightwave Technol. 2006, 24, 38-43.

[25] T. Amemiya; H. Shimizu; Y. Nakano; P. N. Hai; M. Yokoyama; M. Tanaka Appl. Phys. Lett. 2006, 89, 021104.

[26] T. Amemiya; H. Shimizu; P. N. Hai; M. Yokoyama; M. Tanaka; Y. Nakano Appl. Optics 2007, 46, 5784-5791.

[27] W. Zaets; K. Ando Appl. Phys. Lett. 2000, 77, 1593-1595.

[28] V. Zayets; M. C. Debnath; K. Ando Appl. Phys. Lett. 2004, 84, 565-567.

[29] M. Takenaka; Y. Nakano Proc. of IEEE Conference on Indium Phosphide and Related Materials 1999, 289-292.
[30] W. Zaets; K. Ando IEEE Photonics Technol. Lett. 1999, 11, 1012-1014.
[31] M. Tanaka; J. P. Harbison; G. M. Rothberg Appl. Phys. Lett. 1994, 65, 1964-1966.
[32] L. Daweritz; L. Wan; B. Jenichen; C. Herrmann; J. Mohanty; A. Trampert; K. H. Ploog J. Appl. Phys. 2004, 96, 5052-5056.
[33] M. Yokoyama; S. Ohya; M. Tanaka Appl. Phys. Lett. 2006, 88, 012504.
[34] H. Akinaga; K. Tanaka; K. Ando; T. Katayama J. Cryst. Growth 1995, 150, 1144-1149.
[35] T. Amemiya; H. Shimizu; Y. Nakano, Proc. of IEEE Conference on Indium Phosphide and Related Materials 2005, 303-306.
[36] Y. Ogawaa; T. Amemiya; H. Shimizu; Y. Nakano; H. Munekata Proc. of International Symposium on Compound Semiconductors 2007, TuC-P22.

8

GaAs/AlOx Nonlinear Waveguides for Infrared Tunable Generation

E. Guillotel, M. Ravaro, F. Ghiglieno, M. Savanier, I. Favero, S. Ducci, and G. Leo
Laboratoire Matériaux et Phénomènes Quantiques, UMR 7162, Université Paris Diderot France

1. Introduction

New optical sources in the near- and mid-infrared (NIR/MIR respectively) have recently attracted a growing attention for potential applications in telecommunication systems (Yoo, 1996), spectroscopy (Chen et al., 1999; Arie et al., 2002), gas sensing (Lancaster et al., 1999), and quantum information (Gisin et al., 2002; Sergienko & Jaeger, 2003). In this respect, guided-wave frequency conversion is an appealing solution, due to its efficiency, compactness and tunability of the output wavelength. In this process, three guided modes coupled by the material optical nonlinearity exchange power during propagation; provided that their phase-velocity mismatch is absent or cancelled, such interaction allows the efficient transfer of power from the injected pump modes to a new-frequency generated mode, according to energy conservation (Boyd, 2008).

After the demonstration of a few original phase-matching schemes, nonlinear waveguides based on gallium arsenide (GaAs) have carved out a prominent position in the panorama of integrated frequency converters. Among their main assets, in comparison with alternative material systems: a higher nonlinear coefficient, a wider infrared transparency range, and potential monolithic integration with a laser diode pump source. On the other hand, conversion efficiency in such devices is still affected by non-negligible scattering loss, in spite of recent technological developments. Due to this limitation, no optical parametric oscillator has been reported to date in GaAs waveguides, whereas it was demonstrated long ago in lithium niobate ($LiNbO_3$) waveguides (Bortz et al., 1995).

In this chapter we focus on GaAs/AlAs nonlinear waveguides, where phase-matched three-wave mixing can be performed thanks to form birefringence. This is obtained by oxidizing a few AlAs thin layers in the waveguide core, with AlAs being transformed into a low-index non-stoichiometric aluminium oxide (hereafter referred to as AlOx). Although all different $\chi^{(2)}$ processes have been reported so far, these devices especially lend themselves to parametric down-conversion towards mid-infrared, and are promising candidates as compact sources for infrared spectroscopy. In the first section we briefly summarize the state-of-the art of mid-infrared tunable coherent sources, focusing on frequency converters. At present, none of these sources fully meets the main requirements of practical spectroscopic systems. In section two, we describe the phase-matching principle and the design guidelines of GaAs/AlOx waveguides, while the fabrication process and its crucial issues are detailed in section three.

The recent progress in terms of infrared generation efficiency and tunability will be the object of the fourth section. Finally, the fifth section is devoted to work in progress towards the realization of a semiconductor optical parametric oscillator.

2. Scientific context: sources for infrared spectroscopy

2.1 Laser diodes and bulk frequency converters

Mid-infrared spectrum is of strong scientific and technological interest for spectroscopy, as several molecules exhibit their fundamental rotational-vibrational resonances in this region, with absorption spectra highly specific to their atomic structure. As mid-infrared absorption lines are very strong, molecule concentrations as low as parts per-billion can be detected by means of laser-based detectors. In the near-infrared, where only overtone resonances can be excited, absorption peaks are weaker, and the corresponding detection thresholds several orders of magnitude higher. Nevertheless, many measurements are still performed at wavelengths below 2 μm, due to the relative lack of suitable laser sources at longer wavelengths. Infrared spectroscopy requires continuously tunable sources, with compactness and room temperature operation highly desirable. In the mid-infrared, the present state of the art for this type of devices includes diode lasers and sources based on nonlinear frequency conversion (Sorokina & Vodopyanov, 2003).

The main representatives of the former category are antimonide diodes and quantum-cascade lasers. Both are proven to produce significant continuous wave (CW) output power at room temperature in the 2-3 and in the 4-9 μm range, respectively, while maintaining single mode operation and being reproducibly tunable in a manner suitable for spectroscopy. Tuning is typically accomplished by changing either the temperature or the injected current, on an overall range limited to few tens of nanometres.

A versatile alternative to laser diodes is represented by sources based on $\chi^{(2)}$ nonlinear processes. For MIR generation, frequency conversion involves the use of an intense "pump" wave (generated by a laser of angular frequency ω_p), which is coupled to an input wave at ω_s (the "signal"), with an "idler" generated at the difference frequency ω_p-ω_s. The input signal photons can be provided by either a second injected laser beam or quantum noise. In the former case, the interaction is generally performed in a single-pass configuration and is referred to as difference-frequency generation (DFG). The latter process is called parametric fluorescence. Since the single-pass gain is low, it is generally exploited by resonantly enhancing the signal or the idler wave in a cavity containing the nonlinear medium, i.e. an optical parametric oscillator (OPO).

DFG and OPOs are largely employed for the generation of mid-infrared, as they can produce coherent radiation in any temporal format and over a wide spectral range, only limited by the transparency of the nonlinear crystal. For both these processes, one of the main issues is the fulfilment of the phase-matching condition. For the signal and idler waves to be continuously fed by the pump as the beams propagate through the crystal, the three waves must stay unvaried in their initial phase relation, in contrast with crystal dispersion. The classical approach to achieve phase-matched propagation consists in compensating dispersion in a birefringent nonlinear crystal. More convenient quasi-phase matching (QPM) relies on a microstructured crystals where the second-order susceptibility is periodically inverted, so as to keep the three waves into the correct phase relation. This technique, for long time the prerogative of ferroelectric crystals, further reinforced $LiNbO_3$ position as the nonlinear crystal of choice, including for the fabrication of infrared sources. The

combination of high nonlinearity and long interaction length of periodically-poled $LiNbO_3$ has considerably increased the conversion efficiency of DFG and OPO systems based on this crystal, making them accessible to diode laser or fiber laser pumping.
The implementation of QPM had an even stronger impact on GaAs. This semiconductor has a $\chi^{(2)}$ coefficient considerably higher than those of $LiNbO_3$ and of most inorganic crystals. Nevertheless, until the last decade it was not employed for frequency conversion, since it is neither anisotropic nor ferroelectric, and thus not readily suitable to the application of the above phase matching schemes. In 2001, the fabrication of the first effective QPM GaAs bulk crystals, based on lattice-matched heteroepitaxy of GaAs/Ge films on GaAs substrates (Eyres et al., 2001), rapidly opened the way for a series of tunable infrared sources with impressive performances. Among the results that followed from this technological innovation was the first GaAs-based optical parametric oscillator, reported in 2004 (Vodopyanov et al., 2004). Due to the high nonlinearity and wide transparency range of GaAs, this device was continuously tunable between 2.2 and 9 μm (limited only by mirror reflectivity), with a maximum conversion efficiency of 25%. Based on these recent results, GaAs appears to be mature as an alternative to $LiNbO_3$ for the fabrication of infrared sources.

2.2 Integrated frequency converters for the IR generation

In terms of performances, frequency converters show substantial advantages compared to laser diodes. They are suitable for oscillation from CW to femtosecond regimes, offer wider tunability ranges and great flexibility, especially since the advent of periodically poled materials. In addition, the recent demonstration of orientation-patterned GaAs promises to extend the versatility of sources based on periodically poled $LiNbO_3$ to the whole mid-infrared region. On the other hand, frequency converters are based on complex, often cumbersome setups that considerably limit their transportability, preventing their use outside of a laboratory setting. Despite the recent fabrication of miniaturized OPO sources, the degree of compactness offered by room-temperature or Peltier-cooled laser diodes remains unattainable for bulk nonlinear sources.
An intermediate solution between these two families of sources is provided by guided-wave frequency converters, which potentially combine a compactness similar to that of semiconductor lasers with the performances of frequency conversion. In general, nonlinear interactions in waveguides offer several additional features, like compactness, on-chip integration and a variety of design solutions. Moreover, the confinement of the interacting fields on a long propagation distance results in a conversion efficiency orders of magnitude higher compared to their bulk counterparts. On the other hand, the fabrication of waveguides is often technologically more complex than that of bulk crystals. The key issues are scattering losses, which can considerably limit the efficiency of nonlinear processes. In spite of such technological drawbacks, several highly-performing integrated frequency converters have been demonstrated to date. Similar to bulk crystals, periodically poled $LiNbO_3$ has played a major role also for guided-wave generation both in the visible (through frequency up-conversion) (Kintaka et al., 1996), and in the infrared (through down-conversion) (Bortz et al., 1995; Hofmann et al., 1999).
With respect to $LiNbO_3$ and dielectric waveguides in general, GaAs integrated nonlinear devices offer the additional potential benefit of the integration with a pumping source. A tangible evidence of this appeal is given by the large number of integrated frequency converters reported in the last decade. Regretfully, scattering losses have prevented GaAs waveguides from fully developing their potential to date. This is the case of quasi phase

matched AlGaAs waveguides, based on the same inversion technique of bulk orientation-patterned GaAs. The MBE growth of thin AlGaAs multilayers on GaAs/Ge templates has recently allowed the demonstration of guided-wave quasi phase matched frequency doubling (Yu et al., 2005). With a CW 1.55 μm pump, a normalized conversion efficiency η_{norm} = 92 % W^{-1} cm^{-2} was achieved, much lower than the calculated efficiency, η_{norm}=500% W^{-1} cm^{-2}. Such discrepancy arises from high scattering losses, due to waveguide corrugation at the inverted domain boundaries. The resulting attenuation coefficient is of the order of 10 dB/cm, more than two orders of magnitude higher than in GaAs bulk crystals.

At present, a performing alternative to QPM in GaAs waveguides is represented by form birefringence. This kind of integrated frequency converters, based on a strongly birefringent guiding core, were developed in the late 90's at the Thomson CSF (today Thales) laboratories. By embedding in a GaAs guiding layer several low-index AlOx layers, a form-birefringence sufficient to fulfil the phase-matching condition in the infrared was attained (Fiore et al., 1998a). As detailed in the following, AlOx layers result from the selective oxidation of AlAs layers embedded in the structure. Due to design versatility, this approach has since been successfully employed to phase match either down- or up-conversion interactions, with experimental efficiencies η_{norm} up to more than 1000% W^{-1} cm^{-2} and scattering losses lower than 1 cm^{-1} (Moutzouris et al., 2001; Ravaro et al., 2007).

From a technological point of view, in the last decade GaAs/AlOx approach has experienced a strong development. The very first conversion demonstrations suffered from processing immaturity, resulting in high propagation losses. At present, the conversion efficiency for both frequency doubling and down-conversion is comparable to that of $LiNbO_3$ counterparts. Whereas a gap still remains in terms of propagation losses, last advancements paved the way for the accomplishment of a resonant configuration.

3. GaAs/AlOx waveguides for parametric down-conversion

3.1 Form birefringence

A significant difference between TE_0 and TM_0 refractive indices n_{TE0}-n_{TM0}, referred to as modal birefringence, can e.g. be achieved by fabricating waveguides with birefringent materials, like $LiNbO_3$. Nevertheless, even in the case of optically isotropic media, waveguides are slightly birefringent, due to different boundary conditions for TE and TM polarizations. Whereas such "form" birefringence is too weak to compensate dispersion in frequency conversion processes, it can be greatly enhanced by designing a waveguide with several index discontinuities.

In a periodic multilayer stack of isotropic materials, the existence of repeated discontinuities of the refractive index along one direction breaks the original 3 axes rotation symmetry of the constituent media and results in a macroscopic negative uniaxial crystal. With reference to the geometry described in Figure 1, for plane waves in an infinitely extended medium, the ordinary and extraordinary refractive indices are equal to (Born, 1980; Yeh, 1988):

$$n_0^2 = \frac{h_1}{\Lambda} n_1^2 + \frac{h_2}{\Lambda} n_2^2$$
$$\frac{1}{n_e^2} = \frac{h_1}{\Lambda}\frac{1}{n_1^2} + \frac{h_2}{\Lambda}\frac{1}{n_2^2} \qquad (1)$$

where n_i and h_i are the refractive index and the thickness of the ith (i = 1, 2) repeated layer, and Λ the period ($h_1 + h_2 = \Lambda << \lambda$, with λ the optic wavelength). The corresponding form birefringence $\Delta n = n_o - n_e$ depends on the duty cycle and the index contrast.

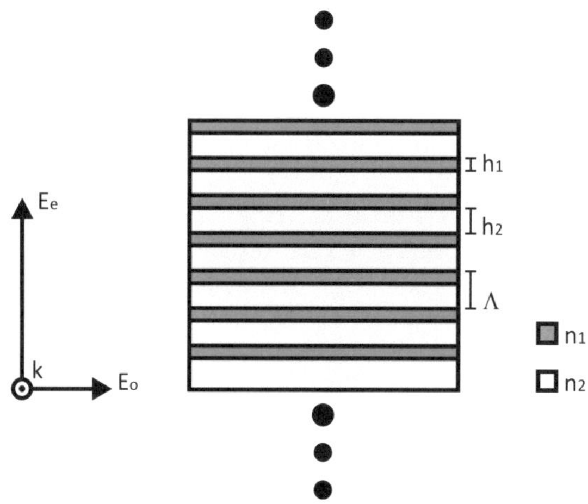

Fig. 1. Scheme of a form birefringent multilayer periodic structure.

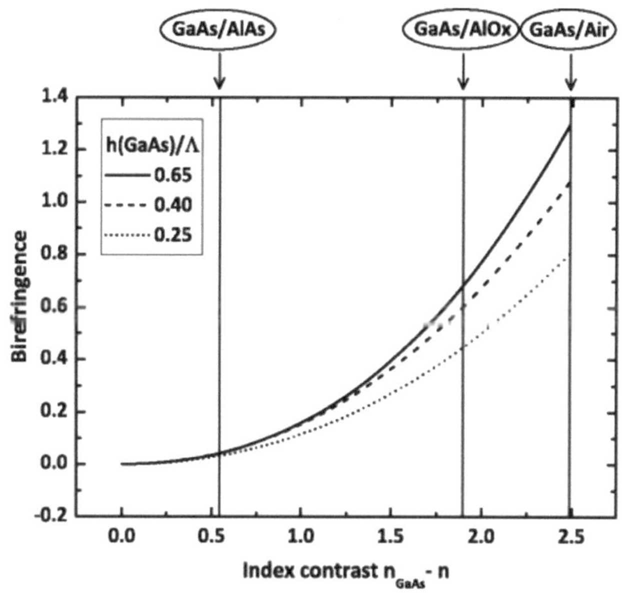

Fig. 2. Form birefringence, in plane-wave approximation, of a multilayer periodic structure of GaAs and a generic material of refractive index n, for three different duty cycles. The index contrasts corresponding to GaAs/AlAs, GaAs/AlOx, and GaAs/air combinations are pointed out.

Form-birefringent phase matching was firstly proposed in multilayer AlGaAs structures, with a birefringence suitable to perform frequency doubling in the mid-infrared, for instance from a CO_2 (λ = 10.6 μm) pump laser (Ziel, 1975; Ziel & Gossard, 1978). As maximum birefringence obtained by GaAs/AlAs index contrast is not sufficient to compensate

dispersion near the forbidden band, near-infrared phase matching is not accessible with these materials. The scenario changed in the early 90's with the demonstration of selective wet oxidation of AlAs layers embedded in GaAs (Dallesasse et al., 1990a). This process produces polycrystalline aluminium oxide layers, with a refractive index of 1.6, whose characteristics will be discussed more in detail in Section 4. The benefit resulting from the obtained index contrast is illustrated in Figure 2, which shows form birefringence for a periodic structure alternating GaAs with a lower index material.
The first achievement of form-birefringent phase matching in an AlOx-based heterostructures was demonstrated with mid-infrared difference frequency generation from two near-infrared guided waves (Fiore et al., 1997). Two continuous-wave pump lasers, a Nd:YAG at 1.32 μm and a Ti:Sapphire at 1.03 μm were end-fire coupled in a GaAs/AlOx waveguide to generate a guided mode at 4.8 μm. Afterwards, a different waveguide design allowed the frequency doubling of a CWdye-laser in the telecom range, with a conversion efficiency of $\eta_{norm} = 4\%\ W^{-1}\ cm^{-2}$, mainly limited by scattering losses (Fiore et al., 1998b).
After a few years of technological development, which mainly produced a significant loss reduction, a similar device was employed to demonstrate parametric fluorescence (De Rossi et al., 2001). In that case a CW Ti:Sapphire laser was coupled into a 3.2 mm-long waveguide and, thanks to a peak efficiency $\eta_{norm} = 1000\%\ W^{-1}\ cm^{-2}$, a parametric fluorescence signal around 2 μm could be detected.

3.2 Waveguide design

The main design criterion for a form-birefringent nonlinear waveguide consists in aiming at the balance between form birefringence and overall (material+waveguide) dispersion for the fundamental modes, at the frequencies involved. Further guidelines followed for the multilayer waveguide design are:

- maximization of the nonlinear overlap integral: in order to improve the efficiency of a frequency conversion process, interacting modes must have similar field distributions with as-high-as-possible peak intensities. In addition, in AlGaAs waveguides, Al content must be minimized, as AlGaAs $\chi^{(2)}$ strongly decreases with Al content.
- minimization of propagation loss: AlOx layers are the main cause for propagation losses in GaAs/AlOx waveguides, due to roughness at the interfaces with contiguous crystalline layers; their number must be kept as small as possible. Scattering reduction requires field intensity at waveguide interfaces with air be minimized as well.

It is worth noticing that the selective oxidation of an AlGaAs multilayer requires the fulfilment of a few conditions: the thickness of layers to be oxidized must be included in a certain range, between a few tens and a few hundreds of nanometres, for technological reasons; in addition, as oxidation concerns not only AlAs but AlGaAs with high Al content, all layers that are not supposed to be oxidized must have an Al content lower than a certain minimum value (Dallesasse et al., 1990a; Dallesasse et al., 1990b)
Dispersion compensation in the near- and mid infrared requires a form birefringence of few tenths (0.2-0.3), lower than the value calculated in plane-wave approximation for the index contrast n_{GaAs}-n_{AlOx} (see Figure 2). This allows increasing the period Λ (and reducing the number of AlOx layers) beyond the long-wavelength range ($\Lambda << \lambda$), in order to limit scattering losses. Thus, the multilayer design for a given frequency conversion process is based on the number of AlOx layers necessary to compensate dispersion for the set of wavelengths concerned. Accordingly, the fabrication of frequency converters operating close to GaAs forbidden band is more critical, due to large number of AlOx layers required.

That is why up- and down-conversion schemes between 0.75 μm and the telecom range have been abandoned, after the first frequency doubling report. Hereafter we refer to GaAs/AlOx waveguides for parametric down-conversion from a 1 μm pump. Such design has led to the most recent relevant results, and is retained as the basis for the fabrication of an optical parametric oscillator.

According to GaAs selection rules for waveguides grown on standard (001) GaAs substrates, second order nonlinear interactions are only possible between a combination of two TE modes and one TM mode. As form birefringence results in a lower refractive index for the polarization orthogonal to the layers plane, i.e. TM, the only possible phase matching scheme in this geometry is a type I with a TM polarized pump. The phase-matching condition for parametric generation at degeneracy reads

$$n_{TM_0}(1\mu m) = n_{TE_0}(2\mu m) \tag{2}$$

which implies that birefringence at 1 μm must compensate TE_0 overall dispersion between 1 μm and 2 μm:

$$\underbrace{n_{TE_0}(1\mu m) - n_{TM_0}(1\mu m)}_{\text{Birefringence}} = \underbrace{n_{TE_0}(1\mu m) - n_{TE_0}(2\mu m)}_{\text{Dispersion}} \tag{3}$$

For a core thickness resulting in single-mode confinement for the pump, the fulfilment of eq. 3 requires the introduction of five thin AlOx layers in the GaAs core. The thickness of the interposed GaAs layer is then set as to precisely tune the phase-matched set of wavelengths. The structure design for 1.064 μm pumping at degeneracy is described in Table 1. Figure 3 shows the corresponding refractive index profile, along with the resulting intensity profiles of the interacting modes at degeneracy. Here, the form birefringence principle is pointed-out: the TM polarization, parallel to the growth direction, is enhanced in the low index layers, due to electric displacement vector conservation at the interfaces. On the other hand, the TE polarization field is continuous on at the same interfaces. As a consequence, the TM effective index is significantly reduced, while the TE one stays almost unchanged.

Lateral optical confinement is then obtained by etching 3 μm wide ridges in $(1\bar{1}0)$ direction. Note that, for such width, the phase-matching condition calculated for the planar structure is just slightly red-tuned. If needed, ridge width can be adjusted to correct a phase-matching wavelength shift due to growth inaccuracy.

For such a device the calculated parametric gain coefficient, normalized to the root of the power, is $g/P^{1/2} = 4.8\ W^{-1/2}cm^{-1}$ at degeneracy. The generated signal and idler wavelengths can then be tuned on a wide spectral region by reducing the pump wavelength with respect

	Composition	Thickness (nm)
	GaAs	Substrate
	$Al_{0.7}Ga_{0.3}As$	1000
	AlOx	33
4×	GaAs	273
	AlOx	33
	$Al_{0.7}Ga_{0.3}As$	1000
	GaAs	30

Table 1. Multilayer structure for 1.064 μm pumping at degeneracy

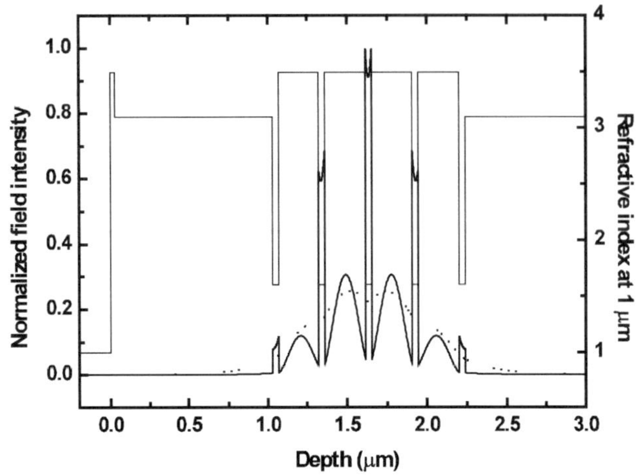

Fig. 3. Vertical structure of the form birefringent parametric generator: refractive index (thin solid line), TM_0 (thick solid line) and TE_0 (dotted line) modal profiles at $\lambda_p = 1.064$ µm and $\lambda_s = \lambda_i = 2\,\lambda_p$, respectively.

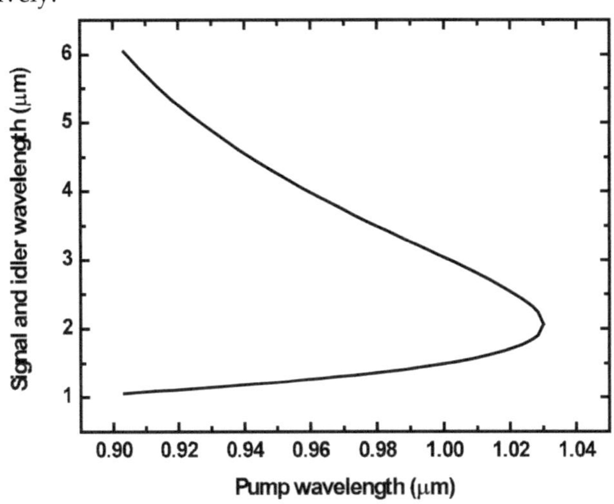

Fig. 4. Calculated parametric generation tuning curve.

to degeneracy (Figure 4). Compared to typical bulk nonlinear crystals, wavelength tuning in this geometry is highly sensitive to pump wavelength. This peculiarity stems from the huge waveguide dispersion at λ_p, much larger than at 2 λ_p, as clarified by the approximated expression for the relationship between λ_p, λ_s, and λ_i around degeneracy (De Rossi et al., 2005a):

$$\lambda_{s,i} = \frac{2\lambda_p}{1 \pm \alpha\sqrt{\frac{\lambda_p^0}{\lambda_p} - 1}} \quad (4)$$

where the "aperture" α of the tuning curve is given by

$$\alpha = \sqrt{\frac{\frac{\partial n_{TM_0}}{\partial \lambda}(\lambda_p^0)}{\frac{\partial n_{TE_0}}{\partial \lambda}(\lambda_s^0)}} \tag{5}$$

In both expressions, the superscript 0 indicates values at degeneracy. Eqs. 4 and 5, which are obtained by neglecting the second-order derivatives of the effective indices, show how a higher $(\partial n_{TM0}/\partial\lambda)/(\partial n_{TE0}/\partial\lambda)$ ratio implies a wider aperture of the tuning curve. A wide aperture α is strongly desirable for a tunable source, especially in the perspective of integrating a pump laser diode, usually exhibiting limited tunability. In addition, waveguide strong dispersion leads to somewhat reduce the down-conversion phase-matching bandwidth, typically much wider than in type II processes, especially at degeneracy.

4. Fabrication of GaAs/AlOx waveguides

Despite high nominal values, the efficiency of GaAs/AlOx nonlinear waveguides has been limited so far by high propagation losses. These arise mainly from scattering at AlOx layers interfaces. Hereafter we briefly describe the origin of this problem and the fabrication improvements we have introduced with respect to the state of the art.

4.1 Growth and lithography

The growth of GaAs/AlAs waveguides is done by molecular beam epitaxy (MBE) for its performances in terms of:

1. precision on the layer thicknesses: this is crucial to achieve a given phase-matching wavelength;
2. abruptness of the layer interfaces: a key issue to limit scattering losses following from oxidation of AlAs layers.

A single narrow ridge is chemically etched for lateral confinement and oxidation. Thanks to the low superficial roughness achieved, the adoption of deeply-etched narrow structures for lateral guidance does not imply a degradation of the waveguide transmission due to scattering at ridge sidewalls.

Initially employed for chemical etching, H_2SO_4:H_2O_2:H_2O solution has been successively replaced with the more performing HBr:CH_3COOH:$K_2Cr_2O_7$. Besides being less selective, the latter reduces under-etching and, above all, further improves etched surface smoothness. A surface analysis at the atomic force microscope has revealed a roughness reduction from 3 to 2 nm rms. This amelioration represents an advantage also for the oxidation process. Indeed, a smoother AlAs/air lateral interface is supposed to improve oxidation kinetics, allowing a better flux of H_2O toward the ridge core and a better release of reaction products.

It is worth comparing the above roughness values with those recently reported for strongly confining AlOx waveguides obtained by e-beam lithography + chlorine plasma etching (Scaccabarozzi et al., 2006). In this case, a 6 nm rms roughness is measured on the ridge sidewalls suggesting that, in spite of recent improvements, plasma-assisted etching does not provide yet the same surface quality as wet etching. The latter, in spite of poor control and reproducibility, remains a reasonable choice when low propagation losses are a priority.

4.2 Selective oxidation

AlAs, or more precisely, AlGaAs with high Al molar fraction gets easily oxidized if exposed to air. (Dallesasse et al., 1990b) The resulting oxide is unstable and causes the material to crumble. This process thus severely affects the performances of AlGaAs devices, such as lasers and more in general waveguides, containing Al rich layers. In the early 90's, an attempt to accelerate this process in a wet atmosphere at high temperature, led to demonstrate the formation of a stable native oxide (Dallesasse et al., 1990a). The so obtained aluminum oxide, AlOx, exhibits a refractive index much lower than that of AlGaAs and it is electrically insulating.

A buried thin AlGaAs layer can be oxidized starting from the etched sidewalls of the waveguide. Oxidation kinetics and, structure and characteristics of the final oxide, are very sensitive to the different parameters involved (thickness of the AlGaAs layers, Al rate, oxidation temperature).

Optical properties of AlOx have been characterized with discrepancies in the published results. AlOx transparency region has been unambiguously demonstrated to extend from deep UV (about 240 nm) up to 10 μm, (Bravetti et al., 1998; Hall et al., 1999) excepted an O-H absorption peak that we identified at 3 μm (Ravaro et al., 2008). However several different values of the AlOx refractive index have been reported. In the NIR these are included between 1.55 and 1.64, with a claimed dependence on the original Al composition. (Knopp et al., 1998; Hall et al., 1999; Durand et al., 2003) This is a major issue for the design of birefringent GaAs/AlOx nonlinear waveguides, as uncertainty on these properties affects the phase-matching wavelength of such devices.

Selective oxidation of GaAs/AlAs waveguides is carried out in a quartz tube oven where samples are heated at 420-430°C while the atmosphere is saturated in water vapor by a 2 l/min flux of wet N_2. This is obtained by previously making dry N_2 pass through a water bubbler stabilized at 70 °C.

This solution and the above novel processing technique have allowed us to further improve the transmission characteristics of GaAs/AlOx waveguides with respect to the best results previously reported. (De Rossi et al., 2001) This has been demonstrated by systematic losses measurements.

4.3 Propagation loss

Compared to gradual index dielectric waveguides, high-index-contrast semiconductor waveguides are significantly affected by scattering losses. Scattering can originate from roughness on either the ridge sidewalls or the epitaxial layers interfaces, as well as from bulk media inhomogeneity.

Scattered light measurements in AlOx waveguides, with a femtosecond NIR OPO, have allowed determining the dependence of propagation losses on wavelength (Venugopal Rao et al., 2003). In the range from 1.3 to 2.1 μm, losses were found to decrease with wavelength, as expected for Rayleigh scattering, confirming the dominant contribution of AlOx/GaAs interface roughness to the attenuation coefficient.

Waveguides propagation loss can be evaluated by the Fabry-Pérot technique (De Rossi et al. 2005b). This resorts to Fabry-Perot transmission fringes in the waveguide to infer the attenuation coefficient independently on input coupling efficiency. Due to the modal reflectivity at input and output facets, the transmission of a waveguide can be treated analogously to that of a Fabry-Perot resonator in plane-wave approximation. In a single-mode waveguide, for a monochromatic input, the transmission is given by the Airy function:

$$T_{FP} = \frac{T^2 e^{-\alpha L}}{(1-R')^2 + 4R'\sin^2(\Phi/2)}\eta \tag{6}$$

where α is the modal attenuation coefficient, L the waveguide length, η the coupling efficiency, T (R) the transmission (reflectivity) of the two facets (supposed to be equal), and $R' = R \cdot \exp(-\alpha L)$. T_{FP} is a periodic function of the round-trip dephasing $\Phi=2k_0N_{eff}L+\Phi_0$, with k_0 the free-space wave vector and N_{eff} the effective index. By tuning the input wavelength or the effective index (by e.g. varying the sample temperature), T_{FP} oscillates between a maximum T_{max} and a minimum T_{min} that depend only on R′. By measuring the contrast of the transmission fringes:

$$K = (T_{max}-T_{min})/(T_{max}+T_{min}) \tag{7}$$

R′ is given by the relation

$$R' = \frac{1-\sqrt{1-K^2}}{K} \tag{8}$$

If the facet modal reflectivity R is known, the attenuation coefficient can finally be calculated as

$$\alpha = \ln(R/R')/L \tag{9}$$

In form birefringent GaAs/AlOx waveguides, current typical lowest propagation loss is in the order of 0.4-0.6 cm^{-1} after oxidation as compared to ≈ 0.1 cm^{-1} before oxidation. Understanding the origin of such losses is thus essential for its reduction. Taking into account the small thickness of the AlOx layers, Transmission Electron Microscopy (TEM) is the best suited instrument to quantify the GaAs/AlOx interfaces roughness.

4.4 Interface roughness characterization by transmission electron microscopy

The fabrication of cross-section specimens for transmission electron microscopy (TEM) is performed with a JEOL Ion Slicer allowing the thinning of one waveguide by sample thanks to 8 keV Ar^+ ion beam with an incident angle of 1.3°. A 2 keV Ar^+ beam at 1.0° final polish ensures minimization of surface amorphous material. The obtained foils are analysed in a JEM 2100F TEM field emission gun TEM operating at 200 kV. Thickness mapping is done for each thin foil, using a Gatan Imaging Filter with 3mm aperture and the two window-based jump ratio imaging technique (Hofer et al., 1995). Exposure time is 1 s for slit widths of 10 eV, with a beam collection angle of 11.0 mrad defined by the objective aperture. For the interfaces characterization we use the high angle annular dark field (HAADF) STEM imaging where the incoherently scattered electrons collected to produce an image are sensitive to variation of the atomic number of atoms in the sample. An example of such image is showed on figure 5.

To accurately measure the amplitude of the roughness, the following procedure is done: (I) binary contrast enhancement of the HAADF image; (II) boundary tracing; (III) line matrix averaging; (IV) gradient removal; and (V) autocorrelation analysis.

Binary contrast enhancement (I) relates to the unique identification of the delineated boundary on either side of which occurs a brightness change for the adjoining of GaAs and AlOx. A median brightness level is identified such that image pixels brighter (higher stored values) than the median will be denoted as 1, and darker image pixels (lower stored values) are denoted as 0, thereby giving a binary image. The median can be deduced in one of two ways; by performing a trace across the image, perpendicular to the GaAs/AlOx interface, to determine the different brightness levels on either side of the boundary, or by determining the median in an image brightness intensity spectrum.

Boundary tracing (II) is next employed to digitally extract the coordinates of the interface through an algorithm that first determines a point at the interface and then follows the boundary, now defined by 0 and 1, in a systematic manner across the image from left to right. The interface coordinates are stored in a separate array.

Such vertical line matrix averaging (III) has been performed with the data being further subject to a linear least squares fit to remove any gradient (IV) displayed. Finally (V), it is possible to calculate the autocorrelation function R(u) of the interface profile defined by the function f(z): $R(u)=\langle f(z)f(z+u)\rangle$. A gaussian fit of the autocorrelation function leads to the mean square perturbation, $\sigma^2=R(0)$, of the boundary and its correlation length L_c. In the present case, values of $\sigma^2 = 0.16\ nm^2$ and $L_c=44$ nm were determined. The non-vanishing thickness of the TEM samples results in a fictitious smoothing of f(z) responsible of a diminution of those parameters evaluated at 39±18% and 20±20% for σ^2 and L_c respectively (Goodnick et al., 1985).

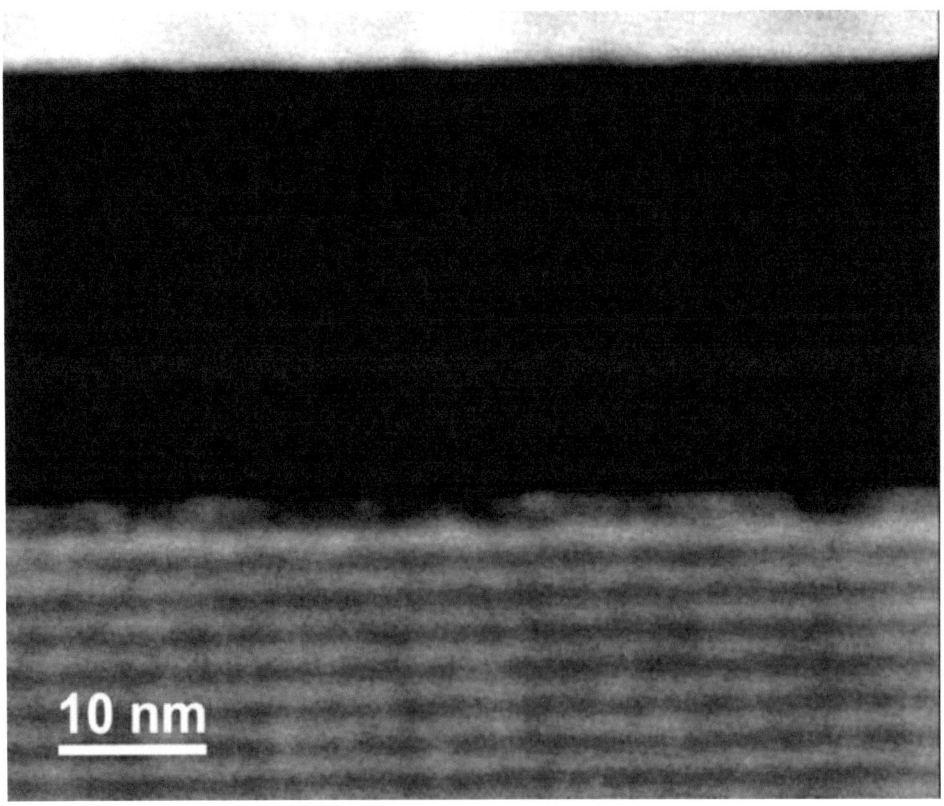

Fig. 5. HAADF image of an AlOx layer (in black) embedded in GaAs (top) and $Al_{0.7}Ga_{0.3}As$ (bottom). Rough interfaces are clearly visible with an enhanced effect at the bottom interface.

To infer Rayleigh scattering loss coefficient from these values, we adapted Lacey and Payne's simple model (Lacey & Payne, 1990) to our multilayer waveguide where the attenuation coefficient is linked to $\tilde{R}$, the Fourier transform of the autocorrelation function R(u), through the expression:

$$\alpha_r = \Phi^2(d)(n_2^2 - n_1^2)^2 \frac{k_0^3}{4\pi n_1} \int_0^{\pi} \tilde{R}(\beta - n_2 k_0 \cos\theta)\, d\theta \tag{10}$$

where n_1 and n_2 are the refractive indices of AlOx and GaAs respectively, k_0 is the wave vector in vacuum, β is the propagation vector, θ is the photon scattering angle as regards as the interface and ϕ(d) is the energy distribution of the propagating mode evaluated at the position of the different interfaces and normalized as follow:

$$\int_{-\infty}^{+\infty} \Phi^2(y)\, dy = 1 \tag{11}$$

Figure 6 illustrates the variation of the loss coefficient with the value of σ^2 and L_c in the case of our GaAs/AlOx waveguides. The star marks the value of the loss coefficient deducted from our TEM observations and the rectangle symbolizes the value of the loss coefficient, taking into account the effect of smoothing occurring through the pass of electrons in the thin foil. Our model gives a value for the propagation loss $\alpha = 0.28 \pm 0.12 cm^{-1}$. This value is in fair agreement with experimental data suggesting that Rayleigh scattering could be the main source of propagation losses.

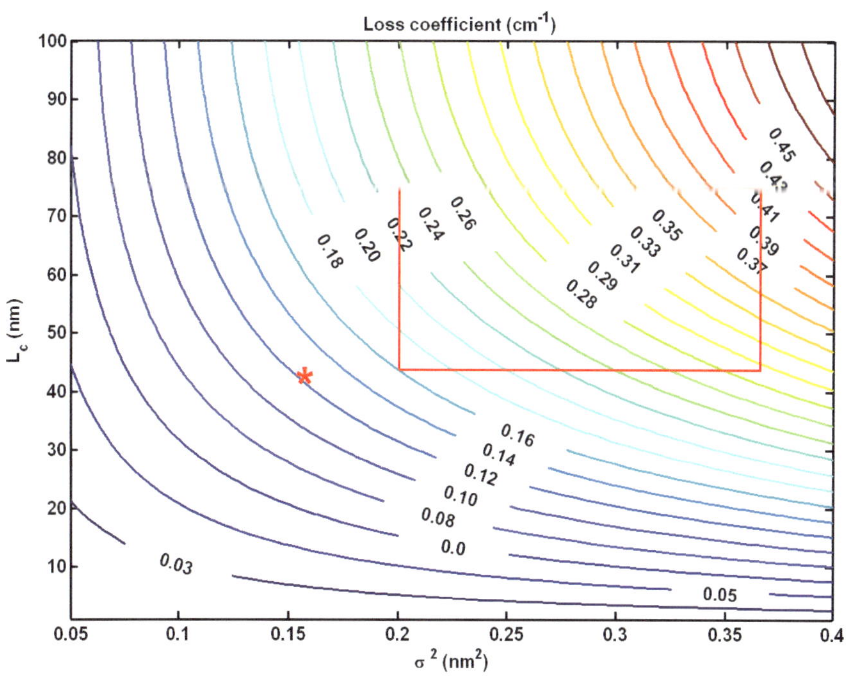

Fig. 6. Variation of the loss coefficient with the mean square value of roughness and the coherent length.

5. Experimental performances

In the first reports of form-birefringent frequency conversion, scattering losses did not allow the full exploitation of the huge normalized conversion efficiency. After the last technological advancements described above, this is now possible in mm long devices, with overall conversion efficiencies comparable to that of $LiNbO_3$ waveguides. Current loss level still prevents the use of cm long devices, which would result in even higher single-pass efficiency and make parametric oscillation more easily accessible.
In the following we present the last advancements of parametric down-conversion in mm long GaAs/AlOx waveguides. In section 5.1 we describe parametric fluorescence results, with a tuning range in excess of 1 µm. In section 5.2 we detail the first parametric amplification demonstration, which, in passing, has led to a precise determination of the parametric gain coefficient.

5.1 Parametric fluorecence

The experiment we report here was performed with a tunable Ti:Sapphire CW pump laser with 40 Ghz linewidth. The beam was TM polarized and coupled with a 60× microscope objective in a GaAs/AlOx waveguide designed for phase-matching at 1.064 µm. Waveguide length was L = 3 mm. Its output was collimated with an identical objective and passed through an high-pass interferential filter before being injected into a spectrometer equipped with an strained InGaAs photo-diode, whose cutoff limited the wavelength detection range at 2400 nm.
Two parametric fluorescence spectra, at degeneracy and 0.2 nm off degeneracy, are reported in Figure 7. Besides the good signal-to-noise ratio, their $sinc^2$ shape demonstrates the high uniformity of the waveguide over a few millimeter length. By acquiring parametric fluorescence spectra for different values of λ_p the tuning curve shown in Figure 8 (crosses) was reconstructed. Idler wavelengths longer than 2400 nm were inferred from measured signal and pump wavelengths, through frequency conservation. A tuning range between 1.69 µm and 2.75 µm has been obtained, limited by loss experienced by the idler mode beyond 2.75 µm. In this spectral region AlOx exhibit an absorption band which is ascribed to an Al hydroxide produced during wet oxidation (Ravaro et al., 2008). However, AlOx layers are expected to be dehydroxylated through a proper thermal annealing, thus giving access to a broader tuning range.
The experimental tuning interval above has been covered by varying λ_p over a 12 nm interval, with degeneracy at λ_p = 1060.5 nm, slightly shorter than the nominal value (λ_p = 1064 nm). These data can be reasonably fitted by a curve (solid line) obtained from effective-index calculations for the planar structure, with the thickness t of GaAs layers as fit parameter (t_{fit} = 271 nm, instead of nominal t = 273 nm). This calculated tuning curve differs from its experimental counterpart for a slightly lower tuning aperture, as determined with Eq. 4: α = 2.14 for the experimental points, versus α = 1.92 for the fitting curve. We ascribe such difference to the narrow lateral confinement of the ridge, which increases the dispersion of the waveguide and thus the aperture of the tuning curve, with respect to the results of 1D modeling.
Parametric tuning can also be obtained by changing the sample temperature, as shown in Figure 9. Here experimental data are compared with a simulated curve based on AlGaAs thermorefractive coefficients (Gehrsitz et al., 2000).

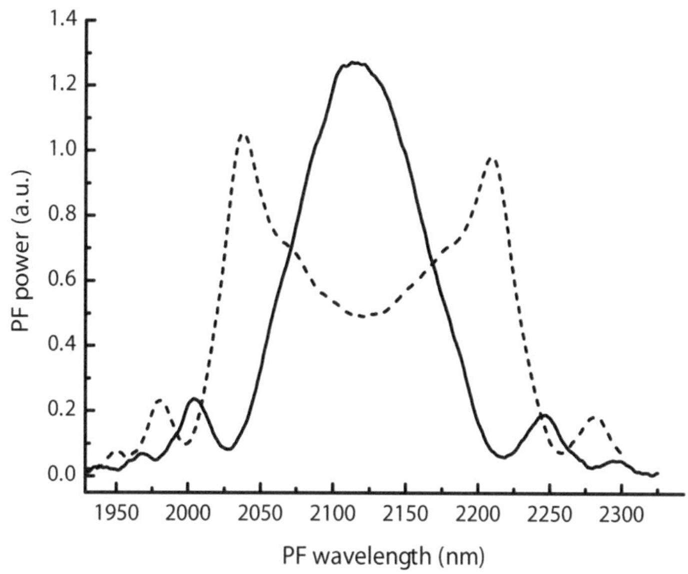

Fig. 7. Parametric fluorescence experimental spectra: at degeneracy (solid line) and 0.2 nm off degeneracy (dashed line).

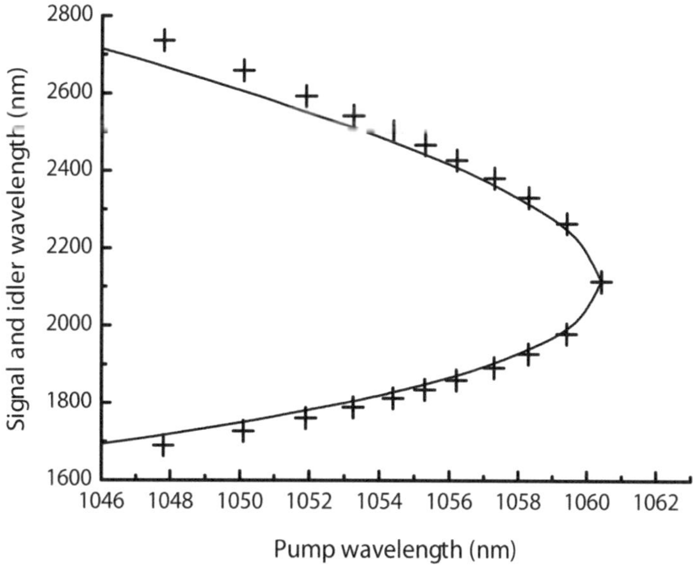

Fig. 8. Parametric fluorescence tuning curve: experimental data (+) and fitting curve for a GaAs thickness t = 271 nm (solid line).

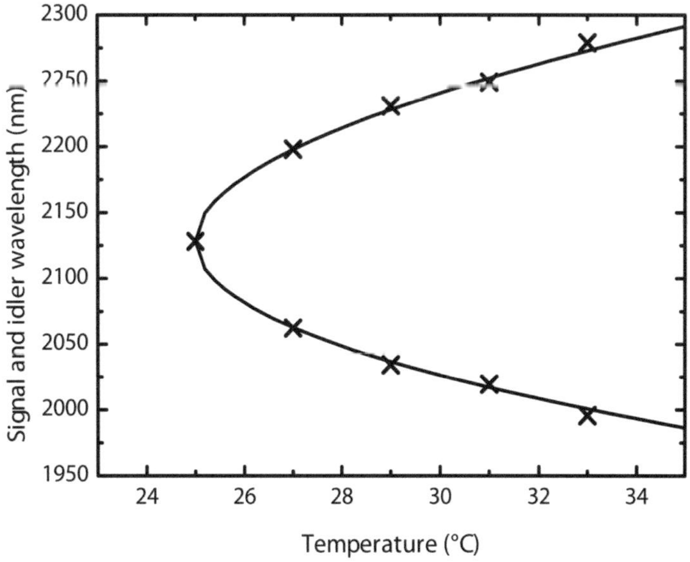

Fig. 9. Temperature tuning for λ_P = 1062 nm: experimental data (crosses) and calculated curve (solid line).

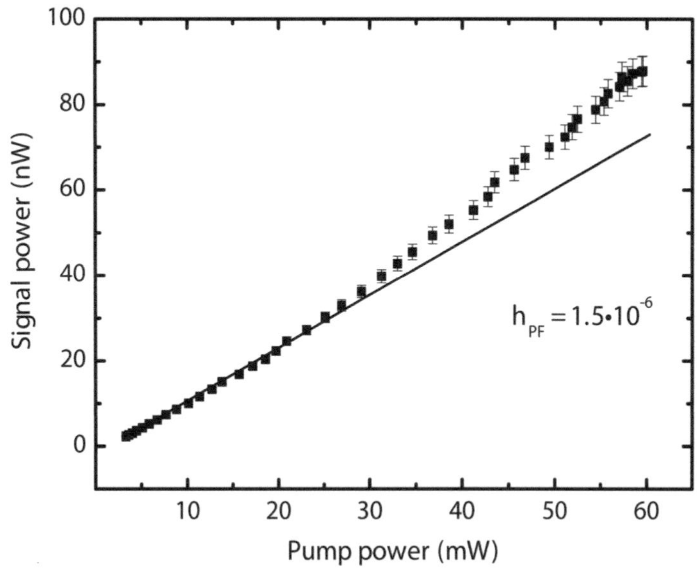

Fig. 10. Generated signal versus pump power: the nonlinear trend of experimental data (squares) is pointed out by the linear fit of the low power points (solid line).

Figure 10 shows the generated signal as a function of the TM coupled pump power P_p at degeneracy, where both vacuum fluctuations effective input and parametric gain coefficient are maximum. The latter can be written as

$$g = \sqrt{8\pi^2 |\Gamma|^2 P_p / (n_p n_i n_s \lambda_i \lambda_s \varepsilon_0 c)} \qquad (12)$$

with

$$\Gamma = \int d_{14}(\rho)\Phi_p(\rho)\Phi_s(\rho)\Phi_i(\rho)d\rho \tag{13}$$

the nonlinear overlap integral of the interacting mode profiles $\Phi_{p,s,i}$, ρ a vector in the transverse plane, ε_0 the vacuum permittivity, c the velocity of light, $n_{p,s,i}$ the effective indices, $\lambda_{p,s}$ the wavelengths, and subscripts p, s, i indicating pump, signal and idler, respectively. Generated signal is expected to grow as the parametric gain at phase-matching $G = \sinh^2(gL)$, thus staying approximately linear in the low gain regime ($gL<1$), when $G \approx (gL)^2$ (Boyd, 2008). Thanks to the high quality of the sample, maximum coupled pump, which is limited by input facet optical damage, reached in this case 60 mW. Accordingly, signal power went beyond such low-gain regime, showing the onset of a nonlinear dependence on P_p and reaching a maximum of 100 nW, with an internal conversion efficiency $P_s/P_p = (1.5\pm0.2)10^{-6}$.

5.2 Parametric amplification

Parametric amplification was demonstrated on a similar sample, with phase-matching at degeneracy shifted at 1030 nm by using slightly thinner GaAs layers in the core (255 nm). As for parametric fluorescence, experiments were performed at degeneracy, where parametric gain and phase-matching bandwidth are maximum. The pump beam was provided by an Yb:YAG mono-mode CW laser slightly tunable around 1.03 µm. The signal was obtained by mixing in a PPLN crystal a part of the same Yb:YAG beam with a narrow-line (1.5 GHz), CW Ti:Sapphire laser around 0.7 µm. At the PPLN output, the so obtained beam at the difference frequency, i.e. at $\lambda_s \approx 2$ µm, is singled-out by means of a Ge filter.

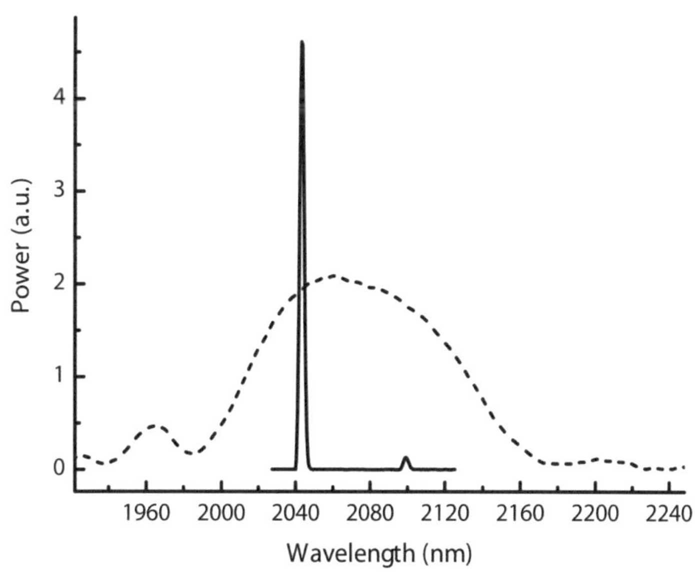

Fig. 11. Waveguide output spectra: amplified signal and idler (solid line) resulting from pump and signal simultaneous coupling, and parametric fluorescence obtained with the only input pump (dashed line).

At first, only the 100 mW (external power) TM polarized pump beam at λ_p = 1036 nm was injected in the waveguide. The resulting parametric fluorescence spectrum was acquired while the sample temperature was finely tuned to exactly set parametric gain at degeneracy, as shown in Figure 11 (dotted line). Then, the Ti:Sapphire wavelength and the PPLN temperature were set in order to produce a signal beam at λ_s = 2046 nm, i.e. well inside the gain band and slightly off its center, so as to spectrally resolve signal and idler outputs. The 500 μW TE signal beam was coupled into the waveguide along with the pump, obtaining the output spectrum reported in the same figure at a different scale (solid line). This shows the amplified signal and a clear idler peak at difference frequency λ_i = 2099 nm. The linewidth of the two peaks is determined by the finite resolution of the spectrometer.
Near degeneracy, where $\lambda_s \approx \lambda_i$ and $G \approx P_i(L)/P_s(L)$ in the low-gain regime, g and G can be directly evaluated by measuring the idler to signal output ratio. In this case, in addition, $P_i(L)$ and $P_s(L)$ can be measured with the same detector, so that their ratio is not affected by calibration uncertainties.
Figure 12 shows the idler-to-signal ratio vs. TM coupled pump power P_p, which is well fitted by a straight line as expected for low-gain. The maximum parametric amplification, obtained for $P_p \approx$ 30 mW, is G = 4.5 %. The slope scaled to the waveguide length L gives the normalized parametric gain coefficient $g/P^{1/2}$= 4.1±0.1 $cm^{-1}W^{-1/2}$.

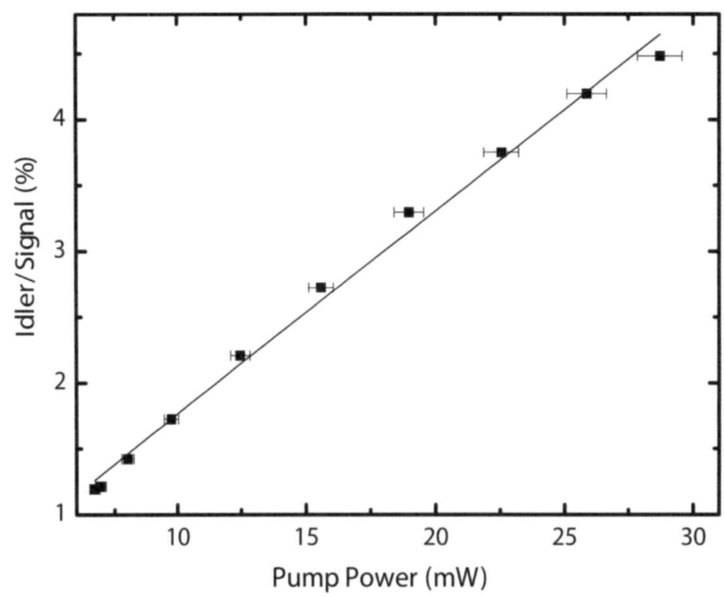

Fig. 12 Output idler-to-signal percentage, i.e. parametric amplification, vs. coupled pump power.

6. Toward an integrated optical parametric oscillator

6.1 Cavity configuration and oscillation threshold

The realization of the first GaAs-based waveguided OPO now faces the difficulties of fabricating a sufficiently resonant cavity. The first choice to deal with is the design of the

OPO cavity. Two common schemes are the singly-resonant OPO (SROPO) and the doubly-resonant OPO (DROPO). In the SROPO the cavity is resonant only at signal (idler) wavelength, i.e. $R_p \approx 0$, $R_s \approx 1$, $R_i \approx 0$ ($R_p \approx 0$, $R_s \approx 0$, $R_i \approx 1$), while in the DROPO both the generated wavelengths are resonating, i.e. $R_p \approx 0$, $R_s \approx 1$, $R_i \approx 1$. As the gain in the active medium is unidirectional, the PM condition is fulfilled only when the three waves are copropagating, signal and idler are amplified in the nonlinear crystal only in one propagation direction.
By equating cavity losses in a round trip and single-pass parametric amplification, the gain threshold for a SROPO results (Sutherland, 1996)

$$(g_{th} L)^2 \approx 2[1 - R_s \exp(-\alpha_s L)] \tag{14}$$

In a DROPO, the threshold is significantly lower:

$$(g_{th} L)^2 \approx [1 - R_s \exp(-\alpha_s L)]\ [1 - R_i \exp(-\alpha_i L)] \tag{15}$$

Since OPOs do not depend on a resonance of the active medium but only on frequency conservation and on the PM condition along the cavity axis, a wide spectral range is attainable, typically limited by the high-reflectivity band of the cavity mirrors.
According to the previous equations, and the typical parametric gain coefficient and propagation losses of our waveguides, the steady-state pump threshold of a SROPO is $P_p^{(th)}$=250 mW, for a waveguide length L=3 mm and a facet reflectivity R = 95%. For a DROPO in the same conditions, the pump threshold is one order of magnitude lower: $P_p^{(th)}$=23 mW. The maximum power we can couple in our waveguide, at present limited to about 70 mW, prevents the adoption of the former configuration. However, as we already mentioned, the standard loss coefficient of our nonlinear waveguides is 0.4-0.7 cm^{-1} which causes the minimum facet reflectivity required to vary. For a symmetric integrated resonator, with equal signal and idler losses, the pump power threshold in a DROPO is:

$$P_p^{(th)} = \frac{1}{\eta_{norm}} \left[\frac{\alpha_p / 2}{1 - \exp(-\alpha_p L/2)} \ln\left(Q + \sqrt{Q^2 - 1} \right) \right]^2 \tag{16}$$

where η_{norm} is the waveguide normalized conversion efficiency and

$$Q = \frac{1 + R_s R_i \exp(-2\alpha L)}{(R_s + R_i)\exp(-\alpha L)} \tag{17}$$

with $R_{s,i}$ the facet reflectivity at the signal and idler wavelength, and α the signal and idler losses. Figure 13 shows the variation of $P_p^{(th)}$ with the facet reflectivity for a typical 3 mm-long waveguide. For the typical values of the propagation losses, the minimum reflectivity required ranges between 84 and 95%.

6.2 Design and fabrication of the integrated mirrors

The solution we have adopted to place our nonlinear waveguide in a DROPO resonator is the deposit of multilayer dielectric mirrors on the facets. However, such deposit on GaAs/AlOx waveguides is made difficult by a 12% contraction of AlAs layers during oxidation. This contraction causes an irregular surface of the waveguide facets (Durand et

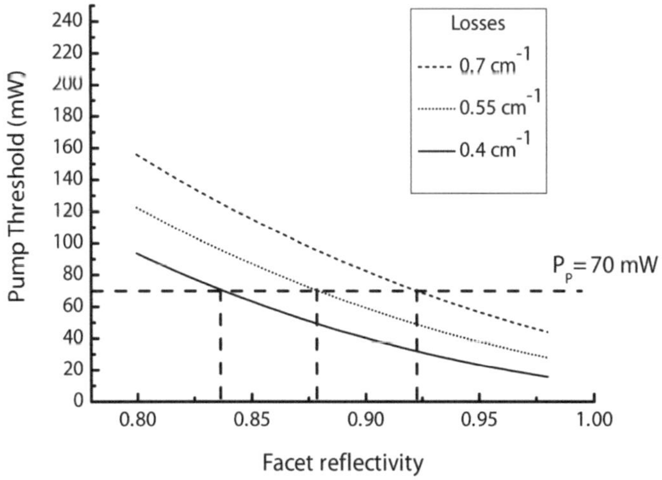

Fig. 13. Calculated pump threshold vs. facet reflectivity, where P_P=70mW corresponds to the waveguides damage threshold.

al., 2003). Moreover, such shrinkage results in a mechanical stress for the whole waveguide structure, making the adhesion of coating more critical.

Accordingly, the choice of the materials to deposit has mainly been dictated by the reduction of the total coating thickness. We have tested two material combinations: Si and low-index Al_2O_3 on the one side and YF_3 and ZnS on the other side.

Three criteria were taken into account for the design of the mirrors. First, we set the signal and idler reflectivity as high as possible. Second, we set the pump reflectivity as low as possible in order to be able to increase the coupled pump power and to prevent a resonant behaviour at λ_p which could affect the OPO stability. Finally the total thickness of the coating must be kept below 3 µm.

The mirrors have been designed with the 1D transfer matrix method (Yeh, 1988). We have optimized the thickness of individual layers with an algorithm for the minimization of a merit function taking into account the three criteria listed previously.

Figure 14 shows the simulated and experimental reflectivity spectrum for both solutions. Experimental spectra, obtained for deposits on substrates, are in good agreement with the calculated ones. $R_{s,i}$ is higher than 95% all over an interval of at least 400 nm around λ=2.0µm and R_p is lower than 15%.

The same types of multilayers have been deposited on the facets of waveguides. For each deposit we noted technologic issues. The Si/Al_2O_3 coating was detached, supposedly because of elements specific to our waveguides and experimental conditions, namely absorption in Si layers or poor adhesion of the coating to the facet. In the case of YF_3/ZnS, the adhesion on the waveguides facets was correct but we noticed the deposition of ZnS all over the samples, tending to form small clusters on the waveguides. Those clusters were responsible of additional Rayleigh scattering from the sides of the waveguides. This problem is mainly due to the evaporation of ZnS that tends to diffuse easily everywhere.

Current research is done on the deposition of other couples of materials. We also investigate the protection of the waveguides during the mirrors deposition and an effort is made on the design of a waveguide with a different geometry to isolate the propagating modes from additional scattering from edges.

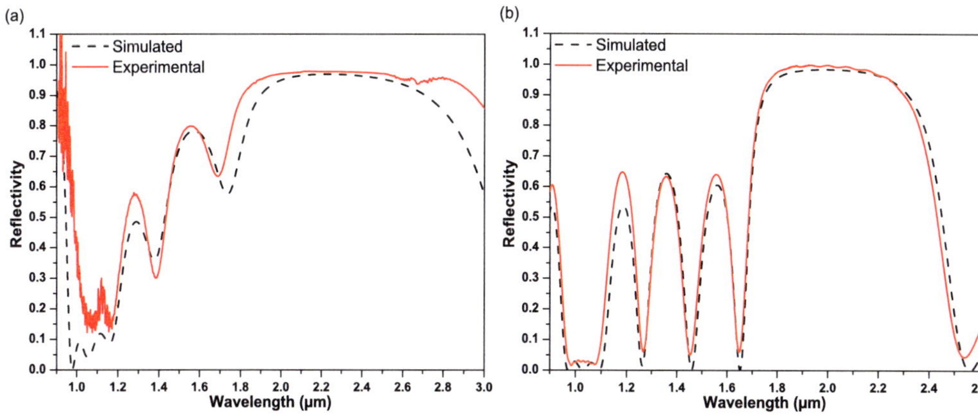

Fig. 14. Calculated and experimental reflectivity spectra for Si/Al_2O_3 (a) and YF_3/ZnS (b) coatings.

7. References

Arie, A.; Fradkin-Kashi, K. & Shreberk, Y. (2002). Frequency conversion in novel materials and its application to high resolution gas sensing. *Optics and Lasers in Engineering*, 37, 2, (February 2002) 159-170, ISSN: 0143-8166

Born, M. & Wolf, E. (1980). *Principles of optics*, Pergamon Press, ISBN: 978-0080264813, Oxford

Bortz, M. L.; Arbore, M. A. & Fejer, M. M. (1995). Quasi-phase-matched optical parametric amplification and oscillation in periodically poled $LiNbO_3$ waveguides. *Optics Letters*, 20, 1, (January 1995) 49-51, ISSN: 0146-9592

Boyd, R. W. (2008). *Nonlinear Optics*, Academic Press, ISBN: 978-0123694706, Boston

Bravetti, P.; Fiore, A.; Berger, V.; Rosencher, E.; Nagle, J. & Gauthier-Lafaye, O. (1998). 5.2–5.6 μm source tunable by frequency conversion in a GaAs-based waveguide, *Optics Letters*, 23, 5, (March 1998) 331-333, ISSN: 0146-9592

Chen, W. D.; Burie, J. & Boucher, D. (1999). Mid-infrared generation by optical frequency conversion and applications to spectroscopy and air monitoring. *Spectrochimica Acta A*, 55, 10, (September 1999) 2057-2075, ISSN: 1386-1425

Dallesasse, J. M.; Holonyak, N.; Sugg, A. R.; Richard, T. A. & El-Zein, N. (1990a). Hydrolyzation oxidation of $Al_xGa_{1-x}As$-AlAs-GaAs quantum-well heterostructures and superlattices. *Applied Physics Letters*, 57, 26, (December 1990) 2844-2446, ISSN: 0003-6951

Dallesasse, J. M.; Gavrilovic, P.; Holoniak Jr., N.; Kaliski, R. W.; Nam, D. W.; Vesely, E. J. & Burnham, R. D. (1990b). Stability of AlAs in $Al_xGa_{1-x}As$-AlAs-GaAs quantum-well heterostructures. *Applied Physics Letters*, 56, 24, (June 1990) 2436-2438, ISSN: 0003-6951

De Rossi, A.; Berger, V.; Calligaro, M.; Leo, G.; Ortiz, V. & Marcadet, X. (2001). Parametric fluorescence in oxidized aluminium gallium arsenide waveguides. *Applied Physics Letters*, 79, 23, (December 2001) 3758-3760, ISSN: 0003-6951

De Rossi, A.; Berger, V.; Leo, G. & Assanto, G. (2005a). Form birefringence phase matching in multilayer semiconductor waveguides: tuning and tolerances. *IEEE Journal of Quantum Electronics*, 41, 10, (October 2005) 1293-1302, ISSN: 0018-9197

De Rossi, A.; Ortiz, V.; Calligaro, M.; Lanco, L.; Ducci, S.; Berger, V. & Sagnes, I. (2005b). Measuring propagation loss in a multimode semiconductor waveguide. Journal of Applied Physics, 97, 7, (March 2005) 073105-1-073105-7, ISSN: 0021-8979

Durand, O.; Wyckzisk, F.; Olivier, J.; Magis, M.; Galtier, P.; De Rossi, A.; Calligaro, M.; Ortiz, V.; Leo, G. & Assanto, G. (2003). Contraction of aluminum oxide thin layers in optical heterostructures. *Applied Physics Letters*, 83, 13, (September 2003) 2554-2556, ISSN: 0003-6951

Eyres, L. A.; Tourreau, P. J. ; Pinguet, T. J.; Ebert, C. B.; Harris, J. S.; Fejer, M. M. ; Becouarn, L.; Gerard, B. & Lallier, E. (2001). All-epitaxial fabrication of thick, orientation patterned GaAs films for nonlinear optical frequency conversion. *Applied Physics Letters*, 79, 7, (August 2001) 904-906, ISSN: 0003-6951

Fiore, A.; Berger, V.; Rosencher, E.; Bravetti, P.; Laurent, N. & Nagle, J. (1997). Phasematched mid-infrared difference frequency generation in GaAs-based waveguides. *Applied Physics Letters*, 71, 25, (December 1997) 3622-3624, ISSN: 0003-6951

Fiore, A.; Berger, V.; Rosencher, E.; Bravetti, P. & Nagle, J. (1998a). Phase-matching using an isotropic nonlinear material. *Nature* 391, 6691, (January 1998) 463-466, ISSN: 0028-0836

Fiore, A.; Sanz, J.; Delobel, L.; van der Meer, P.; Bravetti, P.; Berger, V.; Rosencher, E. & Nagle, J. (1998b). Second-harmonic generation at $\lambda = 1.6\ \mu m$ in $AlGaAs/Al_2O_3$ waveguides using birefringence phase matching. *Applied Physics Letters*, 72, 23, (June 1998) 2942-2944, ISSN: 0003-6951

Gehrsitz, S.; Reinhart, F. K.; Gourgon, C.; Herres, N.; Vonlanthen, A. & Sigg, H. (2000). The refractive index of $Al_xGa_{1-x}As$ below the band gap: Accurate determination and empirical modelling. *Journal of Applied Physics*, 87, 11, (June 2000) 7825-7837, ISSN: 0021-8979

Gisin, N.; Ribordy, G.; Tittel, T. & Zbinden, H. (2002). Quantum Criptography. *Review of Modern Physics*, 74, 1, (January 2002) 145-195, ISSN: 0034-6861

Goodnick, S. M.; Ferry, D. K.; Wilmsen, C. W.; Liliental, Z.; Fathy, D. & Krivanek, O. L. (1985). Surface roughness at the Si(100)-SiO_2 interface, *Physical Review B*, 32, 12, (December 1985) 8171-8186, ISSN: 1098-0121

Hall, D. C.; Wu, H.; Kou, L.; Luo, Y.; Epstein, R. J.; Blum, O. & Hou, H. (1999). Refractive index and hygroscopic stability of $Al_xGa_{1-x}As$ native oxides. *Applied Physics Letters*, 75, 8, (August 1999) 1110-1112, ISSN: 0003-6951

Hofer, F.; Warbichler, P. & Grogger, W. (1995). Imaging of nanometer-sized precipitates in solids by electron spectroscopic imaging. *Ultramicroscopy*, 59, 1-4, (July 1995) 15-31, ISSN: 0304-3991

Hofmann, D.; Schreiber, G.; Grundkötter, W.; Ricken, R. & Sohler, W. (2000) Mid-infrared continuous-wave singly resonant optical parametric oscillator with periodically poled Ti:$LiNbO_3$ waveguide, *Conference on Lasers and Electro-Optics Europe*, paper CDM4, ISBN: 0-7803-6319-1, Nice, September 2000, IEEE

Kintaka, K.; Fujimura, M.; Suhara, T. & Nishihara, H. (1996). High-efficiency $LiNbO_3$ waveguide second-harmonic generation devices with ferroelectric-domain-inverted gratings fabricated by applying voltage. *Journal of Ligthwave Technology*, 14, 3 (March 1996), 462-469, ISSN: 0733-8724

Knopp, K. J.; Mirin, R. P.; Christensen, D. H.; Bertness, K. A.; Roshko, A. & Synowicki, R. A. (1998). Optical constants of $(Al_{0.98}Ga_{0.02})_xO_y$ native oxides. *Applied Physics Letters*, 73, 24 (December 1998) 3512-3514, ISSN: 0003-6951

Lancaster, D. G.; Richter, D. & Tittel, F. K. (1999). Portable fiber-coupled diode-laser-based sensor for multiple trace gas detection. *Applied Physics B*, 69, 5-6, (December 1999) 459-465, ISSN: 1432-0649

Lacey, J. P. R. & Payne, F. P. (1990). Radiation loss from planar waveguides with random wall imperfections. *IEE Proceedings*, 137, 4, (August 1990) 282-288, ISSN: 1350-2433

Moutzouris, K.; Rao, S. V.; Ebrahimzadeh, M.; De Rossi, A.; Berger, V.; Calligaro, M. & Ortiz, V. (2001). Efficient second-harmonic generation in birefringently phase-matched GaAs/Al_2O_3 waveguides. *Optics Letters*, 26, 22, (November 2001) 1785-1787, ISSN: 0146-9592

Ravaro, M.; Le Dû, M.; Likforman, J.-P.; Ducci, S.; Berger, V. & Leo, G. (2007). Estimation of parametric gain in GaAs/AlOx waveguides by fluorescence and second harmonic generation measurements, *Applied Physics Letters*, 91, 19, (November 2007) 191110-1-191110-3, ISSN: 0003-6951

Ravaro, M.; Guillotel, E.; Le Dû, M.; Manquest, C.; Marcadet, X.; Ducci, S.; Berger, V. & Leo, G. (2008). Nonlinear measurement of mid-IR absorption in AlOx waveguides. *Applied Physics Letters*, 92, 15,(April 2008) 151111-1-151111-3, ISSN: 0003-6951

Scaccabarozzi, L.; Fejer, M. M.; Huo, Y.; Fan, S.; Yu, X. & Harris, J. S. (2006). Enhanced second-harmonic generation in AlGaAs/Al_xO_y tightly confining waveguides and resonant cavities. *Optics Letters*, 31, 24, (December 2006) 3626-3628, ISSN: 0146-9592

Sergienko, A. & Jaeger, G. S. (2003). Quantum information processing and precise optical measurement with entangled-photon pairs, *Contemponary Physics*, 44, 4, (July 2003) 341-356, ISSN: 1366-5812

Sorokina, I. T. & Vodopyanov, K. L. (2003). *Solid-State Mid-Infrared Laser Sources*, Springer, ISBN: 978-3540006213, New-York

Sutherland, R. L. (2003). *Handbook of Nonlinear Optics*, Marcel Dekker, ISBN: 978-0824742435, New York

Venugopal Rao, S.; Moutzouris, K.; Ebrahimzadeh, M.; De Rossi, A. ; Gintz, G.; Calligaro, M.; Ortiz, V. & Berger, V. (2003). Influence of scattering and two-photon absorption on the optical loss in GaAs/Al_2O_3 nonlinear waveguides measured using femtosecond pulses. *IEEE Journal of Quantum Electronics*, 39, 3, (March 2003) 478-486, ISSN: 0018-9197

Vodopyanov, K. L.; Levi, O.; Kuo, P. S.; Pinguet, T. J.; Harris, J. S.; Fejer, M. M.; Gerard, B.; Becouarn, L. & Lallier, E. (2004). Optical parametric oscillation in quasi-phase-matched GaAs, *Optics Letters*, 29, 16, (August 2004) 1912-1914, ISSN: 0146-9592

Yeh, P. (1988). *Optical waves in layered media*. Wiley, ISBN: 978-0471828662, New York

Yoo, S. J. B. (1996). Wavelength conversion technologies for WDM network applications. *Journal of Ligthwave and Technology*, 14, 6, (June 1996) 955-966, ISSN: 0733-8724

Yu, X.; Scaccabarozzi, L.; Harris Jr., J. S.; Kuo, P. S. & Fejer, M. M. (2005) Efficient continuous wave second harmonic generation pumped at 1.55 μm in quasi-phase-matched AlGaAs waveguides, *Opt. Express* 13, 26, (December 2005) 10742-10748, ISSN: 1094-4087 van der Ziel, J. P. (1975). Phase-matched Harmonic Generation in a Laminar Structure with Wave Propagation in the plane of the Layers, *Applied Physics Letters*, 26, 2, (January 1975) 60-61, ISSN: 0003-6951

van der Ziel, J. P. & Gossard, A. C. (1978). Optical birefringence of ultrathin AlGaAs-GaAs Multilayer Heterostructures. *Journal of Applied Physics*, 49, 5, (May 1978) 2919-2921, ISSN: 0021-8979

9

Waveguide Photodiode (WGPD) with a Thin Absorption Layer

Jeong-Woo Park
Electronics and Telecommunications Research Institute
Republic of Korea

1. Introduction

Surface illumination photodiode (PD) shows the tradeoff between quantum efficiency and transit time. This is because a thin absorber region is required for a short carrier transit time whereas a thick absorber region is required for high quantum efficiency. In order to achieve good quantum efficiency the absorption region should be ~2μm, which results in a transit time bandwidth of <12GHz. Waveguide photodiodes can overcome this limitation because the thickness of the absorbing region has little effect on the internal quantum efficiency if the absorber region is long enough. WGPDs, in which the quantum efficiency and transit time are decoupled, can overcome this restriction. In this type of devices, the external quantum efficiency is determined principally by the input coupling efficiency because the internal quantum efficiency can close to 100%.

Coupling into WGPDs can be broadly categorized as side-illumination and evanescent coupling. In Figure 1, three types of coupling scheme for WGPDs are shown. Those include a) side illumination type, b) evanescent coupling type, and c) side illumination with a thin absorption/core region. For side illumination, light is focused directly onto the edge of the absorbing layer. With this approach, a responsivity of 0.85A/W and 50-GHz bandwidth has been reported (K. Kato *et al*, 1992). In the report, he used a multimode waveguide in a transverse direction to acquire a higher coupling efficiency than in a typical p-i-n structure. A primary disadvantage of this type of device is poor optical power capability compared to evanescent coupling approach. Evanescently-coupled photodiodes have demonstrated responsivity up to 0.75A/W with a bandwidth of 42GHz (F. Xia *et al*, 2001). As another evanescently-coupled photodiode, an etched short multimode graded index waveguide approach has shown a responsivity of 0.96A/W and 40GHz bandwidth (T. Takeuchi *et al*, 2001). A similar approach that integrates a short planar diluted waveguide with an etched input facet has shown 0.73A/W responsivity and 47GHz bandwidth (M. Achouche *et al*, 2003). The responsivity of 1.02A/W and 48GHz bandwidth has been achieved with a short multimode input waveguide that consists of a diluted waveguide and two optical matching layers (S. Demiguel *et al*, 2003).

In this chapter, a new WGPD with a thin absorption layer will be introduced. Also, methods of design and optimizations for this new type of WGPD are described. A responsivity of 1.08A/W was achieved at 1550nm wavelength, which corresponds to an external quantum efficiency of 86.4% with TE/TM polarization dependence less than 0.25dB. For the same device, the bandwidth was ~40GHz. Also, nonlinearity of the device was characterized.

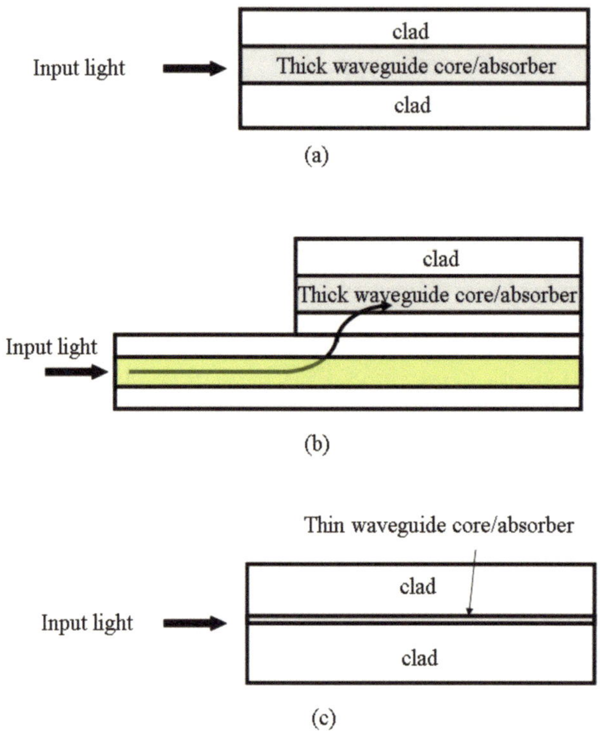

Fig. 1. three types of coupling scheme for WGPDs. a) side illumination, b) evanescent coupling, and c) side illumination with a thin absorption/core region.

2. High responsivity

The guided mode of a waveguide with thin core layer has a larger beam size than that with thick core layer as indicated in Figure 2. This property can be profitably applicable to InP based high responsivity WGPD. Figure 2 shows simulated beam size of guided mode for waveguide structure with InGaAs core/InGaAsP(λg=1.4μm) clad and InGaAs core/InP clad. Beam size is defined at the point, at which field amplitude is reduced down to *1/e* of its maximum amplitude. As shown in Figure 2, for core thickness of less than 0.2μm, guided beam size is enlarged.
The enlarged beam size is well matched with other large sized waveguide, such as optical fiber, silica planar waveguide, or polymer planar waveguide. This property can be applied to overcoming the beam size mismatch between InP based semiconductor waveguides and other waveguides such as optical fiber, silica planar waveguide, or polymer planar waveguide. Using a thin core layer, optical coupling between WGPDs and external waveguides can also enhance a external quantum efficiency, or responsivity.

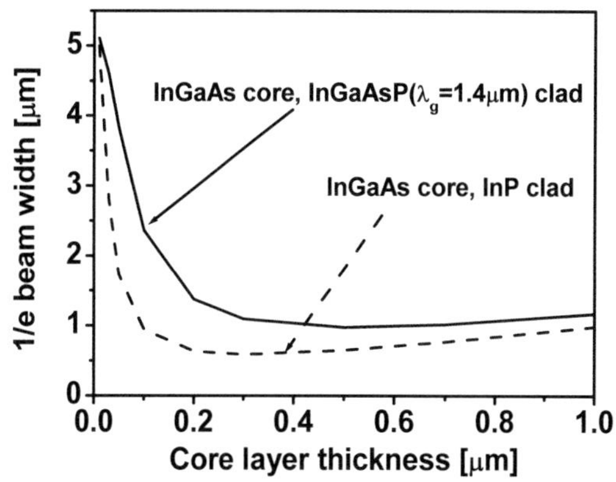

Fig. 2. Two dimensional simulated beam sizes (field amplitude is down to *1/e* of maximum amplitude) for InGaAs core, InGaAsP(λ_g=1.4μm) core and InP clad.

For conform of high coupling efficiency between optical fiber and WGPDs with a thin core layer, four types of WGPDs were fabricated and responsivity for each devices were measured. Figure 3 shows waveguide structures of four different types of WGPDs. Total undoped layer surrounding InGaAs absorption layer was 0.6μm thick for each type of WGPDs. Mesa etching was done past to absorption layer to define deep ridge waveguide. After polyimide passivation and contact opening, Ti/Pt/Au p-electrode was evaporated and rapid-annealed. After Ti/Pt/Au n-electrode evaporation, rapid-annealing were performed. After cleavage, each WGPDs are anti-reflection-coated. Widths of input facet waveguides were 20μm.

Table (I) shows measured responsivity with coupling of lensed fiber and flat-ended fiber. First, responsivity measured at the wavelength of 1550nm, was 0.815A/W for Type (II), which was 300μm long. The calculated vertical mode coupling efficiency, η_v was 65% for Type (II). Horizontal mode coupling efficiency, η_h, is 100% because width of WGPD is wider than that of flat-ended fiber. Thus, total coupling efficiency, $\eta=\eta_h\eta_v$, is 65%. The coupling efficiency of 65% is corresponding to responsivity of 0.81A/W, which well agrees with measured responsivity of 0.815A/W. For Type (II), polarization dependency was less than 0.25dB. This value of less than 0.25dB is originated from different coupling efficiency between TE and TM mode input. Another WGPD with absorption layer thickness of 0.2μm shows similar polarization dependency of 0.25dB. This indicates that 300Å thick absorption layer have a bulk absorption behavior rather than quantum well absorption behavior.

However, Type (I), which has 100Å thick absorption layer thickness, shows quantum well absorption behavior. Figure 4 shows the polarization dependent responsivity curve. For comparison, polarization dependency of Type (II) is drawn together. In Figure 4, x-axis is ε, the parameter on Poincarè sphere, which represents the linear polarization state of input light and y-axis is normalized responsivity with respect to maximum responsivity, in dB unit. The calculated TE/TM difference of coupling efficiency for Type (I) is 0.202dB. Thus, polarization dependency of Type (I), shown in Figure 4, is originated from absorption coefficient difference between TE and TM mode. The comparison of polarization dependence for Type (I) and Type (II) indicates that WGPD with thin absorption layer should have more than 100Å thick absorption layer for polarization independent operation.

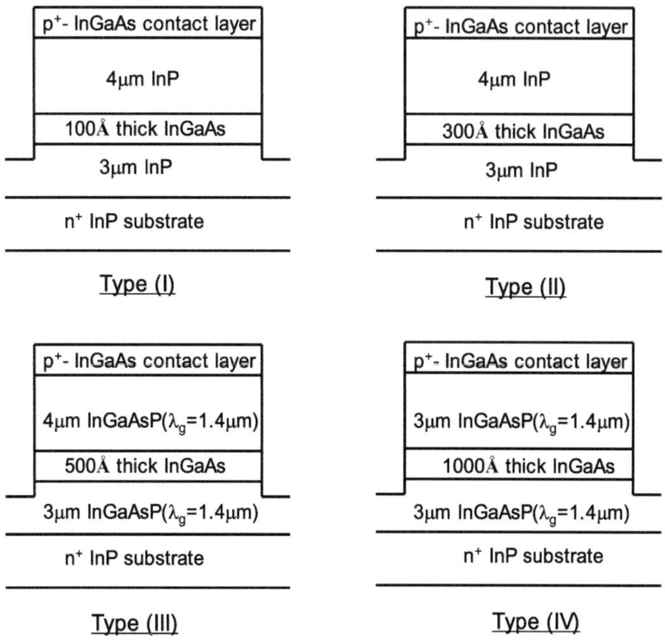

Fig. 3. Tested WGPD structures for high responsivity operation.

PD type	Responsivity (flat-ended fiber)	Responsivity (lensed fiber)
(I)	Polarization dependent	Polarization dependent -
(II)	0.815A/W	1.09A/W
(III)	0.93A/W	1.09A/W
(IV)	0.76A/W	1.08A/W

Table (I) Responsivities for four types of WGPDs

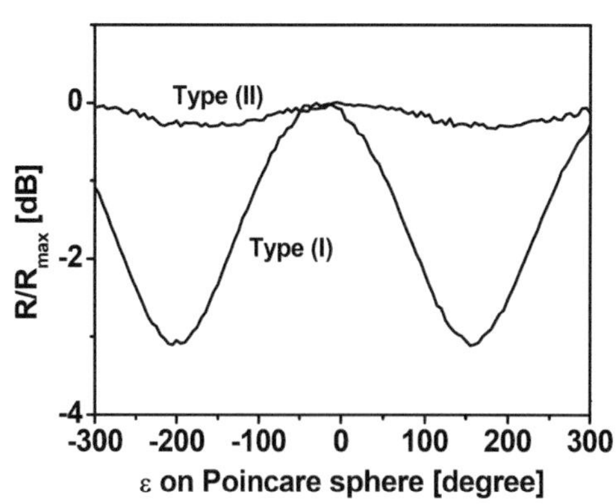

Fig. 4. Polarization dependencies of Type (I) and Type(II).

Another drawback of WGPD with 100Å thick absorption layer is low coupling efficiency, which is contradictory to the simulated value. For type (I), the calculated coupling efficiency, coupled with flat-ended fiber, is 82.6%, which corresponds to responsivity of 0.99A/W for the wavelength of 1490nm. Measured responsivity, however, implies that coupling efficiency of Type (I), when coupled with flat-ended fiber, is 30.6% and maximum responsivity is 0.368A/W for the input wavelength of 1490nm.

To measure the coupling efficiency of Type (I), responsivities of PDs with different lengths were measured for TE input light. Figure 5 shows responsivity values versus PD length. This data was fitted with Equation (1).

$$R = \frac{C \cdot \lambda}{1.24} \cdot (1 - e^{-\alpha \cdot \Gamma \cdot L}) \quad (1)$$

In Equation (1), R, C, λ, α, Γ, and L are responsivity, coupling efficiency, input wavelength, absorption coefficient of absorption layer, confinement factor of guided beam within a absorber, and PD length, respectively.

Fitting results indicate that coupling efficiency is 30.6% and $\alpha\Gamma$ is 0.00511μm^{-1}. Low responsivity for Type (I) was conformed by measuring five PDs. Five PDs show almost same responsivity. Discrepancy between simulated value and measured one can be explained by weakly guiding structure of Type (I). It is estimated that 100Å thick core layer is too thin to support the propagation of light through total waveguide. Even a small perturbation of waveguide structure such as side wall roughness may induce waveguide to be leaky for 100Å thick core layer.

Type (III) shows responsivity of 0.93A/W, when coupled with non-lensed flat fiber. Type (III) has low index difference between clad and core. Thus, guided mode is more spread than the case of high core/clad index difference like Type (II). Calculated beam size of Type (III) is 3.83μm. Compared to the calculated beam size of 2.81μm for Type (II), more enlarged beam size of Type (III) is more similar to mode size of flat fiber, which gives higher responsivity. Polarization dependency of Type (III) is also smaller than 0.25dB, showing bulk absorption property. Comparison of Type (II) and Type (III) shows that small index difference between core and clad is more advantageous, for high responsivity.

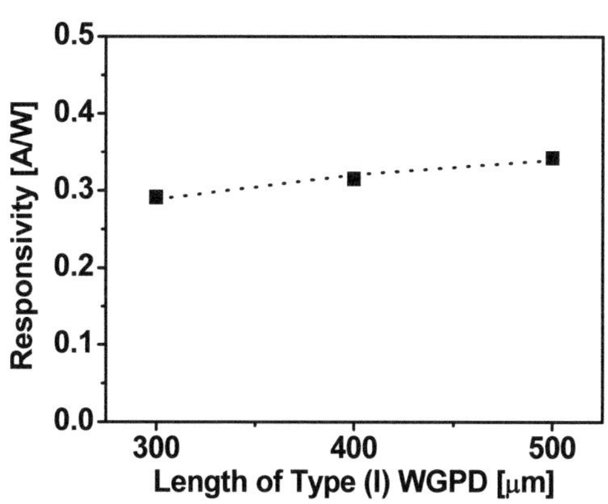

Fig. 5. Responsivity versus PD length of Type (I). Fitting was done by Eq. (1)

Type (IV) with 1000Å absorption layer thickness and 70μm length shows responsivity of 0.76A/W coupled with flat fiber and 1.08A/W coupled with lensed fiber at a wavelength of

1550nm. Calculated guiding mode size of Type (IV) is 2.36μm, which is small compared to 3.83μm of Type (III). Smaller guided mode size of Type (IV), originated from thicker core layer than Type (III), gives more mode-mismatch and smaller responsivity than Type (III).

2. Bandwidth property

To find out time dependent current of photodiode, displacement current should be considered. Including displacement current and photo-generated current, time dependent current of photodiode is given by Equation (2), according to (G. Lucovsky *et al*, 1964).

$$I_{photodiode}(t) = \frac{\frac{1}{L}\int_0^L [I_{drift,e}(x,t) + I_{drift,h}(x,t)]dx}{1 + j \cdot \omega \cdot R \cdot C} \quad (2)$$

In Equation (2), $I_{drift,e}(x,t)$, $I_{drift,h}(x,t)$, R and C are photo-generated electron and hole drift current at (x,t), (series resistance of PD+load resistance) and (photodiode capacitance + stray capacitance), respectively. Transit-time limited response is extracted by developing numerator of Equation (2). To calculate transit-time limited frequency response of WGPDs with thin absorption layer, $I_{drift,e}(x,t)$ shown in Figure 6 should be known first.
Assuming input light can be expressed as $P_o \cdot \exp(j \cdot \omega_m \cdot t)$, where P_o, ω_m are amplitude of input beam and modulation frequency, respectively, electron current at x is given by Equation (3).

$$i_{drift,e}(x,t) = R \cdot P_o \cdot \exp[j \cdot \omega_m \cdot (t - \frac{x}{v_e})] \quad (3)$$

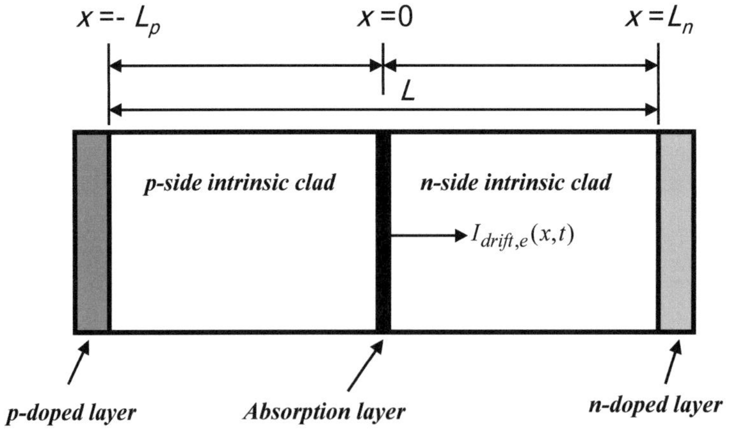

Fig. 6. Configuration for derivation of transit-time limited frequency response of WGPD having thin absorption layer.

In derivation of Equation (3), it is assumed that photocurrent is generated only at x=0. The generated electrons at x=0 drift forward to n-doped region and drift current at x is delayed waveform with respective to current at x=0, with time delay of x/v_e. In Eq.(3), R and v_e are responsivity, electron drift velocity in n-side clad layer, respectively.

Including hole current contribution, the transit-time limited time-dependent photodiode current is given by Equation (4).

$$i_{photodiode}(t) = R \cdot P_o \cdot \exp(j \cdot \omega_m \cdot t) \cdot [\frac{1-\exp(-j \cdot \omega_m \cdot \frac{L_n}{v_e})}{j \cdot \omega_m \cdot \frac{L}{v_e}} + \frac{1-\exp(-j \cdot \omega_m \cdot \frac{L_p}{v_h})}{j \cdot \omega_m \cdot \frac{L}{v_h}}] \quad (4)$$

In Equation (4), v_h is hole drift velocity in p-side clad layer. At optimized condition, electron transit time and hole transit time are equal. This condition can be expressed by $L_n / v_e = L_p / v_h = \tau$. At optimized condition, right most term in Equation (4) can be re-written by Equation (5).

$$\begin{aligned}&[\frac{1-\exp(-j \cdot \omega_m \cdot \tau)}{j \cdot \omega_m \cdot \frac{L_n + L_p}{v_e}} + \frac{1-\exp(-j \cdot \omega_m \cdot \tau)}{j \cdot \omega_m \cdot \frac{L_n + L_p}{v_h}}]] \\ &= \frac{1-\exp(-j \cdot \omega_m \cdot \tau)}{j \cdot \omega_m \cdot \tau}\end{aligned} \quad (5)$$

Transit time limited bandwidth, f_t, is defined as the frequency at which absolute value of Equation (5) is equal to $1/\sqrt{2}$, and can be calculated as

$$f_t \cong \frac{2.8}{2\pi\tau} \quad (6)$$

Including transit time limitation and RC effect, bandwidth of photodiode, f_{3dB}, is given by Equation (7) with an error of less than 5%(K. Kato *et al*, 1993).

$$\frac{1}{f_{3dB}^2} = \frac{1}{f_t^2} + \frac{1}{f_{RC}^2} \quad (7)$$

Figure 7 shows the expected 3dB bandwidth with intrinsic layer thickness variation. Considered structures is Type (IV) of which absorption layer thickness is 1000Å. In calculations, The relative dielectric constant and electron drift velocity of InGaAsP (λ_g=1.4μm) was assumed as 11.16 (S. Adachi, 1982) and 1.5X10^{6}cm/sec (A. Galvanauskas et *al*, 1988). Hole velocity was assumed as the half of the electron velocity. In the calculations, PD length was 70μm and PD width was tapered from 5μm to 1μm. A 70μm length is sufficient for responsivity of more than 1.0A/W for a 3μm mode size fiber. Also, series resistance, Rs and load resistance were assumed as 5Ω and 50Ω.

As can be seen Figure 7 (a), optimized point for maximum bandwidth with pad capacitance of zero, is the point at which RC limited bandwidth and carrier transit-time limited bandwidth are same. At this optimized point, bandwidth can be a 120GHz even though thin absorption layer needs long absorption length of 70μm which is two or three times long compared to typical high-speed WGPDs. When pad capacitance of 10fF is included, however, bandwidth is reduced and optimum point is shfited as can be seen in Figure 7(b).

Based on simulated results of Figure 7 (a), (b), WGPD with a 1000Å thick absorber was fabricated. The thickness of intrinsic layer on n-electrode side and p-electrode side were

0.6μm and 0.3μm, respectively. Width of WGPD was tapered from 5μm to 1μm and length was 70μm. The frequency response of a device was measured using an impulse response. The optical impulse from femto-second laser was applied to WGPD. The impulse response was converted to bandwidth curve using fourier transform. Figure 8 shows the bandwidth response at -3V bias, after the de-embedding the RF loss of the measurement system. The RF losses of measurement system include those of probe, bias tee, cable, and DC block. As can be seen from the Figure 8, bandwidth of ~42GHz was obtained. Hole-trapping at the hetero-interface of i-InGaAsP(λ_g=1.4μm)/i-InGaAs can be a bandwidth limiting factor. However, the bandgap discontinuity at i-InGaAsP(λ_g=1.4μm)/i-InGaAs does not degrade the bandwidth significantly.

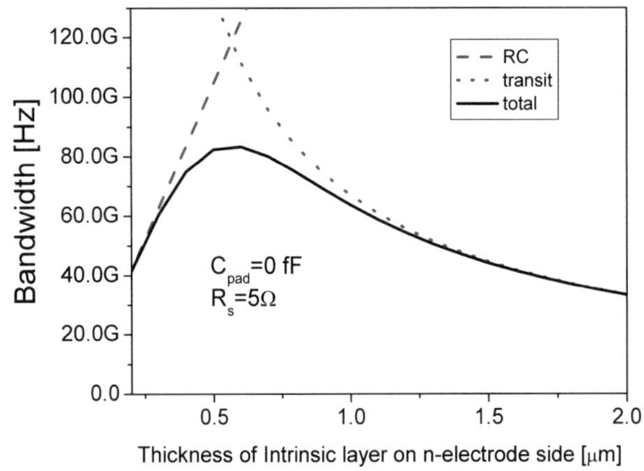

(a)

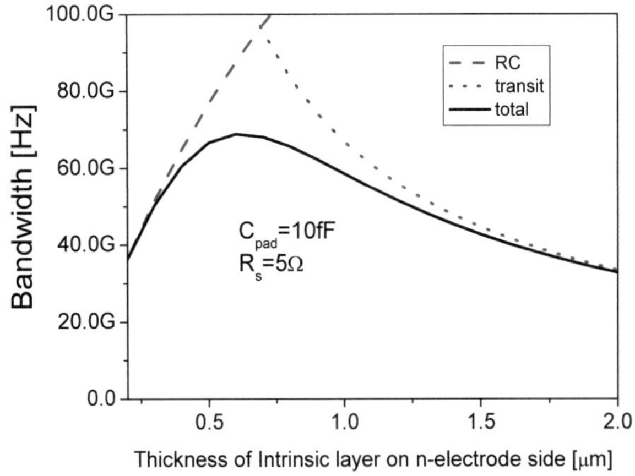

(b)

Fig. 7. RC limited, transi-time limited and total bandwidth traces with variations of thickness of n-side intrinsic layer (a) without consideration of pad capacitance (b) with the pad capacitance of 10fF.

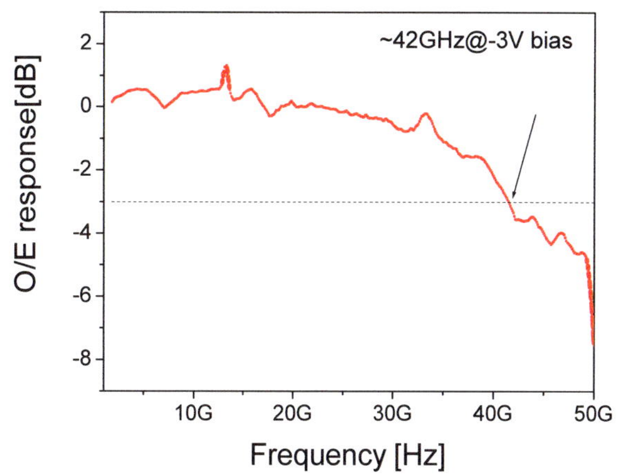

Fig. 8. A measured frequency response of WGPD with a thin absorption layer of 1000Å.

3. Intermodulation distortion properties

In some optical communication systems such as fiber-optic community antanna television (CATV) systems, many optical signals with different modulation frequencies are inputted to a PD. In this case, non-linearity properties of PD should be supressed to re-generate elctrical signals from optical signals without distortions.

When a device shows nonlinear response, input-output relation is represented as shown in Figure 9. An output can be expressed as polymomials of input signal. With this nonlinear relations, supurious outputs of which frequencies are f2+f1, f2-f1, 2f1-f2, 2f2-f1... can be generated when sinusoidal inputs of which frequencies are f1, f2, ..., are applied to device. These supurious outputs should be filtered out not to influence on original signals with

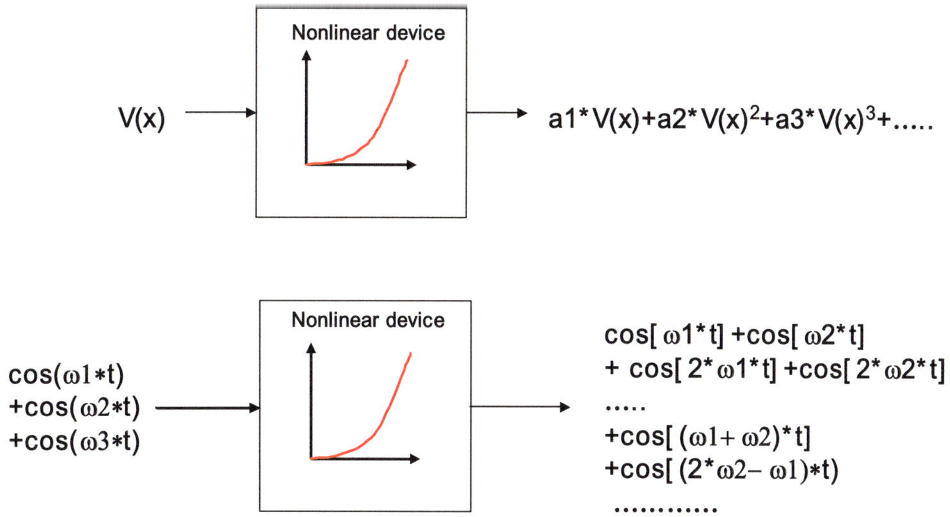

Fig. 9. Supurious signals from nonlinear devices

frequencies of f1, f2,.... As can be seen in Figure 10, however, frequencies of some supurious outputs are close to frequencies of original signal. These supurious signals cannot be filtered out and quality of converted signals from optical to electrical is degraded. The degree of degradations is determined by linearity of PD. The second order intermodulation products of two signals at f1 and f2 occur at f1+f2, f2-f1, 2·f1 and 2·f2. The third order intermodulation products of two signals at f1 and f2 would be at 2·f1+f2, 2·f1-f2, f1+2·f2, and 2·f2-·f1. Among these products, signals at f1+f2, 2·f1-f2 and 2·f2-·f1 are not filtered out. Therefore, to obtain high purity signal among many signals, signals at f1+f2, 2·f1-f2 and 2·f2-·f1 should be supressed when optical-to-electrical conversion occurs at PD. Signals at f2+f1 and f2-f1 are the 2nd order intermodulation distortion (IMD2). Signals at 2·f1-f2 and 2·f2-·f1 are the 3rd order intermodulation distortion (IMD3). The ratio of each intermoulation signal to original signal should be as small as possible and the ratio is expressed with unit of dBc.
The main source of nonlinearity of PD is a space charge induced nonlinearity (K. J. Williams *et al*, 1996), (Y. Kuhara *et al*, 1997). The photo-generated carriers induce space charges in a intrinsic layer of PD. Carrier-dependent carrier velocities associated with a perturbed electric filed due to space-charge and loading effect are main source of photodetector nonlinear behavior. The amount of space-charge generated from photocurrents depends on the power density of incident optical signal. The smaller a density of photo-currents are, the smaller nonlinarity of PD are. To reduce a IMD2 and IMD3, a density of photo-generated carriers should be reduced. WGPDs with thin absorption layer can have a suppressed nonlinearity because thin absorption layer with a long absorption length produce a reduced density of photo-carriers.

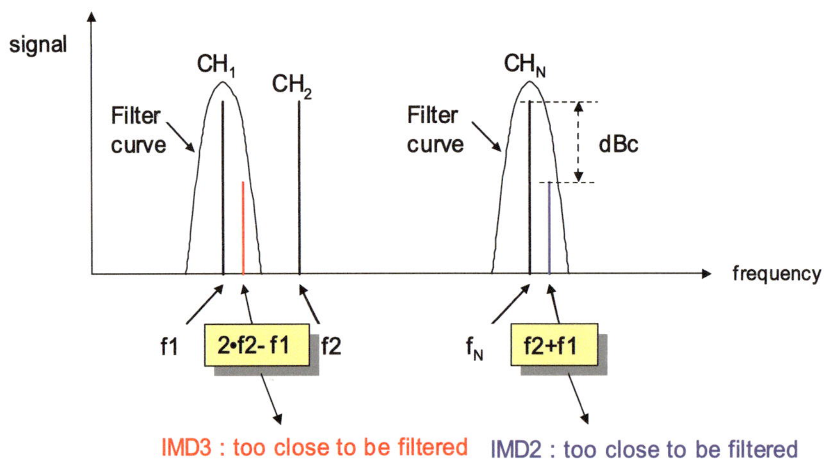

Fig. 10. Intermodulation signals close to original signals. IMD2 and IMD3 signals are too close to original signal to be filtered out

In Figure 11, IMD2 and IMD3 characteristics are presented for a Type (IV) WGPD with width of 10μm and length of 70μm. Its -3dB bandwidth was ~20GHz. The device shows IMD2 of less than -70dBc for a DC photocurrent of 1mA, optical modulation index(OMI) of 0.7 and 50Ω load. Also, IMD3 was less than -90dBc for the same conditions. IMD3 for a voltage range of -6~-8V cannot be measured because IMD3 at that range is too small to be detected within the limit of spectrum analyzer sensitivity.

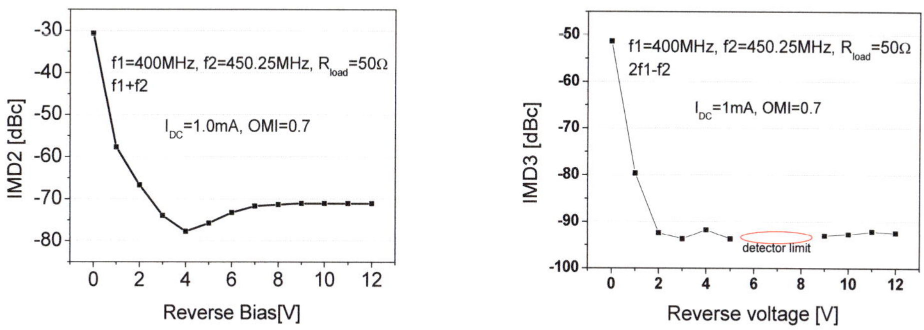

(a)IMD2 (b)IMD3

Fig. 11. IMD2 and IMD3 characteristics of a Type (IV) WGPD

4. Conclusion

A new WGPD with a thin absorption layer was introduced. Methods of design and optimizations for this new type of WGPD were described. Absorber should be thicker than 100Å to obtain a high responsivity and low polarization dependency. A responsivity of 1.08A/W was achieved at 1550nm wavelength, which corresponds to an external quantum efficiency of 86.4% with TE/TM polarization dependence less than 0.25dB. For the same device, the bandwidth of ~40GHz was obtained. The formula for the transit-time limited frequency response of this kind of devices was obtained. With this formula, optimization of frequency response is possible. Also, this kind of devices can show a suppressed nonlinearity.

5. References

K. Kato, S. Hata, K. Kawano, J. Yoshida, and A. Kozen, (1992), *IEEE J. of Quantum Elect.* Vol. 28, No. 12, pp. 2728-2735.

F. Xia, J. K. Thomson, M. R. Gokhale, P. V. Studenkov, J. Wei, W. Lin, and S. R. Forrest, (2001), *IEEE Photon. Tech. Lett.* Vol. 13, No. 8, pp. 845-847

T. Takeuchi, T. Nakata, K. Makita, and T. Torikai, *Proceedings of OFC 2001*, Vol.3, Paper WQ2-1.

M. Achouche, S. Demiguel, E. Derouin, D. Carpentier, F. Barthe, F. Blache, V. Magnin, J. Harari, and D. Decoster, *Proceedings of OFC 2003*, Paper WF5.

S. Demiguel, N. Li, X. Li, X. Zheng, J. Kim, J. C. Campbell, H. Lu, and K. A. Anselm, (2003), *IEEE Photon. Tech. Lett.* Vol. 15, No.12, pp. 1761-1763.

G. Lucovsky, R. F. Schwarz, and R. B. Emmons, (1964) *J. of Applied Phys.*, Vol.35, No.3, pp. 622-628.

K. Kato, S. Hata, K. Kawano, and A. Kozen, (1993), *IEICE. Trans. Electron.*, Vol. E76-C, No. 2, pp. 214-221.

S. Adachi, (1982), *J. of Applied Phys.*, vol.53 , pp. 8775-8792.

A. Galvanauskas, A. Gorelenok, Z. Dobrovol'skis, S. Kershulis, Yu. Pozhela, A. Reklaitis, N. Shmidt, (1988), *Sov. Phys. Semicond.*, Vol.22, pp.1055-1058.

K. J. Williams, R. D. Esman, and M. Dagenais, (1996), *.J. of Lightwave Tech.*,Vol. 14, No. 1, pp.84~96.

Y. Kuhara, Y. Fujimura, N. Nishiyama, Y. Michituji, H. Terauchi, and N. Yamabayashi, (1997), *.J. of Lightwave Tech.*,Vol. 15 No. 4, pp.636~641

10

Resonant Tunnelling Optoelectronic Circuits

José Figueiredo[1], Bruno Romeira[1], Thomas Slight[2] and Charles Ironside[2]
[1]*Centro de Electrónica, Optoelectrónica e Telecomunicacões, Universidade do Algarve*
[2]*Department of Electronics and Electrical Engineering, University of Glasgow*
[1]*Portugal*
[2]*United Kingdom*

1. Introduction

Nowadays, most communication networks such as local area networks (LANs), metropolitan area networks (MANs), and wide area networks (WANs) have replaced or are about to replace coaxial cable or twisted copper wire with fiber optical cables. Light-wave communication systems comprise a transmitter based on a visible or near-infrared light source, whose carrier is modulated by the information signal to be transmitted, a transmission media such as an optical fiber, eventually utilizing in-line optical amplification, and a receiver based on a photo-detector that recovers the information signal (Liu, 1996)(Einarsson, 1996). The transmitter consists of a driver circuit along a semiconductor laser or a light emitting diode (LED). The receiver is a signal processing circuit coupled to a photo-detector such as a photodiode, an avalanche photodiode (APD), a phototransistor or a high speed photoconductor that processes the photo-detected signal and recovers the primitive information signal.

Transmitters and receivers are classical examples of optoelectronic integrated circuits (OEICs) (Wada, 1994). OEIC technologies aim to emulate CMOS microelectronics by (i) integrating optoelectronic devices and electronic circuitry on the same package or substrate (hybrid integration), (ii) monolithically integrate III-V optoelectronic devices on silicon (difficulty since silicon is not useful for many optoelectronic functions) or (iii) monolithically integrate III-V electronics with optoelectronic devices. The simply way to do hybrid integration is combining packaged devices on a ceramic substrate. More advanced techniques include flip-chip/solder-ball or -bump integration of discrete optoelectronic devices on multi-chip modules or directly on silicon integrated circuit (IC) chips, and flip-bonding on IC chips. Although, hybrid integration offers immediate solutions when many different kinds of devices need to be combined it produces OEICs with very low device density. Moreover, in certain cases the advantages of using optical devices is greatly reduced. On the contrary, monolithic integration leads to superior speed, component density, reliability, complexity, and manufacturability (Katz, 1992).

There was been substantial efforts towards monolithical integration of III-V electronics with optoelectronic devices to improve the performance of transmitters and receivers. Approaches to light modulation, light detection and light generation at microwave and millimetre-wave frequencies have been investigated by combining double barrier quantum well (DBQW) resonant tunnelling diodes (RTDs) with optical components such as waveguides (Figueiredo, 2000) and semiconductor lasers (Slight, 2006). These RTD based OEICs can operate as novel optoelectronic voltage controlled oscillators (OVCOs), with potential to simplify clock recovery circuits, improve control of microwave oscillators functionalities, to generate electrical and optical aperiodic waveforms, and as microwave-to-optical subcarrier and optical subcarrier-to-microwave converters for radio-over-fiber systems, where the integration of electrical and optical components in a single chip is a major challenge in order to obtain high reliability, small size and low cost (Sauer et al., 2007).

This chapter reports investigation on resonant tunnelling (RT) based OEICs that demonstrate new functionalities for optical modulators and sources for application in telecommunication systems and signal processing circuits. Section 2 starts with a brief description of DBQW-RTD's operating principle, followed by the presentation of a physics based model of its current-voltage ($I - V$) characteristic, continues with a small-signal equivalent circuit analysis, and ends with an overview of more relevant optoelectronic devices incorporating RT structures. Section 3 describes the integration of DBQW-RTDs within an optical waveguide (OW) towards the implementation of very low driving voltage electro-absorption modulators (EAMs) and optical detectors (OD), with built-in amplifiers, for operation at optical wavelengths around 900 nm and 1550 nm. Section 4 discusses monolithic and hybrid integration of a DBQW-RTD with a laser diode (LD), its operation principle and optoelectronics circuit model used to analyse its modes of operation including optoelectronic voltage controlled oscillator (OVCO), frequency division and multiplication, phase-locking, and the generation of aperiodic, even chaotic, waveforms. The chapter ends with conclusion and acknowledgement sections.

2. Resonant tunnelling diode

Resonant tunnelling diodes (RTDs) are nanoelectronic structures that can be easily integrated with conventional electronic and photonic devices (Davies, 1998)(Mizuta & Tanoue, 1995)(Sun et al., 1998), such as transistors (Mazumder et al., 1998), optical waveguides (McMeekin et al., 1994)(Figueiredo, 2000) and laser diodes (Slight, 2006) with potential to not only reduce power consumption and cost but also increase functionality, speed and circuit reliability, without losing any advantage of using optical devices. They have two distinct features when compared with other semiconductor devices (Mazumder et al., 1998): their potential for extremely high frequency operation up to terahertz and their negative differential conductance (NDC). The former arises from the very small size of the resonant tunnelling structure along the direction of carriers transport. The second corresponds to electric gain which makes possible to operate RTDs as amplifiers and oscillators, significantly reducing the number of elements required for a given function (Mazumder et al., 1998). Functional RTD based devices and circuits span from signal generators, detectors and mixers, multi-valued logic switches, low-power amplifiers, local oscillators, frequency locking circuits, and also as generators of multiple high frequency harmonics (Mizuta & Tanoue, 1995). In this section, the physics of double barrier quantum well resonant tunnelling diodes (DBQW-RTDs) is discussed and analyzed, aiming at its application in high speed optoelectronic converters (rf-optical and optical-rf), such as light emitters, light modulators and light detectors.

2.1 Double barrier quantum well RTD

Resonant tunnelling through double potential barriers was predicted by (Bohm, 1951). Latter, (Iogansen, 1964) discussed the possibility of resonant transmission of an electron through double barriers formed in semiconductor crystals. They concluded that structures with identical barriers show tunnelling transmission coefficients of 1 when the particles incident energy equals the structure resonant energies, however small the transmission through the individual barriers may be (Mizuta & Tanoue, 1995). Figure 1 compares schematically the transmission coefficient $T(E)$ for single and symmetrical double barrier structures. The transmission coefficient lobs broadens with increasing energy because the barriers become more transparent (Davies, 1998).

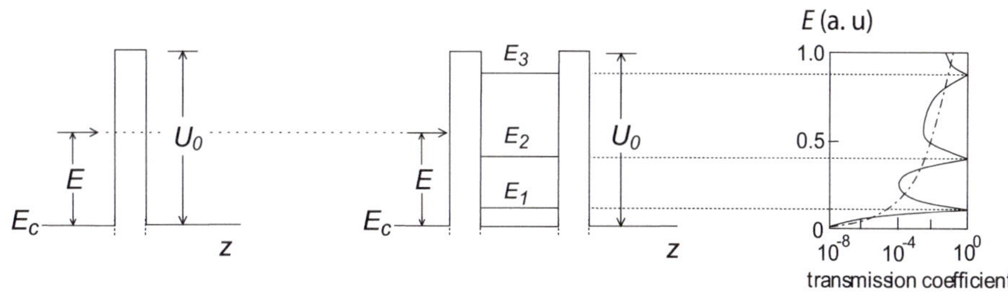

Fig. 1. Single and DBQW transmission coefficients as function of incident carrier energy.

A semiconductor double barrier quantum well resonant tunnelling diode (DBQW-RTD) consists of a low band-gap semiconductor layer (the quantum well, typical 5 nm to 10 nm wide) surrounded by two thinner layers of higher band-gap material (barriers, typical 1.5 nm to 5 nm), both sandwiched between low band-gap *n*-type material layers, typical the well material, as schematically shown in Fig. 2(a) (Mizuta & Tanoue, 1995). The material forming the barriers must have a positive conduction-band offset with respect to the smaller bandgap materials (Weisbuch & Vinter, 1991). When both sides are terminated by highly doped semiconductor layer (the emitter and the collector contacts) for electrical connection the structure is called resonant tunnelling diode (RTD). Figure 2(b) shows a schematic of a *n*-type Al-GaAs/GaAs DBQW-RTD, together with the Γ-conduction band profiles at around zero volts and at the peak voltage. Because finite height of the energy barriers the allowed energy states in the well region become quasi-bound or resonant states, Fig. 2(a), rather than true bound states as it happens with thicker barrier quantum wells (Davies, 1998). In consequence, tunneling of charge carriers through the barriers is strongly enhanced when their energy equals to one of well energy levels, reaching much higher values than the product of the two individual barrier transmission coefficients at the energy values of the system resonant levels, see Fig. 1.

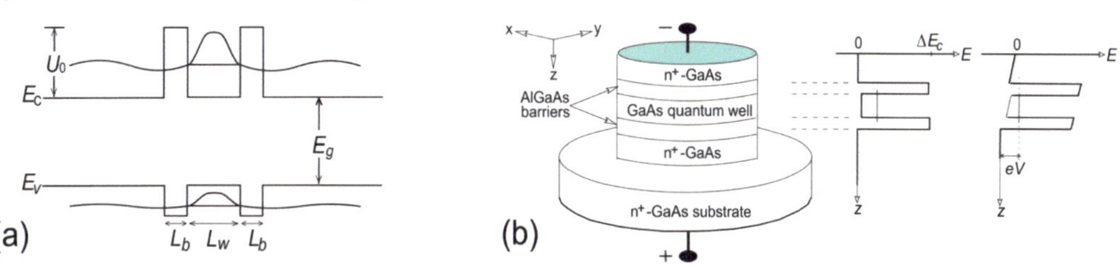

Fig. 2. (a) DBQW semiconductor structure. (b) AlGaAs DBQW structure (left); Γ-conduction band profiles at zero and at the first resonance voltage (right).

Under applied bias, the overall carrier flow through a DBQW-RTD is qualitatively different from that of a single barrier diode since the double barrier structure acts as a band filter to charge carrier energy distribution (Mizuta & Tanoue, 1995)(Sun et al., 1998). This filter action is exploited applying a voltage across the DBQW structure to control the number of carriers that can take part in the conduction through resonant levels. The carrier transmission coefficient maxima shown in Fig. 1 give rise to current-voltage characteristics with regions of strong NDC. The resonant tunnelling phenomenon in AlGaAs DBQW structures was first predicted in 1973 (Tsu & Esaki, 1973), and demonstrated experimentally in 1974 (Chang et al., 1974). In 1983, Sollner et al. demonstrated resonant tunnelling through quantum wells at frequencies up to 2.5 THz (Sollner et al., 1983). Figure 3(a) shows a typical InGaAs/AlAs RTD I -V characteristic. The main carrier flow processes in a DBQW-RTD polarized at the peak voltage (the current first maxima) is schematically represented in Fig. 3(b).

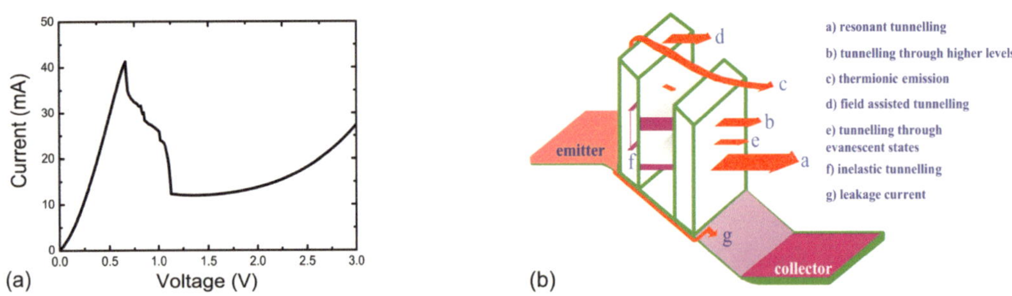

Fig. 3. (a) Typical InGaAlAs RTD I-V characteristic. (b) Current transport mechanisms in DBQW-RTDs at the peak voltage (Sun et al., 1998).

The RTD current-voltage characteristic of Fig. 3(a) can be understood with the help of the Γ-conduction band profile shown in Figs. 2(b) and 3(b) (Davies, 1998). When the applied bias is small, i.e., $V << V_p$ (peak voltage, also referred as resonance voltage), the Γ-conduction band profile is not much affected, remaining almost flat, see Fig. 2(b). The first resonant level is well above the emitter Fermi level, and little current flows. As voltage is increased, the energy of the first resonant level is moved downwards to the emitter Fermi level, leading to an almost linearly current increase with the voltage, the first positive differential conductance (PDC) region, till reaching a local maximum I_p, ideally, at $V \simeq 2E_{n=1}/e$, when the overlap between the emitter electron Fermi sea energy spectrum and the transmission coefficient around the first resonant level reaches a local maximum, as shown in the right side of Fig. 2(b) and Fig. 3(b). A further increase in the applied voltage pulls the first resonant level towards the bottom of the Γ-valley and into the forbidden gap, where there are no longer carriers available to efficiently cross the DBQW. This leads to a sharp current decrease, giving rise to the first negative differential conductance (NDC) portion of the device current-voltage characteristic. At a given voltage, known as the valley voltage V_v, with $V_v > V_p$, the current reaches a local minimum I_v. An additional increase on the bias voltage will further lift up the emitter Fermi level and tunnelling through higher resonant levels or through the top regions of the barriers will lead to new current rise, similar to the classical diode I - V characteristic (Davies, 1998). The resonant tunnelling component dominates at low voltages and the classical diode component takes over at higher voltages. For more details see (Davies, 1998)(Sun et al., 1998). In a circuit, the NDC provides the gain necessary to sustain oscillations (Mizuta & Tanoue, 1995) (Brown & Parker 1996). The

presence of a small inductance in circuit containing an RTD, together with RTD intrinsic capacitance make possible the oscillations at very high frequencies, experimental demonstrated up to 831 GHz (Suzuki et al., 2009). Frequencies never reached by other semiconductor devices: the RTD is currently the fastest purely electronic device.
The most common material systems used to implement RTD devices are III-V compounds such as AlGaAs and InP-based materials.. Si/SiGe RTDs based on Si/SiGe heterojunctions have been demonstrated but the performance is not comparable to III-V RTDs because of the limited band edge discontinuity in both valence and conduction bands. Organic RTDs are currently being investigated (Park et al., 2006)(Ryu et al., 2007)(Zheng et al., 2009).

2.2 RTD based generalized Liénard oscillator

The RTDs inherent high speed operation, up to terahertz frequency, the pronounced nonlinear current-voltage characteristic, wide-bandwidth NDC, structural simplicity, flexible design, relative ease of fabrication, and versatile circuit functionality, make them excellent candidates for nanoelectronic circuit applications. In order to take advantage of the full potential of RTD based devices several attempts have been made to incorporate the full RTD characteristics into circuit simulation packages such as SPICE-like CAD tools (Mizuta & Tanoue, 1995)(Brown et al., 1996)(Sun et al., 1998).
Since a quantum mechanics based model that includes all RTD features is not yet available, a number of empirical models have been advanced (Sun et al., 1998). Most models describe the RTD by small-signal equivalent circuits consisting of a capacitance C, resulting from charging and discharging of electrons of DBQW and depletion regions, in parallel with a voltage depend current source $I = F(V)$, a series resistance R arising mainly from the ohmic contacts and an inductance L due to bond wire connections, Fig. 4. The current source $F(V)$ is usually implemented as polynomial or piecewise functions (Brown et al., 1997)(Sun et al., 1998), which is not satisfactory if a detailed circuit description is needed. More useful RTD non-linear characteristic representations have to consider a wide variety of device structures and the materials available, i.e., the modelled $I - V$ characteristic has to be based as much as possible on the RTD physical parameters such as material properties, layer dimensions, energy levels, dopant concentrations, and the device geometry.

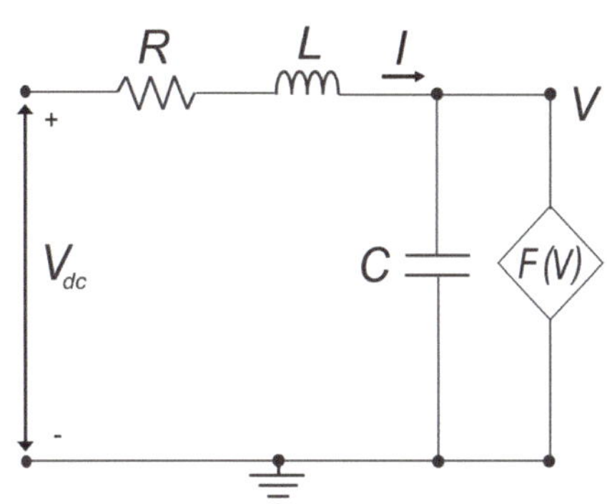

Fig. 4. Electrical equivalent circuit of an RTD represented by a capacitance in parallel with a voltage dependent current source $F(V)$. The inductance L and the resistor R are due to bonding wires and contacts.

The physics based model proposed by Schulman et al. consists of a mathematical function which provides a satisfactory $I - V$ shape characteristic for InGaAs and GaAs RTD based devices (Schulman et al., 1996). The expression obtained contains physical quantities which can also be treated as empirical parameters for fitting purposes. In their analysis the resonant tunnelling current density is expressed within the effective mass approximation (Davies, 1998), which includes nonzero temperature, Fermi-Dirac statistics and the transmission coefficient $T(E,V)$:

$$J_{RT} = \frac{qm^* k_B T \cdot \Delta E_r}{4\pi^2 \hbar^3} \ln\left[\frac{1+e^{(E_F - E_r + qV/2)/k_B T}}{1+e^{(E_F - E_r - qV/2)/k_B T}}\right] \cdot \left[\frac{\pi}{2} + \tan^{-1}\left(\frac{E_r - qV/2}{\Delta E_r / 2}\right)\right] \tag{1}$$

where $E = E_r - qV/2$ is the energy measured up from the emitter conduction band edge, E_r is the energy of the resonant level relative to the bottom of the well at its centre, and ΔE_r is the resonance width. The parameters q and k_B are unit electric charge and Boltzmann constants, respectively. Equation 1 can be rewritten as:

$$J_{RT}(V) = A \cdot \ln\left[\frac{1+e^{q(B-C+n_1 V)/k_B T}}{1+e^{q(B-C-n_1 V)/k_B T}}\right] \cdot \left[\frac{\pi}{2} + \tan^{-1}\left(\frac{C-n_1 V}{D}\right)\right] \tag{2}$$

where the parameters A, B, C, D, and n_1 can be used to shape the curve to match the first PDC region of the measured I -V characteristic, having at the same time a well-defined physical interpretation: A and B are related, among other factors, with resonance width and Fermi level energies, and allow adjustment of the RTD peak current; C and n_1 determine essentially the RTD peak voltage, correlated with the energy of the resonant level relative to the bottom of the well and with the transmission coefficient; finally, D is related to the resonance width ΔE_r.

In order to represent the increasing valley current due to tunnelling through higher resonances or thermal excitation over the barriers, an additional current density component, identical to the classical diode current, the non-resonant term J_{NR}, have to be included:

$$J_{NR}(V) = H\left(e^{n_2 qV/k_B T} - 1\right) \tag{3}$$

Parameters D and H adjustment of adjust the peak to valley current ratio (PVCR) and the peak to valley voltage ratio (PVVR).

Equations 2 and 3 give good estimations of the peak current and the NDC region of current-voltage characteristic. The final form of the RTD current-voltage curve is then given by:

$$I(V) = I_{RT}(V) + I_{NR}(V) = M[J_{RT}(V) + J_{NR}(V)] \tag{4}$$

where the multiplying factor M is used to scale equation 4, in order to take into account the devices area. Figure 5 shows experimental I - V curves of AlGaAs (a), and InGaAlAs (b), RTDs, with the corresponding fit given by equation 4. The fits assumed operation at temperature T =300 K and a multiplying factor M=2×10^{-6} cm^2, with the following parameters: A=1950 A/cm^2, B=0.05 V, C=0.0874 V, D=0.0073 V, n_1=0.0352, H=18343 A/cm^2, and n_2=0.0031 for AlGaAs; A=3800 A/cm^2, B=0.068 V, C=0.1035 V, D=0.0088 V, n_1=0.0862, H=4515 A/cm^2, and n_2=0.0127 for InGaAlAs. Higher values of A and B are used in the InGaAlAs fitting due to RTD higher peak current; parameter D was also slightly larger for the InGaAlAs due to superior PVCR and PVVR. The parameter H was around four times larger in the AlGaAs due mainly to their higher peak voltages.

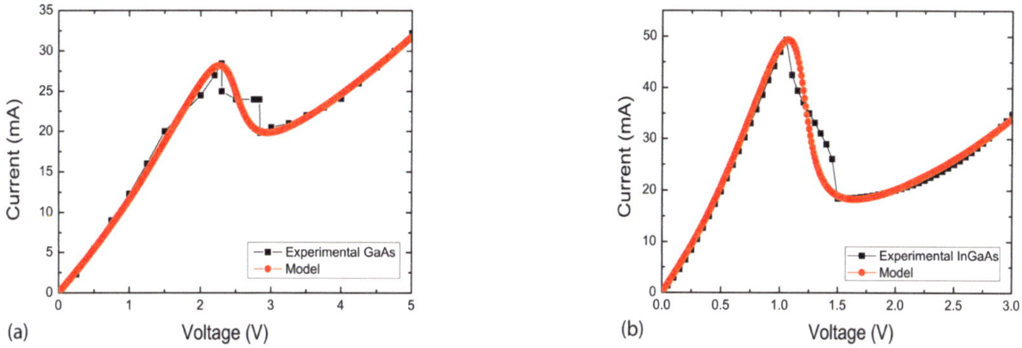

Fig. 5. GaAs/AlAs (a) and InGaAs/AlAs (b) RTD experimental $I-V$s and fittings.

Since the RTD is a voltage-dependent current source device, when incorporated in a resonant circuit and biased in the NDC portion of its $I-V$ characteristic produces oscillations at circuit characteristic frequency (Brown & Parker, 1996). In order to understand the origin of the circuit self-oscillations induced by the RTD we consider the small-signal equivalent circuit of Fig. 4. Typical RTD switching times are in general dominated by the effects of current densities and capacitances, i.e., by the circuit *RC* time constant (Brown et al., 1997) (Brown & Parker, 1996).

A general analysis of a circuit containing an RTD considers the small signal equivalent circuit of Fig. 4, where the RTD non-linear $I-V$ characteristic is represented by a voltage dependent current source $F(V)$, given by equation 4, in parallel with RTD intrinsic capacitance C. Resistor R and inductor L encompasses for the device series resistance and connections inductance, respectively. By applying Kirchoff's laws (using Faraday's law) to the circuit of Fig. 4, the voltage V across the capacitance C and the current I through the inductor L are given by the following set of two first-order non-autonomous differential equations (Slight et al., 2008):

$$\dot{V} = \frac{1}{C}\left[I - F(V)\right] \tag{5}$$

$$\dot{I} = \frac{1}{L}\left(V_{dc} - RI - V\right) \tag{6}$$

After some algebra, we find that the system of Eqs. 5-6 is equivalent to the following second-order differential equation, referred as one of the generalized nonlinear Liénard systems (Slight et al., 2008)(Figueiredo, 1970):

$$\ddot{V} + \left[\frac{R}{L} + \frac{1}{C}\frac{dF(V)}{dV}\right]\dot{V} + \frac{1}{LC}\left[V - V_{dc} + RF(V)\right] = 0 \tag{7}$$

$$\ddot{V} + H(V)\dot{V} + G(V) = 0 \tag{8}$$

"where $H(V) = \frac{R}{L} + \frac{1}{C}\frac{dF(V)}{dV}$ and $G(V) = \frac{1}{LC}\left[V - V_{dc} + RF(V)\right]$. $G(V)$ is a nonlinear force and $H(V)\dot{V}$ is a damping factor.

The circuit of Fig. 4 dc biased in the NDC acts as a relaxation oscillator producing oscillations at a frequency around $f_0(V) \approx \left(2\pi\sqrt{L \cdot C(V)}\right)^{-1}$, the circuit characteristic frequency, whenever the series R is smaller than the RTD operating point negative differential resistance (Brown & Parker, 1996). From the application point of view the wideband NDC of RTD leads to low frequency oscillations instabilities that are detrimental. A most common source of instability arises from the dc source circuitry by introducing in the circuit an equivalent inductance, which together with RTD capacitance leads to oscillations at around few megahertz (Figueiredo, 2000)(Slight, 2006). A method to eliminate these low frequency oscillations and allowing circuit operation at much higher frequency is to place a shunt capacitor across the terminals of the device (Kidner et al., 1990)(Huang et al., 1997). The inductance is now only due to the connection from the shunt capacitor to the RTD.

2.3 Optoelectronic applications of RT structures

Several optoelectronic devices and circuits whose functions depend on embedded resonant tunnelling structures have been proposed and demonstrated, including resonant tunneling light emitting diodes (RT-LEDs) (Van Hoof et al., 1992), vertically integrated semiconductor lasers with RTDs (Grave et al., 1991), resonant tunnelling effect quantum-well lasers (Kawamura et al., 1994), resonant tunnelling injection laser (Capasso et al., 1986), multi-quantum well (MQW) lasers (Kawamura et al., 1987) and photo-detecting (PD) structures (Chen et al., 1991). The nature and the energies involved in the carrier transition induced by the light interaction with the tunnelling layers determine the operation in the optical or in the infrared part of the electromagnetic spectrum. Optical applications such as photo-detection, light emission, optical switching, utilize inter-band transitions (band-gap transitions), whereas infrared applications include intra-band and inter-sub-band photo-detection, and infrared emission. Below is presented a brief summary of the main progress on optical and optoelectronic devices whose functionalities depend of embedded RT structures.

Bistability in the light output of bipolar RT-LEDs has been reported, showing that these devices are capable of ultrafast optical switching and high frequency optical oscillation (Van Hoof et al., 1993). Laser transistors incorporating a resonant tunnelling structure have been reported, with carrier injection or extraction controlled via resonant tunnelling structure, with light output controlled by the collector voltage and achieving higher speed than with conventional semiconductor lasers (Kawamura et al., 1992). Embedding RTs into multi-quantum well (MQW) devices introduces negative differential conductance over wide valley region, which is very effective for getting large voltage switching and high on/off ratio current switching (Kawamura et al., 1988) leading to electro-optic bistability (Chen et al., 1991). Optical bistability in QW lasers integrated with DBQW-RTDs, and a RTD with a MQW modulator/detector based on the p - i(MQW)-n configuration, operating at room temperature, were reported (Kawamura et al., 1994). Clear negative differential conductance and bistability, with high contrast and high sensitivity in resonant tunnelling triangular barrier optoelectronic switch (R-TOPS), which consists of a double barrier resonant tunnelling diode and a triangular barrier phototransistor has been demonstrated (Sakata et al., 1995).

A light pulse incident upon a resonant tunnelling diode produces photo-charges that reduce the series resistance, leading to a shift of the peak and valley voltages which can induce RTD switching and give rise to changes in the current flow (Moise et al., 1995). Optically switched resonant tunnelling diode (ORTD) photo-detectors have been demonstrated (Moise et al., 1997). Phase locking of an oscillating GaAs/AlGaAs RTD to a train of light pulses achieved by direct illumination was reported (Lann et al., 1993), as well as optical switching in resonant tunnelling diode (England et al., 1991) and optical injection locking of the resonant tunnelling oscillator (Kan et al., 2001). The RT structures can be used to implement light-by-light switching (England et al., 1991). Ultra-fast optoelectronic circuits using RTDs and uni-travelling-carrier photodiodes (UTC-PDs) to de-multiplex ultra-fast optical data signals into electrical data signals with lower bit rate and low power consumption has been demonstrated (Sano et al., 1998).

Our work on optoelectronic devices based on the integration of a RTD within an optical waveguide, and on hybrid and monolithic integrations of RTDs with laser diodes is discussed in the remaining sections of this chapter.

3. RTD optical waveguide modulator-photodetector

Novel information and communication technologies relying on microwave/millimetre-wavelightwave interactions are fundamental to the development of applications such as low-cost fibre-optic communication networks, cable television signal distribution, mobile communications, and radio local area networks (Sauer et al., 2007). In this section, electrical active, high speed, highly efficient and low-cost electro-absorption modulators and photo-detectors based on the integration of a RT structure within a semiconductor optical waveguide are described.

3.1 RTD optical waveguide integration

As discussed previously, when the RTD is biased in the valley region most of the applied voltage is dropped across the depletion region formed between the second barrier and the collector contact, Fig. 6(a), where a strong electric field builds-in. Inter-band electro-absorption of light with photon energies close to but smaller than the collector band-gap energy is achieved through the Franz-Keldysh effect (Chuang, 1995). According to the Franz-Keldysh effect the semiconductor material optical absorption band-edge is broadened by the presence of an electric field, resulting in an increase of absorption of light with photon energies smaller but close to the material band-gap (Keldysh, 1958). This effect is used to implement either electro-absorption (EAM) (intensity) modulators (Wakita et al., 1998) or waveguide photo-detectors (Chuang, 1995). However, in typical RTD structures the light is injected perpendicularly to the tunnelling plane, which gives a light interaction (absorption) length well below 100 nm, and thus very small light absorption. This limitation can be easily overcome embedding the RTD into the core of a unipolar semiconductor optical waveguide (McMeekin et al., 1994). A typical waveguide structure is represented schematically in Fig. 6(a), showing also wafer Γ-conduction band-edge and refractive index profiles. This optoelectronic device is called resonant tunnelling diode optical waveguide (RTD-OW). The waveguide refractive index distribution confines light end-fire coupled along the tunnelling layers and the collector depleted region, therefore increasing substantially the light interaction volume along the waveguide length as indicated in Fig. 6(b). The RTD-OW, apart from the light confining layers (the lower refractive index regions - upper and lower cladding layers), corresponds to a DBQW-RTD with thick low doped

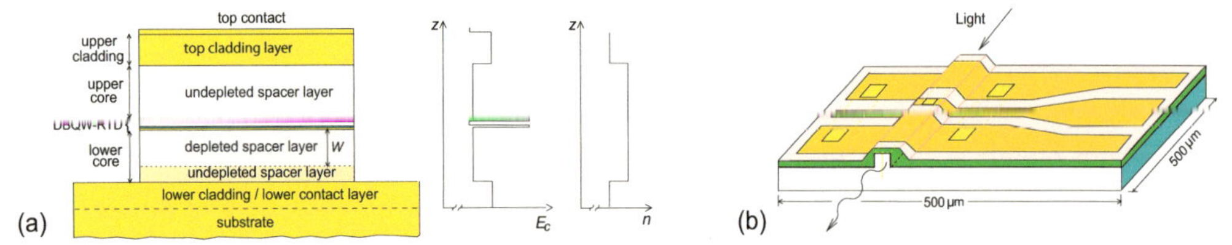

Fig. 6. (a) Diagram of a unipolar resonant tunnelling diode optical waveguide (RTD-OW) wafer structure, and the corresponding Γ-conduction band-edge and refractive index profiles. (b) Ridged waveguide channel configuration.

emitter and collector spacer layers. The presence of the DBQW within the waveguide core modifies the unipolar waveguide linear current-voltage characteristic towards the DBQW-RTD strong nonlinear $I - V$ curve (McMeekin et al., 1994)(Figueiredo, 2000). Moreover, it leads to a non-linear electric field distribution across the collector side waveguide core that is strongly dependent on the bias voltage, due to the electron accumulation close to the emitter barrier and the creation of a depletion region on the collector spacer layer. Since a small voltage can be used to make a RTD operating point to switch between peak and valley regions, the RTD-OW can be employed to implement electro-absorption modulators (McMeekin et al., 1994)(Figueiredo, 2000). A small voltage change results in large modulation of the electric field across the device collector depletion region, resulting, though the Franz-Keldysh effect, in waveguide propagation losses and electro-absorption for photon energies close to but smaller than the waveguide core band-gap energy (Figueiredo, 2000)(Figueiredo et al., 2001).

The RTD-OW electric field distribution dependence on the bias voltage can be understood by considering the Γ-conduction band profile of the collector spacer layer, Fig. 7. Below resonance (first PDC region), the applied voltage is dropped mainly across the DBQW, and the electric field in the collector core is rather small, Fig. 7(a). Any optical loss increase with the applied voltage is mainly due to the thermal effects induced by the current flow, which rise linearly with the current. Above resonance (in the NDC and on the second PDC region), the additional applied bias voltage is dropped mainly across the depleted part of the collector spacer layer, Fig. 7(b), and the electric field magnitude is now much stronger than on the first PDC region, inducing large light absorption. The thermal optical absorption is now much less important because the current flowing through the devices biased on the valley region is significantly lower.

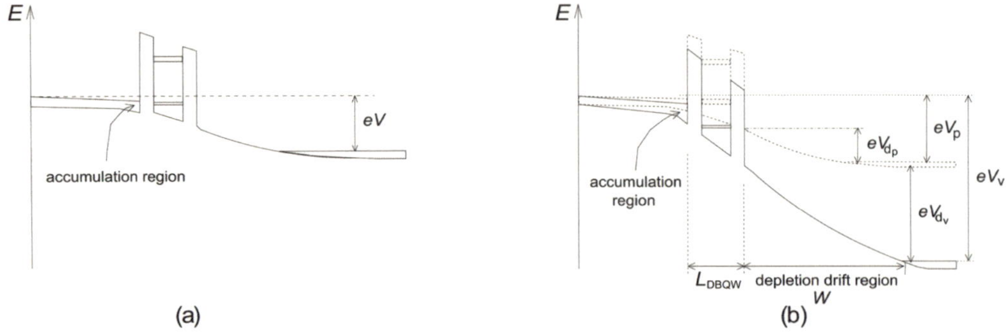

Fig. 7. Effect of applied biased on RTD-OW Γ-band: (a) before the peak and (b) on the valley.

The electric field enhancement ΔE_{VP} induced by the peak to valley switching can be estimated as (Figueiredo, 2000)(Figueiredo et al., 2001):

$$\Delta E_{VP} \simeq \Delta V_{VP} / W_{dep} + (W_{dep} / 2\varepsilon \upsilon_{sat}) \Delta J_{PV} \quad (9)$$

where ΔV_{VP} is the voltage dropped across the depletion region, ΔJ_{PV} is the corresponding current density change, υ_{sat} is the carrier saturation velocity and W_{dep} is the depletion thickness. At a given photon energy the absorption change induced by the electric field enhancement due to the peak to valley switching is given by (Figueiredo, 2000):

$$\Delta\alpha(\hbar\omega, \Delta E_{VP}) = \alpha(\hbar\omega, E_V) - \alpha(\hbar\omega, E_P) \approx \alpha(\hbar\omega, \Delta E_{VP}) \quad (10)$$

where $\alpha(\hbar\omega, E)$ is given, in the weak field approximation, by the Franz-Keldysh effect electroabsorption coefficient (Chuang, 1995)(Keldysh, 1958). The light modulation depth due to the peak to valley switching can be calculated using (Chuang, 1995):

$$R_{VP}(\text{dB}) \approx 4.343 \gamma_f \Delta\alpha(\hbar\omega, \Delta E_{VP}) \ell \quad (11)$$

where γ_f is the optical filling factor which corresponds to the fraction of the optical power guided in the depleted region of the waveguide, and ℓ is the RTD-OW electrically active length, defined by the RTD metal contacts length [see Fig. 6(b)]. The measured Franz-Keldysh effect effective band-edge shift to longer wavelengths can be compared with the value given by theory (Chuang, 1995)(Keldysh, 1958):

$$\Delta\lambda_g \simeq (\lambda_g^2 / hc)(e^2 h^2 / 8\pi^2 m_r)^{1/3} \Delta E_{VP}^{2/3} \quad (12)$$

The measured $\Delta\lambda_g$ gives an independent way to determine the electric field change ΔE_{VP} induced by the peak to valley switching.

As mentioned, RTD-OWs designed to show considerable NDC with a significant portion of the waveguide core being depleted at bias voltages higher than the peak voltage can have their operation point switched between the two I –V PDC regions by small high frequency ac signals (< 1 V). This leads to high speed electric field switching, resulting in high frequency modulation of the waveguide optical transmission loss. In this mode of operation the RTD-OW is called a resonant tunnelling diode electro-absorption modulator (RTD-EAM). In the RTD-EAM the modulation depth depends essentially on the overlap between the electric field in the collector depleted volume and the optical mode. The peak to valley electric field magnitude boost is determined mainly by the NDC region characteristics, ΔV and ΔJ. Figure 8 represents schematically the light absorption on the collector depleted region induced by the Franz-Keldysh effect when the RTD-OW is biased on the valley region, Fig. 8(a), and the change in the absorption coefficient associated with the bistable switching of the device plotted against wavelength, Fig. 8(b).

The device concept was implemented using AlGaAs ternary material system for operation on 900 nm optical window, and InGaAlAs quaternary compound to work on 1300 nm and 1550 nm optical windows, where the optical fibre present zero dispersion and have the lowest losses, respectively. For operation in the 900 nm spectral region, GaAs was used to form the waveguide core and the quantum well; AlAs and AlGaAs were employed to form the barriers and waveguide cladding layers, respectively. For operation at around 1550 nm,

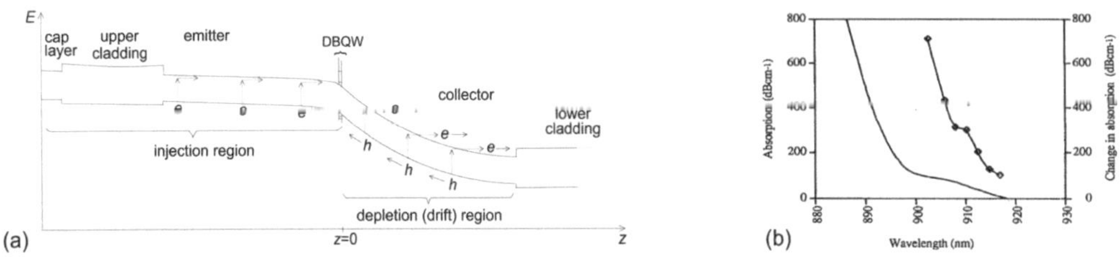

Fig. 8. (a) Schematic diagram of light absorption induced by Franz-Keldysh effect in a RTD-OW biased around the valley point. (b) Change in absorption produced by the change in the voltage characteristic of the NDC pulse plotted with the absorption in dB/cm of bulk GaAs against wavelength (McMeekin et al., 1994).

the InGaAlAs quaternary material system was used to implement the waveguide core and the quantum well, with AlAs and $In_{0.48}Al_{0.52}As$/InP being employed for the barriers and the waveguide cladding layers, respectively. The InGaAsP quaternary compound also allows operation on 1300 nm and 1550 nm optical windows but was not used. A detailed description of the RTD-OW structures implemented can be found in (Figueiredo, 2000). Next we describe the experimental operation of RTD-OW electro-absorption modulators on the optical communication windows around 900 nm and 1550 nm.

3.2 RTD-OW operation as EAM at 900 nm

The RTD-OW operation as an electro-absorption modulator at around 900 nm was achieved by growing the waveguide and DBQW layers using the AlGaAs/GaAs material system on semi-insulating GaAs. The GaAs waveguide core was made 1μm thick to allow easy end-fire light coupling, with *n*-type Si doping concentration of 2×10^{16} cm^{-3}; the cladding layers were made of $Al_{0.33}Ga_{0.67}As$, a direct band-gap compound alloy, with Si doping concentration around 2×10^{18} cm^{-3}. The refractive index difference between the core and cladding layers around 0.224 at 900 nm is sufficiently to obtain efficient light confinement with relatively thin cladding layers. The upper cladding layer thickness was made 300 nm thick, twice the reciprocal of the optical waveguide first mode exponential decaying factor, to keep the device series resistance low. Because the waveguide core and the substrate have similar real refractive indices, the lower cladding layer was made 600 nm thick with Si doping concentration of 2×10^{18} cm^{-3}, to act as an isolation layer separating the core from the substrate, in order to significantly reduce radiation leakage into the GaAs substrate. The DBQW consisted of a 7 nm GaAs quantum well sandwiched between 1.4 nm AlAs barriers. The detailed description and fabrication of AlGaAs/GaAs structures can be found in (Figueiredo, 2000). Figure 9 shows the top view of a RTD-OW die and a packaged device.
When dc biased in the NDC region, all tested devices showed instabilities at around few MHz. These where removed connecting devices to the dc power supply via a wide bandwidth bias-T. In certain cases, a high frequency energy-storage element, such as a coax transmission line, was inserted between the RTD and the bias-T, resulting in a RTD-EAM transmission line relaxation oscillator whenever the cavity characteristic frequency was within the NDC bandwidth (Figueiredo et al., 1999). Typical electrical relaxation oscillations due to a 15 cm long coaxial transmission line are shown in Fig. 10(a). The relaxation oscillations RF spectra show harmonic components up to 15 GHz (Figueiredo, 2000). The free-running oscillation frequency was changeable by varying the optical power coupled

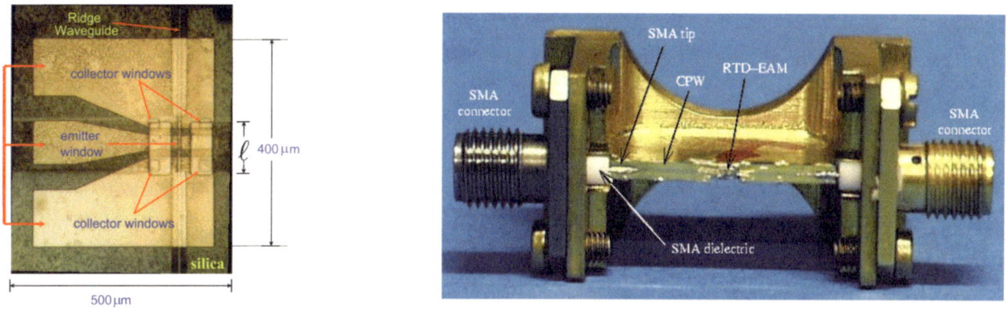

Fig. 9. RTD-OW die top view and a packaged device. The parameter ℓ here represents devices electrical active length, which with the waveguide width defines devices active area.

into the RTD-EAM, as shown in Fig. 10(b); in the cases observed the free-running frequency decreased when the coupled optical power was increased. In a circuit with a free-running oscillation frequency around 470 MHz, a tuning range of 10 MHz was observed. The frequency tuning effect is mainly due to the creation of charge carriers in the depletion region that reduces the device series resistance and moves the operating point through the NDC region, which change the device impedance [mainly the capacitance and the negative differential resistance (NDR)]. In the experiment light from a tunable Ti:sapphire laser emitting at around 900 nm was used; the optical power was kept to few mW in order to avoid damaging waveguide input facet.

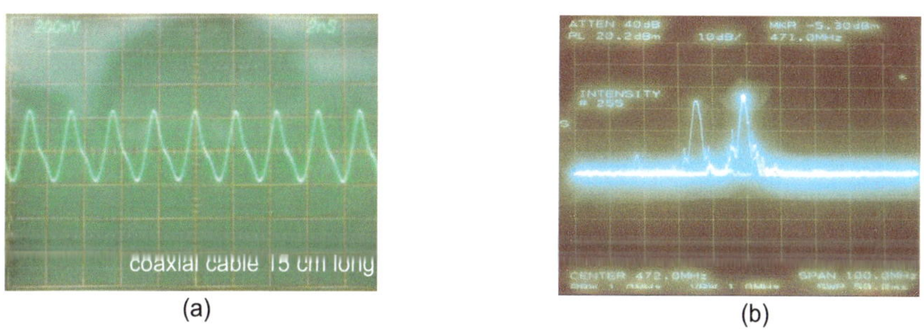

Fig. 10. (a) Self-sustained oscillations in a RTD-EAM connected via a 15 cm long coaxial line. (b) Self-oscillations frequency tuning induced by incident light.

The free-running relaxation oscillation frequency is also affected by the dc bias voltage because of the device intrinsic impedance dependence on the voltage. These behaviours can be used to implement both optical controlled oscillators (OCOs) and voltage controlled oscillators (VCOs). The OCO can be used to optically control microwave oscillators, and will be briefly analyzed when discussing the RTD-OW operation as photo-detector. The VCO behaviour makes possible operating the RTD-EAM as an optoelectronic voltage controlled oscillator (OVCO) since the electric field across the depleted collector region also self-oscillates at the free-running frequency, self-modulating the transmission properties of the waveguide. Before discussing OVCO operation we present electro-absorption response of the RTD-EAM. The RTD-EAM waveguide transmission spectra at zero bias, at slightly below the peak, and just above the valley points, are shown in Fig. 11(a) for devices with active areas around 800 μm^2. (The devices were not dc biased in the NDC region in order to avoid self-oscillation.) As the applied voltage increases from the peak to the valley point, there is a sharp drop in the waveguide transmission at wavelengths in the range 890 nm to 910 nm. The observed Franz-Keldysh absorption band-edge shift was around 12 nm which compares to 9 nm estimated using equation 12, taking in consideration the approximations made (Figueiredo, 2000). Figure 11(b) presents the optical modulation depth as a function of the operating wavelength due to the transition between the two positive PDC regions induced by a square signal with peak-to-peak voltage slight higher than $\Delta V_{VP} = V_V - V_P$. Modulation depth up to 13 dB around 908 nm was achieved. Modulation depths up to 18 dB were observed in waveguides with 400 μm active length and 4 μm wide ridges.

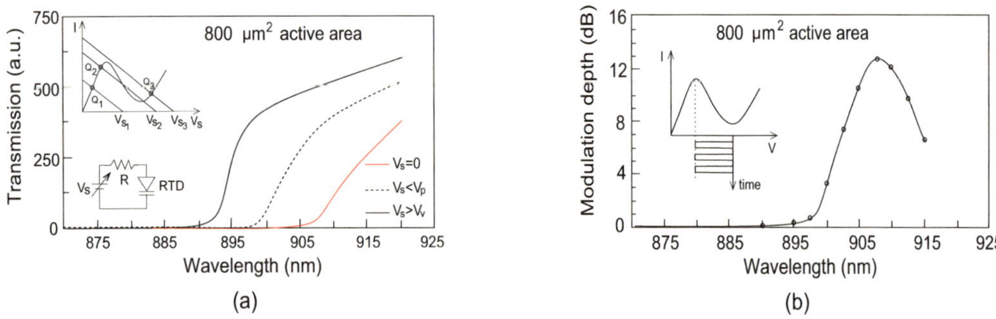

Fig. 11. (a) AlGaAs RTD-EAM optical transmission spectrum at zero volts, around the peak and at the valley region. (b) Modulation depth as function of the operating wavelength due to peak-to-valley switching induced by a square voltage waveform.

Direct modulation was obtained dc biasing the RTD-EAM slightly above the valley point and injecting through a wide band bias-T the rf modulating signals. Figure 12 shows examples of modulation due to 950 MHz and 16 GHz rf signal voltages. In both cases the driving signals amplitude was kept slightly larger than $\Delta V_{VP} \sim 0.4$ V. Optical modulation depths as high as 11 dB were achieved (Figueiredo et al., 1999)(Figueiredo, 2000). The 16 GHz response shown in Fig. 12(b) gives a good estimation of the bandwidth and modulation depth potential of the devices. The modulation efficiency characterized by the bandwidth-to-drive-voltage ratio, defined as the ratio of the operation bandwidth to the operating voltage for at least 10 dB modulation depth, was 40 GHz/V (Figueiredo et al., 1999)(Figueiredo, 2000).

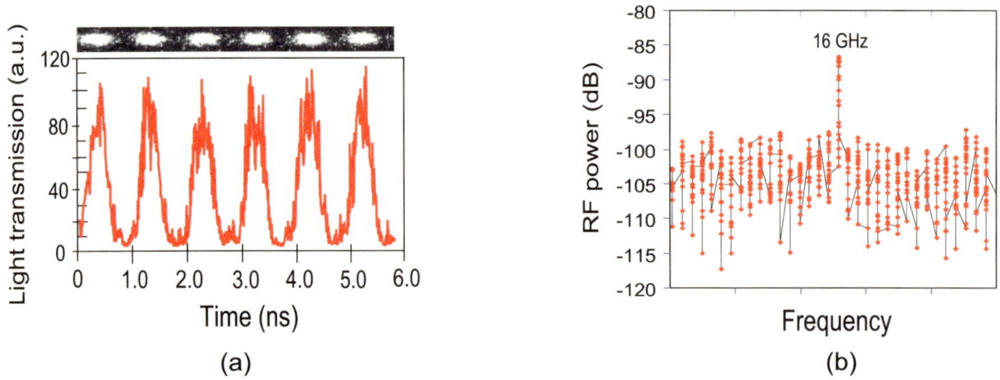

Fig. 12. (a) Direct modulation at around 950 MHz, with modulation depth up to 11 dB (λ=908 nm). (b) Modulator response to a 16 GHz rf signal.

As discussed previously, when dc biased in the NDC region and connected to a bias-T through a coaxial line the RTD-EAM can operate in the self-oscillation mode, producing an optical output modulated by the NDC induced relaxation oscillations, at frequencies determined by the electrical length of the transmission line, as shown in Fig. 13.

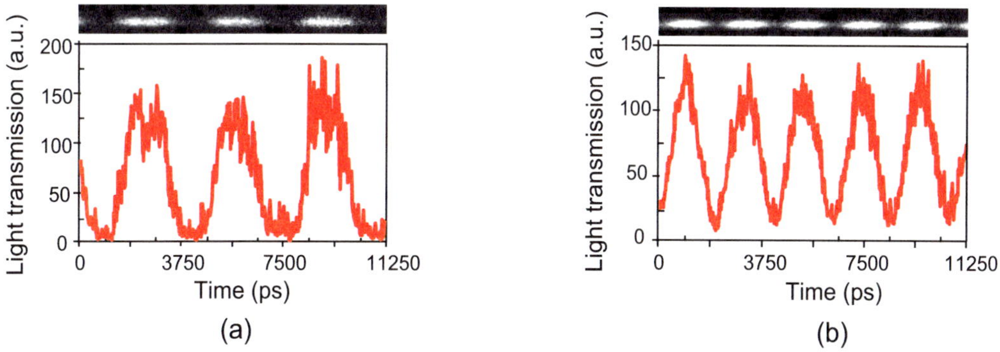

Fig. 13. Optical responses measured with Streak Camera of RTD-EAM transmission line relaxation oscillators with lines 15 cm (a) and 10 cm (b) long.

The AlGaAs RTD-EAM operation modes discussed above can be employed in LANs systems primarily as devices for electrically controlling guided-wave optical signals in the 880 nm to 1100 nm wavelength range such as waveguide intensity modulators, directional couplers and optical switches. The capability to operate in relaxation oscillation mode can be applied in clock extraction circuits, for optical pulse generation and de-multiplexing in optical time division multiplexed systems.

3.3 RTD-OW operation as EAM at 1550 nm

The AlGaAs/GaAs RTD-EAM achieved performances led the work to the demonstration of device concept operation at 1550 nm, where standard single-mode optical fibres have lowest losses (Liu, 1996). For band gap energies between 0.75 eV and 1.439 eV, quaternary alloys lattice matched to InP, which combine In, Ga, Al, and As ($In_{1-x-y}Ga_xAl_yAs$) or In, Ga, As, and P ($In_{1-x-y}Ga_xAs_{1-y}P_y$), can be used (Chuang, 1995)(Figueiredo, 2000). The RTD-OW concept operating at 1550 nm was demonstrated using InGaAlAs lattice matched to InP because phosphorus based heterostructures have lower conduction band discontinuity, which prevents strong localization of electrons in the lower band gap material. Moreover, they are difficult to grow with conventional MBE systems due to the need to handle solid phosphorus and high concentration of phosphorus at its vapour pressure, and also due to the difficulty to control As/P ratio.[1] The InGaAlAs material system shows more favorable material properties such as higher electron mobility, lower effective mass and superior conduction bandedge discontinuity (Figueiredo, 2000). As a consequence it is expected the InGaAlAs RTD-OW shows superior speed and modulation depth performance mainly to the InGaAs RTD higher peak current density and peak-to-valley current ratio, and smaller operating voltage. The In- GaAlAs quaternary system lattice matched to InP allows as well operation at 1300 nm, where standard single-mode optical fibres show zero dispersion (Chuang, 1995)(Figueiredo, 2000).

---

[1] Structures incorporating InGaAsP are usually grown by MOCVD (Bohrer et al., 1993).

The InGaAlAs RTD-OW schematic wafer structure for operation at 1550 nm is shown in Fig. 14, with wafer Γ-valley and refractive index profiles. The core consisted of two $In_{0.53}Ga_{0.42}Al_{0.05}As$ layers (refractive index of 3.56), 0.5 μm thick each, with a band-gap energy around 0.826 eV (absorption band-edge wavelength around 1500 nm), to allow operation at 1550 nm when biased around the peak voltage. The upper cladding was implemented using a layer of $In_{0.52}Al_{0.48}As$, refractive index of 3.24. Because InP refractive index at 1550 nm (3.17) is considerably smaller than the $In_{0.53}Ga_{0.42}Al_{0.05}As$ refractive index, the *n*-type InP substrate acted as lower cladding region. As previously discussed, the upper cladding layer thickness was made 300 nm thick. A detailed description of the wafer structure is given in (Figueiredo, 2000). Most of the RTD-EAMs characterized were 4 μm wide ridges with 200 μm active lengths. Typical current-voltage characteristic of 4 μm × 200 μm InGaAlAs/InP RTD-EAM is presented in Fig. 5(b) (section 2), showing PVCR around 3. These devices showed valley-to-peak voltage differences $\Delta V_{VP} \sim 0.8$ V, with peak-to-valley current density differences $\Delta J_{PV} \sim 10$ kA/cm^2. (Typical GaAs/AlAs devices show $\Delta V_{VP} \sim 0.4$ V and $\Delta J_{PV} \sim 5$ kA/cm^2.)

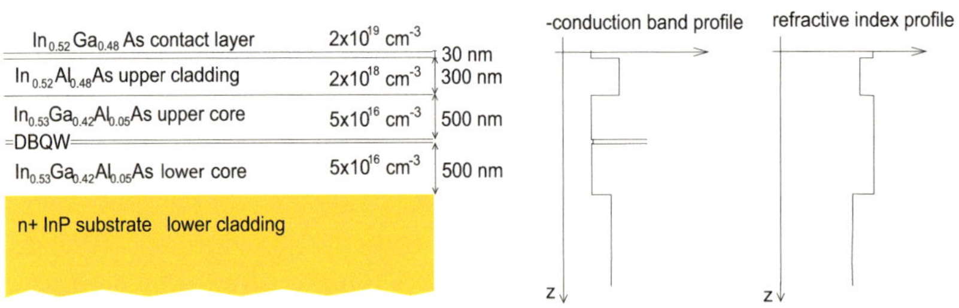

Fig. 14. InGaAlAs RTD-EAM structure, Γ-valley and refractive index profiles.

The devices' frequency response was investigated by on wafer impedance measurements in the 45 MHz to 18 GHz frequency range for all values of bias voltage. The results indicate the InGaAlAs RTD-EAM small signal equivalent circuit consists of a capacitance C in parallel with a non-linear resistor $R_d(V)$, in series with a resistance R_S; the series inductance was found to be negligible (Figueiredo, 2000)(Alkeev et al., 2000). The devices average capacitance C and shunt resistance R around the NDC region were 1 pF and -15 Ω, respectively; the R_S was typical few ohms (less than 5 Ω) ((Figueiredo, 2000)(Alkeev et al., 2000). The device switching time can be estimated as $t_R \simeq 4(\Delta V_{VP}/\Delta J_{PV})C_V$, where C_V is the device capacitance per unit area ($C_V \approx \varepsilon/W_{dep}(V)$). For the devices tested the expected modulation bandwidth was superior to 30 GHz (Figueiredo, 2000).

Following the frequency characterization, the waveguide low frequency electro-absorption response was characterized with no applied voltage, dc biased at slightly below the peak voltage, and on the valley region; the device was not dc biased in the NDC region in order to avoid self-oscillations. Light from a Tunics diode laser, tunable in the wavelength range 1480 nm to 1580 nm was fibre coupled to the waveguide, with the light output fibre coupled to an optical power meter or a high bandwidth photo-detector. The InGaAlAs/InP RTD-EAM waveguide transmission spectrum change due to the Franz-Keldysh effect absorption bandedge broadening induced by peak-to-valley switching is indicated in Fig. 15(a). The measured wavelength band-edge shift was 43 nm, which compares quite well with the estimation of 46 nm, equation 12. The low frequency electro-absorption response showed 5 dB absorption changes induced by 1 mV dc voltage increments, an exceptionally high transmission change per unit of voltage (Figueiredo, 2000). Figure 15(b) shows modulator response as function of the dc bias voltage when driven by 3 GHz voltage signals of amplitude from 1 mV to 100 mV; also represented is the RTD-EAM dc $I - V$ characteristic. The rf photo-detected power increased by about 15 dB when the device dc bias point moved from the peak to the valley region at driving amplitudes as low as 50 mV. An indication the modulator can be driven by very low voltage signals due to its intrinsic built-in electrical amplifier.

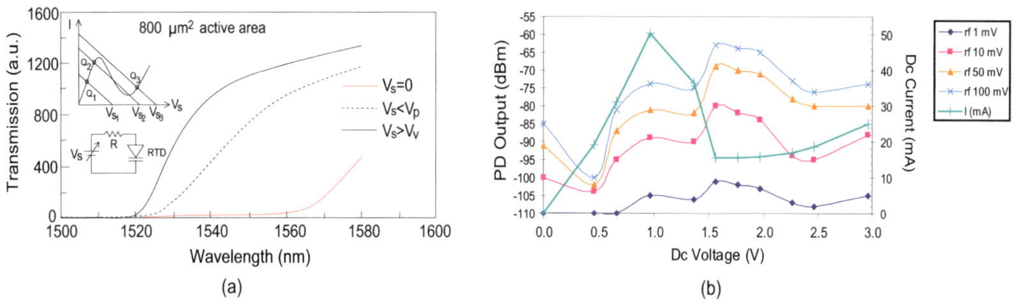

Fig. 15. (a) InGaAlAs RTD-EAM transmission spectrum in the wavelength range 1500 nm to 1580 nm, with the applied voltage as a parameter. (b) Modulator response as function of the dc bias voltage when driven by 3 GHz rf signals, with injected amplitude as a parameter.

RTD-EAM high frequency optical characterisation employed a microwave synthesized signal generator with a maximum output of +20 dBm and an upper frequency limit of 26 GHz (Figueiredo, 2000). Figure 16(a) shows the modulation depth as function of the light wavelength induced by the transition between the two PDC regions produced by a square signal with peak-to-peak voltage slight higher than $\Delta V_{VP} \sim 0.8$ V. The devices were dc biased in the valley region in order to minimize thermal effects and avoid self-oscillations. Modulation depths up to 28 dB were measured on devices with active areas around 800 μm², more than 10 dB superior to the values observed on the AlGaAs/GaAs devices. The modulator response up to 26 GHz driving signals for two power values is shown in Fig. 16(b).

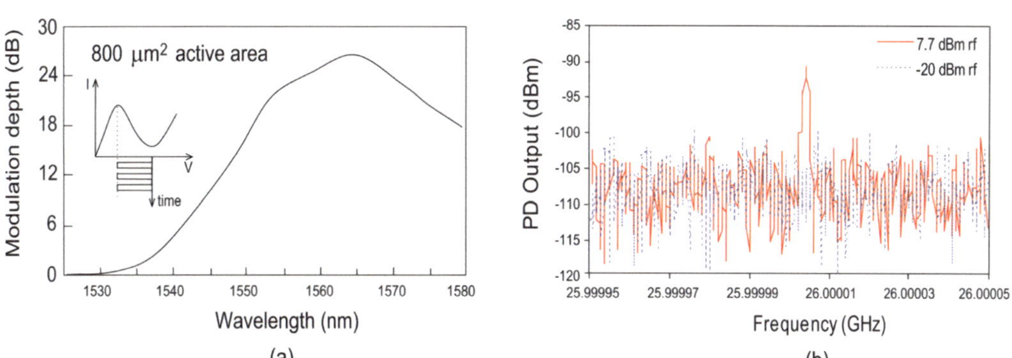

Fig. 16. (a) Modulation depth as function of the wavelength. (b) Spectrum of the 26 GHz photo-detected signal at the modulator driving power of -20 dBm and +7.7 dBm.

The photo-detected power increases more than 10 dB when the driving rf power rises from -20 dBm to +7.7 dBm, an indication the device is capable to achieve modulation extinction ratios higher than 10 dB induced by low power driving signals, less than 10mW, as the consequence of the built-in electrical amplifier. The RTD intrinsic amplifier effect reduces substantially the rf power required for modulation. This on-chip amplification can eliminate the need of an external rf amplifier which is usually required to drive EAMs (Wakita et al., 1998).

3.4 RTD-OW operation as photo-detector at 1550 nm

Light-wave receivers contain photo-detecting devices that convert the light-wave carrier modulation into an electrical signal that needs to be amplified before processing to recover the information signal (Liu, 1996)(Einarsson, 1996). The amplifying circuitry can be the system main penalty in terms of cost and power. We are currently investigating a receiver based on the RTD-OW to take advantage of the RTD intrinsic built-in amplifier.

Because in the RTD-OW the light interaction length is much longer than in conventional RTDs, the RTD-OW will produce substantial inter-band absorption, giving rise to a responsivitygain superior to the one obtained with conventional photo-detectors (Moise et al., 1995). The RTD-OW photo-detection characterization employed light from a Tunics tunable laser diode capable to be directly modulated up to 1 GHz and operate in the mode locked regime at 5 GHz. Figure 17(a) presents the rf power capture level when light modulated at 1 GHz was end-fire coupled to the waveguide. The RTD-OW responsivity-gain increases with the transition from peak to valley voltage, V_p and V_v, by more than 15 dB. Figure 17(b) shows the photo-detected rf power as function of wavelength for dc bias on the peak and on the valley. Photo-detection of mode locked light at 5 GHz showed similar performance.

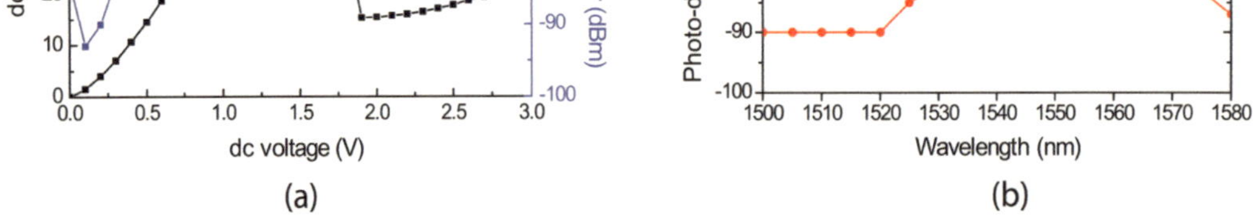

Fig. 17. (a) RTD-OW I - V characteristic and rf power produced due 1550 nm optical signals modulated at 1 GHz. (b) Rf power produced optical signals modulated at 1 GHz as function of wavelength, at DC biased on the peak and on the valley.

When dc biased in the NDC region the RTD-OW self-oscillations lock to the injected light subcarrier, producing electrical signals that emulate the optical subcarrier. We are currently investigating the synchronization between optical subcarriers and RTD-OW free-running oscillations to transfer the information bearing signals such as *Phase Shifted Keyed signals* from the optical to the rf wireless domain without the need of an external amplifier (Romeira[a] et al., 2009).

4. RTD laser diode integration

A light-wave transmitter comprises a driving circuit and a LED or a laser diode which converts the supplied electrical signal containing the information into a light-wave signal. Novel alternatives to traditional laser diode transistor-driver circuits have been proposed based on the integration of a DBQW with semiconductor light sources, since the DBQW layers fit well with the epitaxial layers that make up semiconductor light sources. Furthermore, since the RTD can act as a voltage controlled switch, low voltage digital signals can be employed to switch the RTD between on and off states. It is expected the light sources high-speed modulation characteristics will improve significantly. In what follows we make a brief description of the first monolithic integration of a RTD with an optical communication laser operating at 1500 nm, and give a detailed report on recent advances on the hybrid integrated version operating at 1550 nm optical windows.

4.1 RTD-LD monolithic integration

The first integration of a DBQW-RTD and an optical communication laser operating at around 1500 nm was reported by (Slight & Ironside, 2007). The device consisted of a vertical integration of a DBQW on an InGaAs/InGaAlAs multiple quantum well laser structure. Such integration is straightforward as the RTD section requires only the growth of four to six extra epilayers above a laser structure grown on p–type InP substrate, allowing the RTD to be implemented on the laser junction n–type region. The DBQW was made of a 5 nm InGaAs well and 2 nm AlAs barriers. The devices fabricated were ridge waveguides with the DBQW situated in the ridge between the laser section and the n–type contact, Fig. 18(a). A detailed description of device structure and fabrication can be found in (Slight et al., 2006). The RTD-LD current-voltage characteristic emulates the RTD non-linear I - V curve, hysteresis and bistability (Slight & Ironside, 2007). Figure 18(b) shows a typical RTD-LD optical-voltage characteristic at 130 K, where a hysteresis window is clearly seen; bistable operation was also observed (Slight et al., 2006). The results demonstrate the feasibility of monolithically integrated RTDs with LDs. In order to achieve room temperature operation a new wafer was designed and device fabrication will start soon. Further investigation of the monolithic RTD-LD will include high-frequency operation characterization.

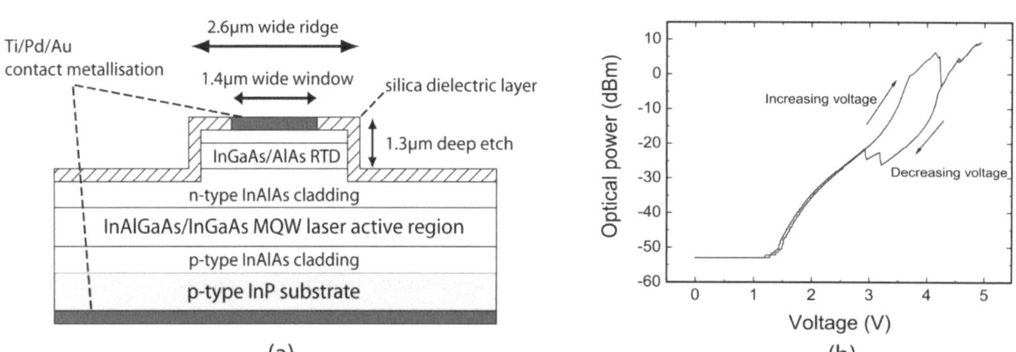

Fig. 18. (a) Cross section schematic of the ridge waveguide RTD-LD. (b) optical-voltage (P –V) characteristic at 130 K, clearly showing bistability and hysteresis.

4.2 RTD-LD hybrid circuit

Once demonstrated the bistable operation of monolithically integrated RTD-LDs the work concentrated on the hybrid integrated circuit (HIC) versions using components similar to the targeted monolithic integrated device. Although without the monolithic expected superior performance, laboratory hybrid RTD-LDs are easy and much less costly to implement, allowing to study both components behaviour separately. The first HICs combined an InGaAs RTD and a commercial prototype laser diode (Slight & Ironside, 2007). The In- GaAs RTD used was fabricated from RTD epi-material originally used in the work described in section 3; the laser diode was a 5 μm ridge wide waveguide designed for continuous-wave (CW) emission at around 980 nm. The RTD and LD were attached to a small copper block using electrically conductive silver epoxy resin, and connected in series through 25 μm diameter gold wire bonding, as schematically represented in Fig. 19(a). Also shown are LD and RTD-LD experimental and PSPICE simulated $I-V$ characteristics, Fig. 19(b) (Slight & Ironside, 2007). The PSPICE code used can be found in (Slight & Ironside, 2007).

The RTD reduces significantly the laser driving circuits' complexity by taking advantage of its high nonlinear $I-V$ characteristic, with the NDC region providing electrical gain to the circuit. The RTD features make possible to operate the RTD-LD as an autonomous OVCO, where the running frequency is fine tuned by the dc bias voltage. Light modulation due to relaxation oscillations at 5 MHz was observed with optical power on/off or extinction ratio up to 31 dB. Moreover, because of RTD bistability the RTD-LD optical output is also bistable, as shown in Fig. 19(c), a feature of particularly convenience for non-return to zero (NZR) digital modulation.

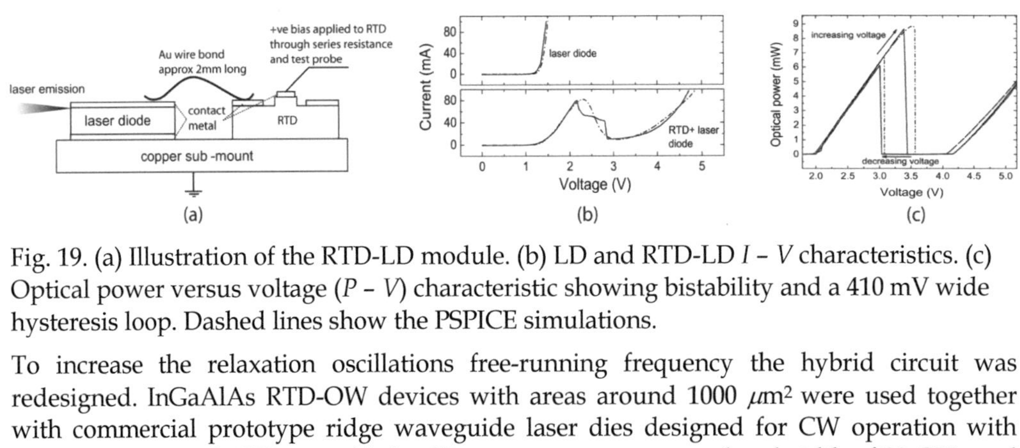

Fig. 19. (a) Illustration of the RTD-LD module. (b) LD and RTD-LD $I-V$ characteristics. (c) Optical power versus voltage ($P-V$) characteristic showing bistability and a 410 mV wide hysteresis loop. Dashed lines show the PSPICE simulations.

To increase the relaxation oscillations free-running frequency the hybrid circuit was redesigned. InGaAlAs RTD-OW devices with areas around 1000 μm^2 were used together with commercial prototype ridge waveguide laser dies designed for CW operation with emission at around 1550 nm with 5 mW average output power, bandwidth of 20 GHz and threshold current I_{th} around 6 mA. The new circuits layouts were mounted directly onto the surface of printed circuit boards (PCBs) containing a 50 Ω copper microstrip transmission line laminated onto the non-conductive PCB substrate. These new improvements on the hybrid RTD-LD circuits lead to some significant breakthroughs: (i) the use of commercial communications laser diodes operating at 1550 nm; (ii) the oscillation frequency went up to for more than two orders of magnitude by solving the instabilities associated to the dc bias circuitry; (iii) demonstration of operation as an autonomous relaxation oscillator in the GHz-range, controlled by voltage; (iv) observation of new operation capabilities induced by injected periodic and phase modulated signals.

In the improved circuits the RTD and LD components were attached directly onto the PCBs using silver epoxy resin and bond wires where used to connect the RTD emitter contact to LD, and the RTD collector contact to the 50 Ω copper microstrip line, as shown in Fig. 20(a). A parallel resistor-capacitor shunt was incorporated as close as possible to the RTD-LD components to reduce the spurious oscillations and to act as a short circuit for the rf signals generated by the RTD-LD. The circuit shunt component values were typically 5 Ω and 3.3 nF. The dc bias and rf injected signals were applied via a wideband bias-T through the resistor-capacitor shunt that also acts as the circuit input port. The circuit electrical output port was defined by the PCB ground plane and the microstrip line, and corresponds to the RTD-LD series terminals as shown in Fig. 20(a). The laser optical output was coupled to a lensed fibre before photo-detection. The light coupling efficiency was estimated from the laser mode profile and single mode fiber characteristics to be around 10 per cent. In Fig. 20(b) are presented the typical *I-V* characteristics of the LD (with the threshold current inset) and of two RTD-LD circuits, I and II, measured without the shunt resistorcapacitor. RTD-LD circuits I and II analysed here have similar PCB layout designs and LD and shunt components. The RTDs used in circuit I and II have approximately the same current peaks, I_p, but different valley currents, I_v, and thus different peak-to-valley current ratios. RTD-LD II was designed to have a lower bond wire length connection between RTD and LD components, which increased its oscillation frequency operation, as discussed below. In both cases $I_{th} < I_v$, which meant that when dc biased in the NDC region, the lasers were working well above threshold current.

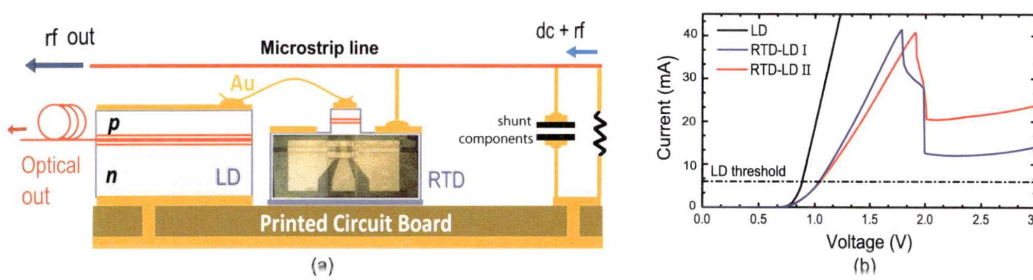

Fig. 20. (a) Layout of the improved hybrid RTD-LD circuit. (b) Current-voltage characteristic of the laser diode and two RTD-LD circuits, showing the RTD NDC is preserved by the RTDLD module.

The RTD-LD circuit of Fig. 20(a) can be represented by circuit electrical layout of Fig. 21(a). When dc biased in or close to the NDC region the laser diode is operating well above the threshold current the laser is well represented simple by its differential resistance. Because its capacitance is much larger than the RTD's, the RTD-LD module equivalent capacitance corresponded to the RTD intrinsic capacitance. This approximation seems reasonable since changing the laser diode did not alter the circuit free-running frequency whenever the lengths of the bond wires used to connect the RTD to the LD were identical. Indeed, the circuit of Fig. 21(a) behaves at rf frequencies like an *RL* circuit connected to the RTD small signal equivalent circuit (a voltage dependent current source $F(V)$ in parallel with the RTD-LD capacitance, as discussed in section 2.2). Its electric behaviour under external perturbation can be studied numerically using the small signal equivalent circuit shown in Fig. 21(b). The lumped *LCR* components of Fig. 21(b) represents the microstrip transmission line and wire bond equivalent inductance, the RTD intrinsic capacitance and the devices equivalent series resistance, respectively.

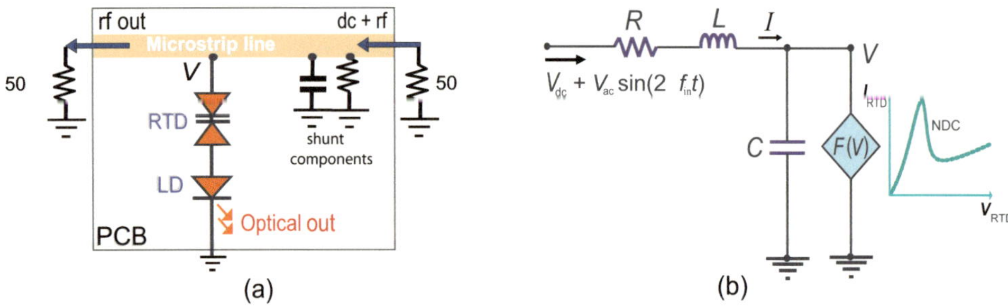

Fig. 21. (a) Electrical schematic of the RTD-LD circuit where V represents the electrical output taken across the RTD-LD. (b) RTD-LD small-signal equivalent lumped circuit. V_{ac} $\sin(2\pi f_{in}t)$ represents an ac injected driving signal.

The maximum operating free-running frequency of circuit RTD-LD I was around 640 MHz, whereas for RTD-LD II the maximum observed free-running frequency was 2.15 GHz (the maximum obtained with the hybrid circuits presented here). The RTD-LD II higher running frequency was mainly due to the smaller inductance achieved with this circuit layout due to the shortening of bond wires length used to connect the RTD to the LD, roughly from 5 mm to less than 2 mm that corresponded to a reduction of the equivalent inductance value from approximately 8 nH to around 1.5 nH. In both circuits the estimated capacitance C was 3 pF. These values when used in the electrical circuit model, Eq. 8, lead to theoretical maximum relaxation oscillation frequencies, given by $1/2\pi\sqrt{LC}$, around 1.03 GHz and 2.37 GHz, respectively.

4.3 RTD-LD optoelectronic model

When dc biased in the NDC region, the circuit of Fig. 20(a) behaves as a classic negative-resistance oscillator (Van der Pol, 1927). Since the circuit of Fig. 21(b) is similar to the circuit of Fig 4, apart from the injected ac driving signal $V_{ac}\sin(2\pi f_{in}t)$, we applied the same procedure, obtaining a second-order differential equation (see section 2.2), commonly referred as one of the generalized forced nonlinear Liénard systems (Romeira et al., 2008)(Figueiredo, 1970):

$$\ddot{V} + H(V)\dot{V} + G(V) = V_{ac}\sin(2\pi f_{in}t) \tag{13}$$

where $G(V)$ is a nonlinear force and $H(V)\dot{V}$ is the damping factor (see section 2.2).
To describe the RTD-LD optoelectronic behaviour we coupled equation 13 to the laser diode single mode rate equations that governs the interrelationship between carrier density and photon density. Assuming the laser oscillates in a single mode and the population inversion is homogeneous, the laser rate equations for photon density S and injected carrier density N are:

$$\dot{N} = \frac{I}{q\vartheta} - \frac{N}{\tau_n} - g_0(N - N_0)\frac{S}{1+\varepsilon S} \tag{14}$$

$$\dot{S} = g_0(N - N_0)\frac{S}{1+\varepsilon S} + \frac{S}{\tau_p} + \beta\frac{N}{\tau_n} \tag{15}$$

where I is the total current through the laser diode given by generalized Liénard's system, Eq. 13, plus the dc bias current; q is the electron charge, ϑ is the laser active region volume, τ_n and τ_p are the spontaneous electron and photon lifetimes, respectively; β is the spontaneous emission factor; g_0 is the gain coefficient; N_0 is the minimum electron density required to obtain a positive gain and ε is the value for the nonlinear gain compression factor. The numerical analysis employed typical parameters of semiconductor laser diodes, as described in (Slight et al., 2008)(Romeira et al., 2008). The coupled system of equations 13-15 has been successfully used to predict the experimental behaviour of RTD-LD electrical and optical outputs.

4.4 RTD-LD optoelectronic voltage controlled oscillator

It is well known that a single-port device that has a negative differential conductance in a portion of its operating range may be used as the basis of a bistable or multistable circuit, and can also be used to form astable circuits (relaxation oscillators), monostable circuits (single-pulse generators), and sine-wave generators (Brown et al., 1997). A simple way to implement a RTD oscillator is to couple a RTD dc biased in the NDC to a resonant tank circuit or a resonant cavity that provides frequency stability (the coupling location in the cavity can serve to partially match its impedance to that of the RTD). Such oscillator corresponds to a relaxation oscillator system since it operates by sequential transitions between unstable states. The RTD-LD circuit of Fig. 20(a), whose circuit schematic is represented in Fig. 21 with the small signal equivalent circuit, operates as a relaxation oscillator when dc biased in the NDC region. The circuit free-running frequency is determined primarily by the round trip time of the ac feedback loop (effective length of equivalent transmission line from the shunt resistor-capacitor to the RTDLD module), in combination with the RTD and the LD parasitics (mainly the inductance from the wire bonding).

The RTD successive switching events (relaxation oscillations) produce sharp current pulses that modulate the laser output yielding sharp optical pulses at the relaxation oscillation fundamental frequency (free-running frequency). Typical RTD-LD self-sustained oscillation voltage output and photodetect optical waveforms are shown in Fig. 22. Figure 22(a) shows RTD-LD I voltage output waveform at free-running frequency around 600 MHz; Fig. 22(b) presents the photo-detected laser optical output modulated by the current relaxation oscillations with an on/off superior to 20 dB.

The pulsed nature of the photo-detected laser optical output shown in Fig. 22 confirms the capacitive character of the current induced by the RTD switching (described in detail in (Brown et al., 1997)). The full width at half maximum (FWHM) of the photo-detected pulses is approximately 200 ps but this measurement is limited by the temporal acquisition resolution of the oscilloscope. Figure 23 shows rf spectra of the electrical and optical outputs of RTD-LD circuits I and II of Fig. 20(b), both dc biased close to the valley region. Figure 23(a) confirms the pulse nature of the current relaxation oscillations with a high harmonic content up to 12th harmonic being measured.

Tuning the dc bias across the NDC region changes the RTD impedance and as consequence tunes the relaxation oscillation frequency making the circuit operate as a voltage controlled oscillator (VCO). Since the current relaxation oscillation waveforms flow through the laser diode, the circuit optical output emulates the current oscillations. The laser output shows the same repetitive switching and harmonic content of the relaxation oscillation current waveforms, making the RTD-LD circuit operate as an optoelectronic voltage controlled oscillator (OVCO). That is, the RTD-LD biased on the NDC region produces electrical and

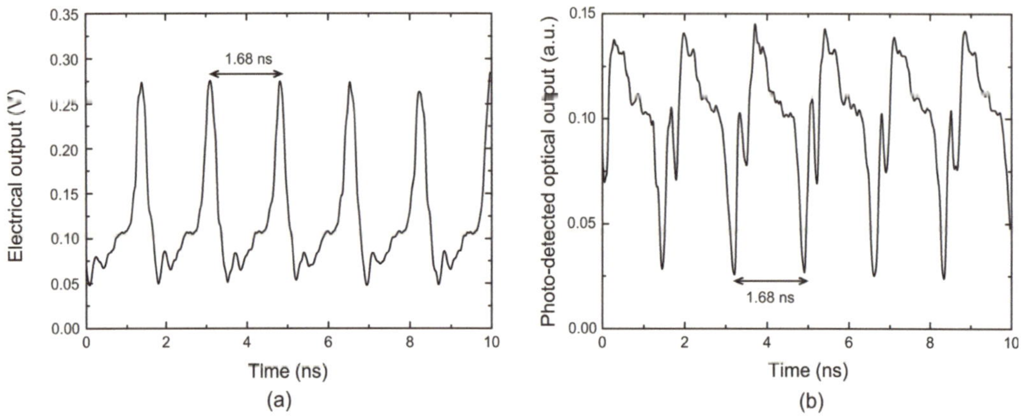

Fig. 22. RTD-LD I relaxation oscillation (a) electrical and (b) photo-detected optical output waveforms at around 600 MHz.

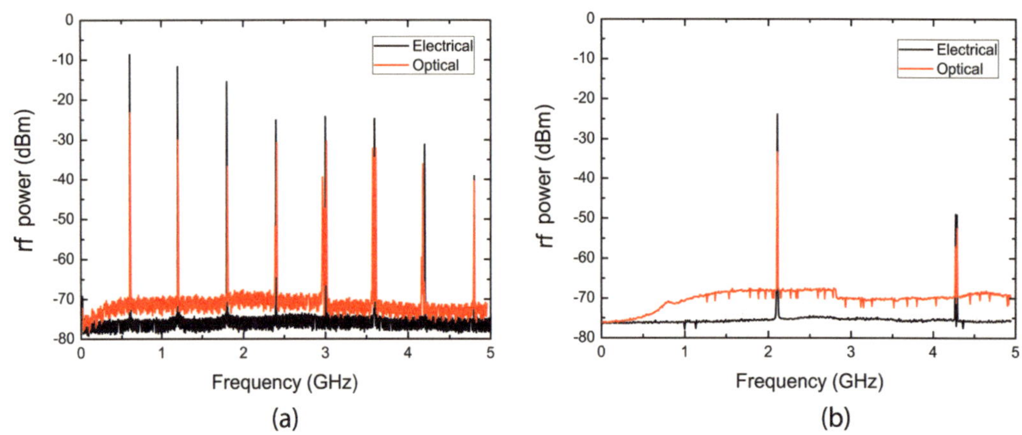

Fig. 23. Electrical and photo-detected optical spectra of free-running oscillations at 600 MHz (a) and 2.1 GHz (b), circuits I and II, respectively.

optical oscillatory signals whose frequency is controlled by the bias voltage quiescent point. Figure 24 shows the frequency response to dc voltage sweep across the NDC region of circuits RTD-LD I and II, whose $I-V$ characteristics are presented in Fig. 20(b).

The oscillation frequency of circuit I changed with the dc voltage from around 500 MHz to 640 MHz, that is, RTD-LD I had a tuning range around 140 MHz, whereas the circuit II oscillate from 1.97 GHz to 2.15 GHz, i.e., RTD-LD II had a tuning range around 180 MHz. Although the dc voltage tuning of circuit I was larger, the tuning sensitivity/tuning performance expressed in tuning range per voltage range was higher for circuit II. In the RTD-LD oscillators analyzed, we found that a linear deviation characteristic is attained considering only voltages close to the peak voltage. The voltage tuning range of circuit I, Fig 24(a), is much larger than the circuit II, Fig. 24(b), as expected from higher PVVR measured in the $I-V$ characteristic. Frequency tuning ranges up to 450 MHz were observed in RTD-LD circuits having NDC widths and $I - V$ characteristics identical to RTD-LD I. Generally speaking, to have a wide dc operating range and therefore large tunability, a wide negative conductance region (large difference between the peak and valley voltage) is required.

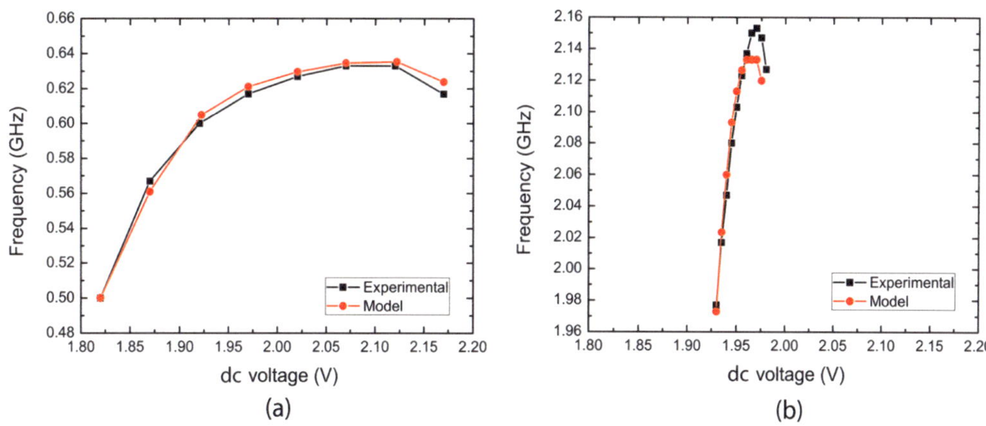

Fig. 24. RTD-LD I (a) and RTD-LD II (b) experimental and simulated frequency tuning responses to voltage sweeping across the NDC regions.

The RTD-LD *optoelectronic* voltage controlled oscillator is a simple way to convert fast, short electrical pulses with low timing jitter and phase noise, into fast, sharp optical pulses.

4.5 Phase-locking

The injection-locking of an electrical oscillator was first described by (Van Der Pol, 1927), and the first locking bandwidth equation for electrically injection-locked oscillators was developed by (Adler, 1946), with a model based on a vacuum tube transistor. The most comprehensive theoretical review of injection-locking solid-state oscillators was given by (Kurokawa, 1973). Most of the characteristic and properties identified by the above authors can be observed with RTD-LD circuits which are much simpler oscillator configuration. When externally perturbed the RTD-LD circuit behaves as a non-autonomous oscillator (Romeira[b] et al. 2009), being a practical demonstration of nonlinear systems theory extensively developed over the last decades (Pikovsky et al., 2001).

Throughout the work, we observed that under appropriated bias and injection conditions the RTD-LD circuit relaxation oscillations lock to low-power injected signals that take over the oscillations, controlling the laser diode output characteristics. To investigate these locking characteristics periodic external signals at microwave frequencies were injected into the circuit. The analysis included the effects of the frequency, signal power level, and injected signal modulation formats. Phase-locking with significant noise reduction to low power signals (below -30 dBm) at frequencies around the circuits' natural frequencies are observed. Figure 25(a) presents rf spectra of photo-detected laser optical outputs when the circuit was free-runing at 600 MHz and when phase-locked to -25 dBm power rf signal also at 600 MHz. The single side band (SSB) phase noise measurement showed the oscillation noise at 10 kHz offset was reduced by about 35 dB due to the phase-locking. For the conditions of Fig. 25(a) the locking range was 1.8 MHz. The frequency locking range increases as the injected power rises, as shown in Fig. 25(b). This behavior is well described by the optoelectronic model presented previously and is represented by the red zone of Fig. 25(b), known as Arnold tongue. Arnold tongues correspond to synchronization regions were locking occurs between two competing frequencies (Pikovsky et al., 2001). When the injected signal frequency becomes out of the oscillator locking range, the circuit generate mixing products of the injected signal and free-running oscillations.

Since the phase of a signal plays an important role in communications, particularly wireless communication, and in the theory of synchronisation, we investigated the effect of phase modulation in the RTD-LD outputs. Figure 25(c) shows circuit response to an injected 600 MHz carrier phase modulated with 1 MHz frequency sub-carrier with phase shift π and $3\pi/2$. As the sub-carrier frequency was varied from 100 kHz up to 2 MHz, the laser output followed the phase modulation of the sine-wave signal subcarrier.

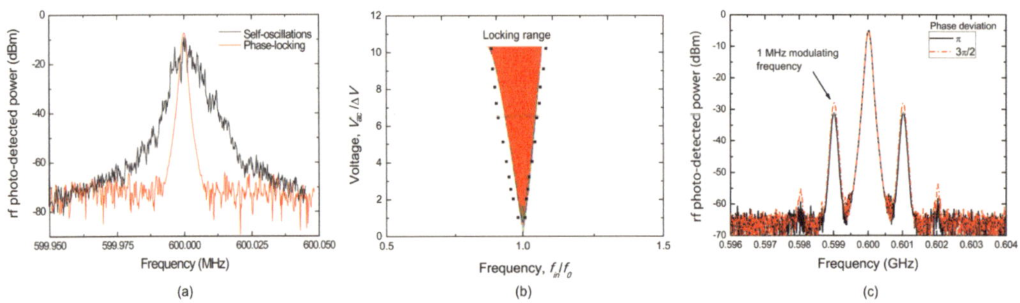

Fig. 25. (a) Rf spectra of photo-detected laser output in free-running mode and when phase-locked to -25 dBm injected signal at 600 MHz frequency. (b) Frequency locking range as function of the injected power. The dotted points are experimental data and the red area (Arnold tongue) was numerically obtained. (c) Rf spectra of photo-detected laser output when phase-locking to a phase modulated 600 MHz sine-wave carrier signal.

The observed phase-locking converts phase differences on shifts in the laser output modulating its intensity. This behaviour can be applied to implement phase shift keying (PSK) digital modulation, which is employed in numerous digital communication systems. The phase-locking capabilities of RTD-LD based relaxation oscillators can also be used for error free timing extraction in optoelectronic circuits.

4.6 Frequency division operation

When the injected signal frequency is out of the oscillator locking range the circuit generates mixing products of the injected signal and free-running oscillations, producing either/ both harmonic and sub-harmonic phase-locking. To investigate the mixing capability of the circuit we analysed numerically the behaviour of the circuit over a range of frequencies to obtain the laser optical output bifurcation diagram of Fig. 26. A bifurcation diagram shows the amplitude peaks heights of output photon density oscillations, S, as a function of the normalized excitation frequency f_{in}/f_0, where f_0 is the free running oscillation frequency. The simulation results show that when the frequency of the injected signal, f_{in}, is successively increased, a stable period-n, n = 1, 2,... is obtained, followed by an unlocked region, then a stable period-(n +1), a new unlocked region and so on (Figueiredo et al., 2008)(Pikovsky et al., 2001). This phenomenon is known as period-adding, where windows of consecutive regions showing frequency division are separated by zones of unlocked, even chaotic, signals. The frequency division regions were obtained experimentally and calculated numerically dc biasing the RTD-LD circuit on the NDC region and varying the frequency of the injected signal from 0.1 GHz to 3 GHz, with drive amplitudes as low as 100 mV. Frequency division regions for constant amplitudes were observed following the period-adding sequence at up to frequency division by 6. In Fig. 26(a) the period-adding is clearly distinguished in a sequence of unlocked (dots) and periodic (branch regions) oscillations, as observed experimentally. Figure 26(b) presents an experimental example of frequency division by 2 when a 0.9 GHz sine-wave was injected.

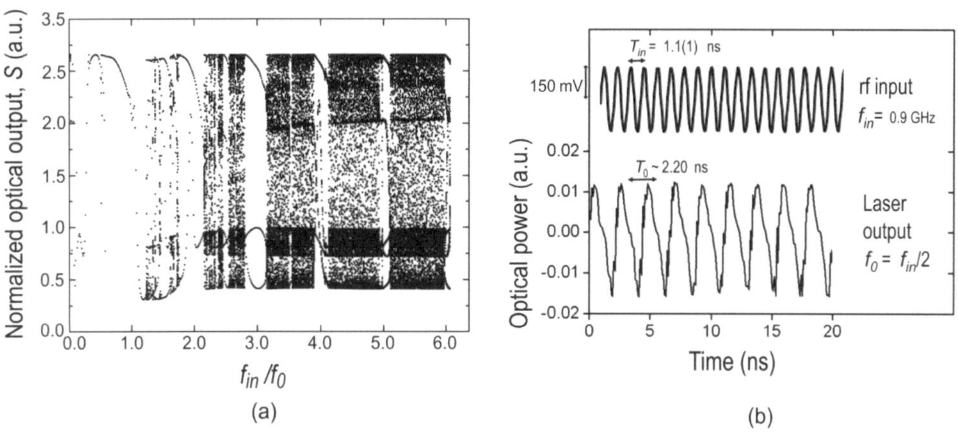

Fig. 26. (a) Calculated bifurcation diagram for V_{ac} = 150 mV up to frequency division by 6. (b) Photo-detected laser output showing frequency division by 2 when a signal with f_{in} = 0.9 GHz was injected into an RTD-LD free-running oscillating at around 0.5 GHz.

Since the sub-harmonic windows appear in limited frequency regions, the RTD-LD circuit can be regarded as an optoelectronic dynamic frequency divider with a selectable dividing ratio.

4.7 Aperiodic and chaotic operation

Electro-optical and all-optical solutions for complex chaos generation have attracted considerable attention in the last decade due to their potential applications in optical chaos communications (Argyris et al., 2005). The use of chaotic carriers allows steganography at the physical layer, which can substantially improve the security of software encryption techniques. The frequency bands corresponding to period multiplication, indicated in Fig. 26(a), are separated by frequency regions where the circuit generates aperiodic signals - chaotic or quasi-periodic output - a direct result from the mixing between free-running oscillation and external injected frequencies (Romeira et al. 2010). An important characteristic of a chaotic signal is its sensitivity to initial conditions. Figure 27 shows an example of a transition to chaos observed in the RTD-LD circuit optical output. The optical waveform presented in Fig. 27(a) is characterized by a series of aperiodic acute peaks (spikes) changing chaotically. Another important characteristic of chaos is demonstrated in the corresponding power spectrum of the time series. Figure 27(b) shows a continuous and broadband spectrum resembling a noisy process with a few dominant frequencies appearing, in this case the rf injected frequency. The results of Fig. 27 are also confirmed numerically by calculating the circuit Lyapunov exponents (Romeira et al., 2001).

This RTD-LD mode of operation provides a simply way to generate and convert electrical chaotic signals into optical sub-carriers that can be transmitted by conventional optical channels. Moreover, the circuit allows direct addition of the message to be transmitted and masked within the chaotic signal.

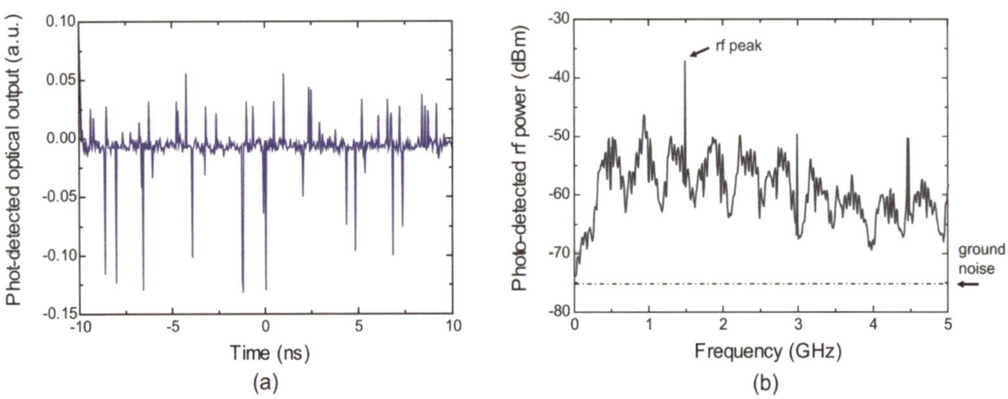

Fig. 27. Chaotic behaviour in the laser output induced by a driving signal of frequency 1.485 GHz and amplitude 793 mV. Optical waverform (a) and corresponding Fourier spectrum (b).

5. Conclusion

As discussed, embedding DBQW-RTDs within semiconductor optical waveguides can lead to the implementation of highly efficient electro-absorption modulators and photo-detectors operating at optical wavelengths around 900 nm and 1550 nm. The presence of the DBQW introduces high non-linearities and NDC regions in the semiconductor optical waveguides current-voltage characteristics, making the electric field distribution across the waveguide core strongly dependent on the bias voltage, which can be used to modulate guided light through the Franz-Keldysh electro-absorption effect. When biased on the NDC region the RTD-OW operates as an optoelectronic voltage controlled oscillator. Electro-absorption modulation up to 28 dB is achieved with high frequency signals as low as 100 mV. The key difference between these RTD-OW electro-absorption modulators and conventional $p-i-n$ electro-absorption modulators is that the RTD-EAM has in essence an integrated electronic amplifier and therefore requires considerably less switching/driving power. Since, the RTD-OWs can also work as photo-detectors with built-in amplifiers, recovering the original transmitted rf signals used to modulated the optical carriers, they can be employed at the base station to convert information from the optical to the rf domains. We foresee that optimized devices can have bandwidths up to 60 GHz.

By integrating a DBQW-RTD with a laser diode low-cost microwave-photonic circuits operating up to 2.15 GHz were implemented. These circuits reduced significantly the driving circuitry of laser diodes. Several optoelectronic operation modes were observed, including optoelectronic voltage controlled oscillator (OVCO), phase-locking, frequency division and generation of aperiodic electrical and optical waveforms. Their simple circuit layout is appropriated for high functional single chip transmitter platforms due to their non-linear optoelectronic characteristics, reduced size and low power consumption. We anticipate that the optimised RTD-LD monolithic integrated versions can operate at much higher frequencies (tens of Gbits), having several advantages when compared to conventional devices currently used in lightwave communication systems.

The RTD-OW and RTD-LD operation as optoelectronic voltage controlled oscillators can be used to simplify significantly clock generation and clock extraction circuits. Due to the nonlinear response to applied voltage the RTD based circuits can work as short optical pulse generators with high repetition rates. At the same time, their integration with other functional devices can be used to encode generated optical pulses. The combination of RTD-OW and RTD-LD functions on a single circuit can be used to incorporate simultaneously rf subcarrier signals into optical carriers and optical subcarrier signals into rf carriers. This is possible due the following simultaneously capabilities: modulation, photo-detection and intrinsic amplification. Thus, the RTD-OW and RTD-LD circuits offer the possibility of implementing very simple microwave/photonics interfaces of cellular network terminal base stations based on radio-over-fiber systems.

Since next generation wireless access picocellular networks will be based on large numbers of short range cells with each office in a building with its own cells and base stations, the RTD based optoelectronic devices offer low cost single chip solutions as microwave/optical interfaces capable of electrical-to-optical conversion of microwave signals into optical subcarriers, taking advantage of the NDC and phase-locking properties of RTD devices. The photo-detecting capabilities allows recovery of the original transmitted rf signals used to modulated the optical carriers sent from the office terminal station to each base station via optical fibre, converting the information from the optical to the rf domain; light generation function is used to transfer the wireless received information bearing signals from the rf domain to optical domain which is then sent from the base stations to the office terminal station via optical fibre.

6. Acknowledgment

Bruno Romeira and José Figueiredo acknowledge the support of the Centro de Electrónica, Optoelectrónica e Telecomunicações, Portugal. This work was also supported in part by the Fundação para a Ciência e a Tecnologia, Portugal, through the grants PRAXIS XXI/BD/2871/94 and SFRH/BD/43433/2008, by the Fundação Calouste Gulbenkian, Portugal, and by Research Networks - Treaty of Windsor Programme 2008/09-U32, Portugal. The authors would like to thank W. Meredith of Compound Semiconductor Technologies Global, Ltd. for providing the laser diodes, and Liquan Wang and Edward Wasige by the fruitful discussions and PCB layout design in the RTD-LD work.

7. References

Adler, R. (1946). A study of locking phenomena in oscillators. *Proc. I.R.E.Waves Electrons*, 351- 357

Alkeev, N. V.; Lyubchenko, V. E.; Ironside, C. N.; Figueiredo, J. M. L. & Stanley C. R. (2000) Super high-frequency characteristics of optical modulators on the basis of

InGaAlAs resonance-tunnel heterostructures. *J. Commun. Technol. Electron.*, Vol. 45, No. 8, Aug. 2000, 911–914

Argyris, A.; Syvridis, D.; Larger, L.; Annovazzi-Lodi, V.; Colet, P.; Fisher, I.; Garcia-Ojalvo, J.; Mirasso, C.; Pesquera, L.; Shore, K. A. (2005). Chaos-based communications at high bit rates using commercial fibre-optic links. *Nature*, Vol. 43, Nov. 2005, 343–346

Bohm, D. (1951). *Quantum theory*, Prentice-Hall Inc., New Jersey

Böhrer, J.; Krost, A. & Bimberg, D. B. (1993). Composition dependence of band gap and type of lineup in InGaAlAs/InP heterostructures. *Appl. Phys. Lett.*, Vol. 63, No. 14, Oct. 1993, 1918

Brown, E. R.; McMahon, O. B.; Mahoney, L. J. & Molvar, K. M. (1996). SPICE model of the resonant-tunneling diode. *Electronics Lett.*, Vol. 32, No. 10, May 1996, 938–940

Brown, E. R. & Parker C. D. (1996). Resonant tunnel diodes as submillimetre-wave sources. *Phil. Trans. R. Soc. Lon. A*, Vol. 354 , No. 1717, Oct. 1996, 2365–2381

Brown, E. R.; Parker, C. D.; Verghese, S. & Geis, M. W. (1997). Resonant-tunneling transmission-line relaxation oscillator. *Appl. Phys. Lett.*, Vol. 70, No. 21, May 1997, 2787

Capasso, F.; Mohammed, K. & Cho A. Y. (1986). Resonant tunneling through double barriers, perpendicular quantum transport phenomena in superlattices, and their device applications. *IEEE J. Quantum Electron.*, Vol. 22, No. 9, Sep. 1986, 1853–1869

Chang, L. L.; Esaki, L. & Tsu, R. (1974). Resonant tunneling in semiconductor double barriers. *Appl. Phys. Lett.*, Vol, 24, No. 12, Jun. 1974, 593

Chen, L.; Kapre, R. M.; Hu, K. & Madhukar, A. (1991). High-contrast optically bistable optoelectronic switch based on InGaAs/GaAs (100) asymmetric Fabry-Perot modulator, detector, and resonant tunneling diode. *Appl. Phys. Lett.*, Vol. 59, No. 13, Sep. 1991, 1523–1525

Chuang, S. L. (1995). *Physics of optoelectronic devices* JohnWiley & Sons Inc., New York

Davies, J. H. (1998). *The physics of low-dimensional semiconductors: an introduction*, Cambridge University Press, Cambridge.

Einarsson, G. (1996). *Principles of lightwave communications*, John Wiley & Sons, New York

England, P.; Golub, J. E.; Florez, L. T. & Harbison, J. P. (1991). Optical switching in a resonant tunneling structure. *Appl. Phys. Lett.*, Vol.58, No. 9, Mar. 1991, 887–889

Figueiredo, R. (1970). Existence and Uniqueness results for Liénards equation. *IEEE Transactions on Circuit Theory*, Vol. CT-17, No. 3, Jun. 1970, 313–321

Figueiredo, J. M. L.; Boyd, A. R.; Stanley, C. R.; Ironside, C. N.; McMeekin, S. G. & Leite, A. M. P. (1999). Optical modulation at around 1550 nm in an InGaAlAs optical waveguide containing an InGaAs/AlAs resonant tunneling diode. *Appl. Phys.Letts*, Vol. 75, No. 22, Nov. 1999, 3443–3445

Figueiredo, J. M. L. (2000). *Optoelectronic Properties of Resonant Tunneling Diodes*, PhD Thesis, Universidade do Porto, Porto

Figueiredo, J.M.L.; Stanley C.R. & Ironside C.N. (2001). Electric Field Switching in a Resonant Tunneling Diode Electroabsorption Modulator. *IEEE Journal of Quantum Electronics*, Vol. 37, No. 12, Dec. 2001, 1547-1552

Figueiredo, J. M. L.; Ironside C. N. & Stanley. C. R.; (2002). Ultralow voltage resonant tunneling diode electroabsorption modulator. *Journal of Modern Optics*, Vol. 49, No. 5, Apr. 2002, 939-945

Figueiredo, J. M. L.; Romeira, B.; Slight, T. J.;Wang, L.;Wasige E. & Ironside C. N. (2008). Selfoscillation and period adding from a resonant tunnelling diode laser diode circuit. *Electronics Letters*, Vol. 44, No. 14, Jul. 2008, 876-878

Gravé, I.; Kan, S. C.; Griffel, G.; Wu, S. W.; Sa'ar, A. & Yariv, A. (1991). Monolithic integration of a resonant tunneling diode and a quantum well semiconductor laser. *Appl. Phys. Lett.* Vol. 58, No. 2, Jan. 1991, 110-112

Huang, C. Y.; Morris, J. E. & Su, Y. K. (1997). Generalized formula for the stability and instability criteria of current-voltage characteristics measurements in the negative differential conductance region of a resonant tunneling diode. *J. Appl. Phys.*, Vol. 82, No. 5, Sep 1997, 2690

Iogansen, L. V. (1964). The possibility of resonance transmission of electrons in crystals through a system of barriers. *Soviet Physics JETP*, Vol. 18, 146

Katz, A. (1992). Indium phosphide and related materials: processing, technology, and devices, Artech House, London

Kawamura, Y.; Wakita, K.; Asai, H. & Oe, K. (1987). Optically bistable operation in InGaAs-InAlAs MQW laser diodes using resonant tunnelling effect. *Electron. Lett.*, Vol. 23, No. 14, 719-721

Kawamura, Y.; Wakita, K. & Mikami, O. (1988). InGaAs/InAlAs bistable multiple quantum well lasers with large on/off ratio using the resonant tunneling effect. *Appl. Phys. Lett.*, Vol.53, No. 16, Oct. 1986, 1462-1464

Kawamura, Y.; Asai, H.; Matsuo, S. & Amaro, C. (1992). InGaAs-InAlAs multiple quantum well optical bistable devices using the resonant tunneling effect. *IEEE J. Quantum Electron.*, Vol. 28, No. 1, Jan. 1992, 308-314

Kawamura, Y.; Asai, H. & Iwamura, H. (1994). Fabrication of resonant-tunnelling optical bistable laser diodes. *Electron. Lett.*, Vol. 30, No. 3, Feb. 1994. 225-227

Keldysh, L. V. (1958). The effect of a strong electric field on the optical properties of insulating crystals. *Sov. Phys. JETP*, Vol. 34, 788

Kidner, C.; Mehdi, I.; East, J. R. & Haddad, G. I. (1990). Power and stability limitations of resonant tunneling diodes. *IEEE Trans. Microwave Theory and Techniq.*, Vol. 38, No. 7, Jul 1990, 864-872

Kurokawa, K. (1973). Injection locking of microwave solid-state oscillators. *Proc. IEEE*, Vol. 61, No. 10, Oct. 1973, 1386-1410

Lann, A. F.; Grumann, E.; Gabai, A.; Golub, J. E.; & England, P. (1993). Phase locking between light pulses and a resonant tunneling diode oscillator. *Appl. Phys. Lett.*, Vol. 62, No. 1, Jn. 1993, 13-15

Liu, M. M.-K. (1996). Principles and applications of optical communications, Irwin Book, London.

Mazumder, P.; Kulkarni, S.; Bhattacharya, M.; Sun, J. P. & Haddad, G. I. (1998). Digital circuit applications of resonant tunneling devices. *Proceedings of the IEEE,* Vol. 86, No. 4, Apr. 1998, 664–686

McMeekin, S. G.; Taylor, M. R. S.; V¨ogele, B.; Stanley, C. R. & Ironside, C. N. (1994). Franz-Keldysh effect in an optical waveguide containing a resonant tunneling diode, *Appl. Phys. Lett.*, Vol 65, No. 9, Aug. 1994, 1076-1078

Mizuta, H. & Tanoue, T. (1995). *The physics and applications of resonant tunneling diodes,* Cambridge University Press, Cambridge

Moise, T. S.; Kao, Y.-C.; Garrett, L. D. & Campbell, J. C. (1995). Optically switched resonant tunneling diodes. *Appl. Phys. Lett.*, Vol. 66, No. 9, Feb. 1995, 1104–1106

Moise, T. S.; Kao, Y.-C.; Goldsmith, C. L.; Schow, C. L. & Campbell, J. C. (1997). High-speed resonant-tunneling photodetectors with low-switching energy. *IEEE Photonics Technol. Lett.*, Vol. 9, No. 6, Jun. 1997, 803–805

Park, T. J.; Lee, Y. K.; Kwon, S. K.; Kwon,J. H. & Jang, J. (2006). Resonant tunneling diode made of organic semiconductor superlattice. *Appl. Phys. Lett.* Vol. 89, No. 15, Oct. 2006, 151114

Pikovsky, A.; Rosenblum, M. & Kurths J. (2001). *Synchronization: A universal concept in nonlinear sciences*, Cambridge Univ. Press,Cambridge

Romeira, B.; Figueiredo, J. M. L.; Slight, T. J.; Wang, L.; Wasige, E.; Ironside C. N.; Quintana, J. M. & Avedillo, M. J. (2008). Synchronisation and chaos in a laser diode driven by a resonant tunneling diode. *IET Optoelectronics*, Vol. 2, No. 6, Dec. 2008, 211-215

Romeira[a], B.; Figueiredo, J. M. L.; Slight, T. J.; Wang, L.; Wasige, E. & Ironside C. N. (2009) Wireless/Photonics Interfaces Based on Resonant Tunneling Diode Optoelectronic Oscillators, *Conference on Lasers and Electro-Optics (CLEO)/The International Quantum Electronics Conference (IQEC)*, paper CTuT4, Baltimore, Maryland, USA, May 31-June 5, 2009, OSA Technical Digest (CD) Optical Society of America

Romeira[b], B.; Figueiredo, J. M. L.; Slight, T. J.; Wang, L.; Wasige, E.; Ironside, C. N.; Kelly, A. E. & Green, R. (2009). Nonlinear Dynamics of Resonant Tunneling Optoelectronic Circuits for Wireless/Optical Interfaces. *IEEE J. Quantum Electron.*, Vol. 45, No. 11, Nov. 2009, 1436-1445)

Romeira, B.; Figueiredo, J. M. L.; Ironside, C. N. & Slight, T. J. (2010). Chaotic Dynamics in Resonant Tunneling Optoelectronic Voltage Controlled Oscillators. to be published in *IEEE Photonics Technol. Lett.* DOI: 10.1109/LPT.2009.2034129

Ryu, S. Y.; Jo, S. J.; Kim, C. S.; Choi, S. H.; Noh, J. H.; Baik, H. K.; Jeong, H. S.; Han, D.W. Song, S. Y. & Lee, K. S. (2007). Transparent organic light-emitting diodes using resonant tunneling double barrier structures. *Appl. Phys. Lett.* Vol. 91, No. 9, Aug. 2007, 093515

Sakata, H.; Utaka, K. & Matsushima, Y. (1995). Diferent types of optoelectronic bistabilities in resonant-tunnelling triangular-barrier optoelectronic switch (R-TOPS). *Electron. Lett.*, Vol. 31, No. 13,Jun 1995, 1098–1099

Sauer, M.; Kobyakov, A. & George, J. (2007). Radio over fiber for picocellular network architectures. *J. Lightw. Technol.*, Vol. 25, No. 11, Nov. 2007, 3301-3320

Sano, K.; Murata, K.; Akeyoshi, T.; Shimizu, N.; Otsuji, T.; Yamamoto, M.; Ishibashi T. & Sano, E. (1998). Ultra-fast optoelectronic circuit using resonant tunnelling diodes and uni-travelling-carrier photodiode. *Electron. Lett.*, Vol 34, No. 2, Jan. 1998, 215-217

Schulman, J. N.; De Los Santos, H. J. & Chow, D. H. (1996). Physics-based RTD current-voltage equation. *IEEE Electron Device Lett.*, Vol. 17, No. 5. May 1996, 220-223

Slight, T. J.; Ironside, C. N.; Stanley, C. R.; Hopkinson, M. & Farmer, C. D. (2006). Integration of a resonant tunneling diode and an optical communications laser. *IEEE Photonics Technology Letters*, Vol. 18, No. 14, Jul. 2006, 1518-1520

Slight, T. J. (2006). *Integration of a Resonant Tunnelling Diode and an Optical Communications Laser*, PhD Thesis, University of Glasgow, Glasgow

Slight, T. J. & Ironside, C. N. (2007). Investigation into the integration of a resonant tunneling diode and an optical communications laser: model and experiment. *IEEE J. Quant. Elec.* Vol. 43, No. 7, Jul 2007, 580-587

Slight, T. J.; Romeira, B.; Wang, L.; Figueiredo, J. M. L.; Wasige E., & Ironside, C. N. (2008). A Liénard oscillator resonant tunnelling diode-laser diode hybrid integrated circuit: model an experiment. *IEEE J. Quantum Electron.*, vol. 44, no. 12, Dec. 2008, 1158-1163

Sollner, T. C. L. G.; Goodhue, W. D.; Tannenwald, P. E.; Parker, C. D. & Peck, D. D. (1983). Resonant tunneling through quantum wells at frequencies up to 2.5 THz. *Appl. Phys. Lett.*, Vol. 43, No. 3, Sep. 1983, 588-590

Sun, J. P.; Haddad, G. I.; Mazumder, P. & Schulman, J. N. (1998). Resonant tunneling diodes: models and properties. *Proceedings of the IEEE*, Vol. 86, No. 4, Apr. 1998, 641-660

Suzuki, S.; Teranishi, A.; Hinata, K.; Asada, M.; Sugiyama H. & Yokoyama H. (2009). Fundamental Oscillation of up to 831 GHz in GaInAs/AlAs Resonant Tunneling Diode. *Appl. Phys. Express*, Vol. 2, Apr. 2009, 054501

Tsu, R. & Esaki, L. (1973). Tunneling in a finite superlattice. *Appl. Phys. Lett.*, Vol. 22, No. 11, Jun. 1973, 562-564

Wada, O. (1994). *Optoelectronic integration: physics, technology and applications*, Kluwer Academic Publishers, Massachusetts

Wakita, K. (1998). *Semiconductor optical modulators*, Kluwer Academic Publishers, London

Weisbuch, C. & Vinter, B. (1991). *Quantum semiconductor structures: fundamentals and applications*, Academic Press Inc., London

Van Der Pol, (1927). Forced Oscillator in a circuit with nonlinear resistance. *Phil. Mag.* Vol. 3, 65-80

Van Hoof, C.; Genoe, J.; Mertens, R.; Borghs, G. & Goovaerts, E. (1992). Electroluminescence from bipolar resonant tunneling diodes. *Appl. Phys. Lett.*, Vol. 60, No. 1, Jan 1992, 77-79

Van Hoof, C.; Genoe, J.; Raymond, S. & Borghs, G. (1993). Giant optical bistable behavior using triple-barrier resonant tunnelling light-emitting diodes. *Appl. Phys. Lett.*, Vol. 63, No. 17, Oct. 1993, 2390-2392

Zheng, T.; Choy,W. C. H. & Sun, Y. (2009). Hybrid Nanoparticle/Organic Devices with Strong Resonant Tunneling Behaviors. *Advanced Functional Materials*, Vol. 19, No. 16, Jul. 2009, 2648-2653

11

Integrated-Optic Circuits for Recognition of Photonic Routing Labels

Nobuo Goto, Hitoshi Hiura, Yoshihiro Makimoto and Shin-ichiro Yanagiya
The University of Tokushima
Japan

1. Introduction

An optical fiber provides enormous capacity of more than tens terabits per second for transmission in photonic networks, whereas packet processing in network nodes will become a bottleneck for large-capacity networking. For realization of large-capacity and high-speed photonic networks, fast optical processing without conversion to electric signal is preferable (Seo et al., 1996; Blumenthal et al., 2000).

Photonic routing has been attracting much interest to overcome the bottleneck of routing function in high-speed networks. In particular, photonic label routing network is expected to provide fast routing of packets at high-bit rate with simple processing. So far, various methods for optical label encoding and decoding have been studied (Kitayama et al., 2000; Goto & Miyazaki, 2005). As one of the nature of light, phase of coherent light has been effectively used in various optical systems, where the interference behavior between multiple signals can be easily used. Using this feature, label recognition techniques have been investigated for photonic routers based upon optical code correlation. However, most of the proposed systems cannot recognize all the binary codes because only the codes that provide enough discrimination between auto-correlation and cross-correlation can be recognized. In addition, in most systems, each optical integrated circuit recognizes only one label (Wada et al., 1999; Takiguchi et al., 2002). Therefore, it is necessary to prepare multiple correlators at each node in order to recognize all the routing labels. On the contrary, for processing of multiple labels, Moriwaki et al. (Moriwaki et al., 2005) and Cincotti (Cincotti, 2004) proposed label recognition systems where self-routing architecture was employed for phase-shift-keying (PSK) labels. On the other hand, Glesk et al. reported a demonstration of optical multiple label recognition for on-off keying (OOK) codes at 250Gbit/s using a self-routing scheme (Glesk et al., 1997). The self-routing of label data stream for label recognition is one of promising methods in label decoding system.

Hiura & Goto proposed a label recognition system for OOK labels (Hiura et al, 2005; Hiura et al, 2007a). Although the proposed system can recognize all the binary-code labels, the system requires many optical switches controlled by optical signals. A similar system was also reported by Kurumida et al. (Kurumida et al., 2006). On the other hand, Hiura et al. also proposed an all-optical passive label recognition system for all the binary codes in binary PSK (BPSK) format (Hiura et al., 2006; Hiura et al., 2007b). The label recognition system consists of a tree-structure connection of passive waveguide components named as asymmetric X-junction coupler (Izutsu et al., 1982; Burns & Milton, 1975; Burns & Milton, 1980), and time gates. The asymmetric X-junction coupler provides a function of wave coupling according to the phase relation between the two incident waves. The asymmetric X-junction coupler has an advantage that the wave coupling behavior does not depend on wavelength because the wave coupling utilizes an adiabatic wave coupling along the X-junction. This feature cannot be obtained with other devices such as 3-dB directional couplers or multi-mode interference (MMI) couplers.

The number of represented labels in PSK format can be increased by employing multiple phases such as four phases as quadri-PSK (QPSK). Optical circuits for detecting QPSK signal have been investigated for receivers in communication systems (Renaudier et al., 2008). When QPSK labels are introduced in label routing system, optical processing of the QPSK labels is required. We have proposed a label recognition circuit for QPSK labels (Makimoto et al., 2008; Makimoto et al, 2009a). The circuit consists of the asymmetric X-junction couplers, Y-junctions and 3-dB directional couplers.

In this article, we describe the principle of label recognition for BPSK and QPSK coded labels in the self-routing scheme. The operation of the label recognition with optical integrated circuits are confirmed by finite-difference beam propagation method (FD-BPM).

2. Photonic label router

Optical labels are used as a routing information in photonic label switching network as shown in Fig.1. A label is attached to the incident packet at an edge router. The label is used to forward the packet to another edge router which is connected to the destination of the packet. At a node, routing switches are controlled according to the label information by referring to a routing table. At first, the label has to be analyzed to find the destination information. Therefore, the label has to be inspected whether the label matches with any of the all labels at the routers.

If the network is designed in a hierarchical structure so that a router at a node in a sub-network is required to find only the packets destined to the sub-network, it is not required for the router to resolve all the labels. Therefore, it depends on the network architecture and routing protocol whether all the labels have to be resolved or only a part of the labels are required to be resolved.

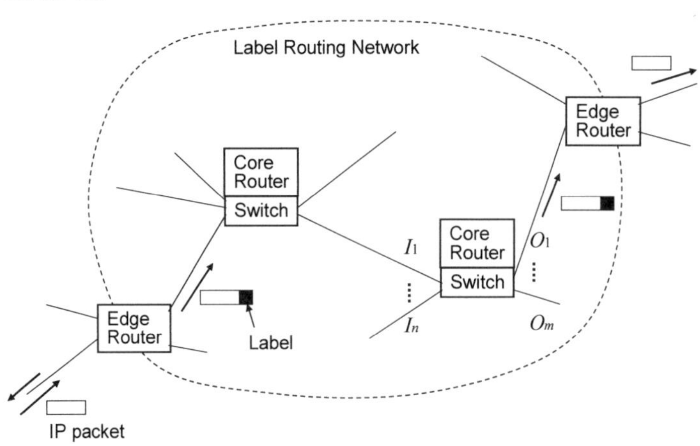

Fig. 1. Label routing network.

3. Optical label and its recognition

Various methods have been investigated to represent routing label information as optical signals, which include coding of the labels in time-domain, in spectral domain, and in their combination. Here we consider time sequential coded pulse train in BPSK and QPSK modulation formats. In these pulses encoded in phase, a reference signal is required to identify their absolute phase. Although differential PSK formats can be alternatives that do not require the reference phase signal, we introduce a reference pulse in advance of the pulses representing an address to identify general PSK address as shown in Fig.2. The electric field of the optical pulse train of an (N+1)-bit label is written as

$$E_{label}(t) = \sum_{i=0}^{N} f_0(t - i\Delta t)e^{j\phi_i}e^{j\omega(t-i\Delta t)}, \tag{1}$$

where $f_0(t)$ denotes the envelope of a pulse with the angular frequency ω, the phase ϕ_i and the pulse period Δt. j is the imaginary symbol of $j = \sqrt{-1}$. The phase is 0 or π in BPSK labels and 0, π/2, π or 3π/2 in QPSK labels. The phase ϕ_0 is assumed to be 0 as the reference pulse, named as identifying (ID) bit in our self-routing label recognition systems. The other phases ϕ_i, i=1,..., N, represent the address.

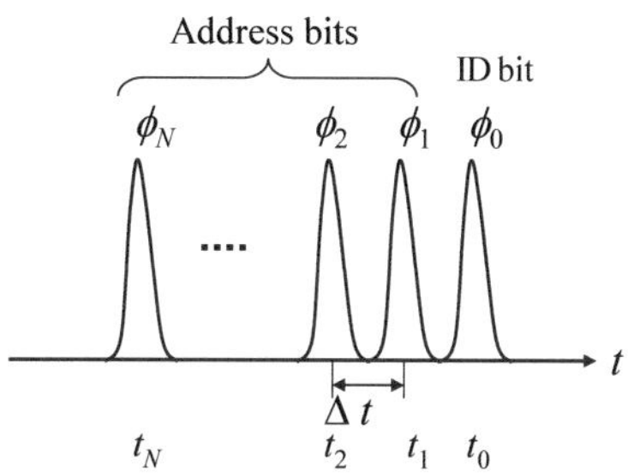

Fig. 2. Label structure in PSK format.

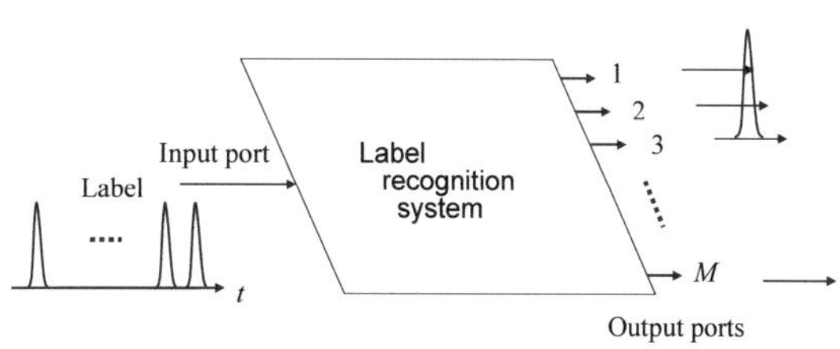

Fig. 3. Label recognition system.

The label recognition is performed by forwarding the ID-bit pulse to an output port corresponding to the destination of the address as shown in Fig.3. The number of the output ports M corresponds to the number of all the coded address, that is, $M=2^N$ and 4^N for BPSK and QPSK codes, respectively. Since all the bits of the label arrive sequentially, we employ a processing circuit to extend all the bits in parallel at a specific time as shown in Fig.4. Equally-divided pulse trains are sent to the input ports of the label recognition circuit through delay elements. The incident electric field of the optical signal to input port I_k, k=1,...,N+1, is delayed by (k-1)Δt as given by

$$E_{label}^{(k)}(t) = \frac{1}{\sqrt{N+1}} \sum_{i=0}^{N} f_0(t-(i+k-1)\Delta t) e^{j\phi_i} e^{j\omega[t-(i+k-1)\Delta t]}. \tag{2}$$

At time t_c, all the (N+1)-bit pulses simultaneously enter to the input ports.

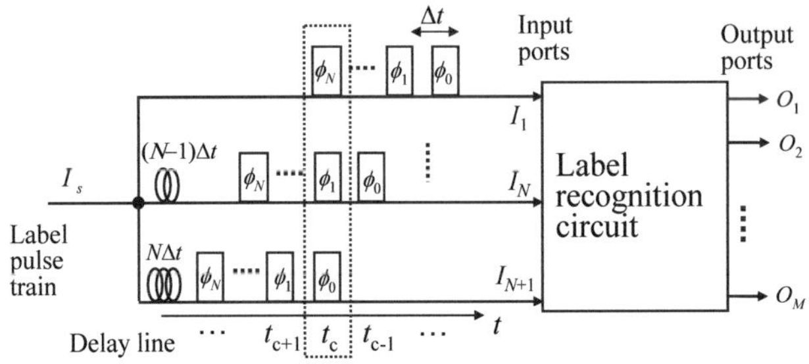

Fig. 4. Label recognition with serial-to-parallel conversion as a pre-processing.

Figure 5 shows the principle of label recognition by self-routing manner. Each bit of the address represents the number of M_0=2 and 4 for BPSK and QPSK, respectively. The nth-stage circuit module forwards the ID bit pulse to the output port corresponding to the nth address bit by using the nth address bit pulse as the control signal. Consequently, the ID bit pulse appears at the destination port among the $M=M_0^N$ output ports in the Nth-stage.

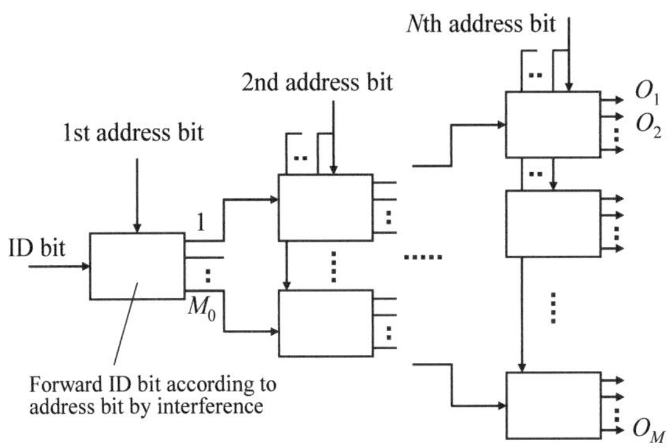

Fig. 5. Tree-structure circuit for label recognition by self-routing manner.

4. Basic devices for recognition circuits

Proposed label recognition circuits consist of passive waveguide devices shown in Fig.6. A 3-dB directional coupler shown in (a) is used to couple the optical incident waves into two output waves. The electric field of the input and output waves, $E_{in}^{(i)}$ and $E_{out}^{(i)}$, i=1,2, are related, by eliminating the common phase shift along the propagation, as

$$\begin{pmatrix} E_{out}^{(1)} \\ E_{out}^{(2)} \end{pmatrix} = \frac{1}{\sqrt{2}} \begin{pmatrix} 1 & -j \\ -j & 1 \end{pmatrix} \begin{pmatrix} E_{in}^{(1)} \\ E_{in}^{(2)} \end{pmatrix}. \tag{3}$$

When an optical wave is incident in only one of the waveguide, the wave is equally divided. The output fields, however, have a phase difference of $\pi/2$. A Y-junction shown in (b) also divides an optical input wave equally into two output ports as expressed by

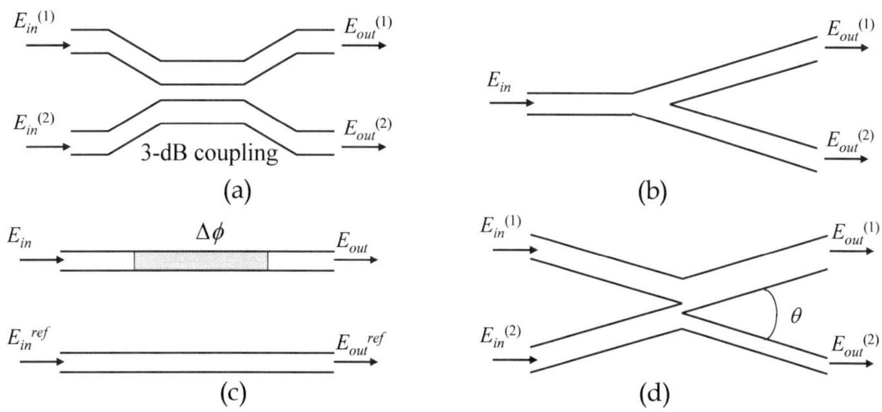

Fig. 6. Basic passive elements for label recognition circuits.

$$\begin{pmatrix} E_{out}^{(1)} \\ E_{out}^{(2)} \end{pmatrix} = \frac{1}{\sqrt{2}} \begin{pmatrix} 1 \\ 1 \end{pmatrix} E_{in}. \tag{4}$$

The two output waves have the same phase. A phase shifter shown in (c) shifts the phase with regard to the reference waveguide. The input-output relation is given by

$$\begin{pmatrix} E_{out} \\ E_{out}^{ref} \end{pmatrix} = \begin{pmatrix} e^{j\Delta\phi} & 0 \\ 0 & 1 \end{pmatrix} \begin{pmatrix} E_{in} \\ E_{in}^{ref} \end{pmatrix}. \tag{5}$$

An asymmetric X-junction coupler shown in (d) is the device whose input and output are a symmetric Y-junction and an asymmetric Y-junction, respectively (Izutsu et al., 1982). When two waves are incident in phase, the output is obtained only at the wider-waveguide port. On the contrary, when two waves are incident in opposite phase, the output is at the narrower waveguide. This input-output relation is given by

$$\begin{pmatrix} E_{out}^{(1)} \\ E_{out}^{(2)} \end{pmatrix} = \frac{1}{\sqrt{2}} \begin{pmatrix} 1 & 1 \\ -1 & 1 \end{pmatrix} \begin{pmatrix} E_{in}^{(1)} \\ E_{in}^{(2)} \end{pmatrix}. \tag{6}$$

It is noted that the output fields have a phase difference of π. The angle θ and the waveguide asymmetry have to be properly designed to realize the ideal routing characteristics. The phenomenon of optical coupling along an asymmetric Y-branch is evaluated by the coupled mode theory (Burns & Milton, 1975; Burns & Milton, 1980). A parameter Ψ to characterize the coupling phenomenon is defined by

$$\Psi = \frac{N_w - N_n}{\theta\sqrt{(N_w + N_n)^2/4 - n_s^2}}, \tag{7}$$

where n_s is the refractive index of the cladding region, N_w and N_n are the effective indices of the wide and narrow waveguides, respectively, θ is the branching angle. The ideal function as given by eq.(6) is expected to be realized in the asymmetric X-junction coupler when $\Psi > 0.44$.

5. BPSK label recognition

The asymmetric X-junction coupler has a function to discriminate the phase of the incident wave in BPSK format. A cascaded connection of the asymmetric X-junction couplers in a tree-structure shown in Fig. 7 can identify two-bit addresses. The symbol X_{ij} denotes an asymmetric X-junction coupler at the *i*th stage. Each of the waveguides numbered by 3 and 4 corresponds to the wider and the narrower waveguide, respectively. The input and output fields are related as

$$\begin{pmatrix} E_{out}^{(1)} \\ E_{out}^{(2)} \\ E_{out}^{(3)} \\ E_{out}^{(4)} \end{pmatrix} = \begin{pmatrix} 1 & -1 & \alpha_2 \\ -1 & 1 & \alpha_2 \\ 1 & 1 & \alpha_2 \\ -1 & -1 & \alpha_2 \end{pmatrix} \begin{pmatrix} E_{in}^{(1)} \\ E_{in}^{(2)} \\ E_{in}^{(3)} \end{pmatrix}, \tag{7}$$

where α_2 is an amplification coefficient of the amplifier placed at input port I_3. We consider two cases having different value of α_2. First we consider the case of α_2=1. When optical two-bit BPSK pulse train is incident, the normalized intensity of the pulse trains appears from four output ports as shown in Fig.8. The maximum output is found at time t_c , and four

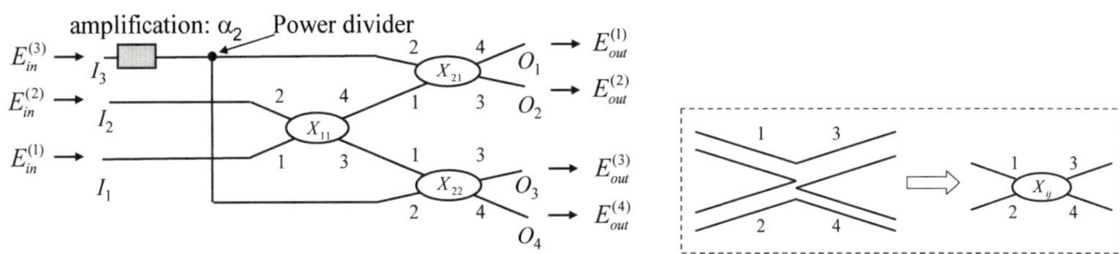

Fig. 7. BPSK label recognition circuit for label length three.

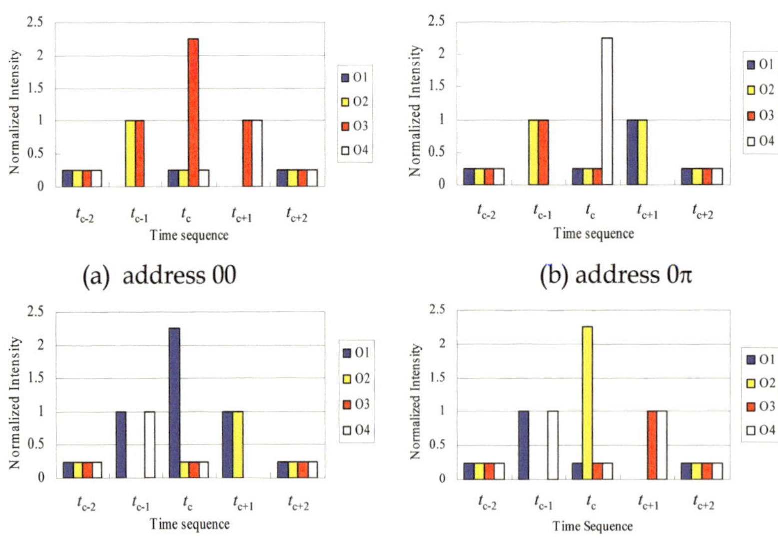

(a) address 00 (b) address 0π

(c) address π0 (d) address ππ

Fig. 8. Time sequential output intensity for four addresses with the circuit of α_2=1.

different addresses have the peak intensity at their corresponding output port. The contrast ratio of the peak intensity to the second largest intensity at time t_c is 9. Since the inputs to the second asymmetric X-junction couplers, X_{2j}, j=1,2, have different intensities, the output from the port whose paired port has the maximum intensity has a non-zero output.
Next, we consider the case of α_2=2. The pulse train from four output ports appear as shown in Fig.9. In this case, the input intensities to the X-junction coupler that has the maximum output are equal, and no output at the paired port. The contrast ratio of the maximum intensity to the second intensity at time t_c is decreased to 4.

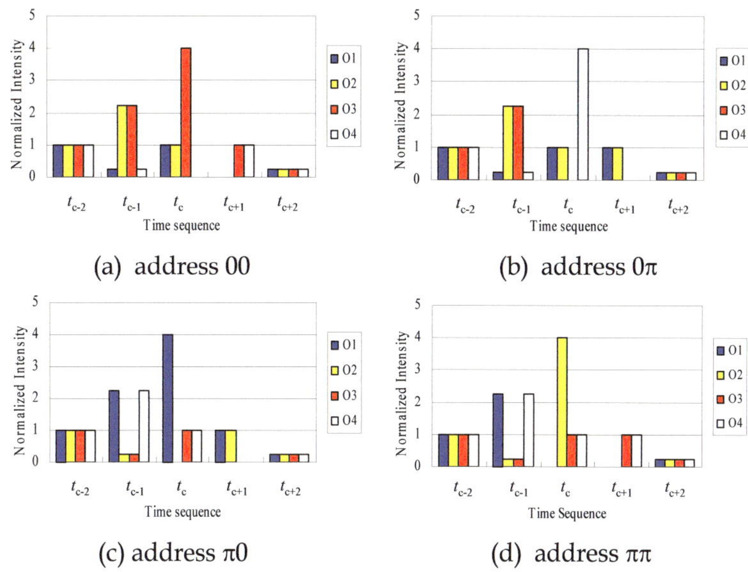

Fig. 9. Time sequential output intensity for four labels with the circuit of α_2=2.

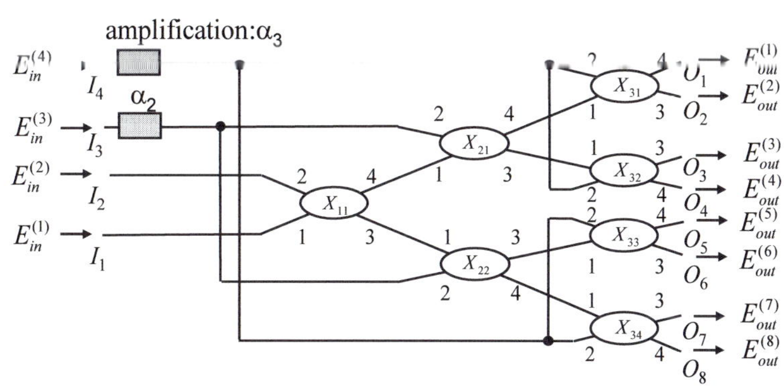

Fig. 10. BPSK label recognition circuit for label length four.

The circuit shown in Fig.7 can be scaled up for label length four as shown in Fig.10. The pulse train coupled in the input port I_4 is amplified by a factor α_3 and divided into four pulse trains for the third-stage X-junction couplers. The contrast ratio of the maximum output intensity to the second largest intensity is decreased as shown in Fig.11. It is found that for the case of α_m=1, m=2,..., N-1, 7-bit addresses can be recognized with the contrast ratio of 2.5[dB]. On the contrary, for the case of $\alpha_m=2^{m-1}$, even 4-bit addresses are difficult to recognize. However, by regarding the null-output port as the port indicating the address, the address can be recognized. The maximum output intensity, $P_{IC,max}$ and the second smallest output intensity $P_{IC,min}$ are plotted in Fig.12. The decrease of $P_{IC,min}$ and the increase of $P_{IC,max}$ limit the recognized address bits. These values are -12dB and 18dB, respectively, for 5-bit addresses.

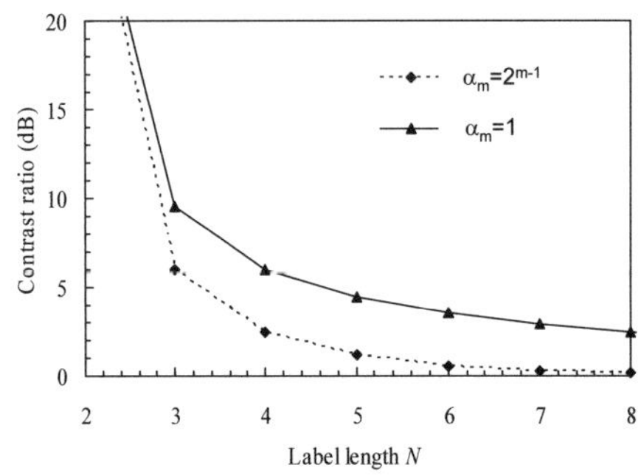

Fig. 11. Comparison of the contrast ratio of the outputs for two cases.

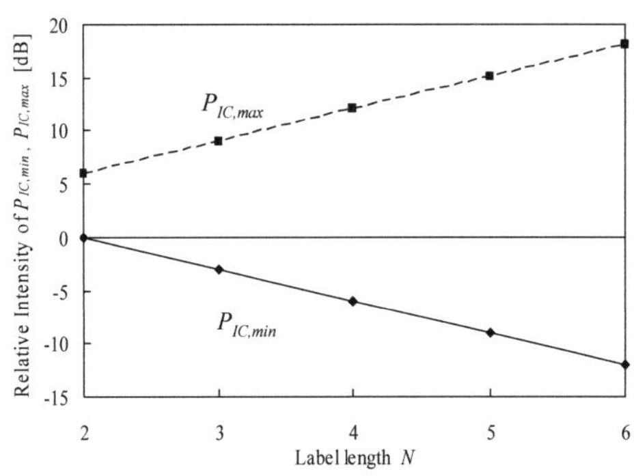

Fig. 12. Relative output intensity of the maximum output and the second smallest output for the second case of $\alpha_m=2^{m-1}$.

The recognition characteristics were verified with computer simulation by FD-BPM. A two-dimensional model of the 2-stage device for label length three is shown in Fig.13. The waveguide circuit is assumed to be made of slab-type Ge-doped silica waveguides. The refractive indices of the core and the cladding regions are 1.461 and 1.450, respectively, at wavelength 1550nm. The core width is W=3μm for the input waveguides 1 and 2, W_3=3.4μm for the wider waveguide 3 and W_4=2.6μm for the narrower waveguide 4. The effective indices are calculated to be N_w=1.4560 and N_n=1.45472 at λ=1550 nm from the dispersion equation of guided waves. The crossing angle θ is 0.004rad. The parameter Ψ of eq.(7) is then evaluated to be 2.56. This value is enough larger than 0.44 and the routing function of eq. (6) is expected.

The hatched waveguides with width W_a=2.9μm and W_b=3.1μm are introduced to compensate the path-length difference. The total device length L is 20mm. The BPM simulation was performed for TE mode. Examples of the results for address "00" and "π0" with α_2=1 are shown in Fig.14, where the intensity profiles is shown. The incident fields to

I_1, I_2 and I_3' are assumed to be e^{j0}, e^{j0} and $(1/\sqrt{2})e^{j0}$ for address "00" and e^{j0}, $e^{j\pi}$ and $(1/\sqrt{2})e^{j0}$ for address "π0", respectively. The results agree well with the theoretical analysis.

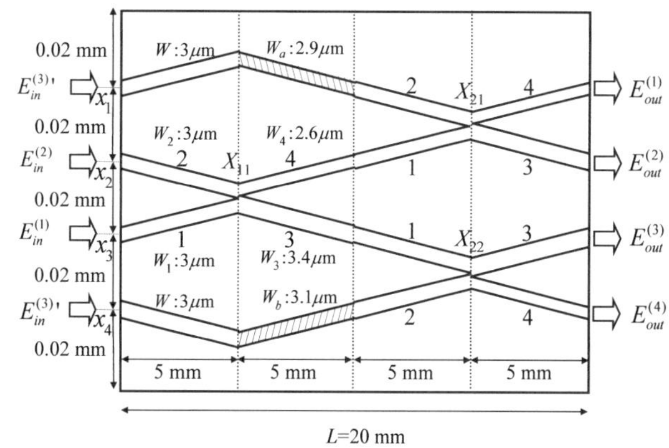

Fig. 13. Two-dimensional waveguide model for FD-BPM simulation.

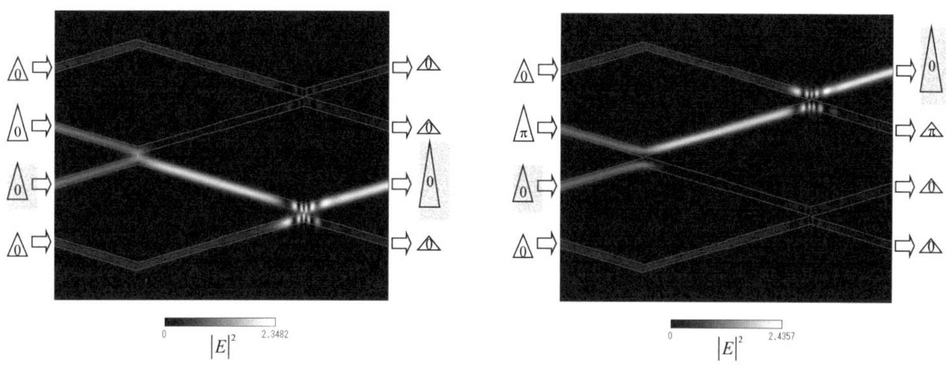

(a) "00": $(E_{in}^{(1)}, E_{in}^{(2)}, E_{in}^{(3)\prime}) = (e^{j0}, e^{j0}, (1/\sqrt{2})e^{j0})$ (b) "π0": $(E_{in}^{(1)}, E_{in}^{(2)}, E_{in}^{(3)\prime}) = (e^{j0}, e^{j\pi}, (1/\sqrt{2})e^{j0})$

Fig. 14. Simulation results for two different labels.

6. QPSK label recognition

By combining the basic waveguide devices shown in Fig.6, we proposed a circuit for QPSK label recognition. Figure 15 shows the basic circuit module which recognizes the phase of an incident QPSK code. This basic module is regarded as a QPSK phase recognition circuit (QPRC). The QPRC consists of a 3-dB directional coupler, two Y-junctions, and an asymmetric X-junction coupler. There are two input ports and four output ports. The electric fields of the optical signals at the output of the two Y-junctions, $E_{Y\text{-}out}^{(i)}$, i=1,…,4, are written using eqs.(3) and (4) as

$$\begin{pmatrix} E_{Y-out}^{(1)} \\ E_{Y-out}^{(2)} \\ E_{Y-out}^{(3)} \\ E_{Y-out}^{(4)} \end{pmatrix} = \frac{1}{2}\begin{pmatrix} 1 & 0 \\ 1 & 0 \\ 0 & 1 \\ 0 & 1 \end{pmatrix}\begin{pmatrix} 1 & -j \\ -j & 1 \end{pmatrix}\begin{pmatrix} E_{in}^{(1)} \\ E_{in}^{(2)} \end{pmatrix}. \tag{8}$$

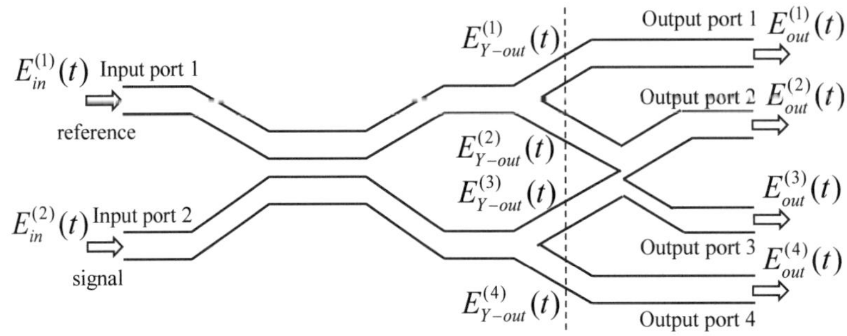

Fig. 15. Phase recognition circuit for one-bit QPSK signal.

After passing the asymmetric X-junction coupler, we obtain the output fields as

$$\begin{pmatrix} E_{out}^{(1)} \\ E_{out}^{(2)} \\ E_{out}^{(3)} \\ E_{out}^{(4)} \end{pmatrix} = \frac{1}{2}\begin{pmatrix} 1 & 0 & 0 & 0 \\ 0 & 1/\sqrt{2} & 1/\sqrt{2} & 0 \\ 0 & -1/\sqrt{2} & 1/\sqrt{2} & 0 \\ 0 & 0 & 0 & 1 \end{pmatrix}\begin{pmatrix} E_{Y-out}^{(1)} \\ E_{Y-out}^{(2)} \\ E_{Y-out}^{(3)} \\ E_{Y-out}^{(4)} \end{pmatrix}$$
$$= \frac{1}{2}\begin{pmatrix} 1 & -j \\ (1-j)/\sqrt{2} & (1-j)/\sqrt{2} \\ (-1-j)/\sqrt{2} & (1+j)/\sqrt{2} \\ -j & 1 \end{pmatrix}\begin{pmatrix} E_{in}^{(1)} \\ E_{in}^{(2)} \end{pmatrix} = \frac{1}{2}\begin{pmatrix} 1 & e^{j3\pi/2} \\ e^{j7\pi/4} & e^{j7\pi/4} \\ e^{j5\pi/4} & e^{j\pi/4} \\ e^{j3\pi/2} & 1 \end{pmatrix}\begin{pmatrix} E_{in}^{(1)} \\ E_{in}^{(2)} \end{pmatrix}. \tag{9}$$

We consider the outputs of the QPRC for each input code. We assume the reference signal $E_{in}^{(1)}(t)$ to be 1, and $E_{in}^{(2)}(t)$ to be four kinds of different phase signals as $E_{in}^{(2)}(t)=1, j, -1$, and $-j$, corresponding to the phase $\phi_2=0$, $\pi/2$, π, and $3\pi/2$, respectively. Table 1 shows electric fields of the optical amplitudes $E_{out}^{(i)}(t)$, $i=1,\dots,4$, at the output ports. The output intensities at the output ports are plotted in Fig.16. The maximum and minimum outputs are obtained at different ports for different phases of $E_{in}^{(2)}$. Thus, the four different QPSK signals can be distinguished with this circuit by identifying the maximum output port or the minimum output port.

Now we consider the output from the QPRC in the system shown in Fig.3. The output intensities at times $t_0-2\Delta t$, $t_0-\Delta t$, t_0, $t_0+\Delta t$ and $t_0+2\Delta t$ are shown in Fig.17, where the phase of the incident QPSK code is assumed as $\phi_1=\phi_2=0$, that is $(E_{in}^{(1)}, E_{in}^{(2)})=(1,1)$. The second largest intensity at $t_0-\Delta t$ is found to be 0.5625 at three output ports. Since this intensity is smaller than the second largest intensity of 0.625 at $t=t_0$, the maximum output port can be recognized by using a threshold device without using time-gating devices.

Input phase ϕ_2	Output amplitude			
	$E_{out}^{(1)}$	$E_{out}^{(2)}$	$E_{out}^{(3)}$	$E_{out}^{(4)}$
0	$e^{j7\pi/4}/\sqrt{2}$	$e^{j7\pi/4}$	0	$e^{j7\pi/4}/\sqrt{2}$
π	$e^{j\pi/4}/\sqrt{2}$	0	$e^{j5\pi/4}$	$e^{j7\pi/4}/\sqrt{2}$
$\pi/2$	1	$1/\sqrt{2}$	$e^{j\pi}/\sqrt{2}$	0
$3\pi/2$	0	$e^{j3\pi/2}/\sqrt{2}$	$-e^{j3\pi/2}/\sqrt{2}$	$e^{j3\pi/2}$

Table 1. Electric fields of the optical outputs for four different phase inputs.

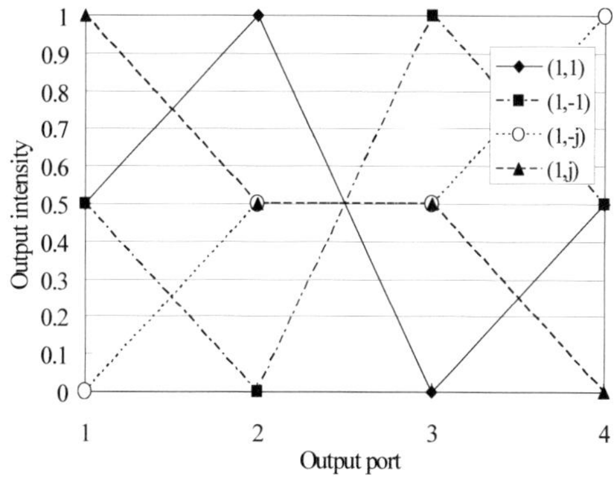

Fig. 16. Output intensities for four different phase inputs.

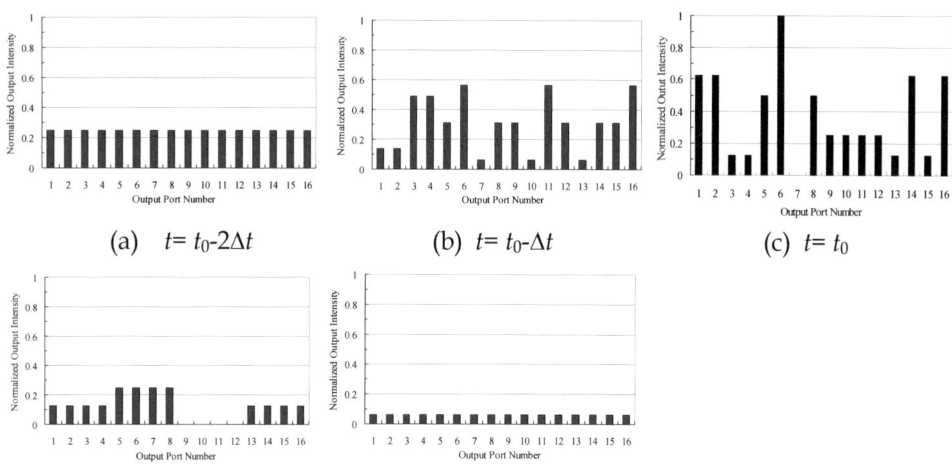

(a) $t = t_0 - 2\Delta t$ (b) $t = t_0 - \Delta t$ (c) $t = t_0$

(d) $t = t_0 + \Delta t$ (e) $t = t_0 + 2\Delta t$

Fig. 17. Output intensities for $\phi_1 = \phi_2 = 0$ at different time instances.

We confirm the principle of the operation with the QPRC module by FD-BPM simulation. The model used in the simulation is two-dimensional waveguide circuit as shown in Fig.18. Optical guided waves propagate in TE mode. The refractive indices of the core and cladding materials are assumed to be 1.461 and 1.45, respectively, at the wavelength of 1.55μm. The basic waveguide width is designed to be 3.0μm to support only the fundamental mode. The wider and narrower waveguides of the asymmetric X-junction coupler are 3.4μm and 2.6μm, respectively, with the branch angle of 0.004rad. The 3-dB directional coupler has the coupling region of 2.1818mm with the gap of 7.2μm. The total length and the width of the QPRC module is 30mm and 100μm, respectively. Figure 19 shows examples of the simulation results for two different kinds of input phases. When the two incident waves are in phase, corresponding to $(E_{in}^{(1)}, E_{in}^{(2)})=(1,1)$, the maximum signal is found at output port 2, whereas no signal is output at the output port 3 as shown in (a). When the input wave in input port 2 has the phase shifted by $\pi/2$, corresponding to $(E_{in}^{(1)}, E_{in}^{(2)})=(1,-j)$, the maximum signal is found at output port 4, whereas no signal is output at output port 1 as shown in (b).

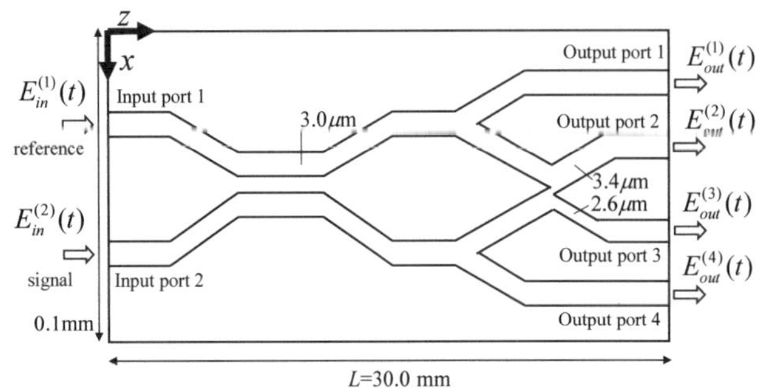

Fig. 18. Two-dimensional model of QPSK phase recogniton circuit for FD-BPM simulation.

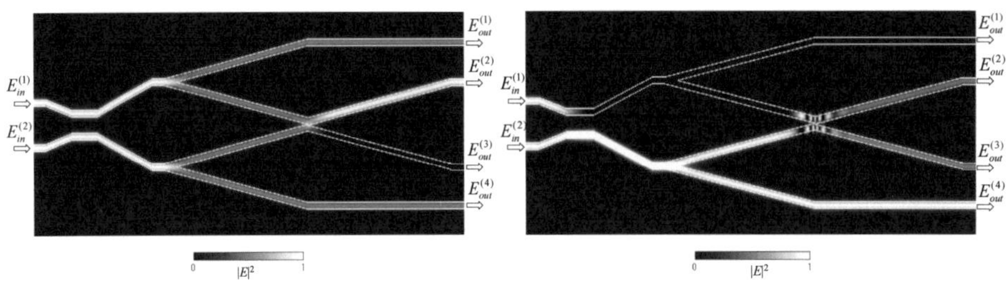

(a) $(E_{in}^{(1)}, E_{in}^{(2)})=(1, 1)$ (b) $(E_{in}^{(1)}, E_{in}^{(2)})=(1,-j)$

Fig. 19. FD-BPM simulation results for two different phase inputs.

Next, we consider a recognition circuits for a two-bit QPSK address. First, we consider a case of two-bit address (N=2). The label recognition circuit for N=2 consists of two-stage connection of QPRC modules in a tree structure as shown in Fig.20. The ID bit pulse and the first bit pulse of the address are incident in the input port 1 and 2, respectively. The four outputs from the QPRC are sent to the input port 1 of the second-stage QPRC modules. The second-bit pulse of the code is four times amplified and then divided into four pulses to be incident in the input ports 2 of the four QPRC modules. In order to perform the same processing as the first-stage QPRC module in the second-stage QPRC modules, the phase of the output signals from the first-stage QPRC module has to be adjusted since the phases of the outputs are not the same as those shown in Table 1. The phase shift circuit is employed to adjust the phase. The outputs of the phase shift circuit $E_{out}^{(i)\prime}$, (i=1,...,4) are written as

$$\begin{pmatrix} E_{out}^{(1)\prime} \\ E_{out}^{(2)\prime} \\ E_{out}^{(3)\prime} \\ E_{out}^{(4)\prime} \end{pmatrix} = \begin{pmatrix} e^{j\Delta\phi_1} & 0 & 0 & 0 \\ 0 & e^{j\Delta\phi_2} & 0 & 0 \\ 0 & 0 & e^{j\Delta\phi_3} & 0 \\ 0 & 0 & 0 & e^{j\Delta\phi_4} \end{pmatrix} \begin{pmatrix} E_{out}^{(1)} \\ E_{out}^{(2)} \\ E_{out}^{(3)} \\ E_{out}^{(4)} \end{pmatrix}, \tag{10}$$

where $\Delta\phi_1=0$, $\Delta\phi_2=\pi/4$, $\Delta\phi_3=3\pi/4$, and $\Delta\phi_4=\pi/2$.

Table 2 shows the amplitudes $E_{out}^{(i)\prime}$ after the phase shift circuit. The maximum output at each output ports can be regarded as an ID bit pulse for recognizing the second-bit pulse of the address. The output intensities from the sixteen output ports for label $(E_{in}^{(1)}, E_{in}^{(2)}, E_{in}^{(3)})$ =(1, 1, 1), are shown Fig.21. It is found from this result that the maximum output intensity of $|E_{out}^{(i)}|^2$=1 is obtained at only one output port, and the minimum output intensity of

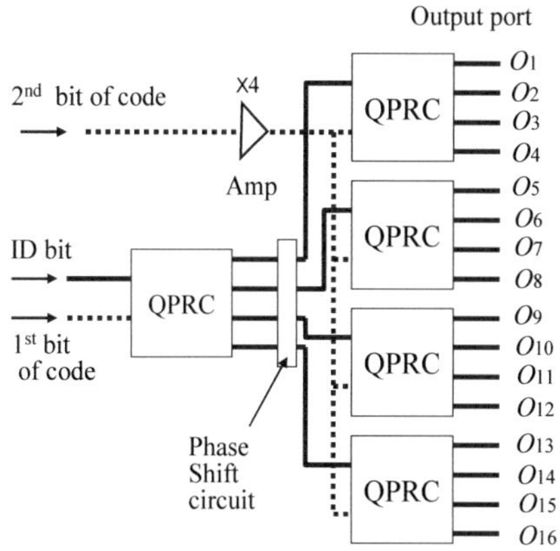

Fig. 20. Label recognition circuit for two-bit QPSK addresses.

Input phase ϕ_2	Output amplitude			
	$E_{out}^{(1)}$	$E_{out}^{(2)}$	$E_{out}^{(3)}$	$E_{out}^{(4)}$
0	$e^{j7\pi/4}/\sqrt{2}$	1	0	$e^{j\pi/4}/\sqrt{2}$
π	$e^{j\pi/4}/\sqrt{2}$	0	1	$e^{j7\pi/4}/\sqrt{2}$
$\pi/2$	1	$e^{j\pi/4}/\sqrt{2}$	$e^{j7\pi/4}/\sqrt{2}$	0
$3\pi/2$	0	$e^{j7\pi/4}/\sqrt{2}$	$-e^{j\pi/4}/\sqrt{2}$	1

Table 2. Electric field of the optical output after phase shift circuit.

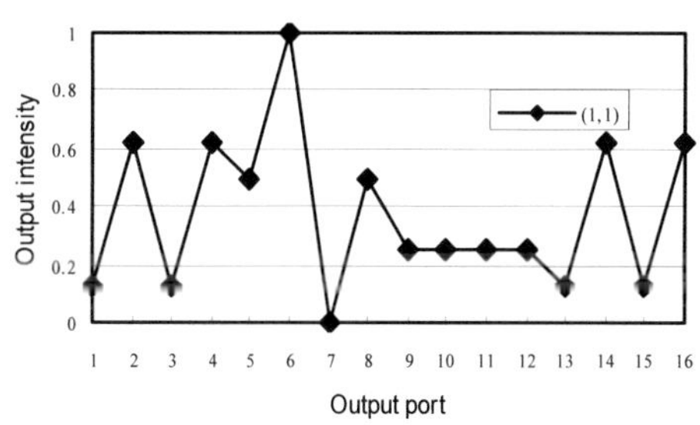

Fig. 21. Output intensities at sixteen output ports for input label $(E_{in}^{(1)}, E_{in}^{(2)}, E_{in}^{(3)}) = (1, 1, 1)$.

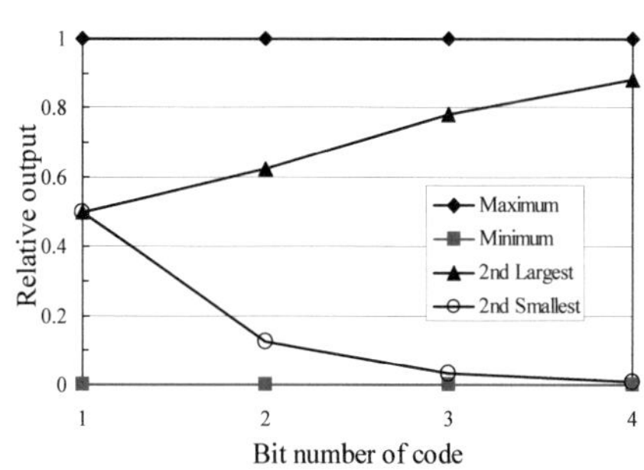

Fig. 22. Relative output intensities of the maximum, 2nd largest, 2nd smallest and minimum outputs.

$|E_{out}^{(i)}|^2=0$ only one of the other output ports. These output ports for $|E_{out}^{(i)}|^2=1$ and $|E_{out}^{(i)}|^2=0$ are different for all the sixteen addresses. The output intensities of the second largest and the second smallest output are 0.625 and 0.125, respectively. By employing threshold devices after the output ports, the output port of the maximum intensity or minimum intensity can be identified. Thus, two-bit QPSK addresses can be recognized with this circuit.

The number N of the code bit to be recognized can be increases by repeating the similar operation with increased stage number. We calculated the output intensities for three-bit (N=3) and four-bit (N=4) addresses. The output intensities corresponding to the maximum, the minimum, the second largest and the second smallest outputs for N=1 to N=4 are plotted in Fig.22. Although refined threshold devices will be required to recognize the addresses for $N > 3$, multiple-bit addresses are recognizable in principle.

7. Conclusion

Label recognition by interfering an identifying pulse with address pulses with integrated-optic circuits was discussed. All the addresses can be identified with a single circuit by self-routing manner. The circuits consist of passive devices such as asymmetric X-junction couplers, Y-junction dividers and directional couplers. Therefore, the circuits can be formed on various substrates. Since the wavelength dependence in these devices is small, the device can be used for optical labels at a wavelength in a wide wavelength range. Although the contrast ratio of the output from the port corresponding to the destination address and that from the other ports decreases as the bit number of the address increases, the contrast ratio can be improved by employing nonlinear post-processing devices such as thresholding devices.

8. References

Blumenthal, D. J; Olsson, B. E.; Rossi, G.; Dimmich, T. E.; Rau, L; Masanovic, M.; Lavrova, O; Doshi, R.; Jerphagnon, O.; Bowers, J. E.; Kaman, V.; Coldren, L. A. & Barton, J. (2000). All-Optical label swapping networks and technologies, J. Lightwave Technol., vol.18, no.12, pp.2058-2075, Dec.2000.

Burns, W. K. & Milton, A. F. (1975). Mode conversion in planar-dielectric separating waveguides, IEEE J. Quantum Electron., vol.QE-11, no.1, pp.32-39, Jan. 1975.

Burns, W. K. & Milton, A. F. (1980). An analytic solution for mode coupling in optical waveguide branches, IEEE J. Quantum Electron., vol.QE-16, no.4, pp.446-454, Apr. 1980.

Cincotti, G. (2004). Full optical encoders/decoders for photonic IP routers, J. Lightwave Technol., vol.22, no.2, pp.337-342, Feb. 2004.

Glesk, I.; Kang, K. I. & Prucnal, P. R. (1997). Ultrafast photonic packet switching with optical control, Optics Express, vol.1, no.5, pp.126-132, Sep. 1997.

Goto, N & Miyazaki, Y. (2005). Label recognition with wavelength-selective collinear acoustooptic switches for photonic label routers, Jpn. J. Appl. Phys., vol.44, no.6B, pp.4449-4454, June 2005.

Hiura, H. & Goto, N. (2005). Proposal of all-optical label recognition using self-routing circuits, Proc. Int. Quantum Electronics Conf. 2005 and the Pacific Rim Conf. on

Lasers and Electro-Optics 2005 (IQEC/CLEO-PR 2005), Tokyo, Japan, CFJ4-3, July 2005.

Hiura, H. & Goto, N. (2006). All-optical label recognition using tree-structure self-routing circuits consisting of asymmetric X-junctions, Proc. of Conf. on Lasers and Electro-Optics, and Quantum Electronics and Laser Science Conf. (CLEO/QELS) 2006, Long Beach, California, USA, JThC73, May 2006.

Hiura, H. & Goto, N. (2007a). All-optical label recognition using self-routing architecture of Mach-Zehnder interferometer optical switches with semiconductor optical amplifiers, IEICE Trans. Electron., vol.E90-C, no.8, pp.1619-1626, Aug. 2007.

Hiura. H; Narita, J. & Goto, N. (2007b). Optical label recognition using tree-structure self-routing circuits consisting of asymmetric X-junctions, IEICE Trans. Electron., vol.E90-C, no.12, pp.2270-2277, Dec. 2007.

Hiura, H.; Makimoto, Y.; Goto, N. & Yanagiya, S. (2008). Optical multiple-wavelength BPSK label recognition with self-touting waveguide-circuit, The 7th Int. Conf. on Optical Internet (COIN2008), Tokyo, Japan, C-16-PM1-2-5, Oct.2008.

Izutsu, M.; Enokihara, A. & Sueta, T. (1982). Optical-waveguide hybrid coupler, Opt. Lett., vol.7, no.11, pp.549-551, Nov.1982.

Kitayama, K.; Wada, N & Sotobayashi, H. (2000). Architectural considerations for photonic IP router based upon optical code correlation, J. Lightwave Technol., vol.18, no.12, pp.1834-1844, Dec.2000.

Kurumida, I.; Uenohara, H. & Kobayashi, K. (2006). All-optical label recognition for time-domain signal using multistage switching scheme based on SOA-MZIs, Electron. Lett., vol.42, no.23, pp.1362-1363, Nov. 2006.

Makimoto, Y; Hiura, H.; Goto, N. & Yanagiya, S. (2008). Proposal of waveguide-type optical circuit for recognition of optical QPSK coded labels in photonic router, OECC/ACOFT 2008, Sydney, ThK-1, July 2008.

Makimoto, Y; Hiura, H.; Goto, N. & Yanagiya, S. (2009a). Waveguide-type optical circuit for recognition of optical QPSK coded labels in photonic router, J. Lightwave Technology, Vol.27, No.1, pp.60-67, Jan. 2009.

Makimoto, Y.; Hiura, H.; Goto, N & Yanagiya, S. (2009b). Wavelength dependence of waveguide-type optical circuit for recognition of optical QPSK labels in photonic router, The 14th OptoElectronics and Communications Conference (OECC2009), Hong Kong, July 2009.

Moriwaki, O.; Kitoh, T.; Sakamoto, T. & Okada, A. (2005). Novel PLC-based optical correlator for multiple phase-modulated labels, IEEE Photon. Technol. Lett., vol.17, no.2, pp.489-491, Feb. 2005.

Renaudier, J.; Charlet, G.; Salsi, M.; Pardo, O. B.; Mardoyan, H.; Tran, P. & Bigo, S (2008) , Linear fiber impairments mitigation of 40-Gbit/s polarization-multiplexed QPSK by digital processing in a coherent receiver, J. Lightwave Technol., vol. 26, no.1, pp.36-42, Jan. 2008.

Seo, S. W.; Bergmann, K. & Prucnal, P. R. (1996) . Transparent optical networks with time-division multiplexing, J. Select. Areas Commun., vol.14, no. 6, pp.1039-1051, June 1996.

Takiguchi, K.; Shibata, T. & Itoh, M. (2002). Encoder/decoder on planar lightwave circuit for time-spreading/wavelength-hopping optical CDMA, Electron. Lett., vol.38, no.10, pp.469-470, May 2002.

Teh, P. -C.; Petropoulos, P.; Ibsen, M. & Richardson, D. J. (2001). A comparative study of the performance of seven- and 63 chip optical code-division multiple-access encoders and decoders based on superstructured fiber Bragg gratings, J. Lightwave Technol., vol.19, no.9, pp.1352-1365, Sep. 2001.

Wada, N & Kitayama, K. (1999). A 10 Gb/s optical code division multiplexing using 8-chip optical bipolar code and coherent detection, J. Lightwave Technol., vol.17, no.10, pp.1758-1765, Oct. 1999.

12

Chip-Scale Programmable Photonic Filters

Duncan L. MacFarlane
The University of Texas at Dallas
United States

1. Introduction

The integrated circuit has driven the electronics industry for half a century of continuous advancement because of three key attributes: an integrated manufacturing platform, a scalable architecture and programmable response. The first attribute, an integrated manufacturing base, provides a cost advantage that stems from a common core of materials science and allows a predictive roadmap for fabrication tool and process advances. The semiconductor manufacturing platform also grants significant reliability and survivability benefits. Secondly, the architecture of integrated electronics has been fundamentally scalable with clear figures of merit - feature size, transistor number per die, clock speed, etc - to demarcate the frontier. The key to this scalability is the ability to regenerate a signal, and historically this stems from the presence of gain in the electronics. Systems with gain are often classified as active, and it is important to note that no sophisticated system or large scale network has evolved that is passive. Finally, programmability provides a third economy, that of manufacturing scale. Because of the power to program microprocessors, DSPs and FPGAs, the development and manufacturing costs of these devices can be amortized over a large number of "niche" applications with medium or small market sizes. The profitable "market of one" is achieved routinely by programmed microprocessors, digital signal processors (DSPs) and field programmable gate arrays (FPGAs). Since the same integrated circuit design may be used in a tremendous number of applications, the fixed costs of design, development and wafer fab can be amortized across disparate small markets. Further, it is quite common to re-program any of these integrated circuits remotely to improve or adapt to their mission.

By contrast, there is the photonics industry, which is today reminiscent of the electronics industry in the 1960s. As an industry, photonics and optics is a heavily fragmented array of customized markets each worth only a few million dollars per year in revenues. We believe that one reason for this is the general philosophy of photonics which concentrates at the component level. In most development work a photonics device is carefully optimized for a specific application, and strategies for multiple markets and design re-use seem to be kept as secondary concerns. Consequently, a photonics module or system is characterized as an assembly of discrete components with very low levels of integration, manufactured with a high manual labor content. These parts are comprised of very different materials that are often epoxied together with glue applied by hand with the tip of an optical fiber. Today, there are few true photonic integrated circuits. Most optical systems are passive. At best, particular components are tunable or adjustable, but not programmable.

However, recent work has shown that the three main lessons of the integrated circuit can be brought to photonics (MacFarlane 2003, MacFarlane & Dowling 2004, Coleman et al 1997). In doing so, an extremely powerful information processor can be developed that will continue to drive improvements in overall processing power in the post CMOS era. The photonic integrated circuit described herein is a versatile, programmable, scalable architecture that will process photonic signals, and provide a sophisticated and practical interface with existing electronics. In particular, two key advances may be leveraged to this end. The first is the use of nanotechnology to provide on chip coupling and routing from one component to another. In the research described here this nanotechnology is integrated in a seamless manner onto the wafer. Second, recent advances in the state of practical technology of photonic devices, components and systems have provided us with the basic tools to design and build photonic filters that are not merely passive but active. This development of the active optical filter is an advance that is analogous to bringing a transistor to an electric circuit comprised of inductors, capacitors and resistors, and promises to provide the same revolutionary impact to photonics. An active filter has power gain as an integral part of its operation. Active filters, with gain, perform better than passive filters because their quality factor, a fundamental figure of merit, is higher. Functionally, active filters with high quality factors can sort or discriminate favoured frequencies with much higher precision than passive filters with low quality factors. A passive filter may only sort, or discriminate, one frequency from other frequencies by attenuating unwanted frequencies. An active filter can use its gain to accentuate, or amplify the desired frequencies in addition to attenuating the unwanted frequencies.

2. Nanophotonic couplers

A wide range of exciting results have emerged in the past several years in the area of nanophotonics (Zachidov et al 1998, Roh st al 2000, Osterback st al 2000, Yeoh et al 2001, Notomi 2002). The nanoscale engineering of optical materials and structures have provided revolutionary precision for traditional tasks and promise significantly new functionality. I see the field of nanotechnology as a key enabler for integrated photonics, and in particular integrated photonics with active elements, or gain. In the simplest application nanophotonic couplers can be used to route or direct optical signals from one component to another, or to several others. These couplers can also be used to multiplex or demultiplex different wavelength lasers on a single chip, direct optical signals to an embedded modulator or detector, and provide input-output coupling onto and off of the chip. The couplers may be

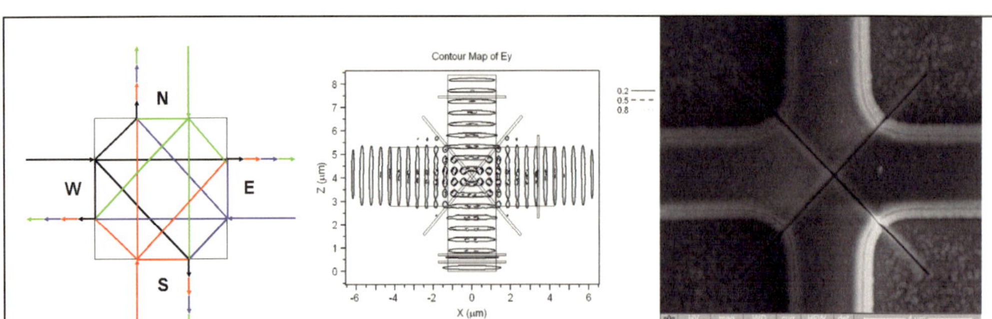

Fig. 1. 4-port nanophotonic coupler concepts

spectrally flat or can provide filter functions including wavelength selectivity and dispersion (chirp) control.

Of particular interest here is the four port nanophotonic coupler shown in Fig. 1. The four port coupler has 4 inputs and four outputs and hence can mix information from four adjacent filter stages. Realization of the four port couple may be accomplished through crossed grating structures or through deep trenches that are etched into the substrate to provide for frustrated total internal reflection.

3. Architecture

The particular class of devices described herein is optical filters. Passive optical filters comprise a market size of approximately $1Bn covering a variety of high tech and commodity applications. Technically passive filters include thin film filters, Fabry Perot etalons, fiber ring resonators, fiber Bragg gratings, interleavers, array waveguides, and star couplers. Applications for many of the more advanced architectures lie in the optical telecommunications arena, however thin film "lattice" structures also find application in antireflection coatings (lenses) and color measurement, lighting and displays. A surprising number of these applications demand precision manufacturing techniques. Producing precision devices with fixed responses for a large variety of applications across many niche markets is an economic and practical challenge. A manufacturer of precision filters for diode laser stabilization recently tripled lead times to 6 weeks to account for expected coating run delays. Such anecdotes are not, unfortunately, uncommon.

The measure of a filter's performance is the quality factor, $Q=\omega/\Delta\omega$. For a bandpass filter, the quality factor may be interpreted as the precision with which one frequency may be separated or distinguished from another. The Q would be the ratio of a frequency divided by an uncertainty in frequency. In general, a filter with a high quality factor is better than one with a low quality factor. A passive filter may only sort, or discriminate one frequency from other frequencies by attenuating or re-directing the unwanted frequencies. An active filter can use its gain to accentuate, or amplify the wanted frequencies in addition to attenuating the unwanted frequencies. This is a physical interpretation for why active filters yield higher Q's, for higher performance.

Schematically the difference between a passive optical filter and an active optical filter can be seen in Fig. 2. Figure 2 shows the signal flow for one stage of an optical lattice filter, a class of filters that includes Fabry Perot etalons and thin film filters. Shown are two mirrors or interfaces labeled k-1 and k, respectively with transmission coefficients and reflection coefficients. The space between the mirrors or the material thickness gives rise to a delay block, labeled $z^{-1/2}$, to invoke the z-transform. Most lattice filters are multistage structures. For example, a modern thin film interference filter may have more than 100 stages. The mathematical treatment of these devices entails unraveling these structures stage by stage, and this process is often called layer-peeling. For this layer peeling, the z-transform technique is a particularly useful design technique that has been pioneered by the author and is now used extensively in engineering practice (MacFarlane and Dowling 1994, Dowling and MacFarlane 1994, Narayan et al 1994, Narayan et al 1995). The same structure is shown in Fig. 2, including a gain block, G. Many of the same design techniques developed for the passive lattice filter may be applied to the active filter.

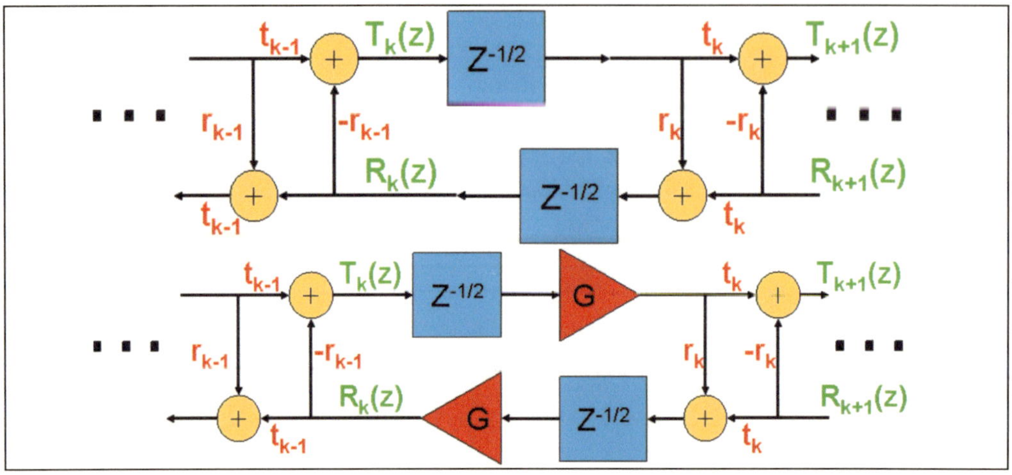

Fig. 2. Lattice filter stages comparing a passive and active filters

In a traditional lattice filter, the flow of signals is forward and backward in one dimension. At the interfaces between stages, or layers, these forward and backward signals combine, with different relevant phases for different frequency components. This delay and superposition action provides for a very rich Infinite Impulse Response (IIR) filter. The lattice structure is closely tied to linear prediction and joint process estimation (Haykin 2002, Proakis and Manolakis 1996), and is generally appreciated in signal processing because it offers desirable stability and robust round-off properties (Haykin 2002, Proakis and Manolakis 1996). However, the surface grating structure provides for significantly more versatile filter architecture. A unique two dimensional flow of signals may be realized if four-port couplers are used.

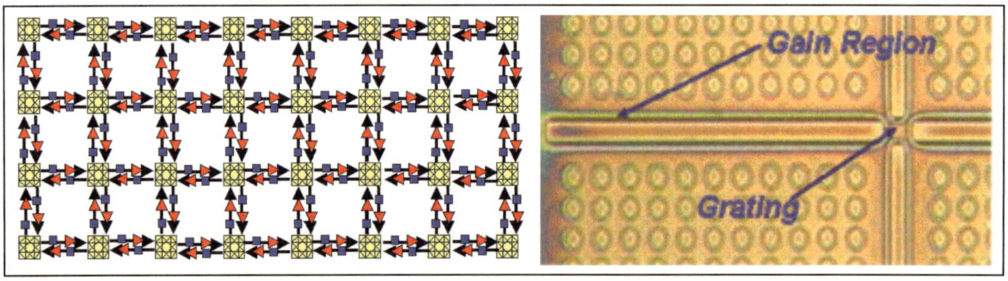

Fig. 3. Two dimensional active lattice filters.

There are significant technical and economic advantages to this two dimensional active lattice filter. Note that the two dimensional architecture optimally uses the two dimensional chip real estate. Operationally, the two dimensional active lattice filter is fully tunable, and allows multiple inputs and outputs. The tunability allows adjustment of the filter response, but also a more basic selection of the type of filter implementation. The device allows for an optimal filter realization for an application. Thus in working with the physical device, an engineer may optimize a design not only with respect to the order of the filter but also with respect to the form, or structure, of the filter. Importantly, completely new filter responses can be programmed on the same device in approximately one nanosecond. The combination of programmability and ease of manufacture provides the optical equivalent of a Field

Programmable Gate Array (FPGA). FPGA economics are extremely attractive because a standardized device can be leveraged across a host of custom, even niche, applications. The fixed costs of development and manufacturing plant can be amortized across a large number of markets, providing economies of scale that are currently unique in the photonics world.

4. Analysis

One strategy for the analysis and design of two dimensional lattice filters follows by extending the 2x2 matrix method for traditional lattices to higher order. In a traditional lattice structure one typically considers a forward propagating signal and a backwards propagating signal which together comprise a vector of rank 2 at any given stage. Movement from stage to stage is governed by a 2x2 scattering matrix. For N stages of a traditional lattice, a total 2x2 matrix may be written as the product of N 2x2 layer matrices. In the thick lattice filter, the 2x2 matrices still hold, albeit their arguments are significantly more complicated. For an MxN two dimensional lattice structure introduced here, one should consider M forward propagating signals and M backward propagating signals which together comprise a vector of rank 2M at any given stage. Movement from stage to stage is then governed by a 2Mx2M scattering matrix, and a total system matrix or rank 2Mx2M may be written as the product of N 2Mx2M layer matrices. However, most layer peeling applications (MacFarlane & Dowling 1994, Dowling & MacFarlane 1994, Narayan et al 1994, Narayan et al 1995, Haykin 2002, Proakis & Manolakis 1996) for analysis and design rely upon the 6x6 scattering matrix that describes the movement from stage k to stage k+1. The conversion of the transfer matrix to the scattering matrix follows by recognizing the 2Mx2M transfer matrix as 4 MxM sub matrices:

$$\begin{bmatrix} \vec{\mathbf{F}}_{k} \\ \vec{\mathbf{B}}_{k-1} \end{bmatrix} = \begin{bmatrix} \boldsymbol{\Psi}_{FF} & \boldsymbol{\Psi}_{FB} \\ \boldsymbol{\Psi}_{BF} & \boldsymbol{\Psi}_{BB} \end{bmatrix} \begin{bmatrix} \vec{\mathbf{F}}_{k-1} \\ \vec{\mathbf{B}}_{k} \end{bmatrix} \tag{1}$$

Two matrix equations may now be written:

$$\vec{\mathbf{F}}_{k} = \boldsymbol{\Psi}_{FF}\vec{\mathbf{F}}_{k-1} + \boldsymbol{\Psi}_{FB}\vec{\mathbf{B}}_{k} \tag{2}$$

$$\vec{\mathbf{B}}_{k-1} = \boldsymbol{\Psi}_{BF}\vec{\mathbf{F}}_{k-1} + \boldsymbol{\Psi}_{BB}\vec{\mathbf{B}}_{k} \tag{3}$$

Equation (3) may be solved for $\mathbf{B}_k$:

$$\vec{\mathbf{B}}_{k} = -\boldsymbol{\Psi}_{BB}^{-1}\boldsymbol{\Psi}_{BF}\vec{\mathbf{F}}_{k-1} + \boldsymbol{\Psi}_{BB}^{-1}\vec{\mathbf{B}}_{k-1} \tag{4}$$

which may be substituted directly into (2) to yield:

$$\vec{\mathbf{F}}_{k} = [\boldsymbol{\Psi}_{FF} - \boldsymbol{\Psi}_{FB}\boldsymbol{\Psi}_{BB}^{-1}\boldsymbol{\Psi}_{BF}]\vec{\mathbf{F}}_{k-1} + \boldsymbol{\Psi}_{FB}\boldsymbol{\Psi}_{BB}^{-1}\vec{\mathbf{B}}_{k-1} \tag{5}$$

The matrix equation follows directly from (4) and (5)

$$\begin{bmatrix} \vec{\mathbf{F}}_{k} \\ \vec{\mathbf{B}}_{k} \end{bmatrix} = \begin{bmatrix} \boldsymbol{\Phi}_{FF} & \boldsymbol{\Phi}_{FB} \\ \boldsymbol{\Phi}_{BF} & \boldsymbol{\Phi}_{BB} \end{bmatrix} \begin{bmatrix} \vec{\mathbf{F}}_{k-1} \\ \vec{\mathbf{B}}_{k-1} \end{bmatrix} \tag{6}$$

with

$$\mathbf{\Phi}_{\mathbf{FF}} = \mathbf{\Psi}_{\mathbf{FF}} - \mathbf{\Psi}_{\mathbf{FB}} \mathbf{\Psi}_{\mathbf{BB}}^{-1} \mathbf{\Psi}_{\mathbf{BF}} \tag{7}$$

$$\mathbf{\Phi}_{\mathbf{FB}} = \mathbf{\Psi}_{\mathbf{FB}} \mathbf{\Psi}_{\mathbf{BB}}^{-1} \tag{8}$$

$$\mathbf{\Phi}_{\mathbf{BF}} = -\mathbf{\Psi}_{\mathbf{BB}}^{-1} \mathbf{\Psi}_{\mathbf{BF}} \tag{9}$$

$$\mathbf{\Phi}_{\mathbf{BB}} = \mathbf{\Psi}_{\mathbf{BB}}^{-1} \tag{10}$$

Equation (6) may now be used to analyze and design the two dimensional lattice filters. While we have found common denominators (poles) and other patterns, the analytic use of Eq. (6) is still an area of research for the proposing team. On the other hand, we have used numerical techniques to study several introductory cases. In Fig. 4, for example, is the transfer function in reflection for a simple 2x2 lattice bandpass filter showing a quality factor of approximately 75,000, which is adequate for the current generation of ITU grid DWDM fiber optic communications systems. Also shown is the tunability of the filter. The surface plot shows that the position of the peak may be shifted by adjusting the gains in the semiconductor optical amplifier, for fixed four directional coupler coefficients. This point is important because it demonstrates that the filter may be adjusted in a useful manner by the user, after fabrication. The tuning range is fairly modest, at approximately 20GHz (for a reasonable 1 THz FSR), and would require changing the current by approximately 4 mA. The tuning range will increase for filters of higher order. The filter theory also shows that the operation of this device will remain stable over this tuning range. In fact the filter theory is particularly convenient in describing stability; so long as the poles of the transfer function remain in the unit circle, no laser oscillation will occur.

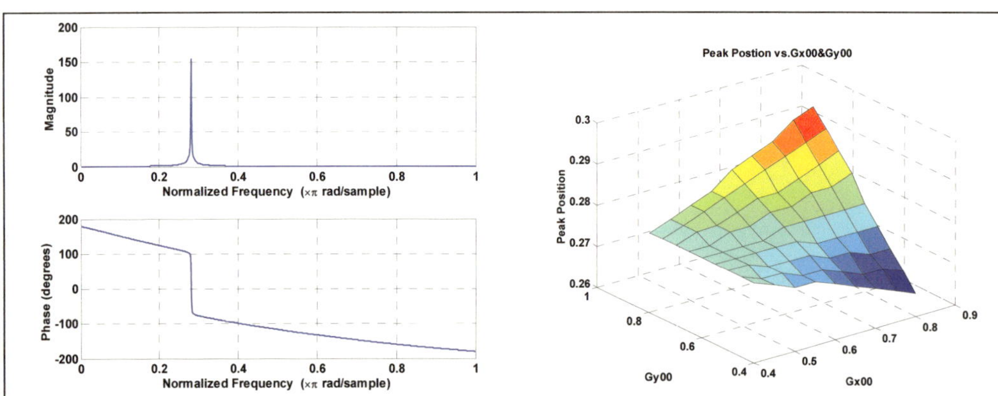

Fig. 4. Predicted bandpass behavior and gain tuning from a tunable two-dimensional active lattice filter.

In fact, there are several approaches to designing stable filters of high Q. In Fig. 10, a relatively "brute force" is taken to obtaining a high Q response with poles lying close to the unit circle. The position of the poles was chosen to account for error in the drive current to be one of the defining factors for the Q. A more thoughtful technique is shown in Fig. 11, wherein the bounded-input, bounded output stability criteria is used to define sheets of stability for a filter using a Newton-Raphson algorithm on the denominator of the transfer function (only one of several surface plots is shown). This stability criterion fixes the poles of a filter response, and provides a filter whose magnitude response is shown in the first Bode plot in Fig. 5. The second Bode plot in Fig 5 is the result optimized by then moving the zeros of the transfer function. The very sharp resonance is therefore determined in a manner that is inherently stable, since the zeros of the transfer function do not affect stability.

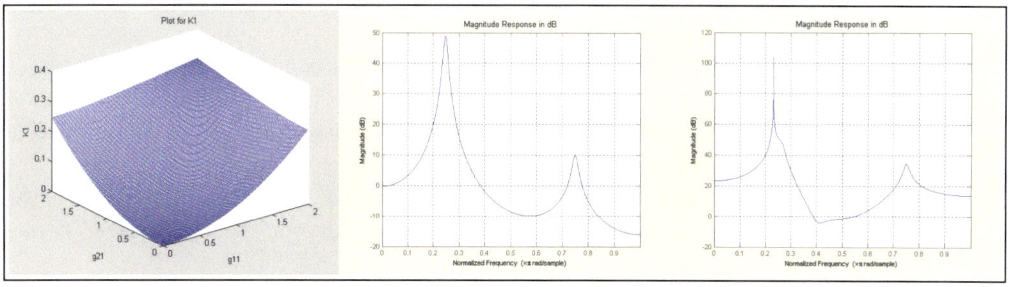

Fig. 5. Predicted bandpass behavior from a tunable two-dimensional active lattice filter. Stability is first set by the roots of the denominator, then the response is optimized by adjusting the remaining degrees of freedom in the denominator. The result is a very sharp filter resonance under fully stable operating conditions.

5. Conclusion

The photonics industry today is at a technically exciting and economically important juncture: The transition from discrete components to early, modest levels of integration. There are early indications of commercial promise for integrated photonics, including dedicated start-ups such as Infinera and Luxtera, and active development groups at large companies such as Intel. The current literature and prevailing views accept most of the basic lessons gleaned from the history of the electronic integrated circuit: the need for an integrated manufacturing platform, the value of chip real estate and overall yield. The role of gain is also well appreciated, especially in driving towards a scalable architecture. But gain also enables programmability, and therefore unlocks huge economic advantages of scale and scope. For example, the power to program microprocessors, DSPs, and FPGAs, allows the development and manufacturing costs of these devices to be amortized over a large number of "niche" applications with medium or small market sizes. The profitable "market of one" is achieved routinely by programmed microprocessors, digital signal processors (DSPs) and field programmable gate arrays (FPGAs). Since the same integrated circuit design may be used in a tremendous number of applications, the fixed costs of design, development and wafer fab can be amortized across disparate small markets. Further, it is quite common to re-program any of these integrated circuits remotely to improve performance or adapt their mission.

6. References

D. L. MacFarlane and E. M. Dowling, "Z-domain techniques in the analysis of Fabry-Perot etalons and multilayer structures," Journal of the Optical Society of America A 11, 236 (1994).

E. M. Dowling and D. L. MacFarlane, "Lightwave lattice filters for optically multiplexed communication systems," IEEE Journal of Lightwave Technology 12, 471 (1994).

V. Narayan, E. M. Dowling and D. L. MacFarlane, "Design of multi-mirror structures for high frequency bursts and codes of ultrashort pulses," IEEE Journal of Quantum Electronics 30, 1671 (1994).

V. Narayan, D. L. MacFarlane and E. M. Dowling, "High speed discrete time optical filtering," IEEE Photonics Technology Letters 7, 1042 (1995).

Duncan L. MacFarlane and Eric M. Dowling, "Active optical lattice filters," U.S. Patent 6,687,461 issued February 3, 2004.

D. L. MacFarlane, "Two Dimensional active optical lattice filters," U.S. Patent filed August, 2003.

Yablonovitch E, Gmitter TJ. Photonic band structure: the face-centered-cubic case. Physical Review Letters, vol.63, no.18, 30 Oct. 1989, pp.1950-3

J.J. Coleman, R.M. Lammert, M.L. Osowski and A.M. Jones, "Progress in InGaAs-GaAs selective area MOCVD toward photonic integrated circuits," IEEE J. Selected Topics in Quantum Electronics 3, 874-884 (1997).

Anvar A. Zakhidov, Ray H. Baughman, Zafar Iqbal, Changxing Cui, Ilyas Khayrullin, Socrates O. Dantas, Jordi Marti, and Victor Ralchenko, "Carbon Structures with three-dimensional periodicity at optical wavelengths," Science 282, 897-901 (1998).

S.D. Roh, T.S. Yeoh, R.B.Swint, A.E. Huber, C.Y.Woo, J.S.Hughes and J.J. Coleman, "Dual wavelength InGaAs-GaAs Ridge Waveguide Distributed Bragg Reflector Lasers with Tunable Mode Separation," IEEE Phot. Tech. Lett 12, 1307-1309 (2000).

Lam CF, Vrijen RB, Chang-Chien PPL, Sievenpiper DF, Yablonovitch E. A tunable wavelength demultiplexer using logarithmic filter chains. Journal of Lightwave Technology, vol.16, no.9, Sept. 1998, pp.1657-62.

R. Österbacka, C. P. An, X. M. Jiang and Z. V. Vardeny "Two-Dimensional Electronic Excitations in Self-Assembled Conjugated Polymer Nanocrystals" Science 287, 838 (2000).

T.S. Yeoh, C.P. Liu, R.B. Swint, A.E. Huber, S.D. Roh, C.Y.Woo, K.E. Lee and J. J. Coleman, "Epitaxy of InAs quantum dots on self organized two-dimensional InAs islands by atmospheric pressure metalorganic chemical vapor deposition," Appl. Phys. Lett. 79, 221-223 (2001).

M. Notomi , "Negative refraction in photonic crystals" Optical and Quantum Elec., 34, 133 (2002)

Mookherjea, S. and Yariv, A. "Coupled resonator optical waveguides", IEEE Journal of Selected Topics in Quantum Electronics (Special Issue on Nonlinear Optics) 8, 448-456 (2002).

Nakagawa, P-C Sun, C-H Chen, Y. Fainman, "Wide-field-of-view narrow-band spectral filters based on photonic crystal nanocavities," Optics Letters, Vol. 27, Issue 3, p.191 (February 2002).

C. Y. Luo, S. G. Johnson, J. D. Joannopoulos and J. B. Pendry , "Subwavelength imaging in photonic crystals," Phys. Rev. B, 68, 045115 (2003)

Minghao Qi, Eleftherios Lidorikis, Peter T. Rakich, Steven G. Johnson, J. D. Joannopoulos, Erich P. Ippen, and Henry I. Smith, "A three-dimensional optical photonic crystal with designed point defects," Nature 429, 538-542 (2004).

Adaptive Filter Theory, Simon Haykin, Prentice Hall 2002

Digital Signal Processing, Principles Algorithms and Applications, John G. Proakis and D. Manolakis, Prentice Hall, 1996.

13

Quantum Dot Photonic Devices and Their Material Fabrications

Naokatsu Yamamoto[1], and Hideyuki Sotobayashi[2]

[1]*National Institute of Information and Communications Technology,*
[2]*Aoyama Gakuin University*
Japan

1. Introduction

Optical frequency resources with wide capacity are required for the construction of photonic transport systems exhibiting high performance and flexibility (Gnauck et al. 2007 & Sotobayashi et al. 2002). The use of ultra-broadbands such as the 1–2-μm wavelength band focuses on photonic communications (Yamamoto et al. 2009a). To utilize a wide wavelength band for photonic communications, novel photonic devices must be developed for each wavelength in the 1–2-μm band. Similar to the conventional wavelength division multiplexing (WDM) photonic transport system shown in Fig. 1, it is well known that several types of optical components of photonic devices such as infrared light sources, optical modulators, optical amplifiers, optical fiber transmission lines, photodetectors, arrayed waveguide gratings, and other passive optical devices are necessary for the construction of photonic transport systems. Photonic transport systems cannot be constructed, if any one component of those photonic devices is not available. Therefore, the development of novel photonic devices in the new waveband is important for the construction of photonic transport and optical communications systems in the all-photonic waveband between 1 and 2 μm. It is expected that ultra-broadband optical frequencies greater than 100 THz can be employed for optical communications. The researches of photonic devices and physics in the all-photonic waveband will help in expanding the usable optical frequency resources for photonic communications. Additionally, the novel photonic devices developed according to the use of the all-photonic waveband can be employed for not only photonic communications devices but also for several scientific applications such as bio-imaging (Yokoyama et al. 2008), environment sensing, and manufacturing.

Figure 2 shows a typical technology map in the all-photonic waveband between 1 and 2 μm (Yamamoto et al. 2009a). Semiconductor device technology is considered to be important for developing active devices in the all-photonic waveband. Generally, InP-based semiconductor devices have been produced for photonic transport systems because conventional photonic networks have been constructed in the C- and L-band (C-band: 1530–1565 nm, and L-band: 1565–1625 nm). The widening of an optical amplifier bandwidth has been intensively studied in the conventional photonic bands of the C- and L-band. However, GaAs-based, Si-based, and SiGe-based semiconductor photonic devices will

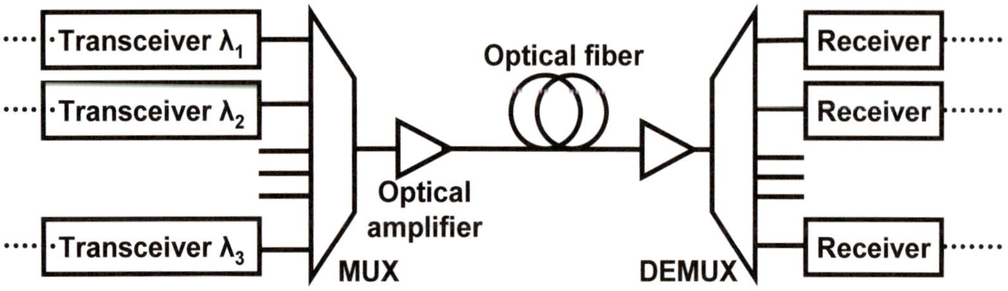

Fig. 1. Schematic image of wavelength division multiplexing (WDM) photonic transport system.

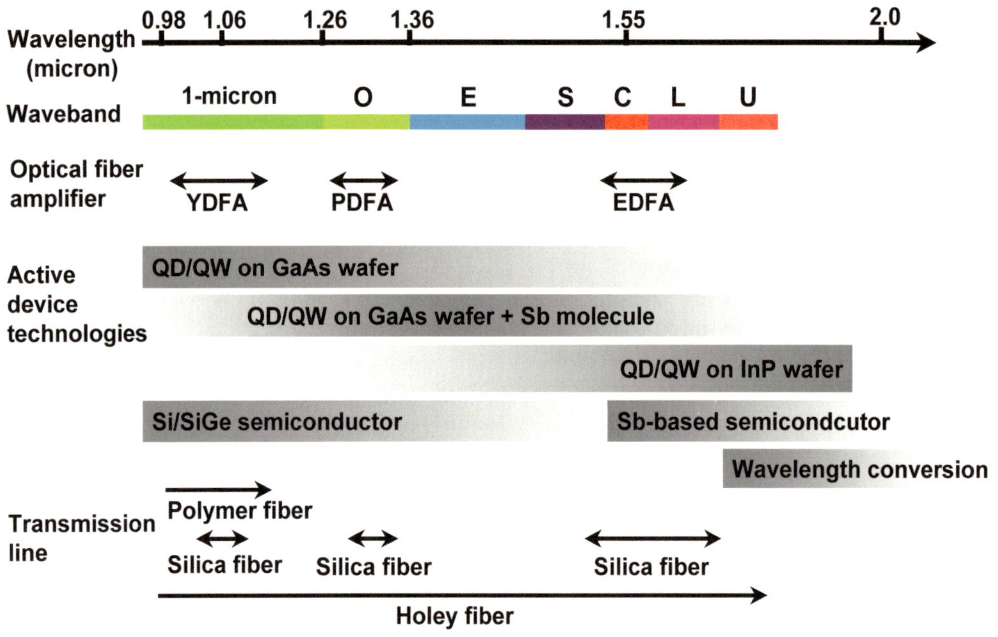

Fig. 2. Technology map and photonic waveband for optical communications. The abbreviations QD and QW denote the quantum dot and quantum well structures, respectively.

become powerful candidates for use in shorter wavelengths such as a 1-μm and O-bands (O-band: 1260–1360 nm) in photonic transport systems (Hasegawa et a. 2006; Yamamoto et al. 2008d; Ishikawa et al. 2009 & Koyama 2009). In particular, high-performance and wide optical frequency band fiber amplifiers (Ytterbium-doped fiber amplifier: YDFA, and Praseodymium-doped optical fiber amplifier: PDFA) can be employed in shorter wavebands (Paschotta et al. 1997). In the ultra-long wavelength band in the 1625–2000 nm and mid-infrared region (>2000 nm), Sb-based semiconductors such as GaSb and InGaSb are useful materials for the development of the photonic devices such as light-emitting diodes, semiconductor lasers, and detectors. Additionally, in this wavelength region, the wavelength conversion technique with an optical nonlinear effect is also employed for constructing light sources. Optical fiber transmission lines are important devices for the construction of photonic transport systems. Ultra-wideband and low-loss photonic transmission lines have been intensively investigated by using holey fiber, hole-assisted fiber, and photonic crystal fiber structures (Mukasa et al. 2007 & 2008). From Fig. 2, it is expected that photonic devices for the all-photonic waveband will be developed by combining GaAs-, InP-, GaSb-, SiGe-, and Si-based semiconductor device technologies. Additionally, implementing nanotechnology for these semiconductor materials is a powerful solution to enhance a usable waveband for semiconductor photonic devices. A quantum dot (QD) is a useful and simple structure for achieving a three-dimensional confinement of electrons and/or holes in the semiconductor (Arakawa et al., 1982). Therefore, the energy levels of the confined electrons and holes can be controlled artificially by controlling the size of the QD structure. It is well known that self-assembled semiconductor QDs exhibit interesting and excellent properties as compared to semiconductor bulk or quantum well structures. The typical properties are as follows: (1) quantum size effect, (2) high confinement efficiency of carriers, (3) desirable quantum levels, and (4) no restrictions on the crystal lattice constant. It is expected that these useful properties improve the device performance. For example, it is possible to fabricate low-threshold lasers (Shimizu et al. 2007), un-cooled lasers (Otsubo et al. 2004; Tanaka et al. 2009), long-wavelength lasers (Ledentsov et al. 2003; Yamamoto et al. 2005 & Akahane et al., 2008), high-power lasers (Tanguy et al. 2004) and ultra-broadband lasers by using QD structures (Rafailov et al. 2007). Additionally, the ultra-broadband semiconductor optical amplifier is also expected to be fabricated by using the QD structure. In this chapter, the development of a semiconductor QD laser and its photonic transport applications are described. The QD structure is considered to be suitable for the development of important devices for the all-photonic waveband.

2. Quantum dot photonic device and optical communications

2.1 Broadband quantum dot laser

The semiconductor QD structures are expected for broadband optical gain materials. Generally, the self assembled semiconductor QD structure is formed on the GaAs or InP substrate under a S-K (Stranski-Krastanov) growth mode by using molecular beam epitaxy (MBE) and a metal-organic chemical vapor deposition (MOCVD) technique. Figure 3(a) shows an atomic force microscope (AFM) image of an InGaAs QD structure fabricated on a GaAs (001) wafer surface. The InGaAs/GaAs QD structure is fabricated by using solid source MBE. The typical height and dimensions of the InGaAs/GaAs QD structure is approximately 4 nm and 20 nm, respectively. It is well known that the density and structure of the QD hardly influence the surface condition before the growth of the QD structure. Therefore, the Sb-molecular irradiation technique (Yamamoto et al. 2008b), Si-atom irradiation technique, and sandwiched sub-nano-separator (SSNS) structure (Yamamoto et al. 2009c) are proposed to enhance the QD density and reduce the giant dot and crystal defect. Figure 3(b) shows a schematic image of the cross-sectional image of the InGaAs/GaAs QD structure embedded in the GaAs matrix with the surface controlling technique of Sb-irradiation. In other words, a high quality QD structure is obtained by using these surface controlling techniques. Therefore, it is expected that the surface controlling techniques employed during QD growth may improve the device performance. The density of the QD structure is estimated to be as high as $5.3 \times 10^{10}/cm^2$. The In composition and deposition amount of the QD structure are controlled in order to tune the emission wavelength. In this case, the In composition and deposition amounts are fixed as approximately 0.5 and 6.0 ML, respectively, in order to fabricate the InGaAs/GaAs QD structure emitting in the 1-μm waveband.

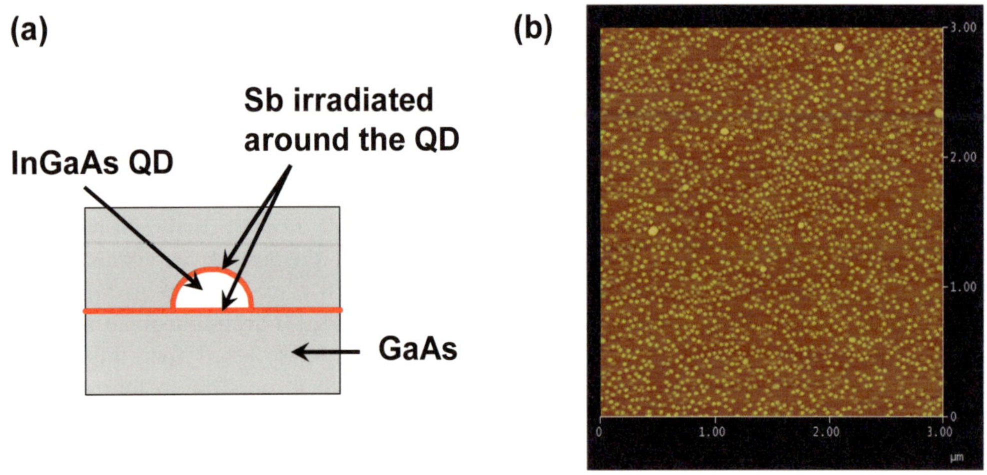

Fig. 3. (a) Schematic cross sectional structure of Sb-irradiated quantum dot structure, and (b) atomic force microscope (AFM) image of the InGaAs/GaAs quantum dot structure on a surface area of 3 μm^2.

Figure 4(a) shows a cross-sectional schematic image of a ridge-type QD laser structure. The fabrication technique employed for a GaAs-based laser device can be applied to QD laser devices on the GaAs wafer. In the core region, multi-stacked QD layers are generally fabricated with a 50-nm spacer GaAs layer. The QD core region is sandwiched by AlGaAs cladding layers with 1- or 2-μm thickness. A growth temperature of the top cladding layer is generally lower than a temperature for the conventional GaAs based laser, because a structure of the fabricated QD is influenced with the high growth temperature of the cladding layer. Figure 4(b) shows a cross-sectional scanning electron microscope (SEM) image of the fabricated QD laser structure. A buried polyimide process and a lift-off technique are carried out for fabricating the ridge-type QD laser diode. The width of the ridge waveguide structure is generally fixed from approximately 2 to 7 μm to achieve a single mode and low-threshold current operations. Naturally, the etching depth of the ridge structure depends on the width of the ridge. A mounted QD laser diode on a chip carrier is applied to photonic transport systems and bio-imaging, because a stable operation of the QD laser diode can be achieved by using the chip carriers. Figure 5(a) shows the mounted QD laser diode on the carrier. This mount technique is similar to the conventional technique used for the GaAs-based laser devices. In other words, wire-bonding and die-bonding techniques are used for the fabrication of the QD laser diode chip. It is important that a large number of fabrication technologies developed for GaAs-based devices are applied to the QD/GaAs device fabrication. Figure 5(b) shows the laser spectra obtained from two types of QD/GaAs lasers. The InGaAs QD/GaAs laser diode emission has a wavelength of 1.04 μm. Additionally, it is clearly observed that other laser emissions have a wavelength of 1.27 μm in the O-band. Laser emission with longer wavelengths can be achieved by using novel optical gain materials such as an Sb-irradiated QD in the well (Sb-DWELL) or InAs/InGaAs QD with SSNS structures (Liu et al. 2003; Yamamoto et al. 2008b, 2009a & 2009c). These emission wavelengths are matched to the ground state of the QD structure. It is well known that ground state lasing must be applied to achieve a low-threshold current density operation of the QD laser. These emission peaks such as 1.04 and 1.27 µm are suitable for the optical gain bandwidth of the YDFA and PDFA, respectively. Therefore, these QD laser devices are highly suitable for photonic transport systems in the 1-µm band and O-band. An emission wavelength of the conventional GaAs-based laser devices has a limitation of up to approximately 1.06 µm. Therefore, it is found that an expansion of the usable wavelength band of the GaAs-based laser diode can be achieved by using QD structures. It is clear that the fabrication of the long and ultra-broad wavelength band (1.0–1.3 µm) light sources can be achieved by combining a novel QD growth technology with the conventional GaAs-based device technology.

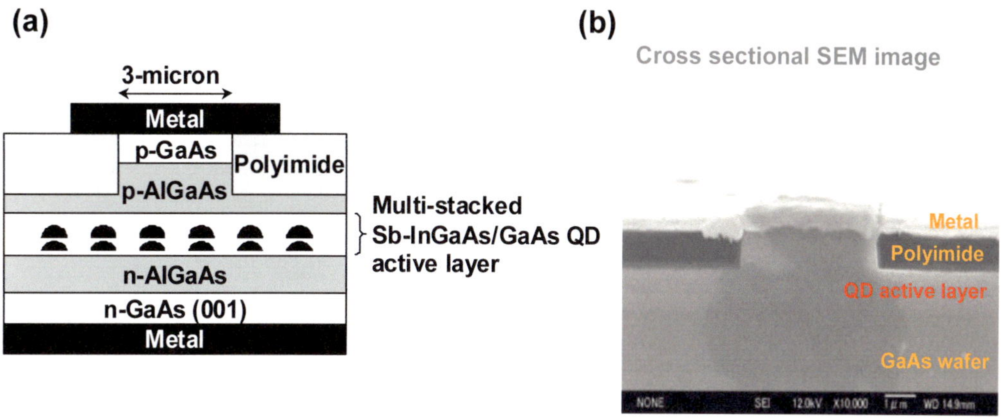

Fig. 4. (a) Schematic cross-sectional image of the InGaAs/GaAs quantum dot laser structure fabricated on a GaAs wafer. (b) Cross-sectional scanning electron microscope image of the ridge-type quantum dot laser.

2.2 Quantum dot wavelength tunable laser

It is expected that an ultra-broadband optical gain will be realized by using QD gain materials. Therefore, the broadband wavelength tunable laser is also achieved by using the QD structures. In this section, one of the QD wavelength tunable laser scheme is introduced. The InGaAs/GaAs QD structure is used to fabricate a wavelength tunable laser in the 1-µm waveband, because the ultra-broadband optical gain can be achieved by using the QD active media as compared to the conventional quantum well (QW) structure. An optical gain material of the InGaAs/GaAs QD laser diode is prepared by using solid-source MBE. A self-assembled QD structure is incorporated by using an Sb-molecule-irradiated InGaAs material on GaAs (001) surfaces together with AlGaAs cladding layers. Here, the emission wavelength corresponding to the QD ground state is tuned in to the 1-µm optical-waveband. Thus, from the MBE-grown QD layers, a 3-µm wide ridge-waveguide laser structure is formed through a standard sequence of GaAs-based semiconductor laser fabrication. The cavity length of the structure is 2 mm. The edge of the laser diode is a cleaved facet. Figure 6(a) shows a schematic configuration of the injection-seeding scheme with the operation wavelength tenability (Yamamoto et al. 2008c & Katouf 2009). A narrow-band optical wedge filter (0.6 nm) is incorporated between the QD laser chip and an external

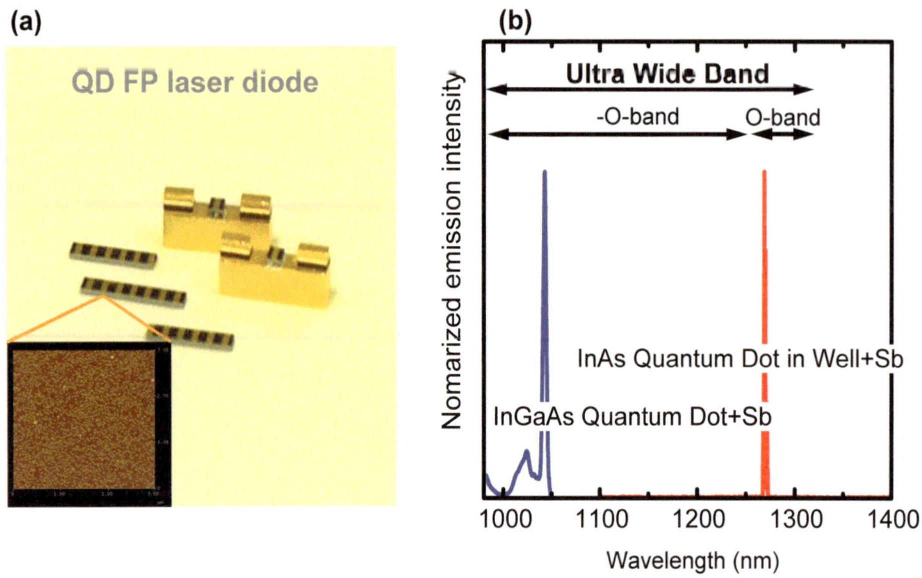

Fig. 5. (a) Photograph of the quantum dot laser diode chip. (b) Laser emission spectrum for the quantum dot laser diode in an ultra-wideband between the 1-µm and 1.3-µm wavelengths.

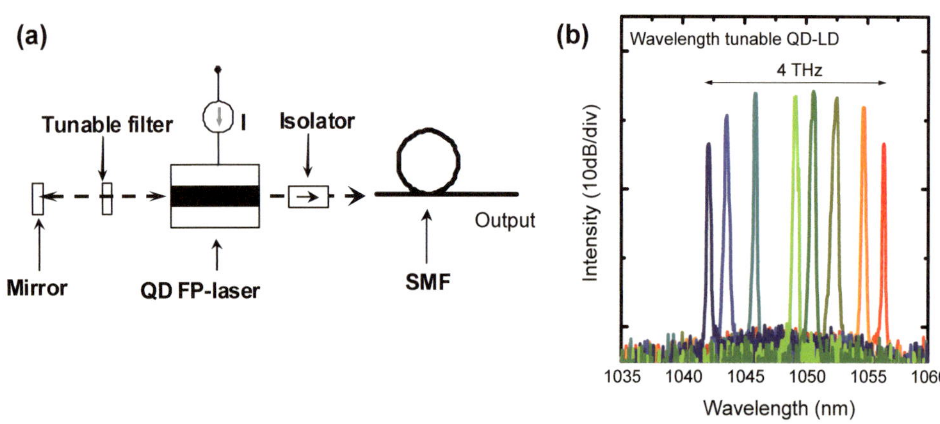

Fig. 6. (a) Wavelength tunable laser constructed with a self-injection seeded quantum dot Fabry-Perot laser. (b) 4-THz tuning range of the 1-µm wavelength tunable quantum dot laser.

mirror, which facilitates the tunability of the emission wavelength. The centre wavelength of the optical filter is controlled by adjusting the light-beam position on the filter. In other words, the wavelength selected by the filter is injected to the laser chip to lock the lasing wavelength of the QD laser diode. The temperature of the laser chip is maintained at 300 K by a thermoelectric cooler stage. The optical output from the laser is coupled to a single-mode optical fiber for the 1-μm optical waveband.

Figure 6(b) shows the typical experimental result of the injection-seeded operation of the QD wavelength tunable laser. The lasing operation is confirmed in a wavelength ranging from 1042 nm to 1057 nm, which corresponds to a broad optical frequency band with a 4-THz bandwidth and consequently to 40 WDM channels with a 100-GHz grid. The tunable frequency of the 4-THz bandwidth is similar to the bandwidth of the C-band. It should be noted that each laser emission peak in Fig. 6(b) is prominent and its optical power level is at least 25 dB higher than that of amplified spontaneous emission. Furthermore, it has been found out that the undulation of the optical output power in the wavelength ranging from 1045 to 1052 nm is 1.0 dB or less. Additionally, all the laser output in the injection-seeding bandwidth can be successfully amplified to up to 10 dBm by using the YDFA. This amplified output power level suggested that broadband WDM photonic transport systems can be feasible with the present devices. On the other hand, a photonic transmission experiment was performed using wavelength tunable QD laser devices. By using the wavelength tunable QD laser for the 1-μm waveband, a 2.54-Gbps error-free transmission with a clear eye opening was successfully demonstrated over the 1-km hole-assisted fiber.

Some wavelength tuning techniques of semiconductor lasers are already proposed, such as conventional techniques of an external cavity scheme and a multi-sectional electrode scheme. These techniques can be simply employed for achieving the broadband tunability width of the QD lasers.

2.3 1-μm waveband photonic transport system

To construct a WDM photonic transport system, the essential photonic devices required are a stable multiwavelength light source suitable for high-speed (>10 Gbps) data modulation, long-distance single-mode transmission optical fiber, wavelength multiplexer (MUX)/demultiplexer (DEMUX), and numerous passive devices. In this section, a 1-μm waveband photonic transport system is demonstrated to pioneer the novel waveband for optical communications (Yamamoto et al. 2008d & 2009b; Katouf et al. 2009). It is considered that a 1-μm waveband QD laser is useful for the optical signal source because a wide optical gain bandwidth can be realized by using the QD structure. Therefore, the QD light source and the photonic transport system are demonstrated. As the QD light source, the generation of a 1-μm waveband optical frequency comb from the fabricated QD optical frequency comb laser (QD-CML) and a method for an optical mode selection for a single-mode operation of the QD-CML are introduced. Additionally, to realize a WDM photonic transport in the 1-μm waveband, a long-distance single-mode holey fiber (HF) and an arrayed waveguide grating (AWG) are also introduced for the transmission line and MUX/DEMUX devices, respectively.

The Sb-molecular irradiated InGaAs/GaAs QD ridge type laser diode was used as the light source for the photonic transport system. The QD laser diode acts as a QD-CML in the 1-μm waveband under high current injection conditions. Figure 7(a) shows the optical frequency comb spectrum obtained from the QD-CML. The frequency bandwidth of the generated optical frequency comb is as wide as ~2.2 THz under a current of few hundred mA. The frequency bandwidth increased with the QD laser current. The free spectral range (FSR) of the optical frequency comb generated from the QD-CML is estimated to be approximately 20 GHz, which is close to the Fabry-Perot mode spacing corresponding to the cavity length.

It is expected that the QD-CML will emerge as an important light source and will have applications as a compact optical frequency comb generator in photonic networks, bio-imaging, etc (Gubenko et al. 2007).

The single- and discrete-mode selections of the QD-CML are important techniques for photonic communications. For applying the single-mode selection technique, an external mirror and a wavelength tunable filter were used for self-seeded optical injection. An optical discrete mode was selected by using the wavelength tunable filter. Figure 7(b) shows an optical spectrum of the single-mode selected QD laser. A sharp peak can be observed at 1047 nm. By using this technique, the side-mode suppression ratio (SMSR) and spectral line width were possibly >20 dB and <0.03 nm, respectively. Hence, the center wavelength of the lasing mode could be selected by controlling the wavelength tunable filter.

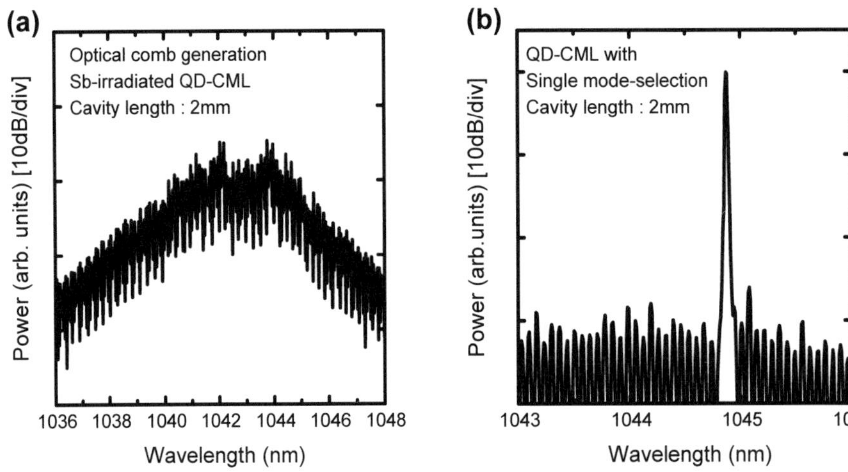

Fig. 7. (a) Optical frequency comb generation from the quantum dot optical frequency comb laser (QD-CML). (b) Optical spectrum of the single-mode selected quantum dot laser.

Figure 8 shows the experimental setup for testing the WDM photonic transmission in the 1-μm waveband (Yamamoto et al. 2009a & 2009b) at 12.5 Gbps. The single-mode selected QD-CML was used as the wavelength tunable non-return to zero (NRZ) signal optical source. The lasing optical mode was selected by using the discrete single-mode selection technique. The selected mode was fitted to the channel spacing (100 GHz) of the AWG device in the 1-μm waveband. The optical signal was amplified by using a YDFA after a 12.5-Gbps and a 2^{15}-1 pseudorandom binary sequence (PRBS) data modulation. The optical signal was passed through the AWG pair. In other words, the AWG pair played the role of a DEMUX and MUX for the multiwavelength optical signal. A single-mode HF was developed for the transmission line in the 1-μm waveband. The dispersion characteristics of the HF were controlled by controlling the size of the holes and their distances from the fiber core (Mukasa et al. 2008 & 2009). The input power to the transmission line was approximately 0 dBm. The transmitted optical signal was amplified again by using a YDFA before the measurements. The optical filters positioned after the YDFAs were used for cutting off the amplified spontaneous emission (ASE) noise in the YDFAs. Figure 9(a) shows the optical spectra measured after a 1.5-km-long HF transmission at four different wavelengths (ch.1:

1042.71 nm–ch.4: 1043.85 nm). Each of the central wavelengths is selected for the 100-GHz channel spacing of the AWG by using the discrete single-mode selection method of the QD-CML. Figure 9(b) shows a typical eye diagram at ch. 2 after transmission. A clear eye opening at 12.5 Gbps is observed after the transmission. Therefore, the 1-μm waveband with a 12.5-Gbps transmission over a long-distance (1.5 km) single-mode HF is successfully

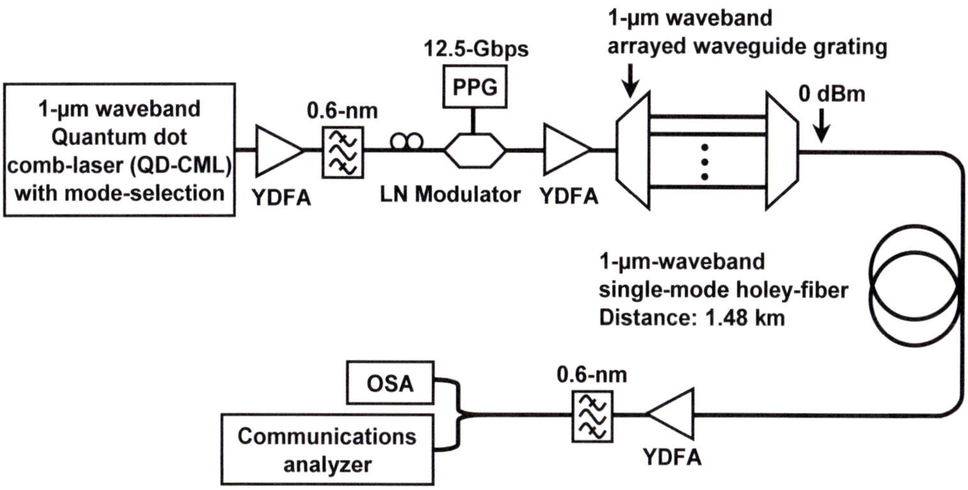

Fig. 8. Experimental set-up for testing the 1-μm WDM photonic transport system. A 1-μm waveband and single-mode selected quantum dot optical-frequency comb laser (QD-CML) was used for the light source.

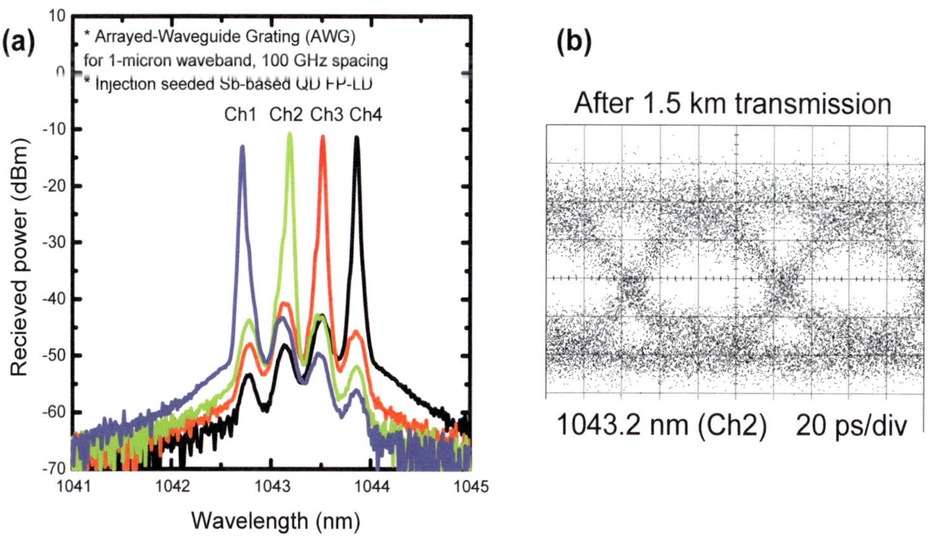

Fig. 9. (a) Optical spectrum of 12.5-Gbps and single-mode selected QD-CML after 1.5-km transmission of the holey fiber. (b) Eye opening of ch.2 after transmission.

achieved at four different wavelengths by using a wavelength-tunable discrete single-mode selected QD laser device. The 1-µm waveband AWG, YDFAs, and other passive devices are also important to construct the 1-µm waveband photonic transport system. From these results, a 12.5-Gbps-based WDM photonic transmission with a 100-GHz channel spacing can be realized in the 1-µm waveband by using the proposed methods. Additionally, it is expected that the QD photonic devices such as a semiconductor laser fabricated on the GaAs wafer will become a powerful candidate to realize an ultra-broadband 1- to 1.3-µm photonic transport system.

3. Quantum dot structure for advanced photonic devices

In this section, novel material systems of a QD structure are introduced for advanced photonic devices. The novel materials of the QD are expected to be used in laser device fabrication, silicon photonics, visible light-emitting devices, etc.

3.1 Long-wavelength quantum dot structure

Sb-based III-V semiconductor materials have very narrow-band gap properties. Therefore, the use of Sb-based III-V semiconductor QD structures (the Sb atoms are included in the QD structure) are expected for producing long-wavelength-emitting devices (Yamamoto et al. 2005 & 2006b). In this section, the Sb-based QD structure fabricated on a GaAs substrate is introduced. However, the fabrication of the Sb-based QD such as an InGaSb QD is difficult under conventional QD growth conditions with the MBE method. To form the high-quality Sb-based QD structure, a Si atom irradiation technique is proposed as one of the methods for surface treatment. Figure 10(a) shows a schematic image of the Si atom irradiation

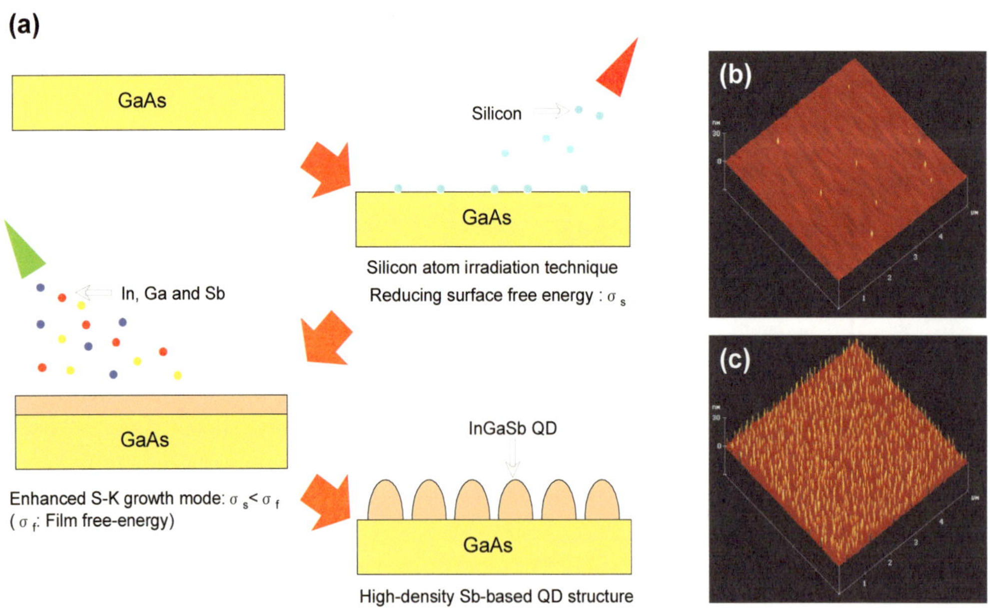

Fig. 10. (a) Schematic image of silicon atom irradiation technique for the fabrication of the high-quality QD structure. AFM images of InGaSb QD structure in a 5 × 5 -µm^2 region on GaAs substrate without (b) and with (c) the Si atom irradiation technique.

technique. Low density Si atoms are irradiated on to the GaAs surface immediately before the Sb-based QD structure growth. It is expected that the surface free-energy may be reduced with the irradiation of Si atoms. Therefore, the density of the Sb-based QD structure is enhanced by using this atom-irradiation technique. Figures 10(b) and (c) show the AFM images of the Sb-based QD structure without and with the Si atom irradiation, respectively. It is found that the QD density with Si atoms is approximately 100 times higher than that without Si atoms. Generally, the QD density as high as $10^{10}/cm^2$ is necessary if the QD structure is used for developing a laser or other photonic devices. Therefore, the optimization of the QD growth conditions such as growth-rate, As-flux intensity, and temperature is also important to obtain the high-quality QD structure. Figure 11(a) shows an AFM image of the Sb-based QD/GaAs structure under the optimized growth conditions. The height, dimension, and density of the Sb-based QD are approximately 7.5 nm, 25 nm, and $2 \times 10^{10}/cm^2$, respectively.

An ultra-wideband emission between wavelengths of 1.08- and 1.48-μm can be successfully realized by using the Sb-based QD/GaAs structure, as shown in Fig. 11(b). The long-wavelength and ultra-broadband emission is also obtained from a light-emitting diode (LED) that contained the Sb-based QD in active regions. From this result, it is expected that ultra-broadband wavelength (>350 nm) light sources may be achieved with the QD structure for the O-, E-, S-, and C-band (Yamamoto et al. 2009a).

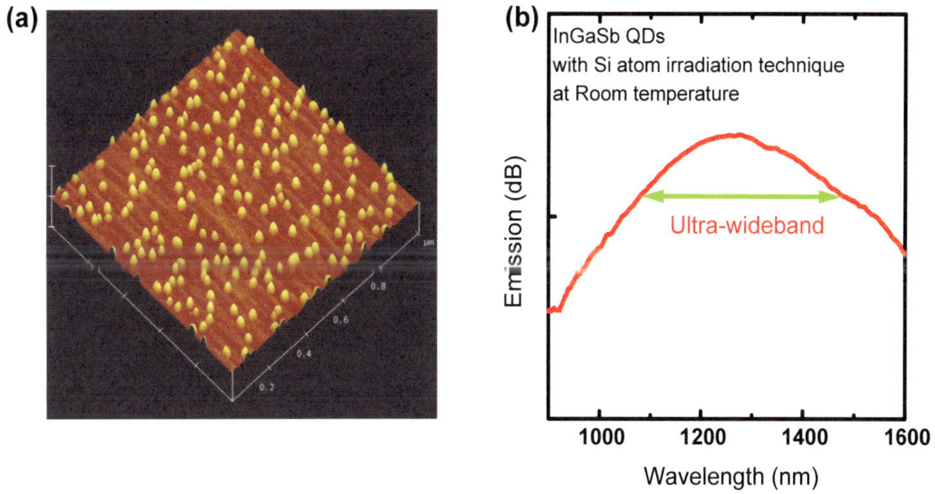

Fig. 11. (a) Atomic force microscope image of high-quality Sb-based QD (InGaSb QD) structure on GaAs surface. (b) Ultra broadband and long-wavelength emission from the Sb-based QD/GaAs structure.

The combination of a micro-cavity structure and the QD structure is a very interesting device structure for the investigation of cavity quantum-electrodynamics (QED). Study on the QED of the QD structure is important for constructing a quantum communications system (Ishi-Hayase et al. 2007 & Kujiraoka et al. 2009). A vertical cavity structure and a photonic crystal structure as an optical resonator are useful for confining the photons (Nomura et al. 2009). Figure 12(a) presents a cross-sectional image of a fabricated vertical cavity structure, which include the Sb-based QD in the cavity. A high-performance diffractive Bragg reflector (DBR) for accomplishing the vertical cavity structure can be simply produced by using an AlGaAs material system. From the Sb-based QD structure in the vertical cavity, a 1.55-µm sharp emission peak, as shown in Fig. 12(b), is successfully observed under the optically pumped condition (Yamamoto et al. 2006a). It is also found that a long-wavelength emission with a 1.52-µm peak can be obtained from the similar QD in the cavity structure at room temperature with a current injection. Therefore, it is expected that the use of the long wavelength QD active media in the semiconductor micro-cavity structure is a very useful and important way for fabricating long-wavelength and multiwavelength vertical cavity surface emitting lasers (VCSELs), resonant cavity light-emitting diodes (RCLEDs), single photon sources, etc.

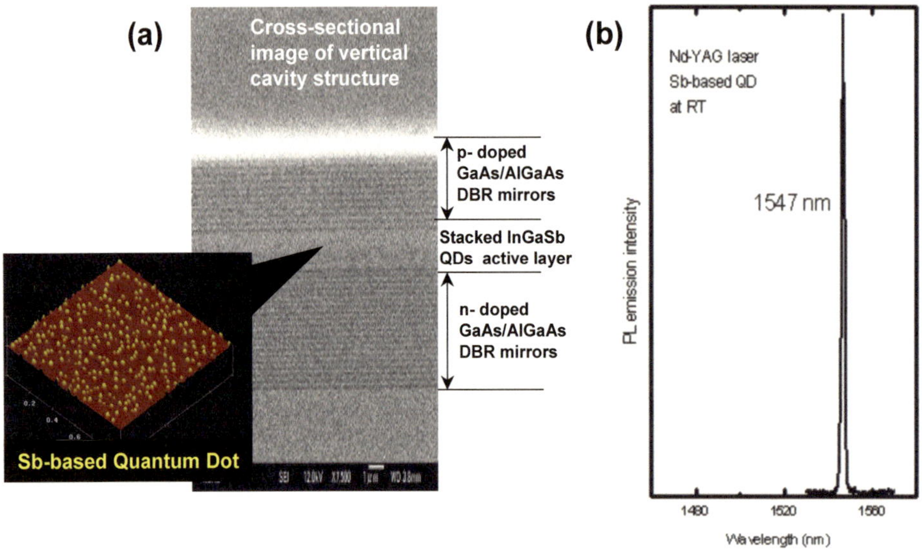

Fig. 12. (a) Sb-based QD in micro cavity structure and (b) 1.55-µm wavelength emission spectrum from optically pumped vertical cavity structure.

3.2 Quantum dot and related materials for silicon photonics

Silicon photonics technology has been conventionally used to fabricate high performance photonic circuits, which have low-power-consumption, are compact, and are relatively inexpensive to fabricate (Liu et al. 2004 & Yamamoto et al. 2007b). Poly-, amorphous-, and crystalline-Si waveguide devices have been developed and their properties have been investigated. An optical gain region must be provided for silicon waveguide structures to enable the fabrication of active devices such as light emitters and optical amplifiers on silicon platforms (Balakrishnan et al. 2006). As one of the candidates of the optical gain media, a III-V semiconductor QD structure on a Si wafer has been investigated. Figure 13 shows the schematic image of the Sb-based QD/Si structure and AFM images of the Sb-based QD structures grown between 400°C and 450°C on Si substrates (Yamamoto et al. 2007a). From the AFM image, it is found that the high-quality and high-density Sb-based QD structure can be obtained under the optimal growth conditions by MBE. Therefore, a high-density (>10^{10} /cm^2) and small-sized (<10 nm) QD structure can be obtained by growing the QDs below 400°C. From this result, it is expected that the nanostructured Sb-based semiconductors with a low-temperature process (<400°C) should become useful materials for complementary metal oxide semiconductor (CMOS) devices compatible with silicon photonics technology (Yamamoto et al. 2008a). Additionally, it is also expected that the nanostructured Sb-based semiconductor will be used for high-speed electro-devices, because the III-Sb compound semiconductor has high-mobility characteristics (Ashley et al., 2007).

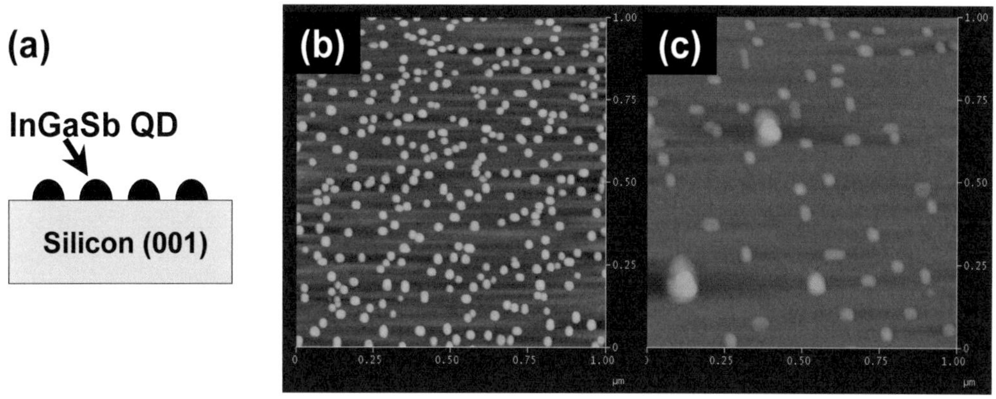

Fig. 13. (a) Schematic image of Sb-based QD structure on Si wafer, and AFM images of the Sb-based QD on Si at (b) 400°C and (c) 450°C.

Compound semiconductors are widely studied for the fabrication of the QD structure because they exhibit an observable quantum size effect in the quantum confinement structure of a relatively large size (approximately few tens of nanometers). On the other hand, a carrier confined structure several nanometers in size, which is generally called a nanoparticle, is necessary when using a silicon semiconductor material. Several techniques have been proposed for the fabrication of the Si nanoparticle as a Si-QD structure (Canham et al. 1990). An anodization method and a photochemical etching method of a Si wafer are proposed for producing the Si nanoparticles (Yamamoto et al. 2001 & Hadjersi et al. 2004). It is known that the Si nanoparticle exhibits a bright visible light emission of red or blue color, and it is considered that this light emission is caused by the quantum size effect of the Si-QD. Figure 14(a) shows a visible emission spectrum from the photochemically etched layers, such as Si nanoparticles (Yamamoto et al. 1999). In addition, electroluminescence devices on a Si wafer are also demonstrated using Si nanoparticles, as shown in Figure 14(b). It is expected that the Si nanoparticle as the Si-QD structure will become a useful material for the visible light-emitting devices with Si-based electric devices (Yamamoto et al. 2000).

4. Conclusion

The quantum dot (QD) structures are intensively investigated as the three-dimensional carrier confined structure. It is expected that the QD structure can act likely as an atom, which has a controllable characteristic of energy levels. The semiconductor QD structure is a very important material for developing novel photonic devices. In this chapter, fabrication techniques and characteristics of novel QD photonic devices such as a broadband QD light

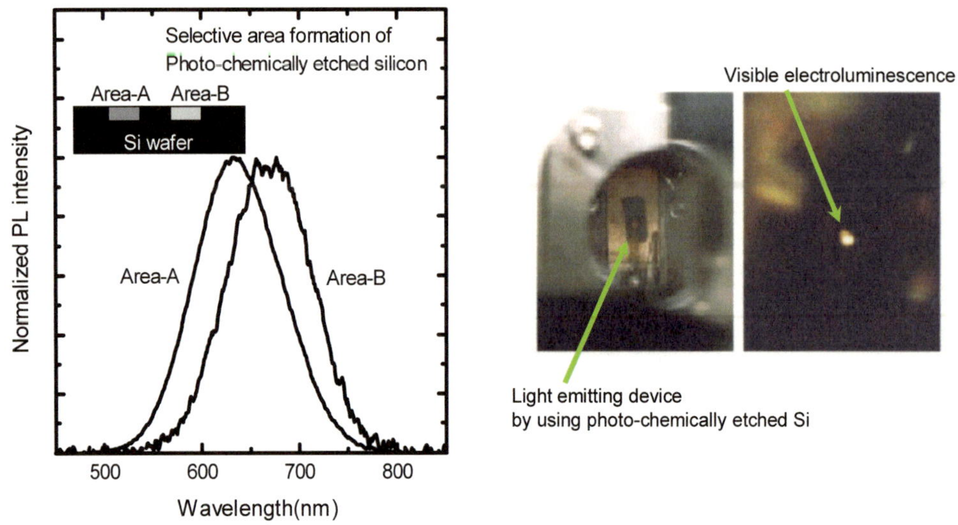

Fig. 14. (a) Emission spectra of photochemically etched layers as Si nanoparticles. The emission colors in areas A and B are observed as yellow and red, respectively. Each layer is formed on the same Si substrate using a selective area formation technique. (b) Visible electroluminescence devices on Si wafer by using the Si-particle as the Si-QD.

source and a wavelength tunable QD laser were explained. The QD light source act in a broad wavelength band between 1-µm and 1.3-µm can be fabricated on the GaAs substrate as a low cost and large-sized wafer by using InAs QD and InGaAs QD structures as an active media. In addition, a fabrication technique of the Sb-based QD structures on the GaAs substrate was demonstrated for the ultra-broadband light source between 1 and 1.55 µm, and the novel photonic devices using the cavity-QED. In other words, by using the QD structure, ultra-broadband optical gain media can be achieved for broadband light-emitting diodes, wavelength tunable laser diodes, semiconductor optical amplifiers, etc. Additionally, the QD structures have interesting opto-electric characteristics compared to the conventional quantum well and bulk materials. It is expected that the QD optical frequency comb laser (QD-CML) can be realized by using the useful characteristics of the QD structure.

Ultra-broadband optical frequency resources in the short wavelength band such as the 1-µm waveband can be used for optical communications. As the 1-µm waveband photonic transport system, over 10 Gbps and a long distance transmission were successfully demonstrated by using high-performance key components such as single-mode QD light sources, long-distance holey fibers, and YDFAs. Therefore, it is expected that the uses of the QD photonic devices enhance the usable waveband for optical communications.

For the silicon photonics, a fabrication technique for the high-quality Sb-based QD structure on a Si wafer was demonstrated clearly. As the other QD structure for the silicon photonics, it is also demonstrated that Si nanoparticles as the Si-QD become candidates for the light-emitting devices on the Si wafer.

It is expected that a fabrication and application of the QD structure will provide a breakthrough technology for the creation of novel photonic devices, improvement in the existing photonic devices, and enhancement of usable optical frequency resources in the all-photonic waveband.

5. Acknowledgments

The authors would like to thank Prof. H. Yokoyama at New Industry Creation Hatchery Center (NICHe) of Tohoku University, Prof. H. Takai at Tokyo Denki University (TDU), Drs. K. Akahane, R. Katouf, T. Kawanishi, I. Hosako, and Y. Matsushima at the National Institute of Information and Communications Technology (NICT) for discussing novel technologies of the quantum dot photonic devices and lasers. The authors are deeply grateful to Drs. K. Mukasa, K. Imamura, R. Miyabe, T. Yagi, and S. Ozawa at FURUKAWA ELECTRIC CO. for discussing broadband transmission lines of the novel optical fibers.

6. References

Akahane, K.; Yamamoto, N., Sotobayashi, H. & Tsuchiya, M. (2008). 1.7-μm Laser Emission at Room Temperature using Highly-Stacked InAs Quantum dots. Proceedings of Indium Phosphide and Related Material (IPRM) 2008, Versailles, 171

Arakawa, Y. & Sakaki, H. (1982). Multidimensional quantum well laser and temperature dependence of its threshold current. *Appl. Phys. Lett.* Vol. 40, 939 (1982) 939

Ashley, T.; Buckle, L.; Datta, S.; Emeny, M. T.; Hayes, D. G.; Hilton, K. P.; Jefferies, R.; Martin, T.; Phillips, T. J.; Wallis, D. J.; Wilding, P. J. & Chau, R. (2007). Heterogeneous InSb quantum well transistors on silicon for ultra-high speed, low power logic applications. *Electron. Lett.*, Vol. 43, (2007) 777

Balakrishnan, G.; Jallipalli, A.; Rotella, P.; Huang, S.; Khoshakhlagh, A.; Amtout, A. & Krishna, S. (2006). Room-Temperature Optically Pumped (Al)GaSb Vertical-Cavity Surface-Emitting Laser Monolithically Grown on an Si(100) Substrate. *IEEE J. Sel. Topics in Quantum Electron*, Vol. 12, (2006) 1636

Canham, L.T. (1990). Silicon quantum wire array fabrication by electrochemical and chemical dissolution of wafers. *Appl. Phys. Lett.* Vol. 57, (1990) 1046

Gnauck, A. H.; Charlet, G.; Tran, P.; Winzer, P.; Doerr, C.; Centanni, J.; Burrows, E.; Kawanishi, T.; Sakamoto, T. & Higuma, K. (2007). 25.6-Tb/s C+L-Band Transmission of Polarization-Multiplexed RZ-DQPSK Signals, *Proceedings of OFC2007*, 2007, Anaheim, CA, PDP19

Gubenko, A.; Krestnikov, I.; Livshtis, D.; Mikhrin, S.; Kovsh, A.; West, L.; Bornholdt, C.; Grote, N. & Zhukov, A. (2007). Error-free 10 Gbit/s transmission using individual Fabry-Perot modes of low-noise quantum-dot laser. *Electron. Lett.*, Vol. 43, (2007) 1430

Hasegawa, H.; Oikawa, Y.; Yoshida, M.; Hirooka, T. & Nakazawa, M. (2006). 10Gb/s transmission over 5 km at 850nm using single-mode photonic crystal fiber, single-mode VCSEL, and Si-APD. *IEICE Electron. Express*, Vol. 3, (2006) 109-114

Hadjersi, T.; Gabouze, N.; Yamamoto, N.; Sakamaki, K. & Takai, H. (2004). Photoluminescence from photochemically etched highly resistive silicon. *Thin solid films*, Vol. 459, (2004) 249-253

Ishi-Hayase, J.; Akahane, K.; Yamamoto, N.; Sasaki, M.; Kujiraoka, M. & Ema, K. (2007). Radiatively limited dephasing of quantum dot excitons in the telecommunications wavelength range. *Appl. Phys. Lett.*, Vol. 91 (2007) 103111

Katouf, R.; Yamamoto, N.; Akahane, K.; Kawanishi, T. & Sotobayashi, H. (2009). 1-μm- band transmission by use of a wavelength tunable quantum-dot laser over a hole-assisted fiber, *Proceedings of SPIE 7234*, p. 72340G, 2009

Koyama, F.; (2009). VCSEL photonics –advances and new challenges-. *IEICE Electronics Express*, Vol. 6 (2009) 651

Kujiraoka, M.; Ishi-Hayase, J.; Akahane, K.; Yamamoto, N.; Ema, K. & Sasaki, M. (2009). Ensemble effect on Rabi oscillations of excitons in quantum dots. *Phys. Stat. Sol. A Applications and Materials Science*, Vol. 206 (2009) 952

Ledentsov, N.N.; Kovsh, A.R.; Zhukov, A.E.; Maleev, N.A.; Mikhrin, S.S.; Vasil'ev, A.P.; Semenova, E.S.; Maximov, M.V.; Shernyakov, Yu.M.; Kryzhanovskaya, N.V.; Ustinov, V.M.; Bimberg, D. (2003). High performance quantum dot lasers on GaAs substrates operating in 1.5 μm range. *Electron. Lett.*, Vol. 39 (2003) 1126

Liu, A.; Jones, R.; Liao, L.; Rubio, D.S.; Rubin, D.; Cohen, O.; Nicolaescu, R. & Paniccia, M. (2004). A high-speed silicon optical modulator based on a metal–oxide–semiconductor capacitor. *Nature*, Vol. 427, (2004) 615

Liu, H. Y.; Hopkinson, M.; Harrison, C. N.; Steer M. J. & Frith R. (2003). Optimizing the growth of 1.3 μm InAs/InGaAs dots-in-a-well structure. *Jpn. J. Appl. Phys.*, Vol. 93 (2003) 2931

Mukasa, K.; Miyabe, R.; Imamura, K.; Aiso, K.; Sugizaki, R. & Yagi, T. (2007). Hole assisted fibers (HAFs) and holey fibers (HFs) for short-wavelength applications, *Proceedings of SPIE 6779*, p. 67790J-1, 2007

Mukasa, K.; Imamura, K.; Sugizaki, R. & Yagi, T. (2008). Comparisons of merits on wide-band transmission systems between using extremely improved solid SMFs with Aeff of 160mm^2 and loss of 0.175dB/km and using large-Aeff holey fibers enabling transmission over 600nm bandwidth. *Proceedings of OFC 2008*, San Diego, OThR1

Nomura, M.; Iwamoto, S., Tandaechanurat, A.; Ota, Y.; Kumagai, N. & Arakawa Y. (2009). Photonic band-edge micro lasers with quantum dot gain. *Optics Express*, Vol. 17 (2009) 640

Otsubo, K.; Hatori, N.; Ishida, M.; Okumura, S.; Akiyama, T.; Nakata, Y.; Ebe, H.; Sugawara, M. & Arakawa, Y. (2004). Temperature-Insensitive Eye-Opening under 10-Gb/s Modulation of 1.3-μm P-Doped Quantum-Dot Lasers without Current Adjustments. *Jpn. J. Appl. Phys.*, Vol. 43, (2004) L1124

Paschotta, R.; Nilsson, J.; Tropper, A.C. & Hanna, D.C. (1997). Ytterbiumdoped fiber amplifiers. *IEEE J. Quantum Electron*, Vol. 33, (1997) 1049–1056

Rafailov, E.U.; Cataluna, M.A. & Sibbett, W. (2007). Mode-locked quantum-dot lasers. *Nature photonics*, Vol. 1, (2007) 395

Shimizu, H.; Saravanan, S.; Yoshida, J.; Ibe, S.; & Yokouchi N. (2007). Long-Wavelength Multilayered InAs Quantum Dot Lasers. Jpn. J. Appl. Phys., Vol. 46 (2007) 638

Sotobayashi, H.; Chujo, W.; Konishi, A. & Ozeki, T. (2002). Wavelength-band generation and transmission of 3.24-Tbit/s (81-channel WDM×40-Gbit/s) carrier-suppressed return-to-zero format by use of a single supercontinuum source for frequency standardization. *J. Opt. Soc. Am. B*, Vol. 19 (2002) 2803

Tanguy, Y.; Muszalski, J.; Houlihan, J.; Huyet, G.; Pearce, E. J.; Smowton, P.M. & Hopkinson, M. (2004). Mode formation in broad area quantum dot lasers at 1060 nm. *Opt. Comm.* Vol. 235, (2004) 387-393

Tanaka, Y.; Ishida, M.; Maeda, Y.; Akiyama, T.; Yamamoto, T.; Song, H.; Yamaguchi, M.; Nakata, Y.; Nishi, K. Sugawara, M. & Arakawa Y. (2009). High-Speed and Temperature-Insensitive Operation in 1.3-μm InAs/GaAs High-Density Quantum Dot Lasers. *Proceedings of OFC 2009*, San Diego, OWJ1

Yokoyama, H.; Sato, A.; Guo, H.-C.; Sato, K.; Mure, M. & Tsubokawa, H. (2008). Nonlinear-microscopy optical-pulse sources based on mode-locked semiconductor lasers. *Opt. Express*, Vol. 16, No. 22, (2008) 17752

Yamamoto, N. & Takai, H. (1999). Blue Luminescence from Photochemically Etched Silicon. *Jpn. J. Appl. Phys.*, Vol. 38, (1999) 5706-5709

Yamamoto, N.; Sumiya A. & Takai, H. (2000) Electroluminescence From Photo-chemically Etched Silicon. *Materials Science & Engineering B69-70*, (2000) 205-209

Yamamoto, N. & Takai, H. (2001). Formation mechanism of silicon based luminescence material using a photo chemically etching method. *Thin solid films*, Vol. 388, (2001) 138-142

Yamamoto, N.; Akahane, K.; Gozu, S. & Ohtani, N. (2005). Over 1.3 μm continuous-wave laser emission from InGaSb quantum-dot laser diode fabricated on GaAs substrates. *Appl. Phys. Lett.*, Vol. 86, (2005) 203118

Yamamoto, N.; Akahane, K.; Gozu, S.; Ueta, A. & Ohtani, N. (2006a). 1.55-μm-Waveband Emissions from Sb-Based Quantum-Dot Vertical-Cavity Surface-Emitting Laser Structures Fabricated on GaAs Substrate. *Jpn. J. Appl. Phys.*, Vol. 45, (2006) 3423-3426

Yamamoto, N.; Akahanea, K.; Gozu, S.; Ueta, A.; Ohtani, N. & Tsuchiya, M. (2006b). Sb-based quantum dots for creating novel light-emitting devices for optical communications, *Proceedings of SPIE 6393*, p. 63930A 2006

Yamamoto, N.; Akahanea, K.; Gozu, S.; Ueta, A.; Ohtani, N. & Tsuchiya, M. (2007a). Growth of InGaSb quantum dot structures on GaAs and Silicon substrates. *Jpn. J. Appl. Phys.*, Vol. 46, (2007b) 2401-2404

Yamamoto, N.; Akahane, K.; Nakamura, Y.; Naka, Y.; Kishi, M.; Sugawara, H.; Kawanishi, T. & Tsuchiya, M. (2007b). Silicon-based photonic network devices and materials, *Proceedings of SPIE, Vol.6775*, (2007) 67750O

Yamamoto, N.; Akahane, K.; Gozu, S.; Ueta, A. & Tsuchiya, M. (2008a). Low-temperature growth of nanostructured InGaSb semiconductors on silicon substrates. *Physica E*, Vol. 40, No. 6, (2008) 2195-2197

Yamamoto, N.; Akahane, K.; Sotobayashi, H. & Tsuchiya, M. (2008b). O-band InAs quantum dot (QD) laser diode with Sb-molecule sprayed Dot-in-Well (DWELL) structures fabricated on GaAs substrates, *Proceedings of OECC2008*, 2008, Sydney, Australia, TuH3

Yamamoto, N.; Sotobayashi, H.; Akahane, K. & Tsuchiya, M. (2008c). Quantum-dot Fabry-Perot laser-diode with a 4-THz injection-seeding bandwidth for 1-μm optical-waveband WDM systems, *Proceedings of ISLC2008*, p. 20, 2008, Sorrento, Italy

Yamamoto, N.; Sotobayashi, H.; Akahane, K.; Tsuchiya, M.; Takashima, K. & Yokoyama, H. (2008d). 10-Gbps, 1-μm waveband photonic transmission with a harmonically mode-locked semiconductor laser. *Opt. Express*, Vol. 16, No. 24, (2008) 19836

Yamamoto, N. & Sotobayashi, H. (2009a). All-band photonic transport system and its device technologies. *Proceedings of SPIE*, Vol. 7235 (2009) 72350C

Yamamoto, N.; Katouf, R., Akahane, K., Kawanishi, T. & Sotobayashi, H. (2009b). 1-μm waveband, 10Gbps transmission with a wavelength tunable single-mode selected quantum-dot optical frequency comb laser, *Proceedings of OFC 2009*, San Diego, OWJ4

Yamamoto, N.; Fujioka, H.; Akahane, K.; Katouf, R.; Kawanishi, T.; Takai, H. & Sotobayashi, H. (2009c). O-Band InAs/InGaAs Quantum Dot Laser Diode with Sandwiched Sub-Nano Separator (SSNS) Structures. *Proceedings of Conference on Lasers and Electro-Optics (CLEO)*, Baltimore, JThE.

14

Silicon Photomultiplier - New Era of Photon Detection

Valeri Saveliev,
National Research Nuclear University
Russia

1. Introduction

More then 50 years Photomultiplier Tubes (PMT's) fills the area of low photon flux detection practically without alternative (Hammamatsu Photonics K.K., 2006), despite the fact that is very well known many disadvantages of this devices.
Concerning modern semiconductor structures for the photon detection, few options were investigated for the detecting of the low photon flux, but main critical problem to develop the semiconductor device was the relative high level of thermal noise of semiconductor detector structure and associated frontend electronics. One of the solutions, overcome this problem is Visible Light Photon Counter (VLPC) (Atac, 1993). This device is semiconductor avalanche structure operated at the temperature of 4K, for the suppression of thermal noise. The results was successful - possibility to detect low photon flux up to single photon, but operational conditions are to complicated to be acceptable for wide area application, cryostat for the 4K temperature up to now is challenge even in the laboratory conditions.
Development of the modern detection structures for the low photon flux Si was initiated at the beginning of 90'th from studies of Silicon Metal Oxide Semiconductor (MOS) structures with avalanche breakdown mode operation for the detecting of single visible light photons [Gasanov et al., 1989]. The results were positive, but strong limitation was the necessity to include external recharge circuits for the discharge the detector structure after charging the MOS structure during the photons detection. Next step was implementation of special resistive layer instead oxide layers, Metal Resistive Semiconductor (MRS) structures, which gives the possibility to recharge the structure after photon detection and in addition to control the breakdown avalanche process by quenching. Such structures had very high and stable amplification characteristics for photons detection, in comparison to conventional avalanche photodetector structures, but limited sensitive area. The idea of Silicon Photomultiplier or more precisely Silicon Photoelectron Multipliers was created for overcoming problem of above mentioned structures as small sensitive area due to nonstability of amplification over large area, low dynamic range, improving the resolution. It was decided create the fine metal resistor semiconductor structure with local space distributed *pn*-junctions (micro-cells) and common output. The result was fascinated, first time clear single photon spectra was detected on the semiconductor structure at room temperature.
Results of study such structures was presented on the 9th European semiconductor conference in 1995 (Saveliev, 1995).

And the first concept of Silicon Photomultiplier was proposed fine silicon structure of avalanche breakdown mode micro-cells with common resistive layer quenching element and common electrodes. Results of this development were presented on the conference Beaune 1999 (Saveliev & Golovin, 2000; Bondarenko et al., 2000).
The goals of next steps were the optimization of the detection structures in particular increasing so called geometrical efficiency - ratio of area sensitive to photons to the total area of the silicon photomultiplier i.e. getting as much as detection efficiency and tuning the optimal operation condition in term of bias and time performance, and generally improve the technological processes. With advanced technology, what became available in the middle of 90th, the micro-cells are positioned as close as possible to each other, the common resistive layer as quenching element was substituted by individual integrated resistors coupled to the individual micro-cells with optimization of position and size. And the modern silicon photomultiplier structures start to be available for the applications (Golovin V. & Saveliev V., 2004).
New problem for optimized structures of silicon photomultipliers was the problem of optical crosstalk in fine detection structure due to light emission during the avalanche breakdown processes in Silicon. The phenomena of light emission from avalanche breakdown process is well known (A.G.Chynoweth & K.G.McKay, 1956). For the Silicon Photomultipliers with tiny space structure of microcells, the probability of detection secondary photons by neighborhood microcells is quite high and should be taking to account. Mainly this problem is affected of area of very low photon flux where the optics crosstalk could significantly change the results of measurement. The solution of this problem was achieved by implementation of modern technology process, physically optical isolation of the micro-cells on the integrated structure level. For the suppression of the optical crosstalk between the micro-cells, the trench structure was implemented around micro-cells as optic isolating elements and filled by optic non transparent material. The latest development in this area brings the very high performance for very low photon flux and created special type of silicon photomultiplier - quantum photo detectors (QPD) (Saveliev et al., 2008).
Silicon Photomultiplier is first semiconductor detector which could not only compete with photomultiplier tubes in term of detecting of low photon flux, but has a great advantages in performance and operation conditions and has great future in many areas of applications such as experimental physics, nuclear medicine, homeland security, military applications and other. Silicon Photomultipliers shows the excellent performance including the single photon response at room temperature (intrinsic gain of multiplication is 10^6), high detection efficiency ~25-60% for the visible range of light, fast timing response ~30 ps. Operational condition are suitable for many applications: operation bias 20-60 V, operated at room temperature as well in cooling conditions, not sensitive to electromagnetic fields. Production on base modern semiconductor technology, compatible with mass production semiconductor technology, compact, typical size of few mm^2 and flexible for assembling of the arrays. In this publication is impossible to eliminate all aspects of the silicon photomultiplier discovery and mainly will emphasise to more common feature to silicon photomultiplier development.

2. Conceptual idea

The main problem of detection of low photon flux or single photon is defined by nature of photons, physics of the photon interaction with matter and processes of converting the results

of interaction to the electric signal, i.e. in mechanism of converting the energy of photons in to the electric signals which is used for utilize by measurement and application systems.
The energy of photons could be estimated by standard expression:

$$E = h\upsilon = hc / \lambda \tag{1}$$

where: h - Plank constant,
v - frequency,
c - speed of light,
λ - wavelength.

This equation gives as example for the 500 nm visible light photons energy of 2.2 eV, it is one of the smallest quant of energy which could be found in nature and detection this quantity or single photon is challenge in many aspects. Moreover, the detection of single photon is interesting as fundamental physics task - study of fundamental quantum nature of light and their characteristics.

The basic principle of the silicon photomultiplier photon detection structure based on the quantum feature of light photon flux as space distributed quanta flux and space distributed array of micro sensors with capability to detect single quant of light – photon by every micro sensor. Main physics process of photons interaction with matter or process converting energy of photons to the other form, in particular charge in semiconductor material is photoelectric effect for the visible range of light. For considering range of light and semiconductor material, this process gives the converting ratio one to one – one photon correspondent energy create one electron-hole pair and this amount of electric charge should be transferred to and measured by electronic system. The basic principle of the photon detecting structure on base semiconductor materials micro sensor, allows utilize the result of photoelectric interaction, is creating the semiconductor structure with possibility of creating region free from charge carriers, depletion area and method of transport the created charge to outside, as example special geometry *pn*-junction (Saveliev&Golovin, 2004). By applying the reverse bias to the structure, between two regions with different type of conductivity forms the depleted area with low concentration of minor carriers and in-build electric field. Process of creating the electron-hole pair due to photoelectric interaction of photon in semiconductor structure and transport of the charges to the output shown schematically on Fig. 1.

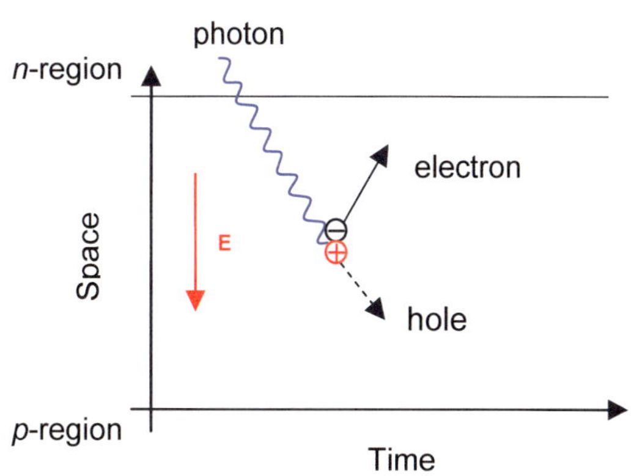

Fig. 1. Process of creating the electron hole pair due to photon absorption in semiconductor materials.

Photon with energy higher than band gap of semiconductor material is absorbed in depleted area with creating of electron-hole pair inside. Carriers, generated by photon, are separated by in-build electric field: electrons drifts to positive enhanced region *n*-region, holes to negative enhanced region *p*-region. Then the carriers are passing though external electric circuit generating the current as measurement signal. As mentioned before for the single photon the value of signal created inside detection volume is extremely low. In terms of measurement electronic system this is equivalent of charge level ~ 10^{-19} C. Registration of such signals is subject of extremely low value of charge as signal, statistical fluctuations of noise of detecting structure itself and and electronic noise generated by the electronic measurement system and is very complicated task.

Electronic noise of measurement system could be characterized in terms of equivalent noise charge for comparison to the charge signal from photon energy converting and represent the equivalent charge generated by the electronic channel in connection to the detection structure. Example of the equivalent noise charge as function of shaping time of discrete high quality frontend electronics system at room temperature presented on the Fig. 2. (Alvares-Gaume L. et al., 2008).

This value of the equivalent noise charge is calculated for discrete high quality frontend electronics. Optimal conditions gives the electronic noise on the level ~10^3 electrons (or elementary charges), it means that the minimal signal which could measured with discrete electronic channel should be higher ~3000 electrons or in term of the photons it is higher ~3000 photons with the 100% detection efficiency of photons.

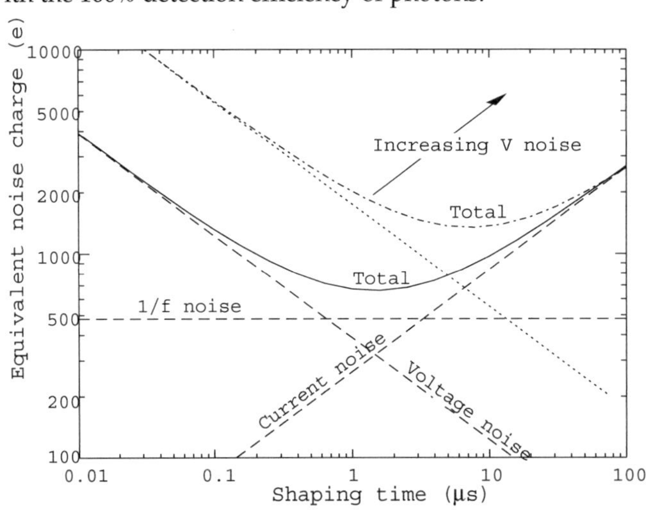

Fig. 2. Equivalent noise charge as function of shape time for discrete frontend electronic

The modern technology of integrated electronics could bring this condition on the level of equivalent noise charge around ~100 electrons at 20 ns shaping time, or equivalent of detection of ~300 photons, but not so many sensors technologies are compatible with integrated electronics on chip. And it is still far from our goal to measure signals range 1 e, which correspondent to single photon.

The way to overcome this problem is provide the internal amplification inside detection structure before transferring the signal to electronic system. The value of the amplification gain should be on range 10^4-10^6, what actually could not be achieved in conventional avalanche photodetection structures due to non stable working point in this region. This is main conceptual idea of the detecting the low photon flux or single photon by the semiconductor strictures, like silicon photomultipliers. Nevertheless to rich the value of intrinsic gain of level 10^6 or more in semiconductor structures is not trivial task in development of silicon photomultipliers.

For remaining, the principle of internal gain of multiplication was realized in the Photomultiplier Tubes - electro vacuum devices, where the electron, created from conversion process of the photon on the photocathode, accelerated by high electric field and multiplied by few stages on the dynode system, due to secondary emission [Hammamatsu Photonics K.K., 2006). The value of the amplification gain is ~10^6, what did the Photomultiplier Tubes as unique device for the detection of very low photon flux. But the high level of statistical fluctuation of multiplication process in photomultiplier tubes don't allowed get the good resolution for the single photon detection.

The amplification in semiconductor structures based on different physical principle. The intrinsic gain in semiconductor structures could be getting by the avalanche processes due to secondary impact ionisation processes (Tsang, 1985). In the high electric field, usually of order $10^5\ V \cdot cm^{-1}$ and higher free carriers in the semiconductors are accelerated and could rich the energy higher then ionization energy of valent electrons. Minimal energy which required for the impact ionisation called threshold ionisation energy. This value is one of the main parameter of the theory of avalanche multiplication in semiconductor materials. To characterise the dynamic of the avalanche processes is used the impact ionisation parameters of the electrons and holes in the semiconductor materials: α - for electrons and β for holes. Those parameters are defined as inverse value of average distance (along the electric field), which is necessary for electrons or holes to produce a secondary ionization and create secondary electron-hole pair. The consequence of secondary impact ionisation interaction gives the avalanche multiplication of the electron-hole pairs and increasing the value of the electric charge correspondent to initial charge created by interaction of photons. Values of α, β, width of high electric field area and carriers injection conditions defined the avalanche multiplication processes in semiconductor photon detection structures. Two types of the avalanche processes could be realized in semiconductor structures. This is strongly depends on value and ratio of impact ionisation coefficients α and β in silicon and on the value of electric field. For the low electric field ~10^{-4}, shown on Fig 3. a, impact ionisation coefficient of holes is much lower and avalanche process created practically by one type of carries - electrons. Avalanche process is one directional and self quenched when carriers is reached the border of depleted area in silicon. This type of avalanche process is usually used

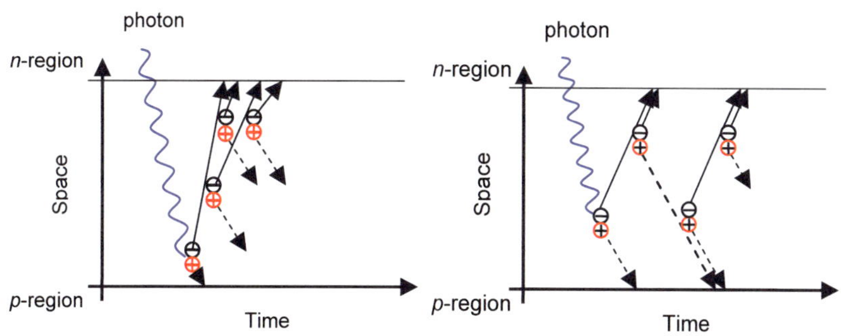

Fig. 3. Two type of avalanche processes in the Si structures, a. -self quenching avalanche process, b. - self sustaining avalanche breakdown process.

in conventional avalanche photo detectors. For high level of electric field in the silicon structure, process shown on Fig. 3.b, impact ionization coefficients coming close to each other and both type of carriers electrons and holes could participate in the avalanche process and create self-sustaining avalanche process, so the curriers rises exponentially with time and reach the breakdown conditions.

In first case the gain of multiplication is limited by thickness of depleted area, second case the gain of multiplication is not limited by the depletion thickness and became infinity even on the limited depleted thickness of silicon, because the different charge carriers undo electric field moving in opposite direction and thickness of amplification region could be just equivalent of length of ionization of electrons or holes under defined electric field. This is gives the possibility to getting the intrinsic multiplication factor enough to get the suitable signal before electronics to detect very small photon flux, up to single photon at room temperature. But avalanche breakdown mode of operation is required special effort for the quenching the avalanche process after initiation by absorbed photon or temperature created electron hole pair inside semiconductor.

The task of getting controlled avalanche breakdown process consist of providing the very high electric field in limited thickness of semiconductor detecting structure to bring the ionization length of electrons and holes less then the depleted thickness of *pn*-junctions, and getting required amplification gain with possibility of control by quenching maechanism.

3. Principle of operation, structure and technology

3.1 Silicon photomultiplier operation principle

Principle of silicon photomultiplier operation is based on quantum nature of light, detecting the space and time distributed photons (photon flux) by the space distributed array of the semiconductor micro sensors - micro-cells, operated in avalanche breakdown mode. Micro-cells are principally designed for detecting single quant of light (photon) with high efficiency. The array of space distributed micro-cells is designed for detecting of the space distributed quanta of light (flux) and sum of the signals from array provide the output signal proportional to the number of incoming photons – measurement of flux. In digital terms – number of micro-cells with avalanche breakdown process gives the number of incoming photons taking to account the detection efficiency.

Operational principle of silicon photomultiplier is based on the controlled avalanche breakdown processes in the silicon microstructure elements – avalanche breakdown micro-cells. Sensor avalanche breakdown micro-cells are special type of planar *pn*-junctions, operated in avalanche breakdown mode, providing the intrinsic multiplication of photoelectrons created by photons, absorbed in the sensitive area of micro-cell.

Above the breakdown voltage the *pn*-junction can be in stable state for a finite length of time, were it does not undergo avalanche breakdown. In this state a single carrier entering the depletion region is enough to initiate avalanche multiplication process and produce a self-sustaining current. The initiation could be as result of incoming photon interaction or termal created carrier inside depleted area. For the stopping of the avalanche breakdown process, the quenching elements are implemented in the silicon photomultiplier for each micro-cell. In case of silicon photomultiplier, the serial resistor for the each sensor micro-cell provides this function. The quenching element acting following way, after the initiation of the avalanche breakdown process by photoelectron or thermal electrons the current is rising in the external circuit and caused the drop voltage drop on the quenching resistor and accordantly of the voltage applied to the *pn*-junction. The process quenching is started when the dropping the voltage on the quenching resistor bringing the voltage applied to the *pn*-junction to value lower then breakdown voltage, and quench the avalanche process. After the micro-cell is quenched, a hold-off time is then necessary to allow any free or stored charge to be swept from the active region of the device, followed by a recharging where the excess bias across the micro-cell is restored.

Important aspect of described process is significant reduction of statistical fluctuation of signal. For silicon photomultiplier structures the amplification factor is defined not by statistic of avalanche processes as in the conventional avalanche photodetectors, but only by *pn*-junction characteristics and quenching circuits. As result, the concept of multiplication noise or access noise is not relevant to the silicon photomultiplier and performance in particular fluctuation of signal is much lower. Output of the micro-cell in process of photon detection is identical charge pulse and overall resolution is defined by identity of characteristics of micro-cell and quenching element. According to this, very important aspect of providing the high performance of silicon photomultiplier is the uniformity of micro-cells characteristics across the sensitive area. This is provided by the modern semiconductor technology, the requirements for the uniformity correspondent to the precision of the charge pulse from different micro-cells detecting the photons across sensitive area and define the single photon detection resolution.

Finally intrinsically the silicon photomultiplier is completely digital device, which produce the number of equivalent charge pulse caused by photon interaction in the space distributed structure of equivalent micro-cells and integrated on the output and correspondent to the number of incoming photons. Future of such devices is providing completely digital signal analysis already inside structure of silicon photomultipliers.

3.2 Silicon photomultiplier structure and technology

Silicon photomultiplier is silicon microstructure consists of:

- large numbers of elementary sensors - array of micro-cells, operated in the avalanche breakdown mode, space distributed with high density on the common substrate, typical size of a few mm^2,
- implemented quenching elements for each micro cell (in present time passive quenching elements - resistor),
- common electrode system, connected individual micro-cells to the common output of silicon photomultiplier.

Schematic view of geometry of silicon photomultiplier and equivalent electric schematics are presented on the Fig. 4, a,b.

On Fig. 4.a, is presented the schematic view of modern silicon photomultiplier with process of photon. The area of detection is divided on the fine space distributed micro cells (light gray square) and consist of the *pn*-junctions (marked as # micro-cells), every micro-cell has the quenching element, located close to the *pn*-junction (dark gray and marked as # Q. elements), the common electrode did't shown on this picture, which connected all quenching resistors to common output electrode, other electrode is on the back side of wafer. On the picture shown also space distributed photons (photon flux) and interaction in the silicon structure - three photons interact in the three micro-cells and initiated the avalanche breakdown processes. On Fig 4, b is presented correspondent electronic schematic of the silicon photomultiplier, shows the *pn*-junctions as array of diodes and quenching elements as serial resistors to the individual diodes. The process of interaction is shown as photons propagation and triggered the correspondent diodes. The electrical connection

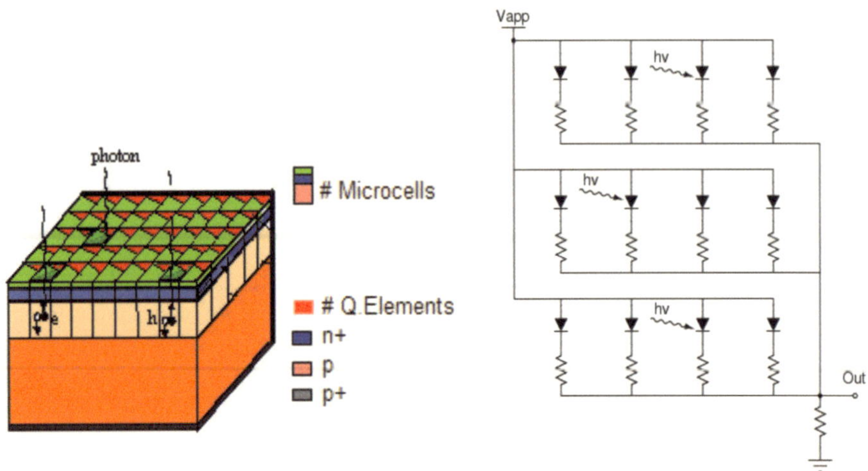

Fig. 4. a) Equivalent schematic of structure of silicon photomultiplier and b) equivalent electronic schematics of silicon photomultiplier

of the micro-cells formed two common electrodes - one for bias and second as signal output connected to the load resistor.

Structure of silicon avalanche breakdown micro-cell, based on the shallow *pn*-junction with so called virtual guard ring is shown on Fig. 5.

Structure of the avalanche breakdown micro-cell consist of silicon substrate (Substrate) with epitaxial layer *p*-type (epi). Avalanche breakdown structure represented by shallow *pn*-junction (n^+p) in silicon epitaxial layer with so called virtual guard ring to prevent peripheral avalanche breakdown processes (the virtual guard ring is formed by special geometry overlapping the n^+-and *p*-type area. To provide possibility to getting high electric field allowed realise the avalanche breakdown mode on the relative thin depletion region is chosen low resistive silicon (epitaxial layer) and additional implantation process to form *p* and n^+ region of *pn*-junction. Heavily doped n+ region connected to electrode an serial quenching resistor. Second electrode is formed on the back side of substrate.

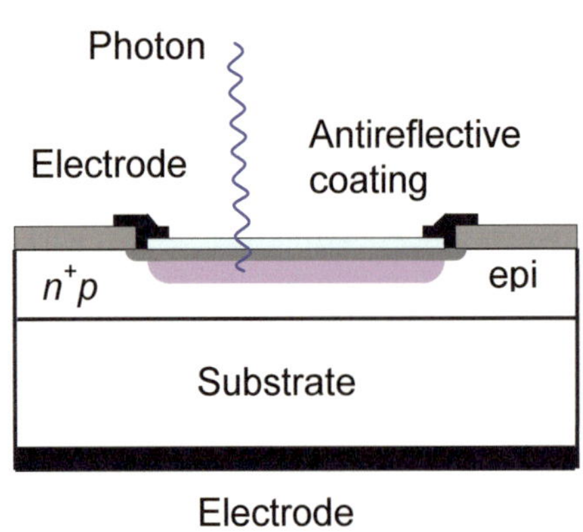

Fig. 5. Schematic schematic of avalanche breakdown micro-cell of silicon photomultiplier *n* on *p* type with virtual guard ring

Other type of avalanche breakdown micro-cell, often used for silicon photomultiplier fabrication is *pn*-junction with physical guard ring, implanted on the periphery of n^+- area. This technology more compatible with standard CMOS technology.
The guard ring in the silicon photomultiplier is important feature of structure, necessary to prevent the intensive breakdown processes in the areas with high electric field and high gradient of electric field caused earlier breakdown in the region of peripheral border of sensitive area and provide more uniform area of avalanche breakdown inside guard ring.
Such type of silicon photomultiplier micro-cell *pn*-junction is called "*n* on *p*" structures. The inverse structure of *pn*-junction, called "*p* on *n*", also is using for the manufacturing of the silicon photomultipliers. Advantage of inverse structures is possibility to increase the short wavelength light sensitivity of silicon photomultiplier (Hamamatsu, 2009)
The quenching elements - passive resistors on base poly-silicon planar technology, doped by implantation to get the correct high resistor value ~ 0.1 - 1 *MOhm* resistors on the limited area (length) of tens of microns. Forming of such elements required high precision because the geometrical characteristic significantly effected to performance of silicon photomultiplier in particular on photon detection efficiency.
The overall topology of silicon photomultiplier is presented on Fig. 6.a,b developed by Kotura Inc. (Kotura, 2009). On Fig. 6, a, is presented the top view of 1 mm^2 silicon photomultiplier with micro-cells size ~30x30 microns. Total number of micro-cells is ~1000 on 1 mm^2 silicon photomultiplier. The typical size of silicon photomultipliers are 1x1 mm^2 up to 5x5 mm^2 without significant changes in performances.

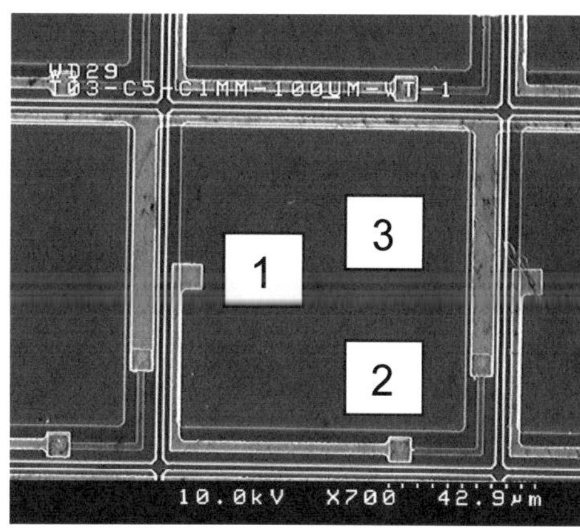

Fig. 6. a,b Micro image of modern silicon photomultiplier , a - overall view 1x1 mm^2, b - detailed view of micro-cell area.

On Fig.6,b is presented microscopic view of single avalanche breakdown micro-cell size ~30x30 microns with visible main elements of structure:

1. sensitive area,
2. quenching element - resistor,
3. part of the common electrode system,
4. optical isolation elements - trenches.

As mentioned before the important feature of the used material, comparison to the conventional silicon avalanche photo detectors is that for the silicon photomultiplier used low resistive silicon material and technologies compatible to the main mass production technology processes as CMOS technology and more important aspect that materials and technology allowed produce the integral device including the sensors and readout electronics on the same substrate. In future the integrated silicon photomultiplier with readout electronics on the chip will dominated on the design and gives unprecedented advantages of such devices.

3.3 Electric characteristics

Fig. 7 shows the typical reverse bias current-voltage (*CV*) characteristics for silicon photomultiplier with sensitive area of 1 mm^2 (Stewart A.G. et al., 2008). The plots shows the current-voltage characteristic in the range of avalanche breakdown at 293 K (room temperature) and at 253K (-20°C).

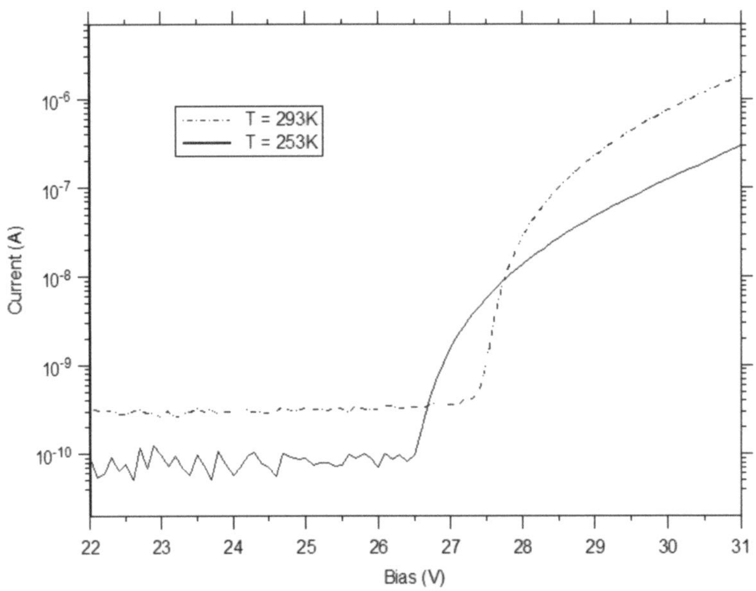

Fig. 7. Current-voltage characteristics of silicon photomultiplier at different temperatures 293 K and 253 K.

Before 26 V the current correspond non avalanche mode of *pn*-junction. At range 27.5 and 26.5 V the currents increase sharply due to avalanche multiplication process. Above avalanche processes started, the current increases by several orders of magnitude and reach the avalanche breakdown conditions, where the current is practically does not depend on the *pn*-junction state and curves follows the resistor behavior of silicon photomultiplier, mainly defined by the resistivity of quenching elements.

The silicon photomultiplier reverse bias current-voltage characteristic is used to determine the breakdown voltage and working point which is expressed in term of overvoltage. As seen from the plot, the breakdown voltage is a function of temperature and has a temperature coefficient of 23mV/°C.

4. Silicon photomultiplier performance

4.1 Single photon detection performance

Single photon detection performance at room temperature is one of the fascination characteristic of the silicon photomultipliers, shows the silicon photomultiplier as an ultimate instrument allowed detection of single photon - single quant of light in detail, that why the special type of silicon photomultipliers has name quantum photo detectors (Saveliev V., 2008). Fig.8.a,b presents the scope signals of the detecting the low photon flux by silicon photomultiplier at room temperature. The both picture shows two signals: top is the signal from silicon photomultiplier and bottom is synchronization signal of special low photon flux light source.

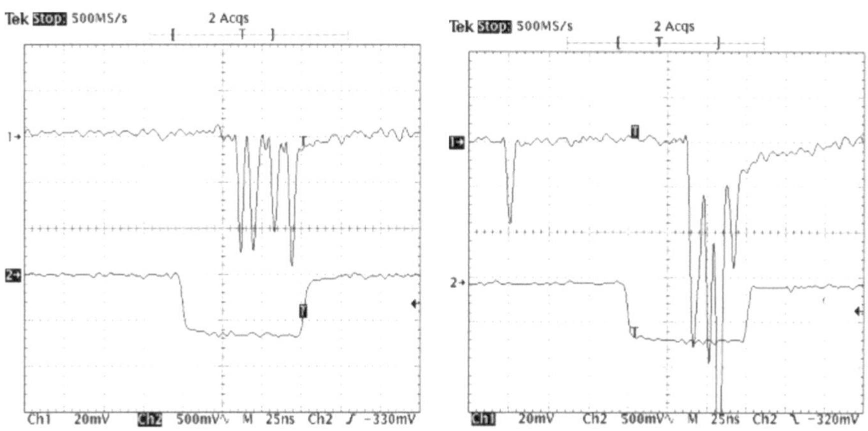

Fig. 8. Single photon detection signals: a) single photon signal, shows the signals from different micro-cells during the light pulse and collected on the output, b) signal with higher intensity of photons which not overlapping in space (signals from different micro-cells) but start overlapping in time and overlapping the electric signal on output of silicon photomultiplier during the photons detection.

On Fig. 8. a. clear visible the four signals each correspondent single photon detection distributed in time from laser diode. Signals are coming from different micro-cells of silicon photomultiplier and summed on the output. The laser diode is flushed in the gate shown on the second signal line. Fig. 8.b shows the signal with higher intensity of photon flux. Signals from different micro-cells started overlapping in time and gives on the output signal amplitudes which correspondent sum of two and more single photon signal, but steel clear seen of formation output signal from the signals of the single photons detected by the different micro-cells.

On the Fig. 9. is presented the signal distribution of the detecting the low photon flux by silicon photomultiplier at room temperature (Stewart A. G. et al., 2008). The signal distribution presented a statistical distributions of detected signals during registrations of low photon flux. Axis are represented correspondently: horizontal - the amplitudes of the signals from phtomultiplier and vertical – statistics of event with particular amplitudes. Clear seen the statistically resolved peaks of particular amplitudes. The resolution of the silicon photomultiplier is enough to distinguish the signals with discrete numbers of photons, which shows the quantum nature of the light, as a collection of discrete quanta with particular energy. The resolution of silicon photomultiplier allowed very precise analysis of the detecting photon flux up to single photon. First peak correspondent to amplitude of noise of electronic channel (pedestal), second peak correspond the amplitude of detecting one single photon, third peak correspond the amplitude of signal of detecting two photons in the same time by different avalanche breakdown micro-cells. Interesting mentioned that second peak is represented the amplitude of single photon detection collected from different avalanche breakdown micro-cells of silicon photomultiplier, - it shows also the high uniformity of characteristics of avalanche breakdown micro-cells around sensitive area of silicon photomultiplier. Clear seen the statistical behavior of the photon flux - Poisson distribution of the overall spectra of detected low photon flux.

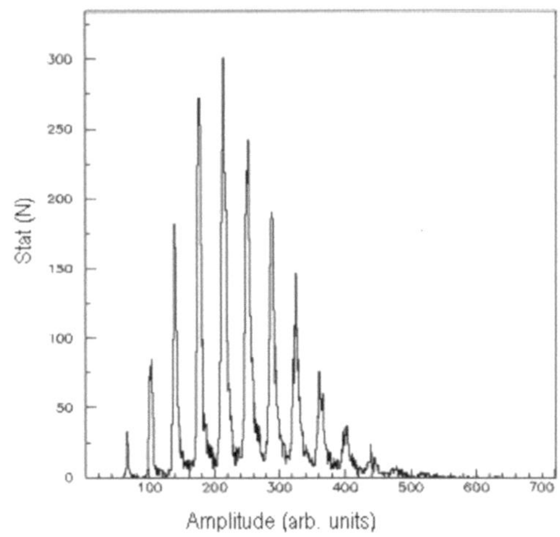

Fig. 9. Signal distribution of low photon flux signal (Poisson statistic of photon flux and peaks correspondent discrete numbers of detecting photons)

4.2 Photon detection efficiency of silicon photomultiplier

Detection efficiency of the silicon photomultiplier is a product of few main factors: quantum efficiency, efficiency of the avalanche process triggering and geometrical efficiency (geometrical filling factor).

(a) Quantum efficiency

The quantum efficiency of silicon photomultiplier is most general characteristic and consistent to general definition of the quantum efficiency of semiconductor detecting structures. Photons illuminated the silicon photomultiplier sensors with energy higher than bandgap are absorbed by the silicon crystal structure of depleted area and created the electron-hole pair, which could be detected as the signal. In some publications the quantum efficiency of semiconductor detectors is defined as number of measured electrons at the output of detector structures to the input flux, but in case photomultipliers more sufficient use the definition from [Tshang W.T (Ed), 1985], the quantum efficiency is the ratio of created electron-hole pairs to the incoming photon flux.

In real detecting structures as silicon photomultipliers part of the photon flux is affected by the reflection on the border of air/sensitive area of detection structure. For the silicon, reflection index on the border air/semiconductor is ~3.5 and correspondent Frenell coefficient for the normal incident photons R=0.3. i.e. losses on the reflection could reach ~30%. This loss could

be efficiently reduced by implementation of the antireflection coating and will be included in the quantum efficiency of silicon photo multipliers (Tsang W.A., 1985).
Main factor which define the quantum efficiency is characteristics of absorption process of photons in the depleted area of detecting structure. A photon flux of intensity $I(\lambda,z)$ will absorbed in silicon according to the Beer-Lambert law given in equation (2), creating an equivalent number of electron/hole pairs.

$$I(\lambda,z) = I(\lambda)e^{-\alpha(\lambda)z} \tag{2}$$

where:

$I(\lambda)$ - initial photon flux,
$I(\lambda,z)$ - photon flux on the distance z fron silicon photomultiplier face
$\alpha(\lambda)$ – optical absorption coefficient,
z - penetrated thickness of silicon.

This process, defined photon attenuation, is a fundamental property of all silicon detectors. I.e. photons entering a silicon layer travel a characteristic distance before giving up their energy to create a photoelectron. This distance is a function of the absorption coefficient, which is defined as the inverse of the distance a photon flux travels in a material before being attenuated by a factor of *e*. It is important parameter for the development of the silicon photomultiplier, because the thickness of depletion layer is relatively thin.
Other aspect of this consideration is that optical absorption coefficients is strong function of wavelength (or energy) of photons for particular semiconductor material, and defined the sensitivity dependency of quantum efficiency to wavelength of detecting photons. Fig. 10. shows the absorption coefficient as function of wavelength of photons in silicon (Tsang W.T., 1985).

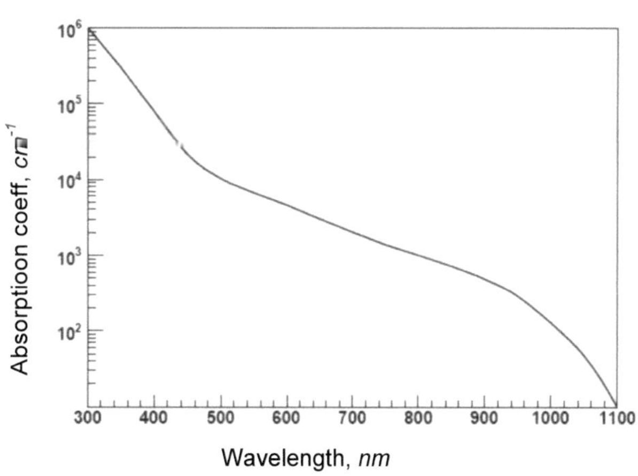

Fig. 10. Absorption coefficient of photons in silicon as function of wavelength.

Cut off at long wavelengths is fundamental limitation and occurs for a silicon photodiode at a wavelength of 1.1 μm where the photon energy is just sufficient to transfer an electron across the silicon band gap. As this wavelength is approached the probability of photon absorption decreases rapidly with increasing wavelength. It will be noted that the absorption coefficient increases with increasing temperature leading to an increase in long wavelength responsivity with temperature. Cut off at short wavelength occurs in silicon photomultiplier through structure feature of micro-cells. The top layer of a micro-cells is formed by an implantation or diffusion process that defines the edge of the depletion layer. Photons absorbed in this layer will not contribute to charge collection process and this effect causes a strong reduction of the sensitivity of photodiodes for photons with a short wavelength which are absorbed close to the surface.

Generally the quantum efficiency of silicon photomultiplier - $\eta(\lambda)$ could be calculated as:

$$\eta(\lambda) = (1 - R)(1 - e^{-\alpha(\lambda)L_d}) \tag{3}$$

where:

R - reflection Frenell coefficient,

$\alpha(\lambda)$ – optical absorption coefficient,

L_d - thickness of depleted area of micro-cell.

Correct design of the sensitive area of the silicon photomultiplier for the maximal quantum efficiency lead provide two main conditions for the vertical layers structure design:

- The top technological layers should be relatively thin, especially if the silicon photomultiplier is developed for the short wavelength region,
- The thickness of sensitive layer (depleted area) should agreed to the condition:

$$\alpha(\lambda)L_d >> 1 \tag{4}$$

where:

a(l) - optical absorption coefficient, which gives the distance over which the photon flux is reduced by a factor *e* ,

L_d - thickness of the depleted area.

In silicon the absorption coefficient for the visible range of light (as example, green light) is ~ 10^4 cm $^{-1}$, and for effective absorption the thickness of the sensitive area of silicon photomultiplier should be order of few microns, this is lead to possibility use the low resistivity silicon material as a basic material of silicon photomultilier structures.

Actually photons could absorbed in other technological layers of the detecting structure and contribute to the charge collection and quantum efficiency, but this effect is second order and will not discussed here (Tsang W. T., 1985).

(b) Efficiency of avalanche process triggering

Not all primary electron-hole pairs succeed in initiating a avalanche breakdown. Although conditions are such that, on the average, the number of carriers in the multiplying region increases exponentially with time, some just start a chain of ionizations that terminates, because of a fluctuation to zero carriers, before it really gets multiplication. The probability that the process of ionizations continues to increase until the whole *pn*-junction is discharged is called the avalanche breakdown probability (P_b). Generally avalanche breakdown probability is function of electric-field profile and the electron-hole ionization coefficients and could be calculated provided the electric field profile and electron, hole ionization coefficient are known [McIntyre, R.J.,1973]. Nevertheless study of silicon photomultiplier detecting efficiency shows that the avalanche breakdown probability in the silicon photomultiplier structures is significantly higher and probably needed to be addition investigation.

(c) The geometrical efficiency (filling factor)

The geometrical filling factor (F) is the proportion of surface area capable of detection single photons to the total area of silicon photomultiplier including technological border. Geometrical filling factor follows from the need to form independent micro-cells, quenching element and electrodes. Some affecting factor on the filling factor is optic crosstalk, which is function of the distance between the micro-cells. Special technology is used to overcome this problem, which will be discussed later.
Finally the photon detection efficiency (PDE) of the silicon photomultiplier could be defined as:

$$PDE = \eta(\lambda) \cdot P_b(V) \cdot F \tag{5}$$

where:

$\eta(\lambda)$ – quantum efficiency of silicon microcell structure,
Pb(V) - probability of avalanche breakdown in silecon microcell structure,
F - geometry filling factor.

Experimental study of photon detection efficiency of silicon photomultipliers is very complicated task, especially in the low photon flux region. To simplify this problem, the experimental photon detection efficiency could be determined by measuring the photon detection probability of individual avalanche breakdown mode micro-cell, identical to those contained within the silicon photomultiplier. The photon detection probability is defined as the quantum efficiency times the avalanche breakdown probability. The photon detection efficiency of silicon photomultiplier is then determined by scaling the photon detection probability with the geometrical efficiency. The micro-cell photon detection probability was measured relative to a calibrated photo detector with monochromator light source. Fig. 11. shows the measured photon detection probability as a function of wavelength at 2 and 4V above the breakdown voltage together with the predicted wavelength response (Stewart A. G. et all, 2008).

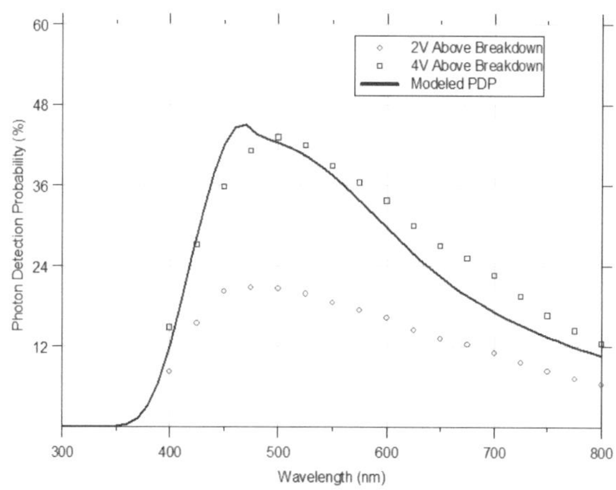

Fig. 11. Photon detection efficiency of silicon photomultiplier as function of wavelength (o - experimental measurements at *2 V* above breakdown, □ - experimental measurement at *4 V* above breakdown, solid line - modeling data)

At $4V$ above breakdown the peak of photon detection probability is 43% and occurs at 500 *nm* wavelength of light. At $2V$ above breakdown the peak of photon detection probability shifts to a slightly lower wavelength and has a peak value of 21%. On the Fig. 11. is presented also the curve of modeled photon detection probability. Model is using the absorption coefficient of silicon and the probability that a photon will be absorbed in the depletion region together with a model of the transmittance of the anti-reflection coating. The depletion layer absorption and anti-reflection coating determine the spectral profile of the photon detection efficiency which is then scaled to fit to the experimental photon detection probability data by using the avalanche breakdown probability as a fitting parameter.
Scaling the micro-cell photon detection probability values with the silicon photomultiplier fill factor gives the photon detection efficiency of silicon photomultiplier. For modern technology the filling factor rich value of 0.6-0.8.

4.3 Multiplication factor

Multiplication factor of silicon photomultiplier is defined by the avalanche process and the characteristic of quenching element. The value of multiplication factor could be precisely calculated from single photon spectra, as mentioned before, the signal is represented the charge generated by the avalanche breakdown process and peaks correspondent to the number of detected photons or photoelectrons. In silicon photomultiplier is possible to get the absolute calibration of the multiplication factor using the position of the single photon peak in the spectra (correspondent value in charge), because this position is exactly correspondent of creating of one electron-hole pair in micro-cell. The multiplication factor is a function of detector bias and temperature. Fig. 12 shows the multiplication factor of silicon photomultiplier increases linearly with over breakdown voltage and is greater than 1×10^6 at 0.7V above the breakdown voltage (Stewart A. G. et all, 2008) .

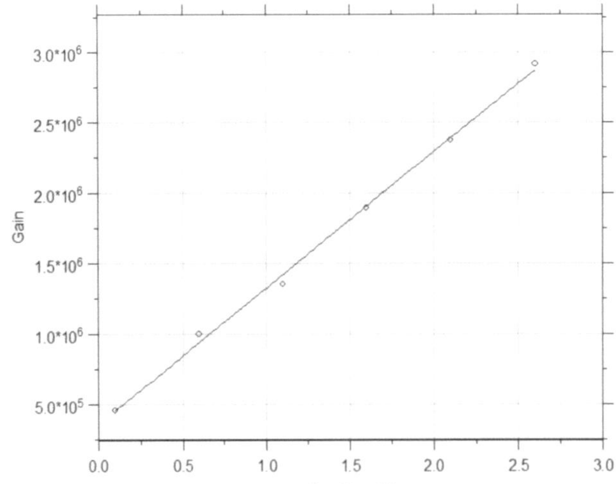

Fig. 12. Multiplication factor of silicon photomultiplier as function of voltage above the breakdown voltage.

The intra and inter micro-cell multiplication factor variation or uniformity of silicon photomultiplier can be determined from the width of the photoelectron peaks on the single photon spectra. The width of single photon peak is defined as combination of the variation of the charge produced by controlled avalanche breakdown process and uniformity of the characteristic of the micro-cells around the total area of silicon photomultiplier, because the single photon peak is collected signals statistically from different micro-cells during photon detection.

4.4 Dynamic range and linearity

The detection of photons by a silicon photomultiplier is a statistical process based on the probability of detecting randomly distributed photons by the limited number of sensitive elements micro-cells. The photon detection efficiency and the total number of micro-cells determine the dynamic range of the silicon photomultiplier. The number of detected photons n_{dph} (number of micro-cells with signal) as function of the number of incident photons can be approximated by the following expression,

$$n_{dph} = N_{mc}\left(1-\exp\left(-\frac{PDE \cdot N_{ph}}{N_{mc}}\right)\right) \tag{7}$$

where:

N_{mc} - total number of micro-cells of silicon photomultiplier,
PDE - photon detection efficiency,
N_{ph} - number of incident photons.

Fig. 13. shows the silicon photomultiplier response (number of pixels with signals) as a function of the number of instantaneous incident photons for an silicon photomultiplier with 620 micro-cells and with a photon detection efficiency of 5% and 10% (Stewart A. G. et all, 2008).

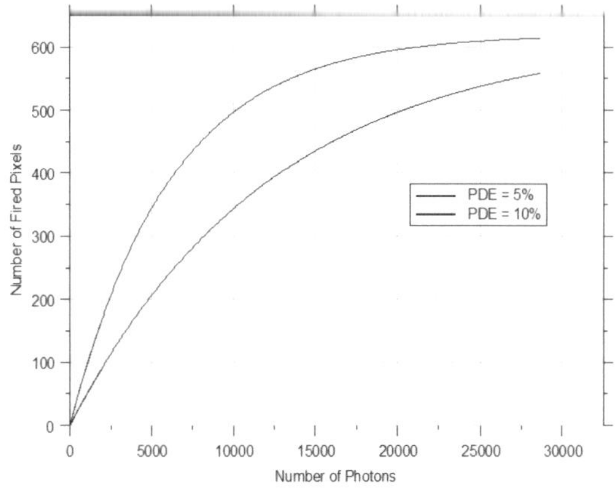

Fig. 13. Dynamic range of silicon photomultiplier – the number of fired micro-cells as function of number of incident photons

The silicon photomultiplier response is linear when the number of incident photons is much less than the total number of micro-cells. The silicon photomultiplier response begins to saturate when the number of pixels fired reaches approximately a quarter of the total number of micro-cells. The figure also shows how the silicon photomultiplier dynamic range and linearity can be extended to handle higher photon fluxes by lowering the operating bias and hence reducing the photon detection efficiency of silicon photomultiplier, in this example, from 10 to 5%. The statistical behaviour of the linearity and dynamic range curves gives the possibility to calibrate this curve to improve the characteristic of silicon photomultipliers. Nevertheless the main way to improve the linearity and increase the dynamics range of silicon photomultipliers is increasing the number of micro-cells.

4.5 Time performance

The time performance of silicon photomultipliers is defined by two parameters: the rising time of the avalanche breakdown signal and the recovery time, which defined by the process of reconstruction the *pn*-junction state after quenching the avalanche breakdown process and recharging through the quenching resistor.

The rising time is defined by the time of creating the avalanche breakdown process and characterized by drift time of carriers under the high electric field. Drift velocity carriers under electric field ~10^5 is limited by scattering and in silicon structures is approximately ~10^7 cms^{-1}, this gives the estimation of the rising time, as example for the thickness of depleted area of 4 microns the rising time is $\sim 30\ ps$.

The timing jitter spectrum, characterized the rising time of the silicon photomultiplier shown in Fig. 14. The jitter histogram is fitted with a Gaussian curve and has a full width on half maximum of 65ps, including the response of measurement system (Stewart A. G. et al, 2008).

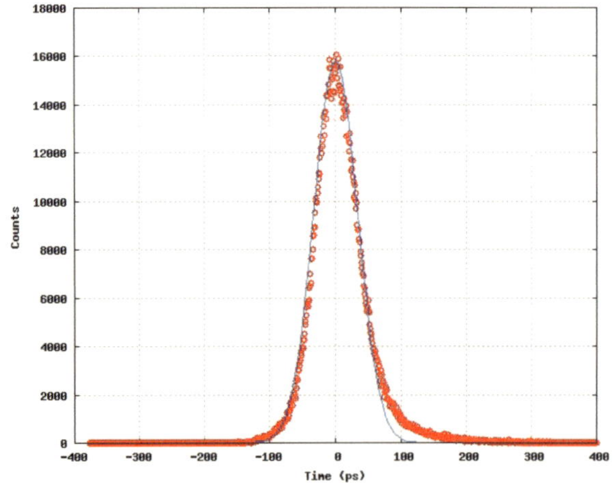

Fig. 14. Time response of silicon photomultiplier (o – experimental measurements, solid line – fit).

The recovery time is defined mainly by recharge process and could be estimated from values of RC - combination of the quenching resistor, diode capacitance and external circuit. The quenching resistor could be tuned in relatively wide range and define the recovery time of silicon photomultiplier in range $\sim 1 - 100\ ns$.

4.6 Noise consideration (dark rate)

One of the main factors limiting the performance and size of the silicon photomultiplier is the dark rate - frequency of the signals with amplitude equivalent of single photon signal, initiated by the thermal electron-hole pair created in sensitive area of silicon photomultiplier. Usually conventional photodetectors is characterized in term of noise by the dark current. In case silicon photomultiplier the very high gain is practically neglected the contribution of the dark current or leakage currents of silicon structure to the output signal, because the leakage current carriers is not affected by the amplification process. Significant for the noise consideration became the processes which included in the processes of multiplication. The thermally generated carriers in the depleted area could also initiate a avalanche breakdown and results in a current pulse that is indistinguishable from a pulse produced by the detection of a single photon. This is particularly important for room temperature operation. The silicon photomultiplier dark rate is the average frequency of the thermally generated avalanches breakdown processes from all micro-cells of silicon photomultiplier. Typical value of dark rate for the modern silicon photomultipliers is in range ~0.1-1 MHz per mm^2.

The results of experimental study of silicon photomultiplier dark rate are shown on the Fig. 15. and Fig. 16., the dark rate signals measured on the threshold correspondent to half of the single photon pulses. Fig. 15. shows the dark rate as a function of over-bias for two types of silicon photomultipliers size 1 mm^2 at room temperature and at temperature -20°C (Stewart A. G. et all, 2008).

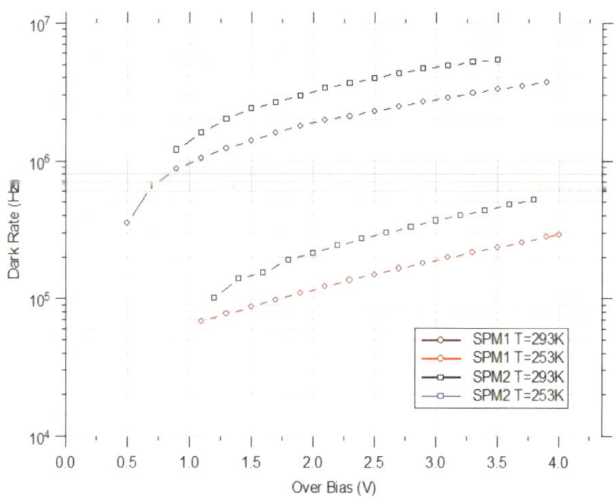

Fig. 15. Dark rate of two silicon photomultipliers as function of voltage over breakdown voltage for two temperatures.

At room temperature the dark rate increases linearly at a rate of 0.95MHz/V for one type and 1.67MHz/V for the another design.

As the dark is thermally activated it is a strong function of temperature. Fig. 16. shows the dark rate of silicon photomultiplier size 1 mm^2 at 2 and 3V above the breakdown voltage (Stewart A. G. et al, 2008).

The plot gives the dark rate thermal activation energy of 433 meV at $2V$ above breakdown and 388 meV at $3V$ above breakdown. These activation energies are close to half the bandgap of silicon (560 meV) and are indicative of thermal generation from a single trap level in the depletion region.
It should be mentioned that the amplitude of the dark rate pulses is equivalent of the single photon signal amplitude, that for many application deal with tens-hundreds photons it could be neglected. For applications with very low photon flux the average dark rate can be measured and subtracted. However, the statistical variation in the dark rate cannot be subtracted and constitutes a noise source that determines the minimum detectable signal.
Dark rate of the silicon photomultiplier is scales as its area, so in practice, the "acceptable" noise dark rate ~ 10^6 limits the maximum designable area around few mm^2.

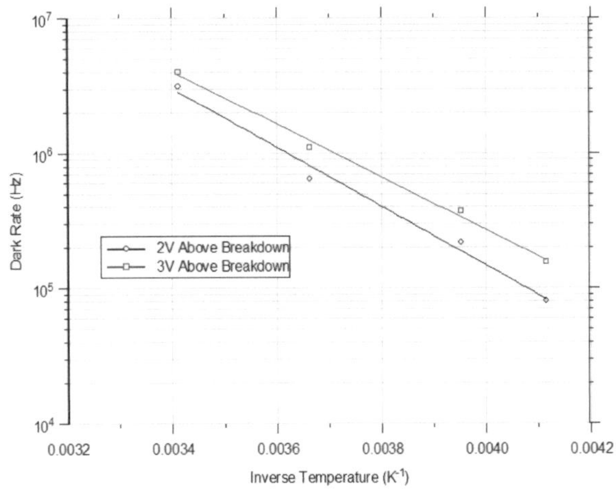

Fig. 16. Dark rate of photomultiplier as function of temperature for two value of voltage over breakdown voltage.

4.7 Optical crosstalk

Generation of the light photons during the avalanche breakdown processes in silicon is well known phenomena (Chynoweth A.G & McKay K.G., 1956). This phenomena becomes very critical in the silicon photomultiplier structures due to very tiny geometrical pattern which allowed the photons created by avalanche breakdown process during the detecting the photon in one micro-cell rich an another micro-cells and give the signal which is not distinguishing from signal of detected outside source photon. The ratio of number of secondary created photons, detected in the silicon photomultiplier to number of detected incoming photons is determinate as optical crosstalk of silicon photomultiplier.
As example the light generation in the silicon photomultiplier is illustrate on the Fig. 17., where shown single micro-cell of photomultiplier under reverse bias (actually the reverse bias is in the high value over breakdown to show the phenomena) (Kotura, Inc, 2009).
The light spots are clear visible in the corner of the avalanche breakdown micro-cell indicate the critical areas of the silicon photomultiplier structures, corners have the two dimensional curvature.

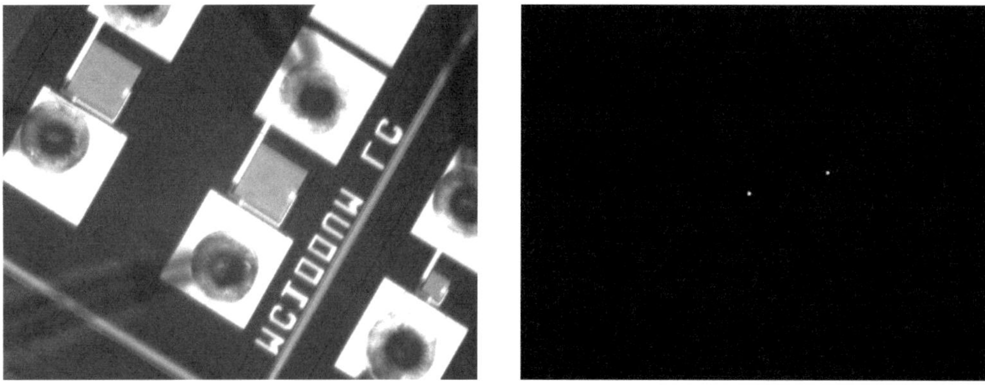

Fig. 17. a,b. Light generation in the silicon photomultiplier micro-cell in regions with high gradient of electric field, a - the micro-cell in external light, b - same micro-cell without external light.

Optical crosstalk of silicon photomultiplier scales as the product of optical generation inside the triggered micro-cell during the avalanche breakdown process, the total geometric cross section for the interaction between two micro-cells and the single photon sensitivity of micro-cells. The optical crosstalk probability is a function of silicon photomultiplier bias voltage and the distance between neighbourhood pixels.

The optical crosstalk probability preliminary can be estimated by the ratio of the count rate at the second photoelectron level to the count rate at the single photoelectron level. The count rate at the single photoelectron level is measured by setting the threshold at half the single photon peak while the rate at the second photoelectron level is measured by setting the threshold at 1.5 times single photon peak height. The measurements were taken at 253K to reduce the primary dark rate. At room temperature were the dark rates are of the order of MHz, there is a higher probability that two uncorrelated dark pulses or a dark pulse and an afterpulse event might coincide to produce a pulse with amplitude greater than the threshold setting for the second photoelectron level. At 253K the primary dark rate is of the order of a few 100kHz and the probability of uncorrelated events occurring simultaneously is minimized.

Fig. 18. a,b. shows the optic crosstalk probability, measured at 253K, for two silicon photomultiplier as a function of voltage over breakdown voltage (left) and the frequency of dark rate as a function of threshold (right) (Stewart A. G. et al., 2008).

The crosstalk probability increases linearly with over-bias and as expected the silicon photomultiplier 2 (SPM2) design has a higher crosstalk probability as the pixel pitch in this design is smaller. Plotting the pulse frequency as a function of threshold position produces a step-like curve. The flat regions of the curve correspond to the frequency of pulses at the single, double, triple photoelectron level. In the example shown the pulse frequency was measured at 253K and at 3V over-bias. At a low threshold the frequency corresponds to the single photoelectron level rate or dark rate of the silicon photomultiplier. As the threshold increases above the peak height of the single Geiger pulses, the rate falls as only pulses at the second photoelectron level, i.e. optic crosstalk, will now be measured. The rate then remains constant until the threshold increases above the peak height of two coincident pulses and then falls to the rate of the third photoelectron level. Pulses at the third photoelectron level are also mostly due to crosstalk.

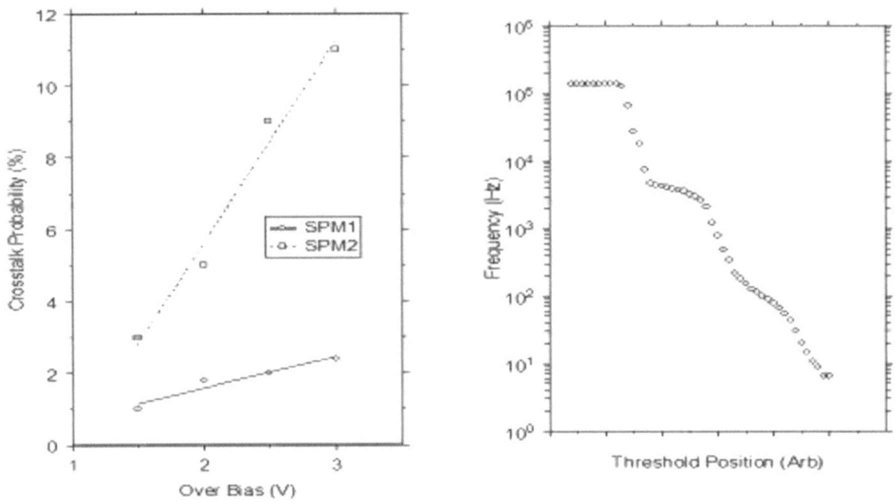

Fig. 18. Optic crosstalk probability of two photomultiplier as function of voltage over breakdown voltage and frequency of dark rate of two photomultipliers as function of threshold.

The optical crosstalk probability is a function of silicon photomultiplier bias voltage and the distance between neighboring pixels. However, the optic crosstalk can be significantly reduced by optically isolating the pixels from each other. This can be achieved by etching a trench between the pixels and filling it with an optically opaque material.

Fig. 19. a,b. shows specific element of the optic isolation, implemented for the quantum photo detector, by the high precision trench technology (Kotura Inc, 2009).

Implementation of trenches in the structure of silicon photomultipliers significantly reducing more then order of magnitude the optical crosstalk, bring on the level less 1% and create new high performance quantum photodetectors for the very low photon flux measurement (Saveliev, V et all, 2008).

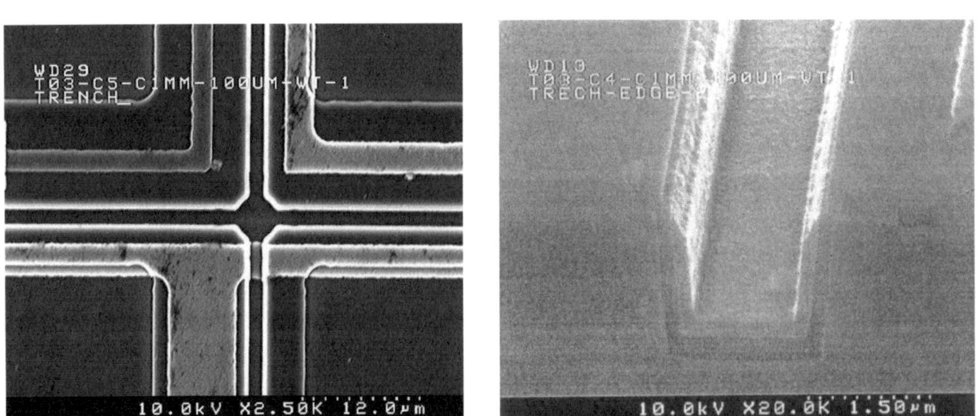

Fig. 19. a,b. micrograf of a -the trench pattern of silicon photomultiplier and b- profile of the trench of silicon photomultiplier

4.8 Afterpulsing

Afterpulsing processes of the silicon photomultiplier are breakdown processes occurs due to release of carriers captured by traps during a previewed breakdown avalanche process. During a breakdown processes a large number of carriers cross the pn-junction and finite probability that a carrier may become trapped by a bandgap state. The carrier may then be released at some time and trigger a new breakdown avalanche event.

Afterpulsing is described in term of probability. The afterpulsing probability is the probability that any gives pulse in followed by a false count or afterpulse. The afterpulsing probability is a function of the mean lifetime of the carrier in the trap. The trap lifetime depends on the trap prpperties, the type of carrier trapped and the temperature. Short trap lifetime result in a low afterpulsing probability since the carriers are likely to be emmited before the diode has fully recovered and the carrier avalanche initiation probabilities are low, Afterpulsing occurring before the diode has fully recovered will also have a lower gain since the diode will not have fully recharged. Preliminary study this effect in silicon photomultipliers shows not so dramatic effect on the performance of silicon photomultipliers.

5. Conclusion

Silicon photomultipliers is novel type of the semiconductor photodetector for the detecting of low photon flux. Already now the technology is developed and suitable for many critical application as medical imaging, and biology, homeland security, optic communications, experimental physics and military applications. Few world well known companies Hamamatsu, Sensl, Kotura are already producing or close to production of silicon photomultiplier type sensors. Near future of silicon photomultiplier development is create the completely digital quantum detector with implemented readout electronics on same substrate. Such solution will be another one step to the new era of quantum detection and will open possibility for many new applications.

6. References

Alvares-Gaume L. et al. (2008). Review of Particle Physics, Particle Detectors. *Physics Letters,* Vol. 667, No. 1-5, 2008 , 281-370, ISSN

Atac, M. (1993). Visible Light Photon Counters for Detecting Very Low Level Light. *ICFA Instrumentation Bulletin,* No 10, 1993 9-11

Bondarenko, G. et all. (2000). Limited Geiger Mode Microcell Photodiode: new results. *Nuclear Instruments and Methods,* A442, 2000 187-192

Chynoweth, A.G. & McKay K.G. Photon Emission from Avalanche Breakdown in Silicon. *Phys Review,* v 102, n2, 1956, 369-376

Gasanov, A. G. et all. (1988). *Soviet Technical. Physics Letters* Sov. 14, 313 (1988)

Golovin, V. & Saveliev V. (2004). Novel Type of Avalanche Photodetector with Geiger Mode Operation. *Nuclear Instruments and Methods,* A518, 2004, 560-564

Hammamatsu Photonics K.K. Photomultipliers Tubes, Basics and Applications, Toshikaza Hakamata et all. (Ed), 2006, 1-309, Hammamatsu Photonics K.K., Electron Tube Division, ISBN, Japan

Kotura (2009) Kotura, Inc., http://www.kotura.com

McIntyre R.J. On the Avalanche Initiation Probability of Avalanche Diodes above the Breakdown Voltage E.D., IEEE Transaction V 20, 7.,1973, 673-641

Saveliev, V. (1995). Avalanche Photodiode on Base of "Needle" Metal Resistor Semiconductor Structures, *Proceedings of New Development on Radiation Detectors, 7th European Symposium on Semiconductor Detectors*, Schloss Elmau, Germany, May 1995

Saveliev, V. & Golovin V. (2000). Silicon Avalanche Photodiodes on base Metal Resistor Semiconductor (MRS) Structures. *Nuclear Instruments and Methods*, A442, 2000 223-229

Saveliev, V. et all (2008). Quantum Photo Detectors with Optical Crosstalk Suppression, *Proceedings of 2008Nuclear Science Simposium, Medical Imaging Conference and 16th Room Temperature Semiconductor Detector Workshop*, Dresden, Germany, October 2008

Sensl (2009). Sensl Technologies Ltd, http://www.sensl.com

Stewart A. V. et all (2008). Performance of 1-mm^2 Silicon Photomultiplier. *IEEE Journal of Quantum Electronics*, Vol. 44, No. 2, 2008, 157-164

Tsang W.T. (Ed) (1985)., *Semiconductors and Semimetals: Lightwave Communication Technology, Part D, Photodetectors,* Tsang W. T. (Ed), 1-309, Academic Press Inc., ISBN :978-0-12-752153-4

15

Optical Mode Properties of 2-D Deformed Microcavities

Soo-Young Lee
Seoul National University
South Korea

1. Introduction

Optical microcavities have a wide range of applications and are used for fundamental studies, such as strong-coupling cavity quantum electrodynamics, enhancement and suppression of spontaneous emission, novel light sources, and dynamic filters in optical communication (Vahala, 2003). As a single-mode, ultralow threshold lasers, dielectric microdisks and microspheres have been studied, since they can support very high-Q whispering-gallery modes (WGMs) (McCall et al., 1992). Basically, strong light confinement of the WGMs in the dielectric microcavities is given by total internal reflection and only small evanescent leakage due to boundary curvature is a possible way of light loss. In a circular microcavity, the WGMs are characterized by two good mode/quantum number (m,l), m is the angular momentum quantum number and l the radial quantum number. Light emission of WGMs is isotropic due to the rotational symmetry of the circle cavity, which is undesired when considered as a light source for device applications.

In their seminal paper (Nöckel & Stone, 1997), Nöckel and Stone have pointed out that a deformation of the cavity leads to partially chaotic ray dynamics and highly anisotropic emission. Optical modes in deformed microcavities do not have good mode/quantum numbers due to absence of rotational symmetry or non-integrability of system. It is also known that the deformed microcavity has rich physics related to quantum chaos, such as chaos-assisted tunnelling, dynamical localization. Only drawback of deforming microcavity would be degradation of Q-factor, but for a slight deformation the degree of degradation is not severe due to the existence of the Kolmogorov-Arnold-Moser (KAM) invariant tori/curves preventing ray diffusion toward the critical angle of total internal reflection from which rays can escape from the microcavity. Some of subsequent works have focused on search for optimal cavity shape supporting optical modes with good emission directionality such as unidirectional emission and directional emission with narrower divergence (Chern et al., 2003; Shang et al., 2008). Other works on microcavities have treated fundamental quantum chaotic features such as scarring phenomena, chaos-assisted tunnelling etc. (Lee, S.-B. et al., 2002; Podolskiy & Narimanov, 2005)

In principle, any microcavity with broken rotation symmetry has directional emission. For a very slight deformation, the emission is the result of tunnelling process, and the emission comes out tangentially at the boundary points with the highest curvature. When the deformation increases enough, the emission directionality is well explained by an ensemble of chaotic ray trajectories. Since a ray trajectory in chaotic dynamics diffuses following unstable manifolds, the structure of unstable manifolds near the critical line for total internal reflection is very important to understand emission directionality of optical modes. Based on this ray dynamical analysis, one can design the shape of microcavity for good emission directionality. Various shapes have been proposed and their optical modes are characterized in terms of the directionality and Q- factor.

On the other hand, the optical microcavity is an analogue of open quantum system. In a 2-D dielectric microcavity, the Maxwell equation governing electromagnetic waves can be reduced to the Helmholtz equation. And the Helmholtz equation is nothing but the Schrödinger equation for a particle moving under a constant potential. Therefore, a ray trajectory and electromagnetic field distributions in the 2-D dielectric microcavity correspond to the particle path and wave functions in billiard problems. In quantum chaos study, billiards have been a paradigm since the classical dynamics varies from regular to fully chaotic depending on the boundary shape and can be easily analyzed compared to other chaotic systems. Therefore, one can study, in the dielectric microcavities, various quantum chaotic features which have been important issues in billiards such as dynamical localization, scarring phenomena etc.

Another important aspect of the dielectric microcavity is its non-Hermitian features. The dielectric microcavities are open systems so that the field inside the microcavity decays with time, therefore, eigen-frequencies of each optical modes have negative imaginary part representing decay rate or spectral linewidth. A non-Hermitian system has complex eigenvalues in general. Therefore, the microcavity is a non-Hermitian system and possesses general non-Hermitian features. The degeneracy point in a non-Hermitian quantum system is called exceptional point (EP) at which two quantum states coalesce, i.e., only one eigenvalue and one eigenstate exist there. In mathematically, the EP is a square root branch point so that the energy surface near an EP has non-trivial topology and geometrical phase. These non-Hermitian features have been treated in dielectric microcavities (Lee, S.-Y. et al., 2008b; Schomerus, 2009).

In Sec. 2 and Sec. 3, several tools for ray and wave analysis on optical modes, such as the steady probability distribution, boundary element method, and Husimi functions, are introduced. Based on the methods, directional emissions from microcavities with various shapes are explained in Sec. 4 and the influence of openness on scarred and quasiscarred optical modes is discussed in Sec. 5. In Sec. 6 the mode-mode interaction along a deformation change and the exceptional point are reviewed. Finally, a brief summary will be given.

2. Ray dynamical analysis

To understand optical mode properties in deformed microcavities, such as internal field intensity pattern and emission direction, it is important to figure out underlying classical ray dynamics. There is a very useful tool to see if ray dynamics is regular or chaotic, it is the Poincaré surface of section (PSOS) that has been widely used in nonlinear dynamics community (Reichl, 1992). This PSOS can be extended to incorporate system's openness for describing open chaotic systems. A useful distribution in phase space made by the incorporation for fully chaotic case is the steady probability distribution (SPD) revealing long-lived ray trajectories in deformed microcavities.

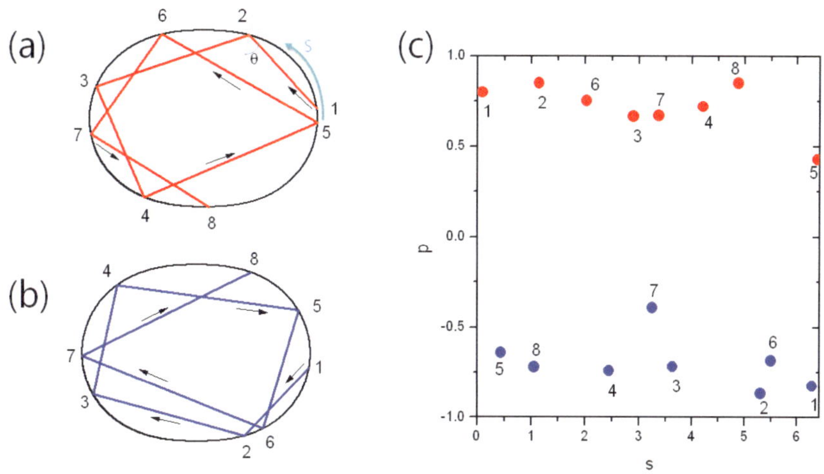

Fig. 1. Correspondence between real trajectories ((a), (b)) in a deformed billiard and a series of points (PSOS) in the phase space (c). s is the boundary coordinate and $p = \sin\theta$, θ is the incident angle.

2.1 PSOS and chaotic transition

Tracing a ray trajectory in a microcavity gives, in general, a useful information for optical mode pattern, e.g., consider a typical Fabry-Perot resonator or a circular microcavity where optical modes are easily matched to underlying ray trajectories. However, when the cavity is deformed, some trajectories are not simple, and they look too complicated to extract some useful information. The most effective way to see whether ray dynamics is regular or chaotic is to depict the Poincaré surface of section (PSOS). As an example, consider a particle trajectory in a 2-D billiard as shown in Fig. 1 (a) and (b). Complete description of the particle movement needs 4 dimensional phase space, e.g., its position (x,y) and momentum (p_x,p_y), but it is very difficult to imagine a trajectory in 4-D space for us living in 3-D spatial space. So, a useful method is to examine only sectional trace. Imagine a 2-D phase space and mark the intersecting points of the trajectory with a 2-D surface. In the billiard problem, the boundary length coordinate (s) of a bounce position and the tangential component ($p = \sin\theta$, θ is the incident angle) is widely used, and they are canonical conjugate coordinates. Then, the trajectory can be represented by a set of points in phase space (s, p) as shown in Fig. 1 (c). In a circular cavity, a ray trajectory would give a straight line in phase space because the incident angle of the ray is invariant, so p is constant. This trajectory can be matched to a WGM. The WGM show a intensity localization similar to localization of the ray trajectory (see Fig. 2 (a)) with incident angle of $\theta=\arcsin(m/\mathrm{Re}[nk_{(m,l)}])$, $k_{(m,l)}$ is the wavenumber of $WGM_{(m,l)}$. As the microcavity is deformed, the rotational symmetry is broken so that the incident angle or angular momentum is not invariant any longer. Accordingly, the PSOS shows some wavy invariant lines and nonlinear resonances. The central points of nonlinear resonances corresponds to a stable periodic orbits and there are unstable periodic orbits between them, as shown in Fig. 2 (b). As the deformation increases more, the chaotic region is getting larger from the vicinity of the unstable periodic orbits. Even in this case, there exists the invariant curves dividing two chaotic regions. The trajectory on an invariant curve cannot go somewhere, i.e., lives forever on the invariant curve. This is important in ray transport because rays in the chaotic region above this invariant curve cannot diffuse into below the curve. This invariant curve thus plays a role of some barrier in ray transport. With more stronger deformation, there are some stable islands and chaotic sea, see Fig. 2 (c). A ray trajectory lying in the islands cannot leave the islands and, on the other hand, a ray motion in chaotic sea shows a fast exponential diffusion in general over the chaotic sea. However the ray existing just outside the islands does not show such fast diffusion, instead, show some power-law diffusion due to the existence of complex small island chains (the red trajectory in Fig. 2 (c)). This kind of phase-space change is typical in non-linear system and is known as KAM scenario.

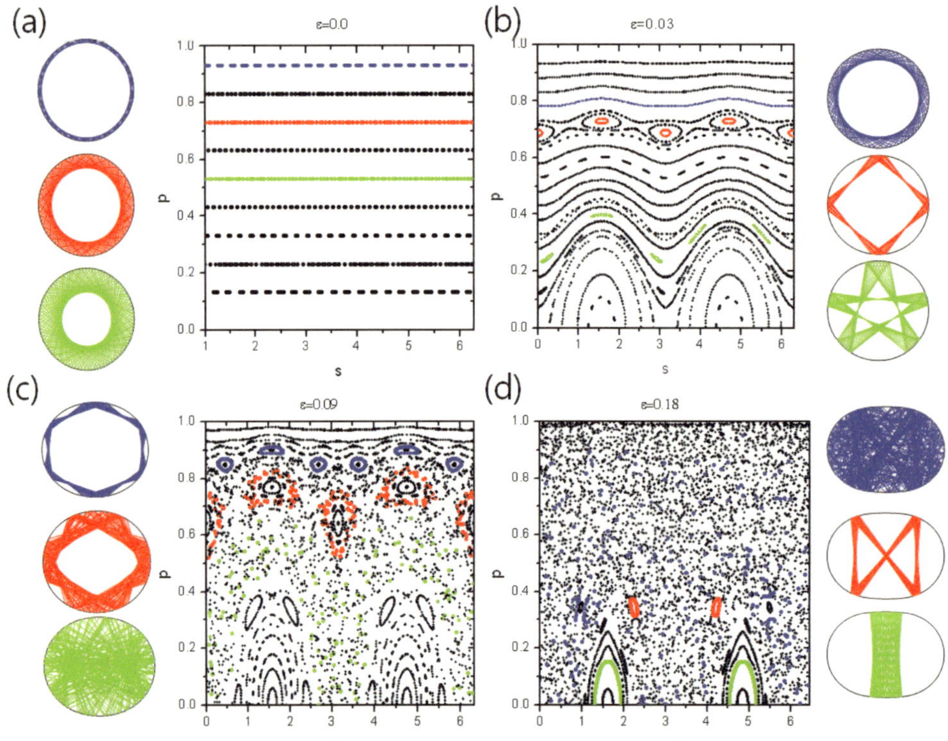

Fig. 2. Change of the PSOS in a chaotic transition. The billiard is qudrupole-deformed, $r = r_0(1+\varepsilon\cos 2\phi)$. (a) $\varepsilon = 0.00$, the circular billiard showing regular dynamics. (b) $\varepsilon = 0.03$, almost regular with some nonlinear resonances. (c) $\varepsilon = 0.09$, some islands in chaotic sea represented by random dots. (d) $\varepsilon = 0.18$, almost chaotic with the robust islands for bouncing-ball trajectories. The blue, red, and green points in the PSOS's correspond to the trajectories shown in deformed billiards in the same color.

2.2 Steady Probability Distribution (SPD)

The PSOS for a fully chaotic system has no specific structure and only random chaotic sea exists in phase space. This implies that the PSOS itself does not give some useful information about optical modes of dielectric microcavities. However, if system's openness is incorporated in the PSOS, one can find some specific structure in the PSOS, revealing both chaotic ray dynamics and openness character.

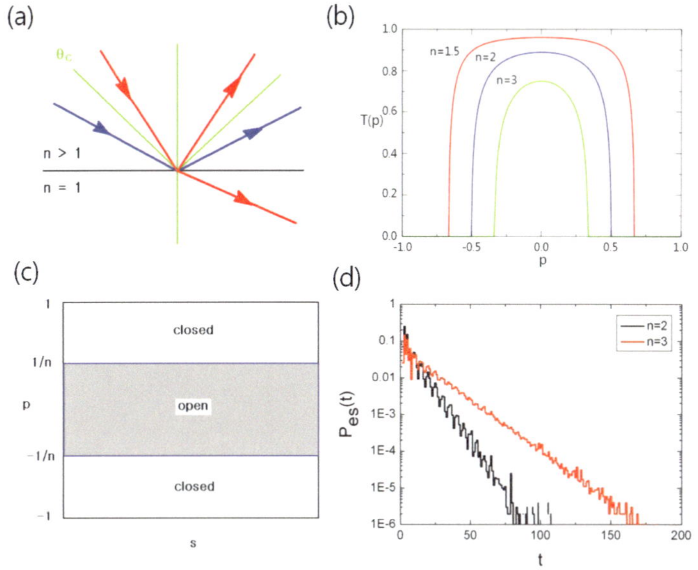

Fig. 3. (a) The incident ray splits into reflected and transmitted rays at the dielectric interface when the incident angle θ is less than the critical angle θ_c (red arrows), and the incident ray would be totally reflected when θ is greater than θ_c. (b) The transimission probability *T(p)* for n=1.5, 2, 3. (c) The open region in phase space where the ray have non-zero probability to escape. (d) The exponential behaviour of the escape time distribution.

As well known in optics, a plane wave incident to an interface between dielectric media of refractive index n and air (n=1) would split into reflecting wave and refracting (transmitting) wave. The reflection probability derived from Fresnel equations is given as

$$R = \left(\frac{n \cos\theta_i - \cos\theta_t}{n \cos\theta_i + \cos\theta_t} \right)^2 \tag{1}$$

where θ_i and θ_t are the incident and refractive angles, respectively, and they are related by Snell's law, $n \sin\theta_i = \sin\theta_t$. The transmission probability is then $T = 1 - R$. As illustrated in Fig. 3 (a), if the incident angle is greater than the critical angle, $\theta_c = \arcsin(1/n)$, the transmission is zero, i.e., total internal reflection takes place. The transmission with an incident angle less than θ_c depends on the refractive index n and incident angle θ_i. This dependence is shown in Fig. 3 (b). The transmission becomes maximum at the normal incident case, $\theta_i = 0$, and the small opening limit is given by $n \to \infty$. In phase space (s, p), the non-zero transmission region is shown as the shaded in Fig. 3 (c).

Imagine a ray trajectory in a deformed microcavity with chaotic ray dynamics. If the incident angle of the ray at a bounce is greater than the critical angle, the ray would be totally reflected. In a chaotic microcavity the incident angle is not invariant, so it becomes soon less than the critical angle. Then, due to the above splitting at interface, part of ray would escape with the transmission probability T and with the reflection probability R the ray would be reflected from the boundary. This openness can be characterized by the exponential behaviour of escape time distribution for a fully chaotic microcavity, as shown for a stadium-shaped microcavity in Fig. 3 (d). This exponential behaviour implies that there are some ray distribution in phase space which are statistically invariant with time (Lee, S.-Y. et al., 2004). This distribution is called the steady probability distribution (SPD). The SPD contains both ray dynamical and open properties of the given chaotic microcavity, and shows the distribution of long-lived ray trajectories in phase space. It is emphasized that the structure of the SPD follows essentially unstable-manifold structure since all ray trajectories diffuse along unstable manifolds. In Fig. 4, three SPDs are shown for a stadium-shaped (a), a rounded triangular-shaped (b), and a spiral-shaped microcavities. In the spiral case, it is clear that only clockwise circulating ray can survive long due to no bouncing from the notch.

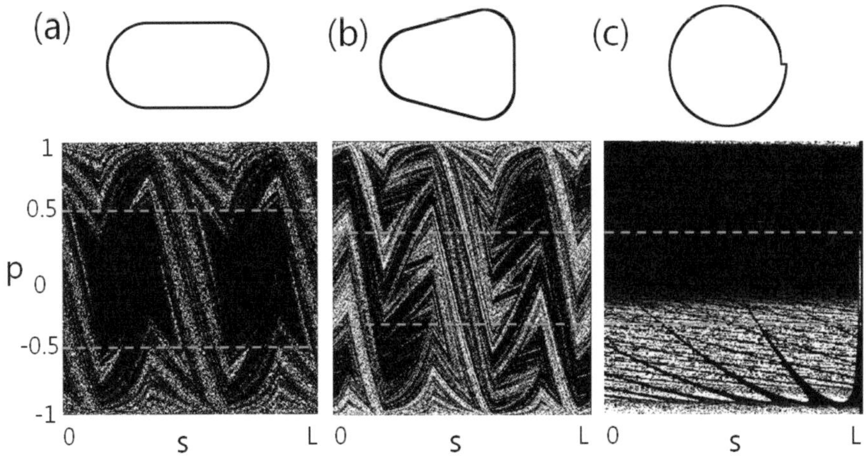

Fig. 4. The steday probability distributions (SPD's) of stadium-shaped (a), rounded triangle-shaped (b), and spiral-shaped (c) dielectric microcavities. The dashed lines denote the critical lines given by $1/n$.

3. Analysis of optical modes

In microcavity experiments, optical mode properties can be characterized by measuring spectra and far field pattern. However, it is very difficult to know some internal field distribution and the role of internal ray dynamics from the both measurements. This gap can be filled by numerical simulation for internal field distributions of optical modes. A popular method for the mode calculation is the boundary element method (BEM), first introduced by Wiersig to mirocavity (Wiersig, 2003). The calculated optical modes can be transformed to some functions in phase space for comparing to the ray dynamics (PSOS or SPD). One of transformed functions is Husimi function which gives more intuitive explanation than another phase space representation, Wigner function.

3.1 Boundary Element Method (BEM)

When a microcavity is slightly deformed, a perturbation method can be used to get optical mode frequencies and internal field distribution (Dubertrand et al., 2008). However, this is not effective for strongly deformed microcavities due to the need of enomous number of basis functions. The boundary element method (BEM) (Wiersig, 2003) is very effective for this strongly deformed case.

Maxwell equations governing electromagnetic waves in a 2-D microcavity can be reduced to the Helmholtz equation

$$\left(\nabla^2 + n(\vec{r})^2 k^2\right) \psi(\vec{r}) = 0. \tag{2}$$

The main idea of the BEM is to discretize a boundary integral equation derived from the Helmholtz equation. To obtain the boundary integral equation, the Green function, the solution of Helmholtz equation with a delta function source, and Green's identity are used. The Green function is given by zeroth-order Hankel function of the first kind

$$G(\vec{r},\vec{r}';nk) = -\frac{i}{4} H_0^{(1)}\left(nk|\vec{r}-\vec{r}'|\right). \tag{3}$$

The resulting boundary integral equation can be written as

$$c\psi(\vec{r}) = \oint_\Gamma ds'\left[\partial_n G(\vec{r},s';nk)\psi(s') - G(\vec{r},s';nk)\partial_n \psi(s')\right] \tag{4}$$

where c is 1 for $\vec{r}$ inside the boundary Γ, 1/2 for $\vec{r}$ on the boundary, and zero otherwise, and ∂_n means the outward normal derivative at the boundary. This integral equation means that if one knows the boundary values of wave function $\psi(s)$ and its normal derivative $\partial_n\psi(s)$, one can then calculate the value of wavefunction $\psi(\vec{r})$ at any point inside the boundary. Note that the Green function inside the microcavity is different a little from the outside Green function due to the different refractive index n.

The unknowns to be calculated are the boundary values of wavefunction $\psi(s)$ and its normal derivative $\partial_n\psi(s)$. By approaching $\vec{r}$ to the boundary and discretizing whole boundary into m elements, one can obtain an equation containing $2m$ unknowns, $\psi(s_i)$ and $\partial_n\psi(s_i)\ (i=1,\cdots,m)$. From Eq. (4) for inside microcavity, one can make m equations by pointing $\vec{r}$ onto m boundary elements, and the other m equations can be obtained by considering outside microcavity where $n=1$ so different argument of the Green function has to be used. Therefore, the discretization leads to a matirix equation and a non-trivial solution exists only when the determinant of the matrix is zero, which is the condition determining the complex wave number k of optical modes. Note that the Green function and its normal derivative are singular at the origin, so a careful treatment is needed to evaluate the diagonal elements of the matrix. The singularity of $H_1^{(1)}(z)$ in $\partial_n G$ can be compensated by a geometrical factor, and the corresponding diagonal terms can be expressed in terms of the curvature of the boundary. Also, the sigularity of $H_0^{(1)}(z)$ in G can be integrated by using the asympotic expression.

Throughout this chapter we focus on TM (transverse magnetic field) polarization where both the wave function $\psi(\vec{r})$ and its normal derivative $\partial_n\psi(\vec{r})$ are continuous across the boundary, and the wavefunction $\psi(\vec{r})$ corresponds to the z component of the electric field when the 2-D microcavity lies on the x-y plane. In the case of TE (transverse electric field) polarization, the wavefunction $\psi(\vec{r})$ represents the z component of magnetic field, and the wavefunction $\psi(\vec{r})$ is continuous across the boundary, but its normal derivative is not, instead $n^{-2}\partial_n\psi(\vec{r})$ is continuous.

Since the boundary curvature appears in the diagonal elements in the matrix, non-analytic boundary points such as an angular cornor, where the curvature is not defined, make a trouble in BEM. This can be overcomed by rounding the angular cornor by a part of circle. Here the radius of circle should be much less than the wave length of the optical mode, because the wave needs not to know the rounding process.

3.2 Husimi functions of optical modes

Sometimes phase-space representation of an optical mode is more useful than internal intensity distribution. If one wants to compare the optical mode to the classical PSOS or SPD in phase space, the transform from wave function in real space to some distribution in phase space is needed. The most popular one is the Husimi function. The Husimi function is determined by the overlap of the wave function with a coherent state of a minimum uncertainty. It constitutes a quasi-probability in phase space, which acquires an intuitive quasi-classical character in the semiclassical limit of short wavelength.
Recently, Hentschel et al. (Hentschel et al., 2003) have derived four Husimi functions applicable to the dielectric microcavity, two are for incident and reflected waves inside the microcavity and the other two are for those outside the microcavity. The incident and the reflective Husimi functions of inside waves are written as

$$H^{inc(refl)}(s,p)=\frac{k}{2\pi}\left|-F\,h(s,p)+(-)\frac{in}{kF}h'(s,p)\right|^2 \tag{5}$$

where $F=\sqrt{n\sqrt{1-p^2}}$ and the components of the Husimi function are given as

$$h(s,p)=\int ds'\psi(s')\,\xi(s';s,p), \tag{6}$$

$$h'(s,p)=\int ds'\partial_n\psi(s')\,\xi(s';s,p), \tag{7}$$

where the minimal-uncertainty wave packet is given by

$$\xi(s';s,p)=\sum_l\frac{1}{\sqrt{\sigma\sqrt{\pi}}}\exp\left[-\frac{1}{2\sigma^2}(s'-s+Ll)-ikp(s'+Ll)\right], \tag{8}$$

where k and L are the wave number inside the cavity and the total length of the boundary, respectively. This packet corresponds to a coherent state centred at (s,p). The aspect ratio factor is taken as $\sigma=\sqrt{\sqrt{2}/k}$.
From the BEM, one can obtain the boundary functions $\psi(s),\partial_n\psi(s)$ of an optical mode $\psi(\vec{r})$, and these can be used for calculating the Husimi function from the above equations. This Husimi function shows details of internal waves of the optical mode such as bouncing positions (s') and incident angles (p'), (s',p') are the phase-space positions of localization pattern of the Husimi function.

4. Emission properties of optical modes

From the viewpoint of applications, directionality of emission has been an important issue many authors have focused on. Various shapes of microcavity have been proposed for optical modes with good directionality. The emission directionality can be understood by ray dynamics and Husimi functions mentioned above.

4.1 Directional emission

The PSOS of a circular microcavity shows simple invariant lines in phase space as shown in Fig. 2 (a), and its optical modes are WGMs with specific mode number *(m,l)*. The $WGM_{(m,l)}$ has $2m$ intensity spots along the perimeter of the circle, and l intensity spots along radial direction, so one can identify *(m,l)* from the internal intensity distribution of the WGM. The Husimi function of the $WGM_{(m,l)}$ reveals wave components with a definite incident angle, $p=\pm\sin\theta=\pm m/\mathrm{Re}[nk_{(m,l)}]$. If the incident angle is greater than the critical angle, the internal wave escapes through tunnelling process. The WGMs showing the tunnelling emission have high quality factor due to the small transmission probability. The tunnelling interpretation comes from the fact that the radial equation of 2-D Helmholtz equation becomes 1-D potential barrier problem. Due to the rotational symmetry the tunnelling rate is invariant about the polar angle, therefore the emission is isotropic. Speaking in detail, the emission is tangential because the incident wave with critical angle would be transmitted tangentially, and far field emission pattern is formed by interference of two tangential emissions tunnelled from oppositely circulating waves.

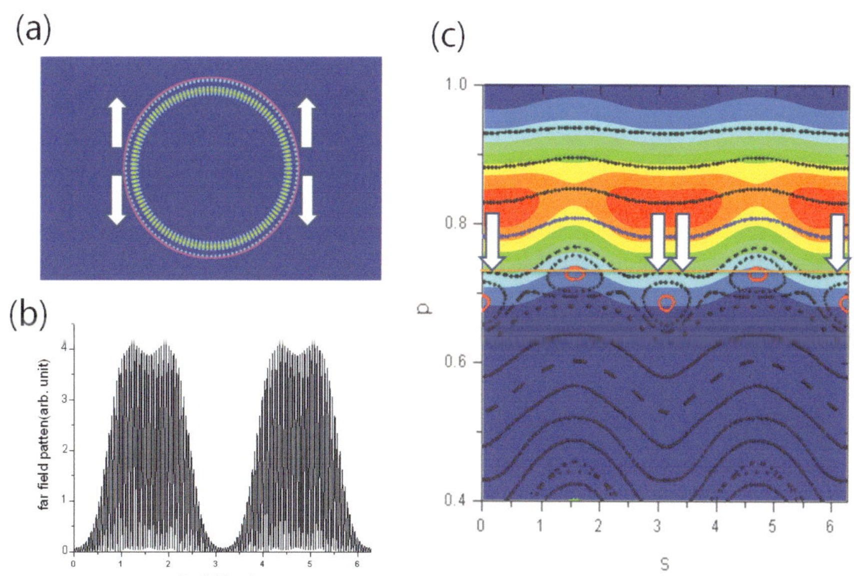

Fig. 5. (a) A WGM in a slightly deformed microcavity, ε=0.03, and the arrows indicate the emission directions that are not clear in the intensity plot due to its negligible intensity compared to the internal one. (b) The far field pattern of emission. This is consistent with the arrows in (a). (c) The Husimi function of the WGM and the PSOS. One can expect that internal wave would tunnel out at the highest curvature points due to the shortest distance to the critical line (the orange line).

When the microcavity is slightly deformed, the tunnelling emission is no longer isotropic. As shown in Fig. 2 (b) and Fig. 5 (c), the straight invariant lines become curved lines showing minimum at the boundary points with the highest curvature. This implies that the tunnelling rate would be maximum at the points as illustrated by arrows in Fig. 5 (c). Then, tunnelling emissions would appear tangential direction at the points with the highest curvature like the optical mode in Fig. 5 (a). This is confirmed by the far field pattern in Fig. 5 (b).

If the deformation increases a little more, the emission direction would have mode dependence because the optical mode can be localized on stable or unstable periodic orbit due to the perturbation originating external coupling (Unterhinninghofen et al., 2008). This mode dependency disappears and chaotic transport, including turnstile transport (Shim et al., 2008), become important when the deformation increases enough. The corresponding ray dynamics is chaotic and the rays diffuse along unstable manifolds, so the structure of unstable manifolds determines the emission directions, which will be discussed in detail when scarred optical modes are discussed.

4.2 Unidirectional emission

Optimal directionality would be unidirectional emission like usual Fabry-Perot laser with two mirrors, one has a perfect reflection and the other does not. In a microcavity, light confinement is not achieved by mirrors, but by the total internal reflection at the cavity boundary. Therefore, one cannot easily expect if the given shape of microcavity can support the uni-directional optical modes or not.

The unidirectional emission is first reported in a spiral-shaped microcavity by Chern et al. (Chern et al., 2003). The key of unidirectionality in the spiral-shaped microcavity is the existence of notch (see Fig. 6 (a)), because every periodic orbits have at least one bounce from the notch (Lee, S.-Y. et al., 2004). Since there are many chances of bounce from the notch, the spiral-shaped microcavity seems to be ideal to support optical modes showing unidirectional emission and the unidirectional lasing emission from the notch is confirmed by other experimental groups (Ben-Messaoud & Zyss, 2005; Tulek & Vardeny, 2007; Kim et al., 2008). However, it is interesting to note that the optical modes showing unidirectional emission have not been reported in BEM calculation.

A rounded triangle-shaped microcavity can support unidirectional optical modes (Kurdoglyan et al., 2004). Some clue of unidirectional emission can be found in the SPD shown in Fig. 4 (b), where the intensity of SPD corresponding to normal incident rays at right vertical boundary is rather strong. The unidirectional optical mode shows WGM-like pattern at the left circular part of boundary, which enables waves to turn their direction without emission to the left direction as illustrated in Fig. 7 (b). However, the unidirectional optical modes would not be high-Q mode, since the incident angles of waves at right vertical boundary are small, implying large transmission probability. Therefore, the unidirectional optical modes are rather leaky.

Some high-Q unidirectional optical modes has been found in an annular microcavity (see Fig. 7 (c)) (Wiersig & Hentschel, 2006). The annular microcavity can support WGM-like high-Q modes and leaky unidirectional modes. These two modes can be superposed under an avoided resonance crossing, and as a result of the superposition the high-Q optical mode has unidirectional emission without severe degradation of quality factor.

Robust high-Q optical mode with unidirectional emission has been found in Limaçon microcavity(Wiersig & Hentschel, 2008). As shown in Fig. 6 (d), initially glancing rays with large incident angle eventually escape at two boundary points with same direction. This can be well explained by the unstable-manifold structure or SPD near the critical line. At the critical lines appreciable intensity of SPD appears only the two positions in phase space corresponding to the two escaping points.

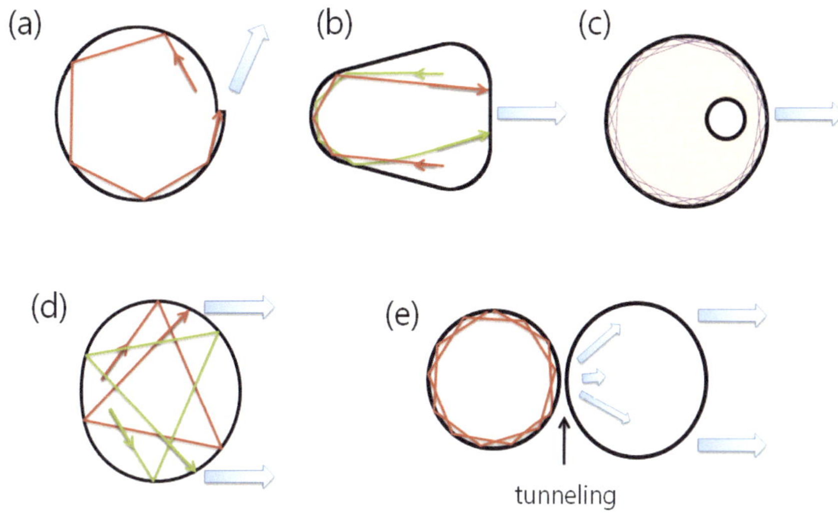

Fig. 6. Various shapes of dielectric microcavities supporting unidirectional optical modes. (a) Spiral-shaped microcavity. (b) Rounded triangle-shaped microcavity. (c) Annular microcavity in which high-Q WGM is coupled to a leaky unidirectional mode without severe degradation of Q factor. (d) Limaçon microcavity, $r=r_0(1+\eta\cos\phi)$, $\eta=0.43$. (e) Coupled asymmetric microdisks.

Recently, it is reported that a coupled asymmetric microdisks (see Fig. 6 (e)) can support unidirectional emission (Ryu et al., 2009). When the left disk in Fig. 6 (e) satisfies the resonance condition and the right disk does not, a WGM appears only on the left disk. In this case, the right disk plays a role of a defect on boundary, so through the contact point waves tunnel out and the circular boundary of right disk makes the tunnelled waves be collimated, resulting in a unidirectional emission.

4.3 Narrow divergent emission

Divergence of emission beam is another important property of optical modes. Narrow divergence of directional emission has been reported in a peanut-shaped ring cavity, made by two optical fibers coated by organic/inorganic hybrid meterials doped by rhodamine B (Shang et al., 2008), the refractive indices of optical fiber and the coating medium are 1.46 and 1.52, respectively. The partial ray trajectories escaped are shown in Fig. 7 (a). The bounced rays at the circular part have incident angles greater than the critical angle so that there is no refractive emission from the circular part, and the rays begin to spread and are collimated to make a parallel beam output with rather large width. This is the key feature of the peanut-shaped microcavity for narrow divergence of output beam. Recall the single slit diffraction where the divergence angle is given by $\theta \approx \lambda / a$, λ is the wavelength and a the opening size of the slit. The larger is the opening size of single slit, the smaller the divergence angle of diffraction. This can be applicable to the peanut-shaped microcavity where the large width of ray beam corresponds to the opening size of the single slit. The corresponding optical mode can be found, it is shown in Fig. 7 (b). Four collimated emission beams come out from rather large part of circular boundary. The far field pattern of this mode shows very narrow divergence angle less than 10 degree (Fig. 8 (c)), comparing with the typical divergence angle, about 40 degree (Chern et al., 2003), of spiral-shaped microcavity. The SPD reveals this property in detail as shown in Fig. 8 (d). In the open region *(-1/n< p <1/n)*, stripe pattern exists and this pattern well overlaps with the red line in Fig 8 (d), indicating rays of parallel emission with far field angle 24 degree. The Husimi function of the optical mode shown Fig. 8 (d) also well overlaps with the red line.

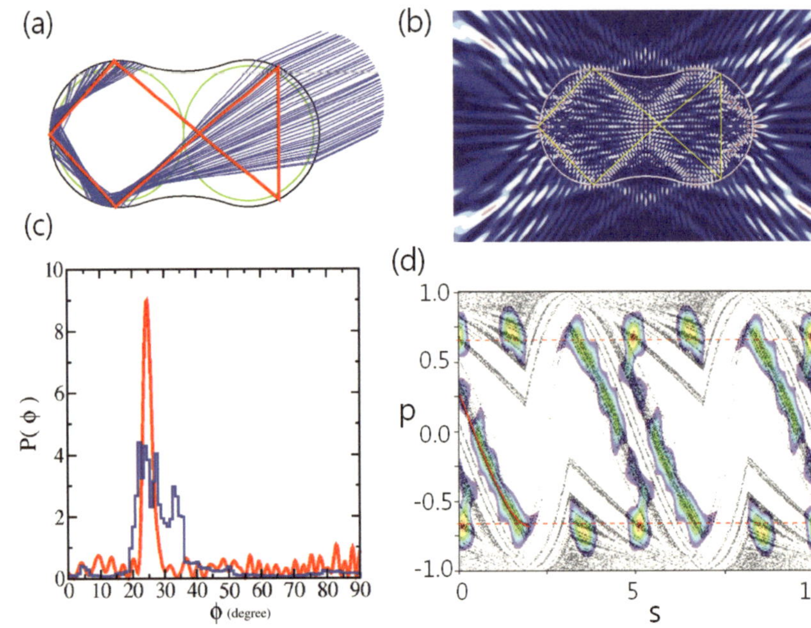

Fig. 7. (a) The ray trajectories forming a collimated broad ray beam. The fish-shaped periodic orbit shows similar bouncing position on the circular part of boundary. (b) The intensity plot of an optical mode showing the narrow divergence. (c) The red line is the farfield pattern of the optical mode shown in (b), which show narrow divergence of less than 10 degree. The blue histogram is the result of SPD. (d) The Husimi function of the optical mode of (b) and SPD (background gray dots). The red line indicates the trajectories with 24 degree farfield angle. The dotted lines are the critical lines for total internal reflection.

5. Scarred and quasiscarred modes

In a stadium-shaped billiard, Heller has found the scarred eigenstate showing enhanced amplitude along a unstable periodic orbit (Heller, 1984). After his work, many authors have studied the scar phenomena in various chaotic systems. The reason of much attention on the scar phenomenon is its counter intuitive aspect. In common sense based on classical chaotic dynamics, showing evenly and randomly distributed PSOS, one would expect quite irregular intensity distribution of eigenfunction. In fact a typical eigenstate has this irregular eigenfunction and this is consistent with the prediction of random matrix theory. Although

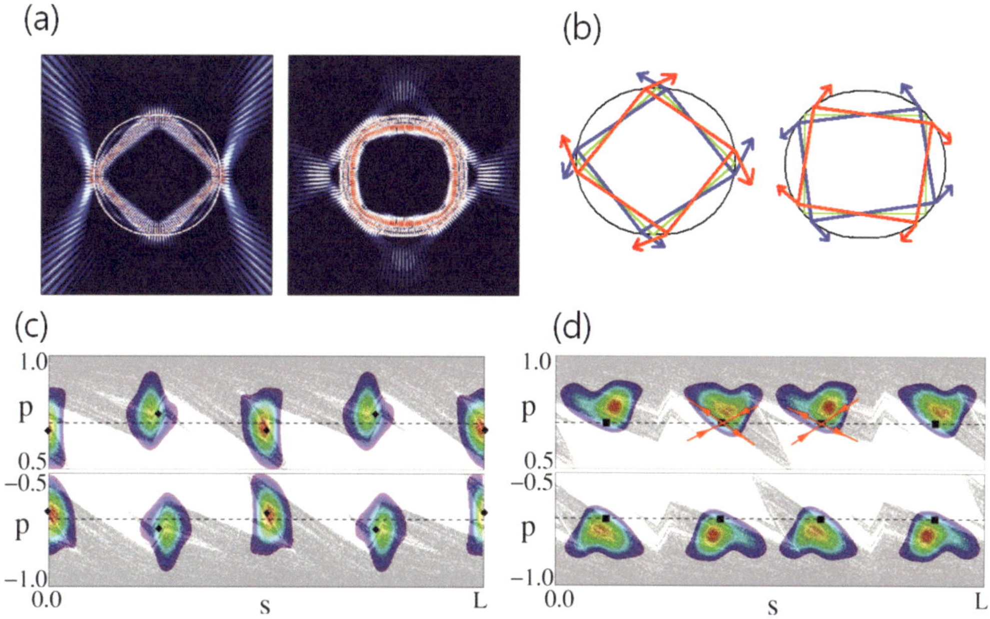

Fig. 8. (a) The scarred optical modes associated with diamond and rectangle unstable periodic orbits in a stadium-shaped microcavity (Lee, S.-Y. et al., 2005). (b) Schematic diagrams of slight pattern rotations depending on the propagating way. (c) The Husimi function of the diamond optical mode with SPD in background. (d) The Husimi function of the rectangle optical mode with SPD in background. The red arrows denote the stable and unstable manifolds emanating from the rectangle periodic orbit. The dotted lines are the critical lines.

only small part of eigenfunctions are scarred in closed billiards, in open systems such as dielectric microcavity the scarred modes can be easily found because the openness suppresses the contribution of long trajectories. The scarred optical modes have been observed in dielectric microcavities (Lee, S.-B. et al, 2002; Rex et al., 2002)

5.1 Scarred modes

The scarred optical modes in a chaotic microcavity have different properties not seen in the scarred eigenfunctions of closed billiards (Lee, S.-Y. et al., 2005). First, consider a scarred optical mode with incident angles comparable to the critical angle. Figure 8 (a) shows scarred modes with diamond and rectangle patterns in a stadium-shaped microcavity with the parallel boundary length *0.2 R*, *R* the radius of circular part. The refractive index is taken as $\sqrt{2}$ so that the critical angle is 45 degree. Note that the incident angles of the diamond and rectangle periodic orbits are about 45 degree. In this case, the emission beams come out from the bouncing positions of the internal periodic orbit pattern.

Although the internal intensity plots of the scarred optical modes looks similar to scarred eigenfunctions of closed billiards, the detail is quite different. From the positions of maximum intensity of Husimi functions for the scarred modes (see Fig. 8 (c),(d)), one can see that the bouncing positions are slightly rotated and the way of rotation depends on the circulating way of internal waves. For example, in the diamond scarred mode the clockwise circulating wave pattern is slightly rotated with positive angle shift and the counter-clockwise one is done with negative angle shift as illustrated in Fig. 8 (b). More interestingly, in the rectangle case, the rotations appear in the opposite way (see Fig. 8 (b)). These intriguing angle shifts are certainly results of openness of the microcavity. At a bounce, partial waves are escaping refractively, and the rest of waves are reflected and then take part in resonance formation, which is known as Fresnel filtering effect (Rex et al., 2002). This is why the scarred patterns do not lie on the exact periodic orbits.

The difference of angle-shift way between diamond and rectangle scarred optical modes can be understood by comparing Husimi functions with SPD. This comparison is shown in Fig. 8 (c) and (d). The black dots indicate the exact positions of diamond and rectangle periodic orbits and the grey dots in the background represent the SPD. One can easily find the maximum intensities of Husimi functions locate on some points near the black dots and the deviations are toward dense part of the SPD. Note that the dense part in SPD corresponds to long-lived ray trajectories, surviving long time in the microcavity. Therefore, it is clear that only long-lived part of waves take part in the resonance formation, as a result, the angle shifts are determined by the structure of SPD near the critical lines.

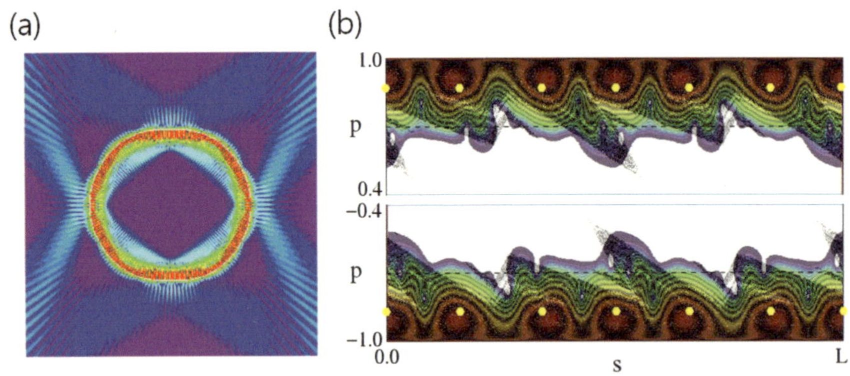

Fig. 9. (a) A high-Q scarred optical mode in logarithmic scale. One can see the mismatch between the bouncing positions of internal pattern and emission positions. (b) The Husimi function of the optical mode shown in (a) with SPD in background. The emission positions coincide with the SPD structure near the (dotted) critical lines (Lee, S.-Y. et al., 2005).

Second, consider a high-Q scarred mode (see Fig. 9 (a)) whose Husimi function is localized far above the critical line due to its incident angles much greater than the critical angle. The emission mechanism in this high-Q case is somewhat different from the previous case. The emission points are not associated with the internal pattern, for example, the hexagonal scarred optical mode in Fig. 9 (a) has four emission points, which is clear only in logarithmic scale. This is also understood by the SPD structure reflecting unstable-manifold structure. The waves forming periodic orbit pattern cannot escape at the bouncing positions where the total internal reflections occur. Instead, the waves would diffuse toward the critical lines and eventually escape by refraction. The wave diffusion would follow underlying ray trajectories moving along the unstable manifolds. This mechanism is confirmed by Husimi function of the high-Q scarred mode. The structure of Husimi function, in logarithmic scale, below the critical line overlaps well on the SPD as shown in Fig. 9 (b).

This result is important because this mechanism can be applied to other high-Q modes in chaotic microcavity. The importance of unstable-manifold structure near the critical line has been stressed by many authors (Schwefel et al., 2004; Altmann, 2009) and experimentally confirmed in liquid-jet microcavity (Lee, S.-B. et al., 2007).

5.3 Quasiscarred modes

Although the openness effect on scarred optical modes has been discussed in the previous subsection, more dramatic effect of openness has been found in the spiral-shaped microcavity (Lee, S.-Y. et al., 2004). As shown in Fig. 10 (a) and (b), scar-like optical modes exist in the microcavity, triangle- and star-shaped localizations are shown in cases of n=2 and n=3, respectively. These modes are named quasiscarred modes because, unlike scarred modes, there is no underlying unstable periodic orbit. It is then obvious that, even though there is no periodic ray orbit, the openness enables waves to make periodic propagation with constructive interference.

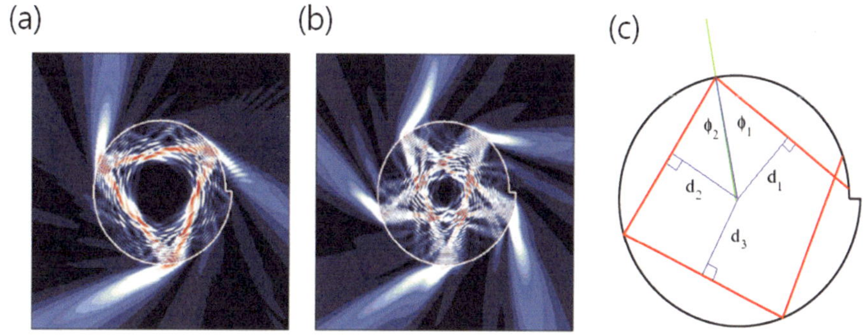

Fig. 10. (a), (b) The quasiscarred optical modes in a spiral-shaped dielectric microcavity (Lee, S.-Y. et al., 2004). The refractive index is 2 in (a) and 3 in (b). (c) A trajectory without bouncing from the notch. The distance between trajectory segment and the origin is getting larger for counter-clockwise circulating ray, $d_1<d_2<d_3$.

Non-existence of triangle- and star-shaped periodic orbits can be easily understood from the ray trajectory shown in Fig. 10 (c). For the counter-clockwise circulating ray without bouncing from the notch, the distance from the origin always increases, so it is impossible to make a periodic orbit without bouncing from the notch.

It is emphasized that many of optical modes at n=2 and n=3 are triangle- and star-shape quasiscarred, respectively (Lee, S.-Y. et al., 2008a). This is a direct evidence of crucial role of openness on the formation of quasiscarred optical modes, noting that the incident angle of internal waves is about the critical angle given by θ_c=arcsin(1/n) (Lee, J. et al., 2008). It is noted that Goos-Hänchen shift, with which reflected wave beam comes out away from the incident beam position, and the Fresnel filtering effects become maximum when the incident angle of wave beam is about the critical angle (Lee, S.-Y. et al., 2005). These unique effects of dielectric microcavities should be important in the quasiscarred-mode formation, i.e., formation of constructive periodic wave propagation without underlying ray periodic orbit. The lasing of quasiscarred modes has been numerically studied (Kwon et al., 2006) using the Schrödinger-Bloch model (Harayama et al., 2003), and recently experimentally confirmed in spiral-shaped microcavity (Kim et al., 2009).

Although the quasiscarred optical modes have been found only in the spiral-shaped microcavity, there is no constraint preventing their appearance in other deformed microcavities with discrete symmetry. To find quasiscarred modes in general deformed microcavities one needs more quantitative criterion for the formation of quasiscarred modes.

6. Avoided resonance crossing and exceptional point

Eigenvalues of a closed quantum system show some variation under change of system's parameter such as deformation of a billiard. An eigenvalue may increase or decrease depending on its mode property. When two eigenvalues encounter each other under a parameter change, they show level repulsion in general unless the two modes belong to different symmetry classes. This repulsive behaviour is called avoided level crossing (ALC) (Takami, 1992). This can be understood as non-zero interaction between two modes, i.e., non-zero off-diagonal terms in a two by two Hamiltonian matrix. Upon this ALC, their internal intensity patterns are exchanged and the gap of ALC is proportional to the interaction strength between two modes. Similar behaviour takes place in the microcavity system, an open system.

In dielectric microcavities, the eigenvalues are complex numbers, i.e., complex wave numbers or complex frequencies with negative imaginary part representing mode decay. Here, the repulsion behaviour called avoided resonance crossing (ARC) has similar property with ALC of closed systems. Upon an ARC in real part of waveunumbers the mode patterns are exchanged, implying the exchange of imaginary part of wavenumbers since the decay rates are characterized by internal mode patterns. At the point of an ARC, the internal mode patterns are given by superposition of interacting two modes like the ALC case. This feature of ARC explains the formation of scar-like localized mode pattern (Wiersig, 2006) and high-Q unidirectional mode in an annular microcavity, a hybrid mode of a high-Q WGM and a low-Q unidirectional modes at ARC (Wiersig & Hentschel, 2006).

The interaction source of ARC might be the classical nonlinearity such as chaotic ray trajectories like ALC in closed billiards. Additionally, in microcavities openness-mediated interaction, called as external coupling, is also possible to be the source of ARC. This external coupling can be observable when the internal coupling caused by classical nonlinear dynamics is almost negligible, like ellipse or rectangle cases, or when the cavity has angular corners at which strong diffraction occurs (Wiersig, 2006). In general chaotic microcavities with smooth boundary, it is thus difficult to find the external-coupling-dominant ARC.

Since the dielectric microcavity is an open system, one can find interesting non-Hermitian features in the eigenvaules and eigenfunctions. The most interesting one of the non-Hermitian properties is the existence of degeneracy point, called exceptional point (EP), where both eigenvalues and eigenfunctions of two interacting modes coalesce into one eigenvalue and one eigenfunction. The exceptional point is a singular point, a square root branch point, so the surface of real part of eigenvalues near the EP shows a non-trivial topology as shown in Fig. 11 (a). At the EP, the transition of ARC to resonance crossing (RC) takes place, and this results in interesting mode interchange during a cyclic parameter change encircling the EP in 2-D parameter space (see Fig. 11 (a)). Therefore, in order to get back to initial modes, one needs double cyclic parameter changes. This topology is equivalent with that of Möbius strip as shown in Fig. 11 (b) where two circulations are need to get back to the starting point. It is also known that the geometrical phase is π for the double cyclic parameter changes, as a result, four cyclic changes are actually required to recover the original mode at starting point (Mailybaev et al., 2005).

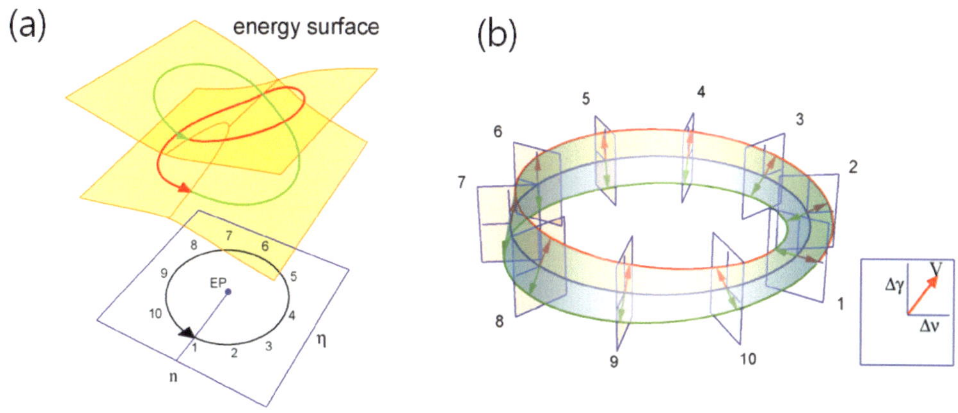

Fig. 11. (a) The topology of energy surface near an EP. (b) Möbius strip made by the difference of eigenvalues of interacting two modes.

In a stadium-shaped microcavity, the exceptional point has been found numerically near an ARC (Lee, S.-Y. et al., 2008b) in a parameter space spanned by a deformation and refractive index. Recently, ARC and RC in microcavity have been experimentally observed in a deformed microcavity made by liquid jet, where some discrete internal parameter, instead of the refractive index, is used and it is expected that the EP would be identified by observing the transition ARC to RC (Lee , S.-B. et al., 2009).

7. Summary

In this chapter, properties of optical modes in deformed dielectric microcavities have been reviewed. Although the ray dynamics in deformed cavity is complicated, through the PSOS one can easily identify the complexity of ray dynamics. The modified PSOS incorporating openness character of dielectric cavity can be characterized by the steady probability distribution (SPD) for fully chaotic case. This distribution reveals combination of unstable-manifold structure and openness character, and it plays a role of classical skeleton for understanding Husimi functions of optical modes supported by deformed microcavities. The directional emission from a strongly deformed microcavity can be well explained by the SPD. Influence of openness changes scarred optical modes to have opposite angular shift of scarred patterns depending on the way of wave circulation, and make it possible to form quasiscarred optical modes without underlying unstable periodic orbit. And the dielectric microcavity can be regarded as an example of non-Hermitian system with complex eigenvalues. The exceptional point (EP), degeneracy point in non-Hermitian systems, can be found in deformed microcavities.

Although much attention has been paid on the microcavity in the past decades and new understandings on optical modes have been achieved, there remain still many challenges. Multi-dimensional tunnelling appears in slightly deformed microcavities. However, there is no quantitative semiclassical theory to treat this tunnelling. Only a perturbation theory, for near integrable microcavity, explains tunnelling emissions (Creagh, 2007). Non-Hermitian properties of optical modes are also important due to their generality applicable to other open quantum systems. The Petermann excess noise factor, a measure of non-orthogonality of eigenstates, is known to diverge at the EP, but its physical implications on the spontaneous emission rate and laser line width are not obvious so far. (Cheng, 2006; Lee, S.-Y. et al, 2008b; Schomerus, 2009)

8. Acknowledgements

This work was supported by BK21 program and KRF Grant (2008-314-C00144).

9. References

Altmann, E. G. (2009). Emission from dielectric cavities in terms of invariant sets of the chaotic ray dynamics, *Phys. Rev. A*, Vol. 79, No. 1, (Jan. 2009) 013830, ISSN 1050-2947.

Ben-messaoud, T. & Zyss, J. (2005). Unidirectional laser emission from polymer-based spiral microdisks. *Appl. Phys. Lett.*, Vol. 86, No. 24, (Jun. 2005) 241110, ISSN 0003-6951.

Cheng, Y.-J. (2006). Spontaneous emission of an atom in a cavity with nonorthogonal eigenmodes. *Phys. Rev. Lett.*, Vol. 97, No. 9, (Aug. 2006) 093601, ISSN 0031-9007.

Chern, G. D.; Tureci, H. E.; Stone, A. D.; Chang, R. K.; Kneissl, M. & Johnson, N. M. (2003). Unidirectional lasing from InGaN multiple-quantum-well spiral-shaped micropillars. *Appl. Phys. Lett.*, Vol. 83, No. 9, (Sep. 2003) 1710-1712, ISSN 0003-6951.

Creagh, S. C. (2007). Directional emission from weakly eccentric resonator. *Phys. Rev. Lett.*, Vol. 98, No. 15, (Apr. 2007) 153901, ISSN 0031-9007.

Dubertrand, R.; Bogomolny, E.; Djellali, N.; Lebental, M. & Schmit, C. (2008). Circular dielectric cavity and its deformations. *Phys. Rev. A*, Vol. 77, No. 1, (Jan. 2008) 013804, ISSN 1050-2947.

Harayama, T.; Davis, P. & Ikeda, K. S. (2003) Stable oscillations of a spatially chaotic wave function in a microstadium laser. *Phys. Rev. Lett.*, Vol. 90, No. 6, (Feb. 2003) 063901, ISSN 0031-9007.

Heller, E. J. (1984). Bound-state eigenfunctions of classically chaotic Hamiltonian system: scars of periodic orbits. *Phys. Rev. Lett.*, Vol. 53, No. 16, (Oct. 1984) 1515-1518, ISSN 0031-9007.

Hentschel, M.; Schomerus, H. & Schubert, R. (2003). Husimi functions at dielectric interfaces: Inside-outside duality for optical systems and beyond. *Europhys. Lett.*, Vol. 62, No. 5, (Jun. 2003) 626-642, ISSN 0295-5075.

Kim, C.-M.; Cho, J.; Lee, J.; Rim, S.; Lee, S. H.; Oh, K. R. & Kim, J. H. (2008). Continuous wave operation of a spiral-shaped microcavity laser. *Appl. Phys. Lett.*, Vol. 92, No. 13, (Apr. 2008) 131110, ISSN 0003-6951.

Kim, C.-M.; Lee, S.-H.; Oh, K.-R. & Kim, j. H. (2009). Experimental verification of quasiscarred resonance mode. *Appl. Phys. Lett.*, Vol. 94, No. 23, (Jun. 2009) 231120, ISSN 0003-6951.

Kwon, T.-Y.; Lee, S.-Y.; Kurdoglyan M. S.; Rim S.; Kim, C.-M. & Park Y.-J. (2006). Lasing modes in a spiral-shaped dielectric microcavity. *Opt. Lett.*, Vo. 31, No. 9, (May 2006) 1250-1252, ISSN 0146-9592.

Kurdoglyan, M. S.; Lee, S.-Y.; Rim, S. & Kim, C.-M. (2004). Unidirectional lasing from a microcavity with a rounded isosceles triangle shape. *Opt. Lett.*, Vol. 29, No. 23, (Dec. 2004) 2758-2760, ISSN 0146-9592.

Lee, S.-B.; Lee, J.-H.; Chang, J.-S.; Moon, H.-J.; Kim, S. W. & An, K. (2002). Observation of scarred modes in asymmetrically deformed microcylinder lasers. *Phys. Rev. Lett.*, Vol. 88, No. 3, (Jan. 2002) 033903, ISSN 0031-9007.

Lee, S.-B.; Yang, J.; Moon, S.; Lee, J.-H.; An, K.; Shim, J.-B.; Lee, H.-W. & Kim, S. W. (2007). universal output directionality of single modes in a deformed microcavity. *Phys. Rev. A*, Vol. 75, No. 1, (Jan. 2007) 011802(R), ISSN 1050-2947.

Lee, S.-B.; Yang, J.; Moon, S.; Lee, S.-Y.; Shim, J.-B.; Kim, S. W.; Lee, J.-H. & An, K. (2009). Quasieigenstate evolution in open chaotic billiards. *Phys. Rev. A*, Vol. 80, No. 1, (Jul. 2009) 011802(R), ISSN 1050-2947.

Lee, S.-Y.; Rim, S.; Ryu, J.-W.; Kwon, T.-Y.; Choi, M. & Kim, C.-M. (2004). Quasiscarred resonances in a spiral-shaped microcavity. *Phys. Rev. Lett.*, Vol. 93, No. 16, (Oct. 2004) 164102, ISSN 0031-9007.

Lee, S.-Y.; Ryu, J.-W.; Kwon, T.-Y.; Rim, S. & Kim, C.-M. (2005). Scarred resonances and steady probability distribution in a chaotic microcavity. *Phys. Rev. A*, Vol. 72, No. 6, (Dec. 2005) 061801(R), ISSN 1050-2947.

Lee, S.-Y.; Rim, S.; Ryu, J.-W.; Kwon, T.-Y.; Choi, M. & Kim, C.-M. (2008a). Ray and wave dynamical properties of a spiral-shaped dielectric microcavity. *J. Phys. A: Math. Theor.*, Vol. 41, No. 27, (Jul. 2008) 275102, ISSN 1751-8113.

Lee, S.-Y.; Ryu, J.-W.; Shim, J.-B.; Lee, S. W. & An, K. (2008b). Divergent Petermann factor of interacting resonances in a stadium-shaped microcavity. *Phys. Rev. A*, Vol. 78, No. 1, (Jul. 2008) 015805, ISSN 1050-2947.

Lee, J.; Rim, S.; Cho, J. & Kim, C.-M. (2008). Resonances near the classical separatrix of a weakly deformed circular microcavity. *Phys. Rev. Lett.*, Vol. 101, No. 6, (Aug. 2008) 064101, ISSN 0031-9007.

Mailybaev, A. A.; Kirillov, O. N. & Seyranian, A. P. (2005), Geometrical phase around exceptional points. *Phys. Rev. A*, Vol. 72, No. 1, (Jul. 205) 014104, ISSN 1050-2947.

McCall, S. L.; Levi, A. F. J.; Slusher, R. E.; Pearton, S. J. & Logan, R. A. (1992). Whispering-gallery mode microdisk lasers. *App. Phys. Lett.*, Vol. 60, No. 3, (Jan. 1992) 289-291, ISSN 0003-6951.

Nöckel, J. U. & Stone, D. (1997). Ray and wave chaos in asymmetric resonant optical cavities. *Nature*, Vol. 385, No. 6611, (Jan. 1997) 45-47, ISSN 0028-0836.

Podolskiy, V. A. & Narimanov, E. E. (2005). Chaos-assisted tunneling in dielectric microcavities. *Opt. Lett.*, Vol. 30, No. 5, (Mar. 2005) 474-476, ISSN 0146-9592.

Reichl, L. E. (1992). *The transition to chaos*, Springer-Verlag, ISBN 3-540-97753-8, New York.

Rex, N. B.; Tureci, H. E.; Schwefel, H. G. L.; Chang, R. K. & Stone, A. D. (2002). Fresnel filtering in lasing emission from scarred modes of wave-chaotic optical resonators. *Phys. Rev. Lett.*, Vol. 88, No. 9, (Feb. 2002) 094102, ISSN 0031-9007.

Ryu, J.-W.; Lee, S.-Y. & Kim S. W. (2009). Coupled nonidentical microdisks: Avoided crossing of energy levels and unidirectional far-field emission. *Phys. Rev. A*, Vol. 79, No. 5, (May 2009) 053858, ISSN 1050-2947.

Shim, J.-B.; Lee, S.-B.; Kim, S. W.; Lee, S.-Y.; Yang, J.; Moon, S.; Lee, J.-H. & An, K. (2008). Uncertainty-limited turnstile transport in deformed microcavities. *Phys. Rev. Lett.*, Vol. 100, No. 17, (May 2008) 174102, ISSN 0031-9007.

Schomerus, H. (2009). Excess quantum noise due to mode nonorthogonality in dielectric microresonators. *Phys. Rev. A*, Vol. 79, No. 6, (Jun. 2009) 061801(R), ISSN 1050-2947.

Schwefel, H.; Rex, N.; Tureci, H.; Chang, R. K.; Stone, A. D.; Ben-Messaoud, T. & Zyss, J. (2004). Dramatic shape sensitivity of directional emission patterns from similarly deformed cylindrical polymer laser. *J. Opt. Soc. Am. B*, Vol. 21, No. 5, (May 2004) 923-934, ISSN 0740-3224.

Shang, L.; Liu, L. & Xu, L. (2008). Highly collimated laser emission from a peanut-shaped microcavity. *Appl. Phys. Lett.*, Vol. 92, No. 7, (Feb. 2008) 071111, ISSN 0003-6951.

Shinohara S. & Harayama, T. (2007). Signature of ray chaos in quasibound wave functions for a stadium-shaped dielectric cavity. *Phys. Rev. E*, Vol. 75, No. 3, (Mar. 2007) 036216, ISSN 1539-3755.

Takami, T. (1992). Semiclassical interpretation of avoided crossing for classically nonintegrable system. *Phys. Rev. Lett.*, Vol. 68, No. 23, (Jun. 1992) 3371-3374, ISSN 0031-9007.

Tulek, A. & Vardeny, Z. V. (2007). Unidirectional laser emission from p-conjugated polymer microcavities with broken symmetry. *Appl. Phys. Lett.*,Vol. 90, No. 16, (Apr. 2007) 161106, ISSN 0003-6951.

Unterhinninghofen J.; Wiersig, J. & Hentschel M. (2008). Goos-Hänchen shift and localizatioin of optical modes in deformed microcavities. *Phys. Rev. E*, Vol. 78, No. 1, (Jul. 2008) 016201, ISSN 1539-3755.

Vahala, K. (2003). Optical microcavities, *Nature*, Vol. 424, No. 6950, 839-846, ISSN 0028-0836.

Wiersig J. (2003). Boundary element method for resonances in dielectric microcavities. *J. Opt. A: Pure Appl. Opt.*, Vol. 5, No. 1, (Jan. 2003) 53-60, ISSN 1464-4258.

Wiersig J. (2006). Formation of long-lived, scarlike modes near avoided resonance crossings in optical microcavities. *Phys. Rev. Lett.*, Vol. 97, No. 25, (Dec. 2006) 253901, ISSN 0031-9007.

Wiersig J. & Hentschel M. (2006). Unidirectional light emission from high-Q modes in optical microcavities. *Phys. Rev. A*, Vol. 73, No. 1, (Jan. 2006) 013802(R) (2006), ISSN 1050-2947.

Wiersig J. & Hentschel M. (2008). Combining directional light output and ultralow loss in deformed microdisks. *Phys. Rev. Lett.*, Vol. 100, No. 3, (Jan. 2008) 033901, ISSN 0031-9007.

16

Practical Continuous-Wave Intracavity Optical Parametric Oscillators

Dr David J M Stothard
University of St. Andrews
United Kingdom

1. Introduction

The mid-infrared spectroscopic region (~1.5-5μm) is one of ever increasing importance. Many hazardous, contraband or otherwise important molecules and compounds exhibit their peak rotational and vibrational absorption features over this wavelength range and so can be readily detected and identified through the use of spectroscopic techniques. There is an urgent requirement, therefore, for high spectral purity, compact and wavelength-flexible optical sources operating over this range. Laser based spectrometers operating at visible or near-infrared wavelengths offer a combination of unprecedented resolution and ease of use due to their extremely high spectral brightness and tunability. Mid-infrared laser-based spectroscopy is, however, far less developed (even though this spectral range is arguably of more scientific importance) due to a severe lack of suitable continuous-wave (cw), broadly tunable laser sources. Whilst this area has attracted intense research interest over the past decade, current state-of-the-art mid-infrared laser systems are still not poised to address this shortfall. Quantum-cascade, difference-frequency mixing techniques and lead-salt diodes produce very low output power, limited tunability, poor spatial mode quality, require liquid cryogens, or a combination of these.

Fig. 1. The generation of long wavelength light through parametric frequency down-conversion. Here, $\nu_p=\nu_s+\nu_i$.

The use of nonlinear optical techniques to convert the output of laser systems operating at too short a wavelength, but otherwise exhibiting meritorious characteristics (e.g. high efficiency, robust design, etc) to the low frequency, mid-IR band of interest has received considerable interest since the invention of the laser in the early 1960s. Such nonlinear devices are called *optical parametric oscillators* (OPOs) and they operate by dividing the energy of an incoming, high energy pump photon into two lower energy photons (denoted the signal and idler); the energy (and hence, frequency) of which add up to that of the pump (see Fig. 1). One of the simplest incarnations of this device is the externally-pumped, or *extra-cavity, singly-resonant OPO* (ECOPO) - (see Fig. 2(a)). Here, a nonlinear optical crystal is placed within an optical cavity exhibiting high finesse at one of the down-converted waves (most usually, the signal wave). Once pumped hard enough, the parametric gain overcomes the round-trip loss experienced by the resonant wave and the OPO reaches threshold: down-conversion from the incident pumping wave to signal and idler begins. Crucially, as the parametric process is not limited to a particular electronic or vibrational transition (as in the case of a laser), the tuning range of the down converted signal and idler waves are limited only by the transparency of the nonlinear dielectric material in which they are generated. Hence it is possible to realise devices which exhibit very broad tunability in the down-converted signal and idler waves even if the pumping laser is not itself tunable (although pump-laser tunability does enable an additional tuning mechanism).

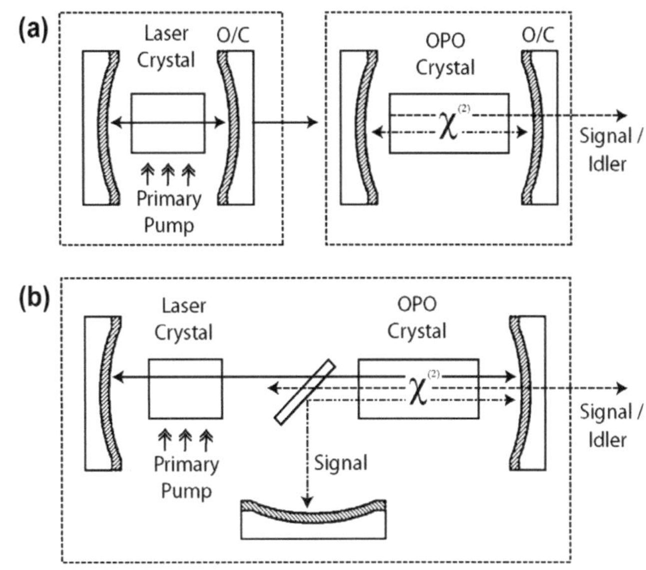

Fig. 2. Externally-pumped (a), and intracavity (b) optical parametric oscillators (ECOPO and ICOPO). Note that in both of these geometries, the optical cavity in which the nonlinear crystal resides is resonant at only one of the down-conveted waves (i.e. either signal *or* idler).

It is the large offset pumping power needed before downconversion begins (the *threshold* pumping power) which is the main objection to the widespread implementation of the ECOPO. Before the advent of long interaction length periodically-polled nonlinear crystals exhibiting comparatively large nonlinearity-interaction length products, threshold pumping powers were on the order of many tens of watts – therefore precluding their use with all but the most powerful cw pump lasers. When one takes into account the primary pumping power required to excite the pumping laser gain medium then the overall efficiency picture of these devices looks even bleaker. This has changed with the introduction of the aforementioned periodically-poled nonlinear materials, most notably the now-ubiquitous periodically-polled $LiNbO_3$ (PPLN) crystal. This brought threshold pumping powers down to the 3-5W level, i.e. within the reach of moderately powered cw laser systems. Overall "wall-plug" efficiency is however still very poor, though, unless ECOPOs are operated well above (~2-3x) threshold (more of this on section 2.1). The highly efficient production of multiple-watt output in the down-converted signal and idler fields is therefore perfectly possible (and indeed has been amply demonstrated (Bosenberg, Drobshoff et al. 1996)) in the ECOPO geometry *but* very poor efficiency results when the output is in the 10s-100s mW region (i.e. the device is operated closer to threshold). This is problematic as many (if not most) of the potential applications of a broadly tunable mid-IR source only require moderate power levels. In addition to this, for many industrial, medical, forensic and field uses, high efficiency, highly compact devices (i.e. battery powered, air-cooled) are a must.

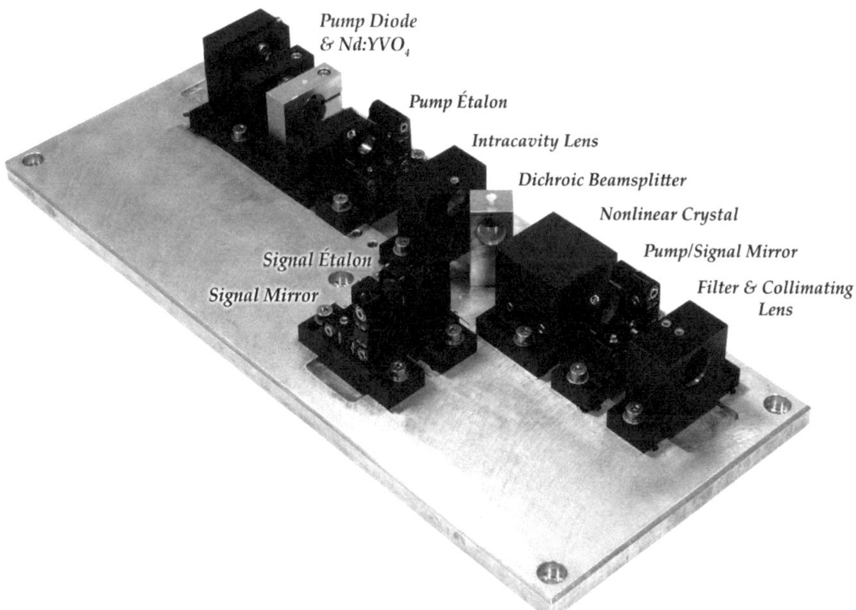

Fig. 3. A well-engineered, all solid-state miniaturised cw-ICOPO. This device consumes just ~10W electrical power, can deliver >500mW in the down-converted optical fields and requires no forced cooling. The function of the various components is discussed later in the text.

An elegant solution to this problem comes through taking advantage of the very high circulating field found within the (high-finesse) cavity of a laser. If one replaces the lasers' output coupling mirror with a high reflector, very high (10's W) circulating fields can result even when pumped at low (100's mW) levels. Placing the OPO *inside* the laser cavity (see Fig. 2(b)) then gives the parametric process access to this high field and the OPO comes to threshold at very much lower primary pumping powers than is the case with the ECOPO, thus obviating the high primary pumping power threshold requirements associated with that geometry. This, the *intracavity optical parametric oscillator* (ICOPO) enables the realisation of extremely compact, highly efficient devices which can exhibit high output powers in the down converted waves (100's mW) when pumped with only very modest (1's W) primary (i.e. diode-laser) pumping sources. An important consequence of the unprecedented down-conversion efficiency afforded by the intracavity approach, coupled with the robust operating nature of the singly-resonant design, is the possibility of realising battery / field operable systems as the need for large frame pumping lasers, forced water cooling and high cost is eliminated. A photograph of such a system is shown in Fig. 3. Here, for just 3W of primary pump power from the integrated diode laser pump module, 300mW and 150mW of

broadly tunable signal and idler power are delivered. Because of the very high efficiency exhibited by the ICOPO, no forced air or water cooling is required. The device consumed <10W electrical power, making it ideal for battery power, portable or remote operation.
From a power and efficiency point of view, then, the cw-ICOPO represents an excellent solution to the problem of inadequate spectroscopic laser-source coverage over the mid-IR range. Unfortunately, there is a particular problem associated with the intracavity approach which has to date severely hampered its widespread implementation. The practical application of very narrow linewidth (sub MHz), diode pumped ICOPOs requires continuous wave output and therein lies a serious limitation inherent in the underpinning physics of the ICOPO. This is due to the impact of the OPO upon the transient dynamics of the Neodymium-based pump lasers in which to date they have been operated. Clearly, maintaining a diode pumped, all solid-state parent laser is highly desirable and hence the majority of ICOPO research has been predicated upon the use of Neodymium (Nd) based laser gain media. Whilst exhibiting many excellent characteristics ideally suited to this technology, their long upper state lifetime (compared to the decay time of the laser and signal waves in their respective cavities) leads to unpredictable and prolonged bursts of relaxation oscillations when used in consort with the intracavity technique. Such behaviour has an unacceptable impact on the frequency and amplitude stability of the down-converted waves and has to date precluded the Nd-based CW ICOPO from having lived up to its considerable potential.
In this chapter we will explore the design criteria for the realisation of practical intracavity cw-OPO systems, with a particular emphasis on overcoming their susceptibility to the spontaneous onset of relaxation oscillations. We shall begin with a comparison between the operating characteristics of cw intracavity OPOs and their externally-pumped counterparts (without becoming bogged down in a turgid foray into nonlinear optical theory (Oshman & Harris 1968)), and the design rules which must be fulfilled in order to realise optimal operation in the intracavity regime. These rules will be applied and tested by then considering the design and realisation of a real-life system previously reported in the literature; the steps taken in order to maximise the chances of successful operation of the device will be reviewed. The discussion will then move on to the vexing problem of relaxation oscillations which occur in the intracavity context; this will be investigated with the aid of a simple numerical model showing how and why they occur. The remainder of the chapter will then describe two examples of state-of-the-art diode laser pumped, cw-ICOPOs which are designed to obviate the problem of relaxation oscillations without losing any of the significant advantages which the intracavity technique confers.

2. The power characteristics of optical parametric oscillators

Much has been written on the principles underpinning the operation of OPOs and we shall avoid repetition here. For a theoretical and analytical thorough discussion of the physical processes underpinning these devices the reader should refer to (Ebrahimzadeh & Dunn 1998). In this section we shall describe the different operating regimes of both intra- and extra- cavity OPOs and examine those best suited to each geometry. Finally, we shall briefly discuss a strategy for operating the ICOPO under optimal efficiency conditions.

2.1 Power characteristics and the advantage of the intracavity technique

It is a common misconception that the ICOPO is somehow fundamentally superior to the ECOPO in terms of conversion efficiency, due to its much lower external threshold pump power requirements. Whilst this is certainly true at lower powers, where the ICOPO is capable of efficient output when the ECOPO would not even be able to achieve threshold, at higher pump powers we shall see that the ECOPO is also capable of exhibiting excellent conversion efficiency. The crucial disadvantage of the ECOPO is its large offset threshold pumping power requirement.

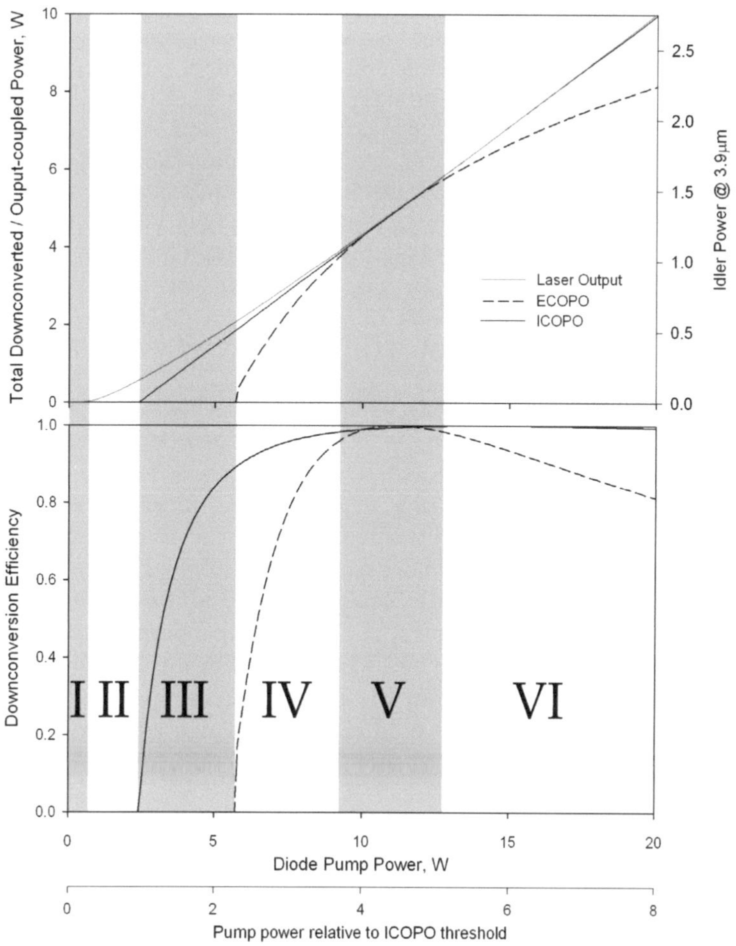

Fig. 4. Down-conversion characteristics of IC- and EC-OPOs

Even with high quality, modern nonlinear crystals exhibiting a high nonlinearity and interaction length, the finite cavity round trip loss for the down converted wave sets the minimum attainable ECOPO threshold in the region of ~2-5W, which would require at least 5-10W of primary optical diode pump power *simply to reach threshold*. However, once above threshold the down conversion efficiency (that is, the fraction of incident pump power down converted to longer wavelengths) rapidly increases to the point at which 100% down conversion efficiency is achieved once the ECOPO is pumped ~2.5 times above its threshold level (Ebrahimzadeh & Dunn 1998). A good example of this is (Bosenberg, Drobshoff et al. 1996) where ~93% of the incident 1μm pumping power was down converted into signal and idler power. ECOPOs have enjoyed something of a revival in recent years due to the

availability of high power, high spatial and longitudinal mode quality fibre lasers and the drop in cost of their associated diode laser pumping modules. The requirement to operate these devices 2-3 times threshold, and the limitations in nonlinear crystal interaction length / nonlinearity and finite signal round-trip loss, still results in the requirement for many 10's W electrical power required in order to operate these devices efficiently. Such a requirement precludes the realisation of the ECOPO in compact, low power designs.
The various operating regimes in which the devices can be operated are shown graphically in Fig. 4, where the output power characteristics of the parent pump laser, an ECOPO and an ICOPO are contrasted. In this model, a typical parent laser is assumed (i.e. Nd:YVO_4, pumped by an 808nm laser, ~2% round trip parasitic loss) and the linear loss effects of the intracavity OPO components is ignored. A note on nomenclature: "down-conversion efficiency" and "down-converted power" refer to the total power converted through the parametric process, i.e. both idler *and* signal. In general, only the longer wave idler is of interest and none of the signal is usefully extracted (although this need not be so - output coupling of the signal field is perfectly possible if this wavelength is also required). Therefore, a second axis has been added in the figure to indicate the total idler power obtained from the device, taking into account the quantum defect between the diode pump and generated idler field wavelengths.
So that the performance of each pumping geometry can be better compared, the threshold condition of the ICOPO and ECOPO in the model have been tailored such that maximum efficiency in either case occurs at the same pumping power (in this example, at about 11.5W). In reality this means artificially increasing the threshold of the ICOPO (by modelling the pump and signal field with only a very weak focus in the nonlinear material); real-world ICOPOs exhibit OPO threshold at far lower pumping powers than shown here - as little as a few hundred mW (Stothard, Ebrahimzadeh et al. 1998). We can see that Fig. 4 has been separated into 6 'zones' of operation. The first and second are merely below and above laser threshold, respectively. The ICOPO comes to threshold at the beginning of zone III, still well before threshold occurs in the ECOPO. In zone IV, ECOPO operation is achieved but the down converted power is still significantly less than in the case of the ICOPO. Clearly, if the available pump power were limited to the range ~2.5-8W then the ICOPO is obviously the superior choice in terms of the amount of mid-infrared light generated. As the down-conversion efficiencies in either case become optimised (i.e. near unity), the total down- converted power is comparable in each case (zone V) and there is little to differentiate between the two devices in terms of performance. In order to optimise for maximum overall efficiency, both devices would be operated in this zone. As the pump power is increased beyond the optimum operating condition (zone VI), the efficiency in each case drops (markedly so in the case of the ECOPO). Here, back conversion of the signal and idler takes place. In practice, one would not operate either device in this zone; in order to obtain very high output powers and maintain optimal efficiency the threshold of each OPO would be *increased* such that optimal down conversion (zone IV) occurs at the required operating point. The crucial advantage of the ICOPO over the ECOPO is that in a practical device, zone V can be achieved at very much lower primary pumping levels, whereby a combination of very high efficiency and moderate down-converted output power is possible. In the ECOPO, high efficiency is only achievable at ~2.5x threshold. As this threshold is locked at relatively high powers by the finite parametric gain / signal wave loss product (~2-5W of incident pumping power), high efficiency only occurs when very high powers are being obtained. For clarity, we summarise these operating regimes in tabular form.

Zone	Operating Regime	Notes
I	Laser below threshold	
II	Laser above threshold	
III	ICOPO above threshold	If diode pump power is limited then ICOPO performance clearly superior over these zones
IV	ECOPO above threshold	
V	Down conversion approaches 100%	Little to differentiate between devices in terms of down-conversion performance
VI	Over pumping	Would never operate either device here in practice

Table 1. Summary of the operating 'zones' depicted in Fig. 4

In the above treatment, the linear loss of the intracavity OPO components placed within the parent pump laser is ignored. In the case of the ECOPO, the pump is only used on a single pass and so linear loss effects, to a first approximation, have little impact upon performance. However, placing lossy components within a laser cavity has obvious consequences in terms of laser performance. With reference to Fig. 2(b) we see that the two additional components which the laser cavity must tolerate are the dichroic beamsplitter and nonlinear optical crystal. Clearly, these components must be antireflection coated in order to minimise loss at the pump wavelength. It is particularly important to secure the finest coatings available upon the nonlinear crystal and inner surface of the beamsplitter as these need to be specified at three separate wavelengths. However, coating techniques have now matured to the point at which such advanced coatings are generally obtainable, particularly in devices which do not require broad tuning of the OPO (and, hence, broad-band AR/HR coatings). For a well established nonlinear crystal such as PPKTP or PPLN, absorption at the pump wavelength is negligible and so the additional round trip loss of the ICOPO components can be as low as ~3-5% at the pumping wavelength. Significant crystal-induced loss is only encountered, and is therefore problematic, when the intracavity technique is used in conjunction with lossy nonlinear materials, such as ZGP pumped at ~2μm. In this case, care has to be taken that the round trip loss of the laser cavity accommodating such lossy components does not impact too heavily on the attainable circulating field and, hence, obviate the advantage that the intracavity technique confers. The use of such crystals is beyond the scope of this chapter.

2.2 ICOPO efficiency optimisation

Unlike the case of a laser, minimizing the point at which the ICOPO comes to threshold (in terms of the primary pump power from the laser-diode) does not necessarily bring about the highest output (or efficiency) at the maximum available pump power. This is because the nonlinear parametric process acts as the output coupler for the laser, and so for a given pumping power one requires that the OPO operates in such a way that it behaves as an optimal output coupler for the pump cavity (i.e. is operating in zone V (Fig. 4) for a given primary pumping power). Therefore, the threshold level of the OPO, in terms of external pumping power, is a function of both laser threshold and the external pumping power at which the device is to be optimised. If the OPO comes to threshold too quickly, then at the maximum available primary pumping power the laser will be over coupled, hence reducing the down-converted power obtained. For a particular value of laser threshold and maximum available primary pumping power, optimum down-conversion efficiency occurs when the condition

$$P_{th}^{OPO} = \sqrt{P_{th}^{L} \cdot P_{in}} \qquad (1)$$

is met (Colville, Dunn et al. 1997), where P_{th}^{L} and P_{th}^{OPO} are the primary pump powers at which the laser and OPO, respectively, reach threshold, and P_{in} is the primary pumping power at which the device is to be optimised. When operated in this regime, the ICOPO acts as an optimum output coupler to the parent pump laser and maximum conversion of primary pump to down-converted power is achieved (this power being equal to that extractable from the pump laser under optimal output-coupling conditions with the OPO components accommodated within the pump cavity but with down-conversion suppressed).

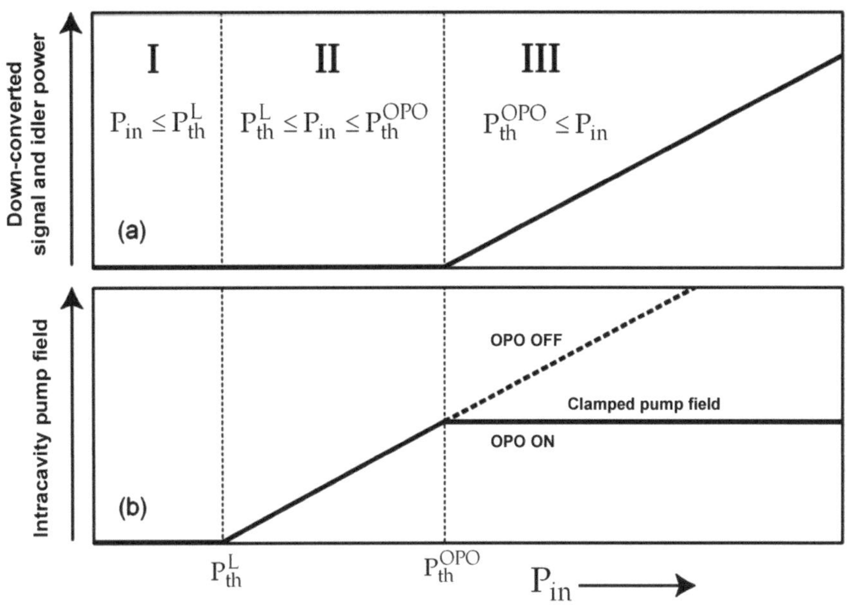

Fig. 5. (a) Linear output power of OPO once above threshold and (b) clamping effect of the ICOPO upon the circulating field (Turnbull, Dunn et al. 1998)

Whilst a very low value of P_{th}^{OPO} is highly desirable when the available primary pumping power is limited, reduced down-conversion powers are experienced when higher power pump sources are used as the system is operated too many times above threshold (because of the aforementioned over coupling of the pump field). Due to the high pumping fields available when using the intracavity technique, coupled with the low parametric thresholds enabled by long interaction-length, high-nonlinearity periodically-poled crystals, a choice can therefore be made when optimising the performance of the device either for maximum down-converted power or minimising parametric threshold in terms of primary pump power. Both of these cases are considered in a practical system later on in section 5.2.

Once above threshold, the parametric oscillator acts like an optical zener diode and 'clamps' the circulating field at the OPO threshold value, as shown in Fig. 5(b). Increased pumping power is then transferred from the laser gain medium population inversion, through the circulating field into increased power in the signal and idler waves, which grow linearly. When characterising the performance of an ICOPO, it is often well worth measuring the quality of the pump-field clamping above OPO threshold as the primary diode pump power is increased. Good clamping is indicative of a well designed pump and signal cavity which is either free of (or robust in the presence of) any dynamic thermal effects which may be present within the laser gain medium and nonlinear optical crystals. Significant thermal lens effects manifest themselves in poor clamping of the pump field and a non-linear relationship between primary pumping and down-converted power. We shall see examples in the following section of how to calculate the circulating field required to bring the OPO to threshold, and experimental observations of the pump-field clamping effect.

Let us now take these simple design rules and see how they are applied when planning, constructing and characterising a system on the optical bench.

3. Designing a cw-ICOPO

In this section we shall take a specific example of a previously demonstrated ICOPO system reported in the literature (Stothard, Ebrahimzadeh et al. 1998) and walk through the process of realising such a device, ensuring that the first-time experimentalist will maximise his or her chances of success - by which we primarily mean at least getting the OPO above threshold. Here we will assume the experimenter has access to readily available pumping sources, Nd laser gain media and nonlinear crystals.

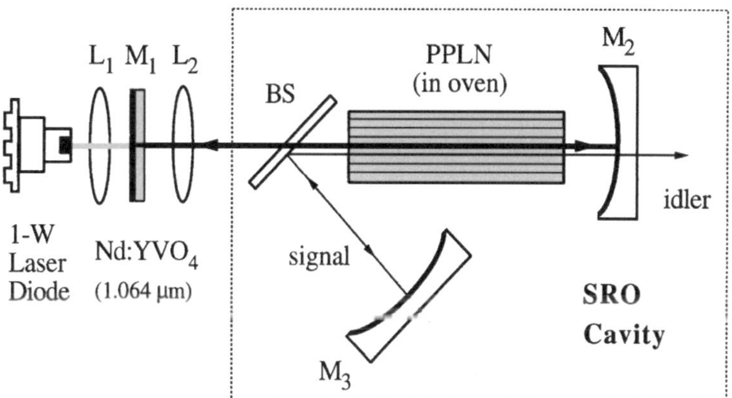

Fig. 6. A simple cw PPLN- Nd:YVO_4 ICOPO (Stothard, Ebrahimzadeh et al. 1998)

Our requirement is that the device, once constructed, will operate comfortably above threshold, delivering 10's mW of tunable power in the down-converted waves. Steps to circumvent the onset of relaxation oscillations will not be addressed in this discussion; here we will restrict ourselves to simply realising a low threshold, high efficiency device. In particular, we will consider the practical design choices which were taken in order to realise the first ICOPO based upon Nd as reported in (Stothard, Ebrahimzadeh et al. 1998), and use the physical parameters as used in that case. A schematic of that device is shown in Fig. 6.

The system was pumped by a c-packaged, temperature stabilised diode laser capable of delivering just 1W of optical power into the rear face of a 1% doped Nd:YVO_4 laser crystal. The laser cavity was defined by a highly reflective (at 1.064μm) coatings applied directly to the outer-most facet of the laser gain crystal and mirror M2. All of the components within the cavity were anti-reflection coated, such that the round trip loss experienced by the pump field was ~3%. Mirror M2 was also coated to be highly reflecting at the signal wavelength, as was M3 and the dichroic beamsplitter BS, thus defining the signal cavity. Due to the limited diode pump power available (only 1W), a crystal exhibiting a high nonlinearity / length product (more on this in the following section) was required in order to minimise parametric threshold, and so a 50mm long PPLN crystal was procured. This was placed within an oven to avoid the effects of photorefractive damage.

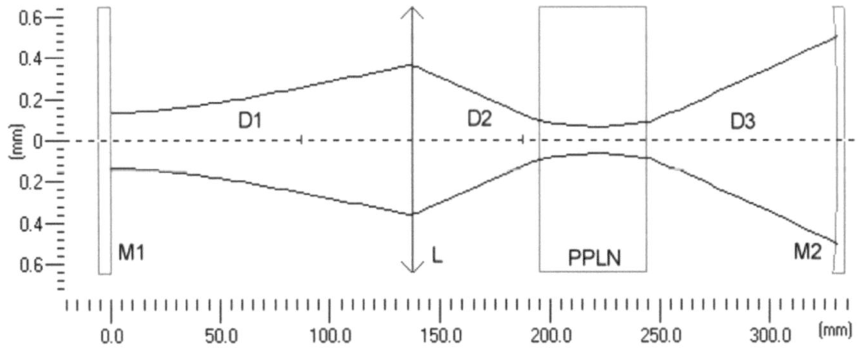

Fig. 7. Stability simulation of the pump cavity. Note that the beamsplitter has no focal power and is therefore omitted. Its optical length is encorporated into distance D2.

The cavity was modelled and its pump mode diameter, as a function of cavity position, is shown in Fig. 7. It is important that the cavity remain stable over a wide range (~50mm → ∞) of thermally-induced (by the diode pump) radius of curvatures modelled in mirror M1. Note the somewhat large distance D3 between the PPLN crystal and M2; this was set by the mirror substrate radius of curvature available at the time of the experiment (200mm). Such a long distance and the use of a relatively weak focal-length mirror results in a somewhat "loose" cavity, more susceptable to the effects of thermally-induced lensing prevalent in the PPLN crystal. A better solution is to use a substantially shorter curvature mirror, perhaps 25mm, placed close in to the PPLN crystal. This has the added advantage of increasing the free-spectral range of the pump cavity: helpful when trying to line-narrow the pump field.

3.1 Parametric gain and threshold

Clearly, it is of crucial importance that the OPO exhibits a threshold pumping requirement that is significantly less than the circulating pumping field available within the cavity of the pump laser, so ensuring that the threshold pumping level can comfortably be reached and exceeded. Let us examine the physical parameters which effect this level, and the steps which can be taken in order to minimise it.

When pumped by a polarized laser beam exhibiting sufficient spectral and spatial coherence, a nonlinear optical crystal designed for use in an OPO will exhibit fluorescence (i.e. gain) over its phase-matched bandwidth in much the same way that a laser crystal will exhibit gain over its gain-bandwith (albeit by a different physical process). This gain is given by (Vodopyanov, 2003)

$$G = \frac{P_{out}}{P_{in}} - 1 = \sinh^2(\Gamma\ell) \tag{2}$$

where ℓ is the length of the nonlinear crystal and Γ is the gain increment given by

$$\Gamma^2 = \left(\frac{d_{eff}^2}{n^3}\right)\frac{2\omega_s\omega_i I_{pump}}{\varepsilon_0 c^3} = \left(\frac{d_{eff}^2}{n^3}\right)\frac{8\pi^2 I_{pump}}{\lambda_s\lambda_i\varepsilon_0 c} \tag{3}$$

Here, I_{pump} is the power density of the laser mode within the crystal, ω_s, ω_i, λ_s & λ_i represent the signal and idler angular frequency and wavelength, d_{eff} is the effective nonlinearity of the nonlinear crystal and n^3 is the product of the nonlinear material refractive index at the three transmitted wavelengths. Note that the factor d_{eff}^2/n^3 is referred to as the *figure of merit* (FOM) and indicates that a high nonlinearity alone does not necessarily yield high gain: it is moderated by ever-increasing refractive index. This is particularly important at longer signal and idler wavelengths where transparency issues mandate the use of semiconductor-based nonlinear crystals whose refractive indices are significantly larger than their phosphide- or arsenide-based counterparts. At low gains ($\Gamma\ell \le 1$, as is experienced in the cw-regime), equation (2) approximates to

$$G_{cw} \approx \Gamma^2\ell^2 \tag{4}$$

And therefore, when properly phase-matched, the single pass gain has a quadratic dependence upon $\Gamma\ell$. The full expression describing the parametric gain experienced as a function of circulating pump power P_{circ}, when the OPO is placed within the cavity of the pump laser, is then

$$G_{cw} = \left(\frac{d_{eff}^2}{n_p n_s n_i}\right)\frac{4\omega_s\omega_i\ell^2 P_{circ}}{\varepsilon_0 c^3\pi(\varphi_p^2 + \varphi_s^2)} \tag{5}$$

Where the refractive index at each of the propagating waves is now explicitly stated, as is the radii of the confocally-focussed pump and signal beams, φ_p and φ_s. This waist radius is given by

$$\varphi_\lambda = \sqrt{\frac{\lambda\cdot\ell}{2\pi}} \tag{6}$$

Note the factor of 2 increase in (5) over (2); this is a consequence of the signal field experiencing gain on each pass of the pumping field, which is of course travelling in both directions through the nonlinear crystal on each round-trip of the pump cavity. Threshold occurs when the circulating pumping field is sufficiently powerful that the parametric gain exceeds the round-trip loss experienced by the resonated down-converted (in this case, the signal) wave:

$$G_{cw} \ge \alpha_{cav} \tag{7}$$

Where α_{cav} is the round-trip loss of the signal cavity. Finally, therefore, we define P_{th} as circulating pumping field (*not* the threshold diode pump power) at which the OPO comes to threshold and re-arrange (5) to give

$$P_{th} = \frac{n_p n_s n_i \varepsilon_0 c^3 \pi(\varphi_p^2 + \varphi_s^2)}{4\omega_s\omega_i\ell^2 d_{eff}^2}\cdot\alpha_{cav} \tag{8}$$

This relation, then, lets us examine the various parameters we can influence in order to attain parametric threshold for the minimum of circulating pump field and, hence, primary pump power. It also reminds us that we are always limited by the material properties of the crystals available to us and the wavelengths over which we wish the device to operate, and illustrates why advances in this field often go hand-in-hand with the development and improvement of new nonlinear materials.

Clearly, in order to obtain the lowest possible threshold we need to maximise the denominator of (8) which means utilising a nonlinear material which offers the largest $d_{eff} - \ell$ product. This is why, given the very modest primary pump power used in this experiment, the nonlinear material PPLN was selected: this crystal exhibiting a then unprecedented 17pm/V nonlinearity and available in lengths as long as 50mm. It is also clearly crucial to minimise the signal cavity round trip loss α_{cav}. When procuring the optical coatings applied to the beamsplitter and signal cavity mirrors it is wise to place most emphasis on the best specification at the signal wavelength. The coating applied to the inner face of the beamsplitter, which must be anti-reflecting at the pump wavelength and broad-band highly reflecting at the signal, is particularly challenging for coating manufacturers. When specifying this coating, it is often helpful to encourage the coating engineer to let the incidence angle and polarisation of the pump and signal waves 'float' in his or her modelling calculations (if these parameters are not fixed by other demands placed on the system design), thereby giving him or her the freedom to maximise the performance of this challenging coating. Typically, one can conservatively expect the round-trip loss of the signal cavity to be ~2-5% (i.e. $\alpha_{cav} \approx 0.02 - 0.05$).

The chosen length of the crystal, along with the desired signal and idler wavelengths, fixes the confocal beam waist radius of the two resonant beams, as given by (6). The refractive index of PPLN at the three different wavelengths is calculated using Sellmeier equations (which shall be addressed in the following section). For the particular case under discussion, where the pump, signal and idler wavelengths were ~1.0, 1.5 and 3.6μm respectively, and a signal cavity round trip loss estimated to be 4%, we find upon solving (8) that parametric threshold occurs when ~3.5W is circulating within the pump cavity.

We now need to assess whether this intracavity field can be comfortably reached and exceeded with the available pumping power. With knowledge of the gain parameters of the laser gain medium and cavity (upper-state life time, stimulated cross-section, pump mode intensity, parasitic loss, etc.) the relation between the primary diode pumping power and the circulating pump field can be accurately modelled. However, it is often more straight forward to simply measure the output power of the laser through a well-chosen output coupler and then infer the intracavity field. For instance, with mirror M2 in Fig. 6 removed and replaced with an (optimal) 5% transmissive output coupler, 510mW of power at the pump wavelength was extracted. This indicates that 10W of field was circulating within the cavity, easily enough to bring the OPO to threshold when the laser is tolerating the additional loss of the output coupler. When highly reflecting mirror M2 was replaced, we estimated that the circulating field increased above 20W - enough to place the OPO well above threshold.

In marginal threshold cases it is possible to lower the threshold requirements of the OPO by increasing the intensity of the resonant fields within the nonlinear crystal. This is achieved by reducing the spot sizes of the pump and signal waists φ_p and φ_s. This however results in less optimised operation at higher pumping powers and can have practical consequences such as mode aperturing at the facets of the crystal, increased susceptibility to the effects of thermal lensing (and, in extreme cases, optical damage) but is a useful trick to try when out of other options.

3.2 Phase-matching and tuning

Much has been written about phase matching in nonlinear optical processes and for the sake of space it will not be repeated here save for a brief overview. Most applications to which the ICOPO will be turned will require the production of a specific idler and, hence, signal wavelength pair. Many applications (e.g. spectroscopy) also place both a coarse and fine tunability requirement on the device. For energy conservation, the signal and idler wavelengths are related to that of the pump by the relation

$$\nu_p = \nu_s + \nu_i \tag{9}$$

or, more usefully,

$$\frac{1}{\lambda_p} = \frac{1}{\lambda_s} + \frac{1}{\lambda_i} \tag{10}$$

This, however, implies an infinite combination of signal and idler wavelengths for a given pumping wavelength. How does one successfully achieve device operation at the required signal and idler wavelengths?

The particular signal and idler frequency pair that is generated is governed by the *phase-matching* criterion of the nonlinear optical crystal employed. The efficient flow of power from the pumping wave into signal and idler waves only occurs when the three waves (pump, signal and idler) are travelling at the same speed (i.e. are in phase) within the nonlinear medium. When this condition is satisfied then the process is said to be phase-matched. Clearly, this criterion cannot be met in isotropic media due to linear refractive dispersion and so more subtle phase-matching schemes are called for. Phase-matching has traditionally been achieved by using bi-refringent crystals through which the waves were propagated at an appropriate angle and polarisation with respect to the crystallographic axis such that the respective refractive indeces experienced by the different wavelengths were equal, satisfying the condition

$$\frac{n}{\lambda_p} - \frac{n}{\lambda_s} - \frac{n}{\lambda_i} = 0 \tag{11}$$

Unfortunately these angles of propagation rarely coincided with that which the optimal nonlinearity of the material was encountered, leading to low overall nonlinear coefficients. In addition, tuning of the signal and idler waves was often achieved through rotation of the crystal angle, thus leading to complicated mechanical designs required to keep the optical cavity stable whilst crystal rotation took place. This changed with the advent of periodically poled nonlinear media where the phase-matching criteria could be "engineered" into the material by periodic inversion of the crystallographic domains (as shown in Fig. 8(a)), thereby making the generated signal and idler wavelengths simply a function of polling period (and crystal temperature). This enabled the somewhat cumbersome tuning mechanisms associated with conventional bi-refringently phase-matched devices to be dispensed with. The axis of propagation could also now be chosen in order to access the highest material nonlinearity. We shall only concern ourselves with these *quasi-phase-matching* (QPM) schemes in this discussion as all of the devices described in this chapter utilised this method of phase matching

The period Λ of the domains (often called the *grating period*, but not to be confused with diffraction gratings) written within the nonlinear crystal is chosen such that it takes up the 'slack' in the phase-mismatch so that phase-matching is achieved:

$$\frac{n_p(\lambda_p,t)}{\lambda_p}-\frac{n_s(\lambda_s,t)}{\lambda_s}-\frac{n_i(\lambda_i,t)}{\lambda_i}-\frac{1}{\Lambda(t)}=0 \qquad (12)$$

Note that the refractive index of the material is a function of both wavelength and temperature. Due to thermal expansion of the nonlinear crystal as its temperature is varied, the grating period is also somewhat dependent upon temperature. It is the dependence of these parameters on wavelength and temperature, along with the need for the conservation of energy, which enables the OPO to be tuned by crystal temperature as well as and pump wavelength. Recently more advanced grating patterns have been demonstrated where the grating period varies linearly across the lateral axis of the crystal. In this, the so-called fanned grating design (Fig. 8(b)), the phase-matching condition is therefore a function of crystal position and very rapid tuning of the signal and idler can be achieved by translating the crystal through the circulating pumping field.

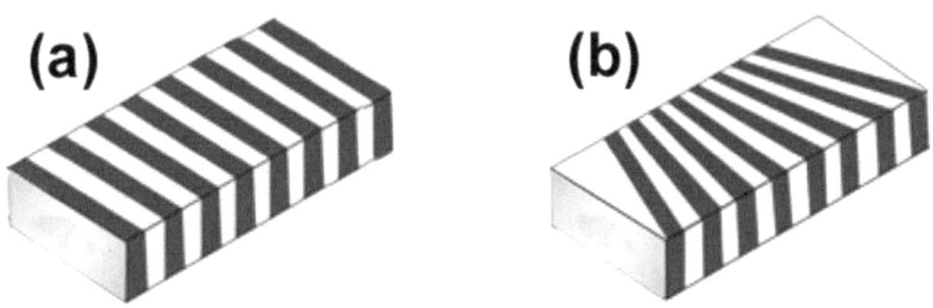

Fig. 8. Periodically-poled nonlinear crystals with (a) single and (b) fanned grating designs.

Relation (12) is solved by using empirically-derived Sellmeier equations which relate the refractive index of a particular material to the wavelength of light propagating within it. Modified Sellmeier equations also include temperature-dependence terms in order to enable the modelling of temperature tuning of the phase-matched condition. Not only is the format of each Sellmeier equation (and the constants used) specific to a particular nonlinear material, it is also often specific to the method of crystal growth used during manufacture. Whilst most commonly used nonlinear materials are very well characterised and their Sellmeier equations are available in the literature, it is often prudent to contact the crystal manufacturer and either ask which Sellmeier equations best describes their material, or better still let them calculate the required grating period in order to phase-match for the desired signal and idler wavelength pair at the required temperature.

The Sellemeier equation and its coefficients describing the PPLN nonlinear crystal used in this particular experiment is described in (Jundt 1997) and the accuracy with which it was able to predict the refractive index of the PPLN crystal and, hence, the phase-matched signal and idler wavelength pair for a given material temperature and grating period is shown in Fig. 9. The PPLN crystal used in the experiment had eight discrete grating zones of different polling periods written within it and so Fig. 9 comprises eight pairs of signal and idler curves, each particular pair corresponding to a different grating zone.

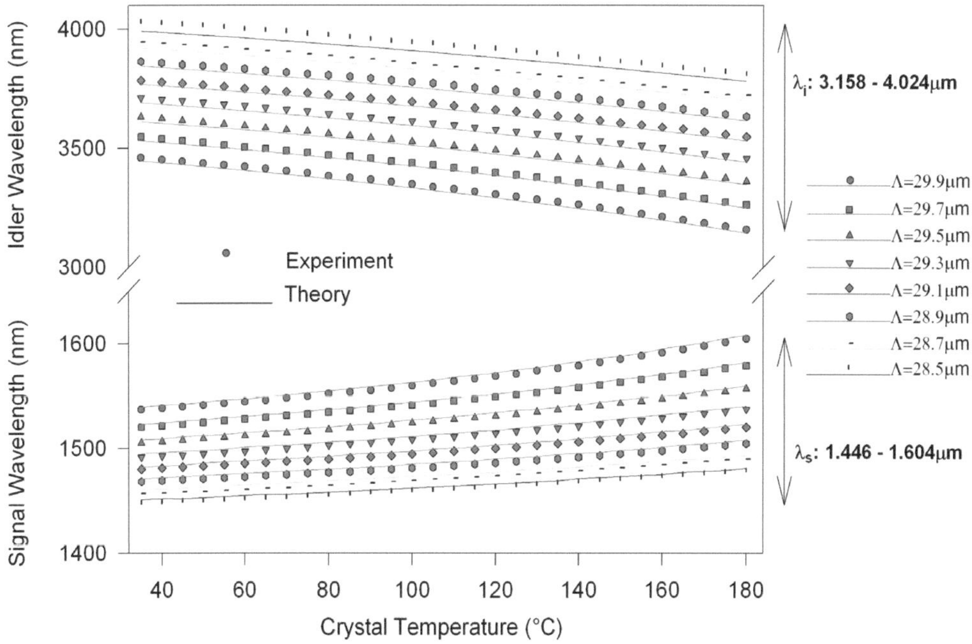

Fig. 9. Predicted and measured tuning of the signal and idler wavelengths

An accurate determination of the anticipated signal and idler wavelengths and tuning ranges is important, not only from the point of view of the end application of the device, as this information must be first determined before specifying the centre-point and bandwidth of the coating pertaining to the idler and, of particular importance for the reasons outlined above, the signal wavelength. The threshold pumping power requirement often rises substantially at the extremes of the tuning range as signal cavity round trip loss creeps in at the edge of the coating bandwidth. On condition that it does not compromise overall performance, it is often prudent to specify a coating bandwidth exceeding the tuning range over which the parametric process is expected to phase-match in order to obviate this effect.

3.3 Performance evaluation and optimisation

The down-conversion performance of the device is indicated in figure Fig. 10, where the extracted idler is shown as a function of increased primary diode pump power as is the circulating pump field both in the presence and absence of down-conversion. The laser and OPO threshold occurred at a diode pump power of 69 and 310mW respectively. In this latter case, 5.2W of circulating pump power was present, a figure somewhat larger than the anticipated threshold field of 3.5W. This is accounted for by sub-confocal focussing of the pump and signal fields resulting in reduced field intensity. Whilst this leads to the increase in threshold pump power, the cavity resistance to thermal lensing effects within the PPLN crystal was significantly reduced leading to more robust performance of the device. Despite this increase, the primary advantage of the ICOPO approach is still clear. In order to bring the OPO to threshold in an extra-cavity system, 5.2W of power from the pumping laser would be required, which itself would therefore require ~10W of primary optical pumping power. We achieve the same here for just 310mW of primary pump power - a significant drop indeed. The robust nature of the system is evident from the both linear relationship between the circulating field and pump power in the absence of parametric down conversion and the excellent clamping of the pump field once the OPO is above threshold. It is worth comparing the measured performance of the device as indicated in Fig. 10 with the theoretical behaviour shown in Fig. 5.

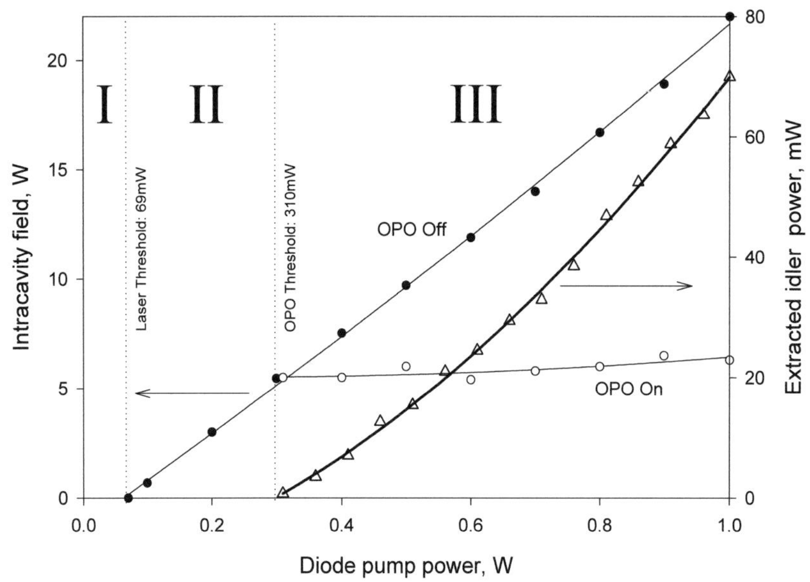

Fig. 10. Extracted idler (triangles) and pump-field with (open circles) and without (closed circles) operation of the OPO as primary diode pump power is varied. The idler wavelength was 3.66μm.

The slightly super-linear nature of the extracted power is a consequence of a thermally-induced increase in the focal power induced in the Nd crystal reducing the mode size (and hence, increasing intensity) of the pump field within the PPLN crystal at higher primary pump powers

At the maximum pump power of 1W the device delivered 70mW of tunable idler through M2. In order to calculate the total down-converted power (that is, the total signal and idler power generated) we need to take into account the quantum defect between the signal and idler waves and for the fact that the idler is generated in both directions within the PPLN crystal (the 'other' direction being lost within the system). The total down-converted power is therefore

$$P_{DC} = 2 \cdot P_i \cdot \left(1 + \frac{\lambda_i}{\lambda_s}\right) \tag{13}$$

This, for an idler power of 70mW and an idler wavelength of 3.6μm, corresponds to a total down-converted power of 476mW from the pump wave into the signal and idler. Recall that when mirror M2 was replaced with an optimal output coupler for the pump cavity, 510mW of power at the pump was obtained. We can therefore take the *down-conversion efficiency* of the device (that is, the fraction of the total obtainable power which is down-converted) to be 476/510 = 93%. A down-conversion efficiency of unity can only be achieved when the OPO is optimally output coupling the pump field through the parametric effect, which is achieved when relation (1) is satisfied. For a laser threshold and operating pump power of 69 and 1000mW respectively, the optimal OPO threshold is then 250mW - slightly less than is the case in this system. As we have said, the stability of the cavity has been improved by slightly defocusing the pump (and signal) waists within the PPLN crystal which has raised the OPO threshold to this non-optimal level. The resulting improvement in performance, however, makes this slight drop in overall efficiency a price worth paying.

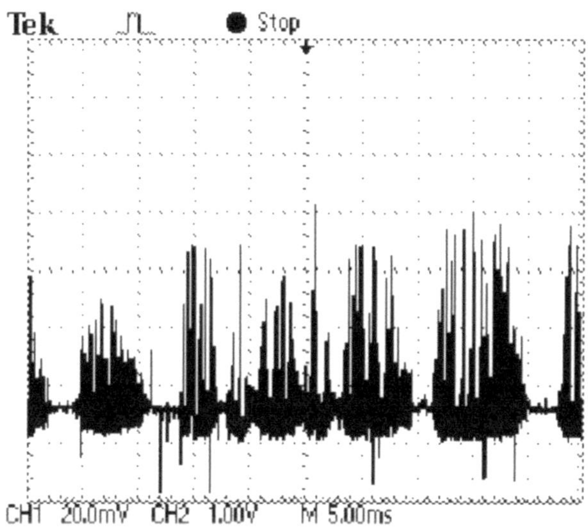

Fig. 11. Spontaneous and long-lived bursts of relaxation-oscillations manifesting themselves on the circulating pump field.

Finally, we turn our attention to the transient stability of the device which was measured by directing the small amount of pumping field reflected off of the rear face of the beamsplitter onto a fast photodetector. An example of the resulting trace is shown in Fig. 11. In the absence of any external perturbation mechanism the pump (and hence signal and idler) fields exhibited spontaneous and very long-lived bursts of high frequency relaxation oscillations. This resulted in ~100% modulation of the extracted idler field and erratic longitudinal mode hopping of the pump field, both of which are most undesirable in the context of high resolution spectroscopy and renders the device unsuitable for all but mean power, "crude" mid-IR applications. This is regrettable as in all other respects this system displays the very many highly desirable characteristics as discussed in sections 1 & 2,, such as very high efficiency, broad tunability, compact geometry, etc. In order to release the potential of this technology, a solution to the problem of relaxation oscillations is crucial. Let us now focus on the nature of these oscillations, the physical processes underpinning their behaviour and some real-life strategies for their elimination.

4. Transient dynamics and the origin of relaxation-oscillations

Whilst laser systems based upon Nd-doped gain media readily exhibit relaxation oscillations when substantially perturbed from their steady-state they do not, on the whole, display any stability problems when left to their own devices. Why is it, then, that a perfectly stable and well-behaved Nd laser should suddenly become prone to spontaneous burst of erratic and long-lived relaxation-like oscillations when an OPO is placed within its cavity?

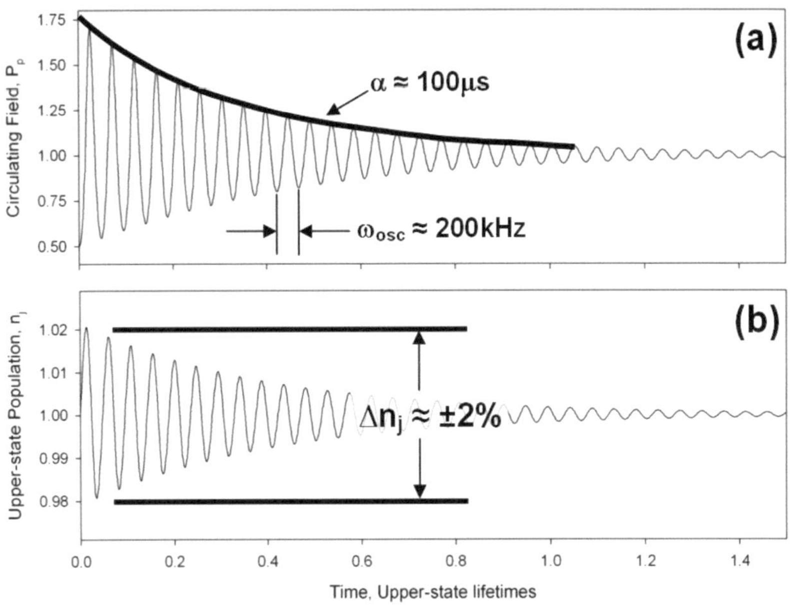

Fig. 12. Modelled relaxation-oscillation oscillations for a simple Nd-based laser. Photon (a) and upper-state (b) population evolution over time after the steady state is perturbed.

Experimental observation indicates that the relaxation oscillations displayed by ICOPOs are different in three key ways to those displayed by the parent pumping laser. They (a) are very long lived, (b) have a far higher oscillation frequency and (c) occur spontaneously. In this section we will explore the origins and the nature of relaxation oscillations in cw-ICOPOs, and thereby find potential strategies for their elimination.

Relaxation oscillations in laser systems are caused by interplay between the energy stored in the upper state of the lasing transition within the gain medium (n_j) and that stored within the optical cavity of the laser due to the circulating pump field (P_p). In the ICOPO, a third parameter, the energy stored within the signal cavity due to the circulating signal field (P_s) is also introduced. As all of these parameters are cross-coupled, it is useful to investigate their behaviour by constructing a set of coupled rate equations which can then be solved numerically with different starting parameters. In the case of the ICOPO, the three coupled rate equations might be as follows:

$$n_j' = \frac{1}{\tau_u} \cdot \left[1 + k - n_j - \left(k \cdot P_p \cdot n_j\right)\right] \tag{14}$$

$$P_p' = \frac{P_p}{\tau_p} \cdot \left[\frac{n_j \cdot \sigma_j}{1+k} - 1 - \frac{(\sigma_j - 1 - k) \cdot P_s}{1+k} \right] \qquad (15)$$

$$P_s' = \frac{P_s}{\tau_u} \cdot [P_p - 1] \qquad (16)$$

We say "might" as there is some freedom in the construction of these equations; it is up to the investigator to choose the way in which the model is normalised. In this particular set of expressions, n_j, P_p and P_s are assumed to be unity in the steady-state. The cavity photon lifetimes of the pump and signal cavities are given by τ_p, τ_s and the upper-state lifetime by τ_u. These parameters are normalized to the laser gain medium upper-state lifetime and so τ_u is always unity. The factors k and σ_j refer to the pumping levels of the laser and OPO with respect to their threshold conditions: σ_j is simply the number of times above threshold at which the laser is operated, and (k+1) is equal to the number of times at which the OPO is operated above its threshold level.

Let us initially consider the dynamics of the laser operating in the absence of down-conversion. In practice this could easily be achieved by blocking the signal cavity without effecting the operation of the laser. Once running, the laser steady state is 'plucked' by instantaneously halving the circulating field. The numerical model is then allowed to run until the laser returns to the steady state. The results of this simulation are shown in Fig. 12. In this model the pump cavity photon lifetime was assumed to be 20ns and the upper-state lifetime 100μs (i.e. τ_u=1; τ_p=0.0002). We can see from the figure that once purturbed, the laser returns to its steady-state in about 100μs (i.e. about an upper-state lifetime), and has an oscillation frequency of about 200kHz. This frequency is determined by the mean of the upper-state and pump cavity photon lifetimes as:

$$\omega_{osc} \propto \sqrt{\frac{1}{\tau_u \cdot \tau_p}} \qquad (17)$$

Thus we see that the very long upper-state lifetime (compared to that of the pump cavity photon) moderates the oscillation frequency to the relatively low ~100's kHz range. This is crucial as this relatively low oscillation frequency gives the upper-state population time to "respond" to the variations in the circulating field. Upon close inspection of Fig. 12, it is clear that the circulating field and upper-state population are in quadrature-phase. It is a combination of this phase difference, and, crucially, the modulation depth of the upper-state population which is the primary damping mechanism returning the system to its steady-state after a perturbation event. In the case of the system modelled above, the ~±2% modulation depth of n_j is sufficient to return the system to its steady-state in about an upper state lifetime.

We now re-run the model using the same laser parameters, but this time in the presence of intracavity parametric down conversion. Here the system is simulated operating at 2 times OPO threshold (k=1) with a signal cavity photon lifetime of 40ns (i.e. τ_s=0.0004). Fig. 13 shows the transient dynamics of the system after perturbation.

It is obvious that the inclusion of parametric down-conversion within the cavity of the laser has had an enormous impact upon the transient dynamic behaviour of the system. Even after 20 upper-state lifetimes has passed (i.e. 2ms) the oscillations have still yet to damp away.

Clearly, any system exhibiting such oscillations having triggering mechanism on the order of or less than this time would display semi-continuous oscillatory behaviour. The time scale used in Fig. 13 is such that the individual oscillations are not resolvable. These can be seen by re-plotting the figure over just a few fractions of an upper-state lifetime, as shown in Fig. 14 (a). We see now that as well as a sharp decline in the damping of the oscillations (giving rise to the very long oscillation events depicted in Fig. 13), the inclusion of the OPO within the parent pump laser has led to a very substantial increase in the oscillation frequency.

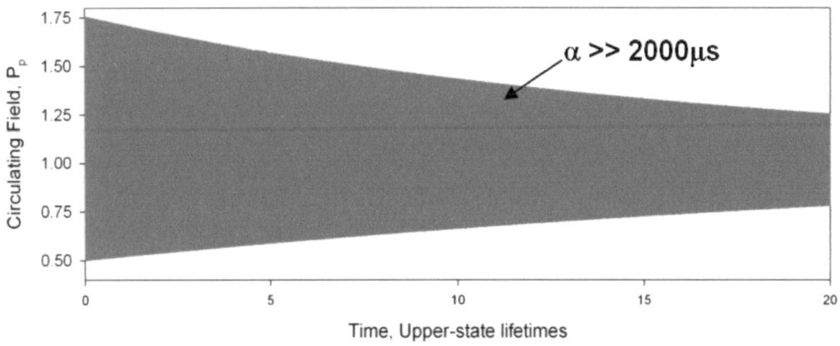

Fig. 13. Modelled relaxation-oscillation oscillations for a perturbed Nd laser based ICOPO. Note that the individual oscillations are not visible on this time scale.

We previously stated that in the case of the laser operating in the absence of the OPO the oscillation frequency was given by the upper-state and pump cavity photon lifetime (17). Now that the OPO is included in the system the processes giving rise to oscillations have shifted down a tier from upper-state population → pump cavity power to pump cavity power → signal cavity power, and therefore the oscillation frequency is now independent of the upper-state lifetime. It is now solely governed by the pump- and signal- wave cavity photon lifetimes:

$$\omega_{\text{osc}} \propto \sqrt{\frac{1}{\tau_{\text{p}} \cdot \tau_{\text{s}}}} \tag{18}$$

The relatively long lifetime of the upper-state, compared to that of the pump- and signal-cavity photon lifetimes, is now no longer able to moderate the oscillation frequency and the very short lifetimes of τ_p and τ_s result in such high, order-of-magnitude increased oscillation frequencies.

Whilst the influence of the upper-state lifetime upon the oscillation frequency has been removed in the ICOPO, its population response to the changing circulating pump field is still the primary damping mechanism for the system. This is problematic as the very high oscillation frequencies have a serious impact upon the ability of the upper-state population to respond to the rapidly changing circulating field. This is somewhat analogous to the charge on a large capacitor being unable to track a high-frequency signal placed across its plates.

The resulting upper-state population is shown in Fig. 14(b). The modulation depth has now been sharply reduced by two orders of magnitude to just ±0.02%, i.e. just 1 part in 10^4. This explains why the oscillations, once induced, carry on for so long – there is very little damping within the system. It also shows why the oscillations are so easily triggered as only very small excursions in the upper-state population are necessary to begin an oscillation event. Such an excursion could be caused by an acousto-mechanically induced longitudinal mode-hop of the circulating pump field or small modulation in the primary pumping field. The long upper-state lifetime (compared to the oscillation period) precludes the use of a feedback system between the circulating pumping field and external primary pump power modulation as a mechanism to improve the damping in the system as it acts as an "optical capacitor"; even 100% modulation of the primary pumping diode scarcely effects the upper-state population on the time scale of the oscillation periods indicated in Fig. 14.

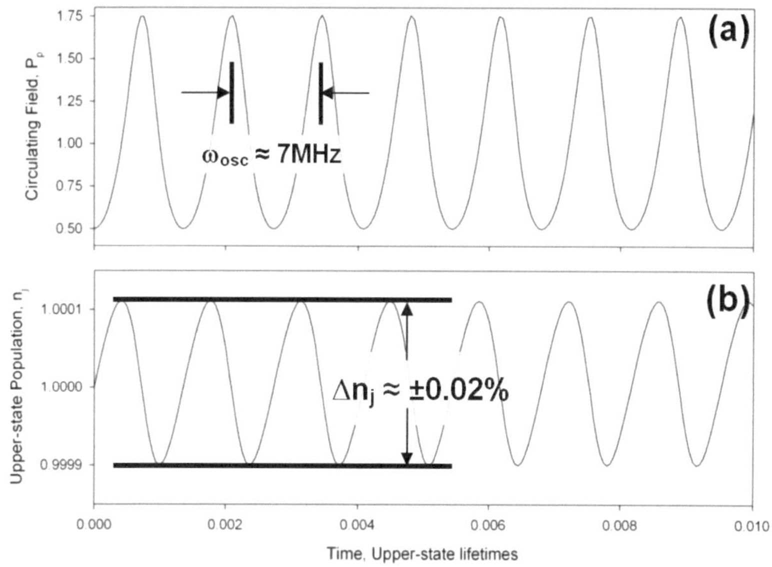

Fig. 14. Modelled results as in Fig. 13 but also showing the upper-state population and with a much enlarged time axis.

To summarise: we have shown, through the use of a simple numerical model based upon a set of three coupled rate-equations, that it is (a) the high oscillation frequencies associated with ICOPOs, brought about by high-speed energy transfer between the pump and signal wave cavities via the parametric process, coupled with (b) the weak damping induced onto these high frequency oscillations by the very long upper-state lifetime of the laser gain medium which leads to both high susceptibility to, and such long bursts of, relaxation oscillations in Nd-ICOPOs. In the next section, we examine ways in which the problem of relaxation oscillations can be circumvented by examining two state-of-the-art cw-ICOPOs which both display excellent transient stability.

5. Examples of practical, relaxation-oscillation free cw-ICOPOs

In the previous sections we have touched upon the various parameters governing the operation of cw-ICOPOs and looked in some detail at the physical processes underpinning the poor transient dynamic behaviour of these devices. We now turn our attention to two practical examples of state-of-the-art diode-pumped cw-ICOPOs which have been engineered with particular emphasis on the elimination of the relaxation oscillations which have to date severely restricted this technology from reaching its full potential.

5.1 Relaxation-oscillation suppression through second-harmonic generation

We have seen that a combination of the long upper-state lifetime of Nd-doped laser gain materials coupled with the high-frequency oscillations resulting from rapid power flow from the pump, through the parametric process, to the signal cavity leads to very weak damping and, hence, long oscillation times. One potential route to break this cycle is the inclusion of an additional damping mechanism within the system. Such a scheme could be introduced actively by utilising a fast detector to monitor the circulating pump or signal field strength, and then somehow instantaneously feed this back onto the loss of the respective cavity. Fig. 15 shows such a technique implemented in the schematic ICOPO depicted in Fig. 2(b).

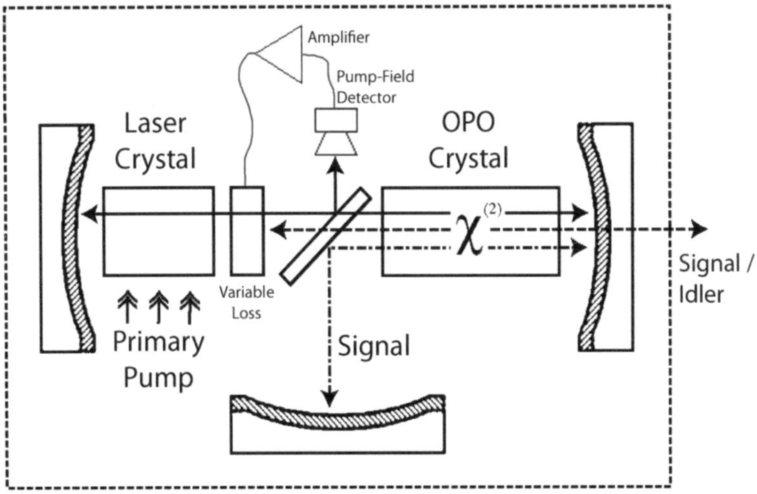

Fig. 15. Potential active scheme to eliminate relaxation-oscillations via an active damping mechanism

In this device, the small amount of circulating pump field which reflects off of the (anti-reflection coated) beamsplitter is directed towards the fast photodetector. The signal from this detector is then amplified and buffered before being fed into a loss mechanism; perhaps an acousto- or electro-optic modulator. This loss mechanism modulates the loss of the pump cavity and, hence, the cavity photon lifetime thus forming a negative feedback loop which preferentially suppresses the high intensity peaks of the oscillations and significantly reduces their damping time. Numerical modelling of this technique indicates that the dynamics of the system are modified to the point at which the oscillation decay time is potentially even shorter than those exhibited by the laser in the absence of the parametric process. In practice, though, the finite bandwidth of the feedback loop described here would introduce an unacceptable phase lag between the peaks of the oscillations and the modulation of the intracavity loss, a problem exacerbated by the high frequency of the oscillations and their higher-harmonic spectral components (>10MHz), the requirement for very fast electronic amplification and, in the case of an acousto-optic modulator, the acoustic wave front time-of-flight between the transducer and the pump field beam within the modulator medium.

A very attractive alternative scheme based upon this philosophy (i.e. pump-cavity photon lifetime dependence upon pump-field intensity) is facilitated by placing a nonlinear frequency-doubling (or second-harmonic generation (SHG)) crystal within the pump-wave only section of the pumping cavity, as shown in Fig. 16. Such a scheme is a successful oscillation damping mechanism as the second-harmonic conversion efficiency (and hence its 'loss' within the pump cavity) is proportional to the *square* of the power propagating within it. The higher intensity peaks of the oscillations are therefore preferentially attenuated compared to the troughs. This scheme has many highly desirable characteristics in the context of relaxation-oscillation suppression, most notably that the process of up-conversion is instantaneous (i.e. there is no phase lag between the increase in pump intensity and the corresponding drop in pump cavity photon lifetime) and, particularly attractive in the context of practical devices, is entirely passive.

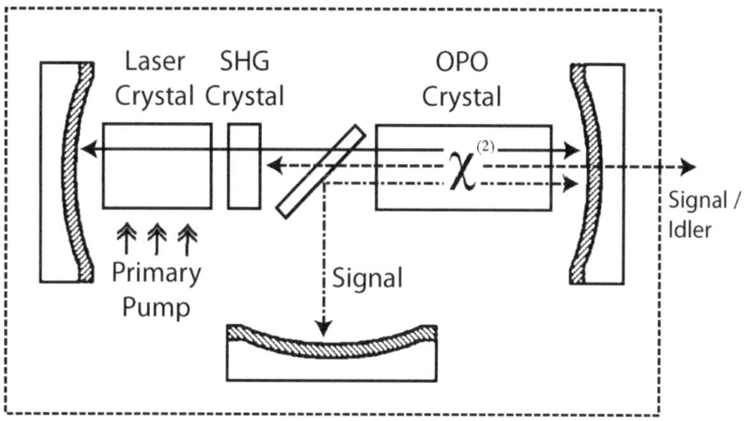

Fig. 16. CW ICOPO with SHG relaxation-oscillation suppression.

In order for this scheme to operate correctly, it is important that a small amount of the circulating pumping field is lost to up-conversion even in the steady state. Clearly, as SHG has no threshold associated with it (unlike the parametric down-conversion), some frequency doubling will always be present as long as the pump laser is above threshold. This "bias" loss level, and indeed the amount of SHG which occurs in the presence of relaxation oscillations, can be controlled by variations in (a) the length of the SHG crystal, (b) the choice of focussing of the circulating pump field within the SHG crystal, (c) the choice of SHG crystal and (d) variation in the up-conversion phase-matching condition. In practice it is variations in this latter parameter which is most straight forward to implement as it can be varied in real time (in the case of temperature tuning of the phase-matched condition) without otherwise impacting upon the operation of the pumping laser. The clamping effect of the OPO upon the circulating field results in comparatively low levels of green light generation, as the field will typically be held at ~2-4W by the OPO. When the scheme is put into effect and the signal cavity is blocked, thus disabling the operation of the OPO, the rise in green generated power is significant.

We can model the effect of up-conversion upon the transient dynamics of the device by incorporating an additional term into the rate equation describing the fractional change in pumping field (equation 15) and re-running the model.

$$P_p' = \frac{P_p}{\tau_p} \cdot \left[\frac{n_j \cdot \sigma_j}{1+k} - 1 - \frac{(\sigma_j - 1 - k) \cdot P_s}{1+k} - \delta P \cdot P_p \right] \tag{19}$$

Equation (19) shows the modified pump-field rate equation with the additional term (the last term in the square parenthesis). The parameter δP relates to the magnitude of the SHG effect and it is this parameter which is varied in the model in order to gauge the efficacy of the scheme. This is an empirical value which is determined experimentally by observing the resulting damping time and the drop in down-converted power. We can see that once equation (19) is factored out the rate at which power is lost from the pumping cavity is proportional to δP but to the *square* of P_p. In order to evaluate the impact of the SHG on ICOPO transient dynamics, with particular emphasis on damping time, we re-run the modified numerical model with the same parameters which produced the very long damping time shown in Fig. 13.

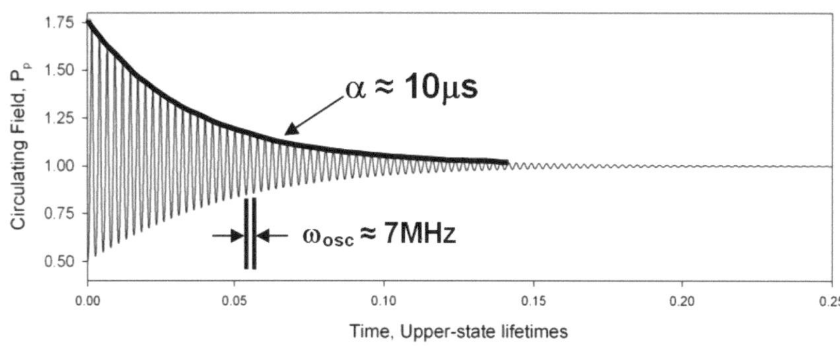

Fig. 17. The modelled effect of SHG on ICOPO transient stability

It is immediately obvious that the inclusion of the SHG crystal has had an enormous impact upon the damping time of the system. As the frequency of the oscillations is governed only by the pump and signal cavity photon decay times, the former of which is not significantly altered by the very low level of SHG, the resulting oscillation frequency is almost the same as the system operating in the absence of up-conversion. We have seen that the laser, operating in the absence of parametric down-conversion, demonstrates a high level of stability when free running and returns to the steady state in about an upper-state lifetime (~100μs) once perturbed. The modelled ICOPO, with the SHG stabilisation implemented, demonstrates an even shorter damping time than this - returning to the steady state in only a *tenth* of an upper-state lifetime. It is when the dynamics of the ICOPO with and without the inclusion of the SHG stabilisation are compared that the efficacy of the technique is thrown into particularly sharp releif. A comparison of Fig. 13 and Fig. 17 shows that the inclusion of the SHG crystal within the cavity of the pumping field reduces the damping time by a factor of ~50,000 - a reduction of well over four orders of magnitude. Clearly, this approach is very effective from the point of view of damping time - but what impact does circulating field lost to frequency up-conversion have upon the total down-converted power? The model is re-run for different levels of δP (i.e. changing the magnitude of the up-conversion effect); the resulting damping time and drop in down-converted power is shown in Fig. 18.

In order for the system to operate with an adequate level of stability (comparable to that of the laser operating in the absence of frequency up- or down-conversion), we require a level of δP which corresponds to a damping time on the order of or less than an upper-state lifetime. This region is indicated by the shaded area in Fig. 18. We can see that a reduction of damping time to between a tenth and a whole upper-state lifetime has only a marginal penalty in down-converted power - approximately 5% - which is perfectly acceptable when one considers the stability advantage which the SHG process confers.

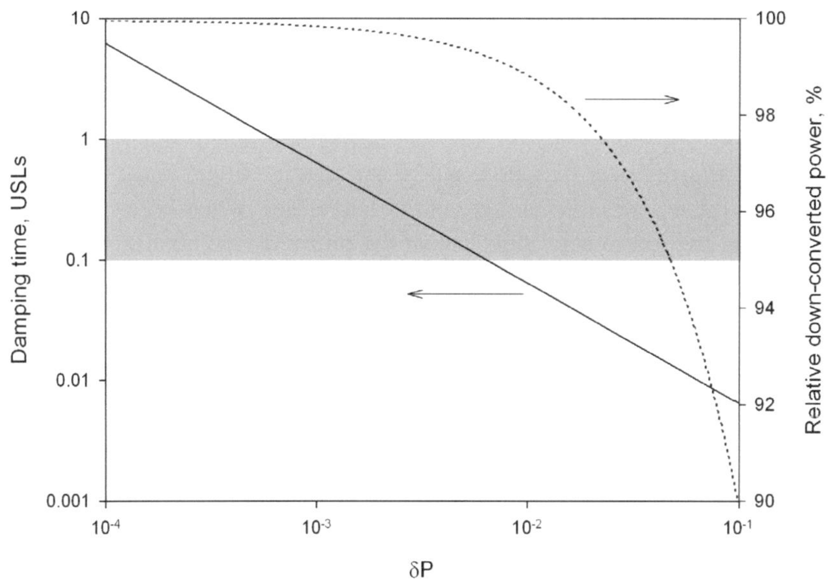

Fig. 18. The impact of increased δP upon damping time and down-converted power. The shaded region corresponds to the zone in which one would prefer to operate.

The strategy outlined above indicates a very promising route to the elimination of relaxation oscillations in the context of the ICOPO - but how well do these modelled results mirror the performance of an actual device? A particular embodiment resulting from the above discussion is illustrated in Fig. 19 (Stothard & Dunn 2009), and is essentially the same design as the system outlined in section 3 (Stothard, Ebrahimzadeh et al. 1998) with the addition of an LBO SHG crystal in the pump field only arm of the device.

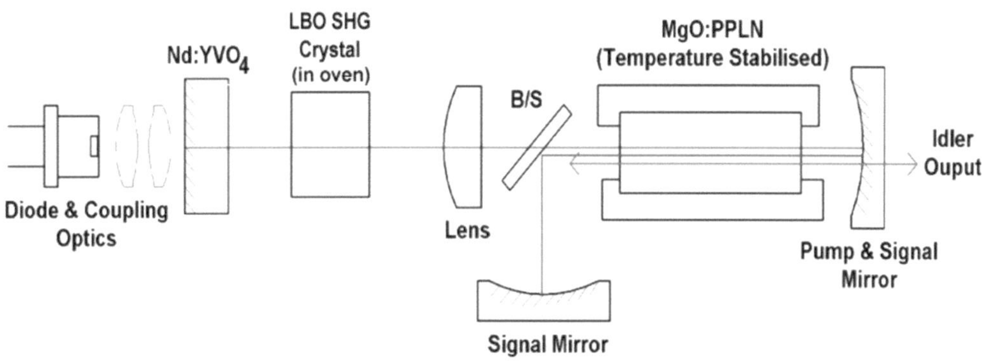

Fig. 19. Experimental configuration of the ICOPO with SHG relaxation oscillation suppression (Stothard & Dunn 2009).

Up-conversion was achieved with an antireflection-coated (for the pump field and its second-harmonic), 2×2×20mm^3 LBO crystal placed in the pump field only section of the cavity between the Nd:YVO_4 crystal and the intracavity lens. The LBO crystal was placed within a temperature stabilized oven (not shown in the diagram) in order to maintain it at the correct temperature for phase matching. The pump mode was weakly focused in this part of the cavity with a beam diameter of ~300μm at the centre of the LBO crystal. Whilst perhaps not the most obvious choice, LBO (as opposed to, say, KTP) was chosen as the doubling crystal as it exhibited type I phase-matching and therefore did not exhibit any bi-refringence into the pumping cavity. This is crucial: such bi-refringence can perturb the polarisation of the circulating pump field and cause the device performance to severely degrade. In addition to this, the efficiency of the SHG, and its impact on the transient stability of the system, could be varied simply by control of the crystal temperature, i.e. without changing the round-trip linear loss of the pump cavity. Finally, the large aperture ($2\times2mm^2$) made it easy to accommodate the near-collimated pump mode without the risk of aperturing. A superior, and perhaps more elegant solution, would be to implement a hybrid MgO:PPLN crystal featuring a fanned grating (as shown in Fig. 8 (b)) for rapid control of the signal and idler wavelengths with a non-fanned SHG grating integrated into a small portion at the end of the crystal. Changing the temperature of the crystal would therefore vary (or optimise) the SHG up-conversion and the signal and idler waves would be tuned by translating the crystal laterally through the pumping field. As well as improving the simplicity of the device, such a scheme would also result in four less surfaces for the pump-wave to encounter on each transit of the pump cavity.

In order to gauge the effect of the SHG, regular bursts of relaxation oscillations were induced into the system by rectangular-wave modulation of the diode-laser pumping current. The frequency of the perturbation was 1 kHz with a duty cycle heavily biased towards the diode laser active state. It was important to maintain this very high duty cycle in order to keep the mean thermal load within the Nd:YVO_4 unchanged from the steady-state case. Variations in this parameter are undesirable as it can destabilize or at least modify the stability of the pump cavity by altering the effective radius of curvature of the lens induced into the Nd crystal, thereby changing the threshold conditions of the pump laser and OPO. In order to minimize this effect in the case of external modulation of the diode pump power, the drive current was modulated on a ~90% duty cycle; the mean power delivered to the Nd:YVO_4 crystal thereby changing very little.

The transient dynamics of the system was measured in three modes of operation: with both SHG and the OPO suppressed (in the former by temperature tuning the crystal well away from the phase-matching condition, in the later by blocking the signal cavity); with the OPO operational but no SHG (i.e. operating as a straight-forward ICOPO with no suppression) and finally with operation of both the SHG and parametric process. Measurement of the system response to perturbation from the steady-state in these three regimes was achieved by monitoring the leaking pump field through the signal and idler mirror with a fast photodetector, and the traces obtained under the three operating regimes outlined above are shown in Fig. 20.

As expected, the laser operating in the absence of parametric down-conversion (Fig. 20(a)) exhibits standard relaxation-oscillation behaviour with a damping time of the order of the upper-state lifetime and an oscillation frequency of a few hundred kHz. The anticipated and deleterious impact of OPO operation is shown in Fig. 20(b); both the oscillation frequency and the damping time have risen considerably – although the damping time has only risen by a factor of four or five (and not by the orders-of-magnitude which the model would predict). This is due to parasitic SHG within MgO:PPLN crystal, which cannot be "switched

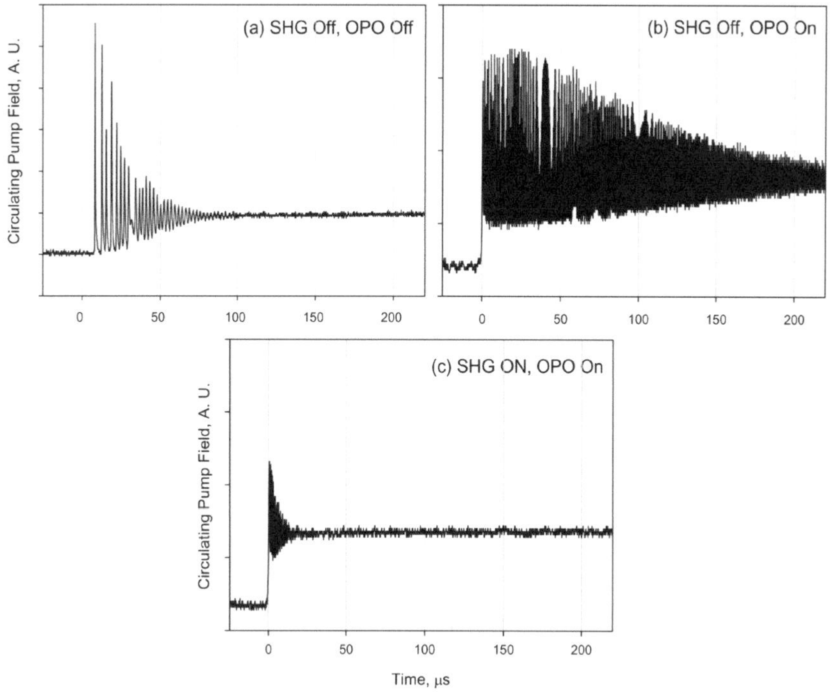

Fig. 20. Transient dynamics of (a) the laser alone, (b) the ICOPO without SHG and (c) the ICOPO with SHG.

off", having a damping effect on the oscillations and also the representative nature of the curve shown in Fig. 20(b) which was very erratic and fluctuated from damping time slightly shorter than that indicated to many 100's μs. Finally, SHG is enabled and the resulting response is shown in Fig. 20(c). Thankfully, the up-conversion has had the intended effect and the damping time of the oscillations has been reduced to even less than that of the simple laser - in this case, about 0.15 upper-state lifetimes (~15μs). Although not very clear on the axis scales used in Fig. 20, the oscillation frequency has not been significantly affected by enabling the SHG, as expected. When the modulation of the external pumping diode was ceased, and the device allowed to operate unperturbed in its steady state, it exhibited superb amplitude stability on both the short (100's μs) and long (10's s) timescale, thereby proving the efficacy of this technique.

We can empirically deduce a value of δP resulting from the LBO crystal nonlinearity, interaction length and choice of pump mode focussing by measuring the damping time of the oscillations and comparing them to the curve shown in Fig. 18; doing so indicates a figure of $\delta P \approx 0.005$. Whilst our attention is turned to this figure, and armed with an approximate value of δP, we can also anticipate the drop in down-converted power the SHG oscillation suppression induces - in this case, in the region of 4%. Confirming this drop experimentally is a simple process of measuring the extracted idler power both in the absence and presence of SHG; this is shown as a function of primary diode laser pumping power in Fig. 21. At maximum pumping power, the inclusion of SHG upon the total down-converted power is just ~3%; in close agreement with the figure predicted by the model and an entirely acceptable price to pay when one considers the huge improvement in transient stability.

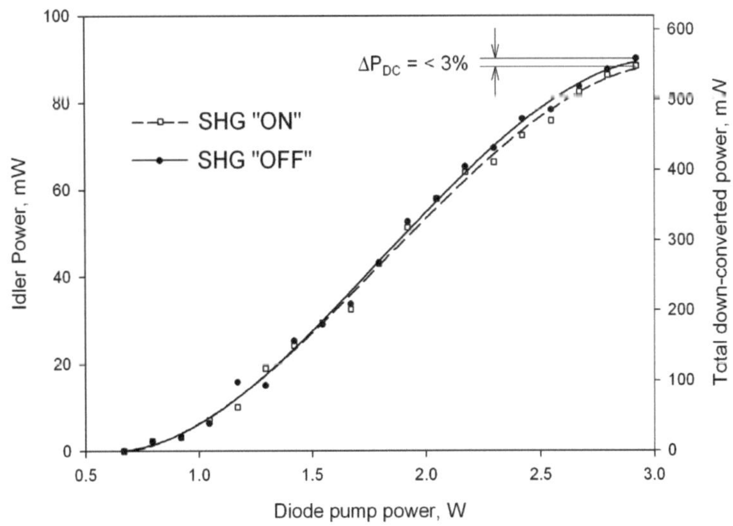

Fig. 21. SHG impact on down-converted power.

5.2 Utilising short upper-state lifetime materials

During our discussion in section 4 we concluded that the susceptibility of the ICOPO to spontaneous bursts of long-lived, high frequency oscillations was a consequence of the long upper-state lifetime exhibited by Nd compared to that of the pump and signal photons within their respective cavities. In this final section we shall discuss the elimination of ICOPO relaxation oscillations by devising a system such that the upper-state and cavity photon lifetimes are comparable by the use of laser gain materials which exhibit very short upper-state lifetimes.

The pioneering early cw-ICOPO work (Colville, Dunn et al. 1997; Turnbull, Edwards et al. 1997; Edwards, Turnbull et al. 1998) was carried out using argon-ion pumped Titanium:Sapphire (Ti:S) lasers, and as such did not experience any problems with the transient instability which have plagued similar systems predicated upon Nd lasers. This is, of course, due to the short upper-state lifetime exhibited by Ti:S of ~3μs (approximately two orders of magnitude less than Nd:YAG) in addition to the longer cavity photon lifetimes resulting from the somewhat lengthier cavities used in these systems. Assuming a pump cavity photon lifetime of 20ns and 50ns in the case of a Nd:YAG and Ti:S laser, respectively, then the ratio of the two lifetimes is 60:1 in the case of Ti:S and 10000:1 in the case of Nd:YAG - a very substantial difference which accounts for the absence of relaxation oscillations in these early Ti:S-based devices. A short upper-state lifetime enables the upper-state population to be more responsive to fluctuations in the circulating pump field power and therefore can heavily damp the system even in the presence of the very high frequency oscillations which result from the very high speed power flow from the pump to the signal-wave cavity. ICOPOs based upon Ti:S are in fact excellent optical sources for use in very high resolution spectroscopic applications due to their inherent stability and the ease with which they can be made to operate on a single longitudinal mode; it is only their somewhat cumbersome pumping requirements (i.e. multiwatt, high spatial mode quality green laser radiation and their associated cost and cooling requirements) which has impeded their wide-spread implementation. We therefore require a gain medium which has the combined advantage of being straight forward to implement in all solid-state, diode pumped designs, as is the case with established gain media such as Nd, whilst exhibiting an upper-state lifetime comparable to or preferably shorter than that of Ti:S.

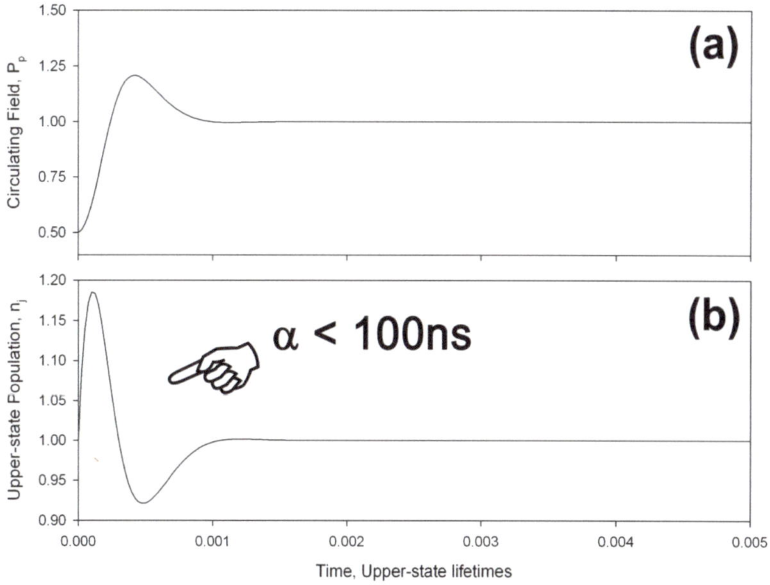

Fig. 22. Modelled transient response of an ICOPO operated internal to a VECSEL. Note that in order to ease comparison with previous figures, the time scale is in units of Nd:YVO_4 upper-state lifetime (100μs).

The semiconductor-based gain chips employed in optically-pumped, vertical extended-cavity, surface emitting lasers (VECSELs) are just such media. These materials have received much attention in research laboratories over the last few years due to their numerous meritorious characteristics. Their broad gain bandwidth has made them ideal candidates as a basis for very short pulse width, mode-locked laser systems and the ability to engineer their band-gap rather than rely upon a particular material electronic transition has enabled them to become particularly successful as intracavity SHG sources of visible radiation. For some specific examples of VECSEL devices and an overview of the technology, see (Kuznetsov, Hakimi et al. 1999; Tropper, Foreman et al. 2004; Kim, Cho et al. 2007; Maclean, Kemp et al. 2008); here we shall restrict ourselves to a brief overview of the characteristics pertinent to the realisation of a viable cw-ICOPO device.

Of course, the most important characteristic which VECSEL gain media exhibits in the context of the current discussion is their extremely short upper-state (or, more accurately, carrier) lifetime - on the order of 1ns. This leads to extremely rapid and deep modulation of the upper-state (or carrier) population when the circulating pumping field is perturbed, returning the system to its steady-state in only a few pump cavity transit times. We can confirm this by re-running the model which produced the curves shown in Fig. 13 and Fig. 14 with the same parameters except for a VECSEL-like laser gain medium upper-state lifetime. The resulting exceptionally high damping levels are evident in Fig. 22, where the system returns to the steady-state in just 0.001 Nd:YVO_4 upper-state lifetimes - that is just 100ns. This represents a reduction in damping time of some *five orders of magnitude* when compared to the system predicated upon Nd. Such performance, if translated into a practical device, would result in excellent amplitude stability, exceeding even that of the Ti:S devices discussed above.

The very thin active region within the VECSEL chip reduces spatial hole burning significantly when compared to the case of the Nd-based laser systems, where absorption depths can be 100's μm. This significantly eases pump-wave longitudinal mode control to the point that single longitudinal mode operation can be realised in standing wave geometries. As the linewidth of the pumping wave is taken up by that of the non-resonant idler, taking steps to operate the pumping cavity in a single frequency makes the VECSEL-based ICOPO an ideal source for very high resolution spectroscopic applications. The second consequence of the very thin gain region is slightly more pragmatic: the elimination of thermally-induced lensing within the pump cavity. From a laser-physicists' perspective, one of the most problematic and vexing parameters one has to evaluate when designing a laser cavity is gauging the magnitude of the thermal lens induced into the gain media. The magnitude of this is sensitive to many inter-related parameters within the system, such as doping concentration, circulating field, extraction efficiency, spot sizes, etc. An incorrect assignation of this parameter renders the position and diameter of the waist within the nonlinear down-conversion crystal ambiguous and can even lead to unexpectedly unstable cavities. It also makes the cavity geometry a function of pumping power and it is often the case that power in / power out curves are skewed by the cavity changing its stability criterion at low pumping levels. All of these issues do not occur when utilising VECSEL gain media as the gain region is too short for any significant thermal lens to form. The essentially two-dimensional geometry of the active region also simplifies mode matching between the circulating pump field and the pumped volume as the divergence of the incident diode-pump radiation has only to spatially overlap with the resonant intracavity mode over a ~1μm depth. Lastly, the broad gain-bandwidth of VECSELs, typically 5-10nm, also enable pump tuning of the OPO phase matched condition. Whilst altering the parametric phase-matching condition is the primary tuning mechanism within the system, the added facility of pump-wave tuning is attractive as it can lead to very broad and simple tuning mechanisms when the device is operated under certain circumstances.

An OPO located within the cavity of a VECSEL has recently been demonstrated (Stothard, Hopkins et al. 2009), and is shown in Fig. 23. The cavity is similar in philosophy to that outlined in section 3 except that the gain medium is now a InGaAs-based VECSEL chip and a bi-refingent filter (BRF) is included within the pump wave only section of the cavity. The system was pumped by a fibre-coupled diode laser bar capable of delivering 8.5W onto the surface of the VECSEL. The BRF was included to stop erratic hopping of the pump-field wavelength as the laser tries to avoid the loss introduced into the cavity once the OPO is significantly above threshold.

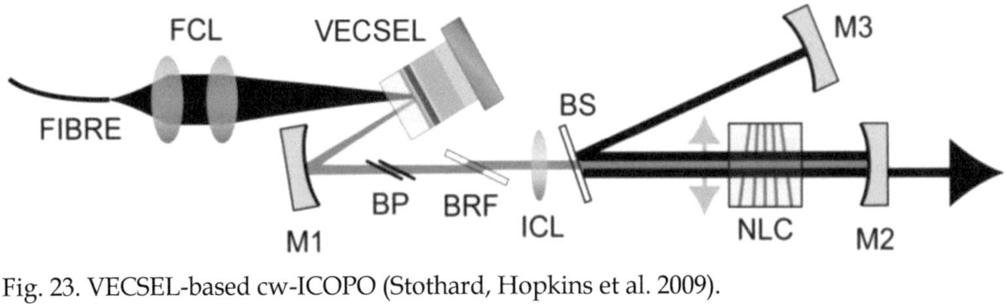

Fig. 23. VECSEL-based cw-ICOPO (Stothard, Hopkins et al. 2009).

This must now be considered as the gain-bandwidth exhibited by VECSEL gain media is so much broader than that of Nd-doped materials. As the BRF is only effective when there is polarisation-dependant loss within the cavity, it is placed into the cavity at Brewster's angle. This however does not induce enough discrimination against unwanted polarisation (and therefore unwanted wavelengths) due to the competing loss of the OPO. Once significantly above threshold the OPO can output-couple significant amounts of power from the circulating pump field into the signal and idler waves. This introduces sufficient round-trip loss such that the laser can operate more effectively by operating against the Brewster surfaces of the BRF alone and rotate its polarisation to a position where the OPO efficiency is substantially diminished. In order to clamp the pump-wave polarisation, and therefore its wavelength, two additional Brewster plates (BP) were included into the cavity. Heat extraction is a significant factor in the successful implementation of VECSEL systems (Kim, Cho et al. 2007; Maclean, Kemp et al. 2008) and so the gain chip was bonded to an uncoated diamond substrate which, due to its very high thermal conductivity, rapidly sinks heat away from the surface of the VECSEL to the copper block into which the assembly was housed. This heat spreader is within the pump cavity and so also acts an étalon for the pump field. A chopping wheel was placed within the pumping cavity in order to investigate the dynamics of the device after disruption from steady-state operation of the device. The resulting behaviour of the circulating field is shown in Fig. 24, where the dynamics of a ICOPO based upon $Nd:YVO_4$ but with otherwise similar operating parameters is shown for comparison.

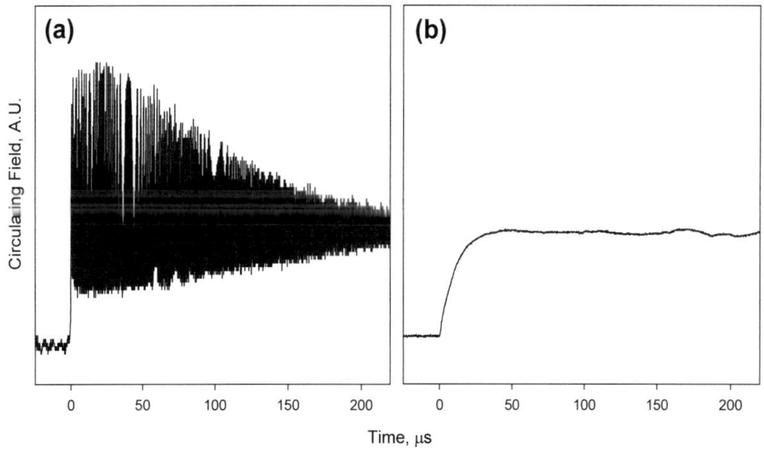

Fig. 24. Recovery transient dynamics of a chopped ICOPO pumped internal to (a) an $Nd:YVO_4$ and (b) VECSEL laser.

It is immediately obvious that that the VECSEL displays excellent transient stability: there is no trace of oscillatory behaviour at all. The finite rise-time of the trace shown in Fig. 24(b) is simply due to the reveal time of the chopping wheel. When the wheel was removed the pump, signal and idler all exhibited excellent amplitude stability. In order to quantify this free-running stability, the amplitude spectrum of the pump field fluctuations was obtained via the use of an RF spectrum analyser, as shown in Fig. 25 where again a comparable trace obtained form the Nd system is shown for comparison.

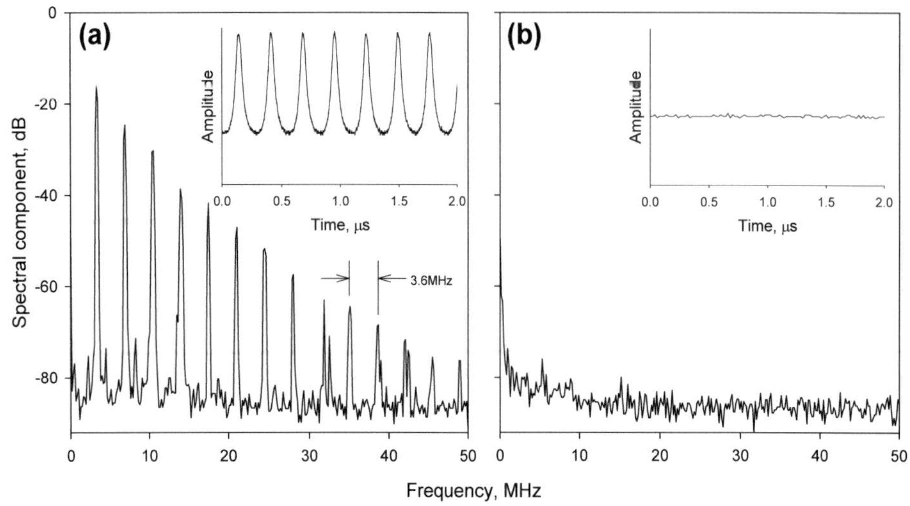

Fig. 25. Amplitude spectrum of the circulating pump field in (a) the Nd:YVO_4- and (b) SDL-based ICSRO. The insert to each panel shows the temporal behaviour of each respective pump field.

The trace shown in Fig. 25 (a), as measured from the Nd-based ICOPO, was obtained during a spontaneous oscillation event as is evident from the characteristic wave form shown in the amplitude-space inset trace. The frequency spectrum shows a large peak at the fundamental oscillation frequency (~3.6MHz) with the higher-order harmonic components tailing off beyond. This is in sharp contrast to the result obtained from the VECSEL system which, as the inset to Fig. 25(b) shows, displayed true cw behaviour. The frequency-space trace only displays spectral content at very low frequencies (not resolvable on the axis scale used in the figure) – a consequence of acousto-mechanical noise within the environment in which the system was operated. Such noise could be easily remedied through the use of established cavity length and amplitude stabilisation techniques, or the implementation of a more robust mechanical design.

We conclude this section by examining the down-converted power characteristics of the device. The considerably higher circulating pump powers available in this device compared to those discussed previously in this chapter, brought about by the enhanced power of the primary pump diode laser, meant that the device could be operated many times above OPO threshold. As we have previously seen, operating the device in this mode results in over-coupling of power from the pump circulating field and hence reduced down-converted powers at maximum primary pump power. In order to optimise the efficiency of the device at full pump power it is therefore necessary to raise the threshold of the OPO such that, for a given laser threshold and primary pumping level, power relation (1) is satisfied. The device could therefore be optimised in one of two regimes: for minimum OPO threshold or maximum obtained power in the extracted idler field at full primary pump power. Both of these cases were demonstrated with the VECSEL ICOPO under discussion, and their down-converted power performance is shown in Fig. 26. In both cases it can be seen that as expected, once the OPO comes above threshold the power in the intracavity circulating pump field clamps at (or near) its OPO threshold value, indicative of a robust optical geometry for the pump-wave and signal-wave cavities. Fig. 26(a) indicates the performance of the device when optimised for low parametric threshold. Here, laser and OPO threshold are achieved at diode pump powers of 1W and 1.4W respectively. At this latter pumping power a circulating pump field of 3.1W was present within pump cavity.

That OPO threshold is achieved at a power level well within the reach of very modest pumping sources once again shows the great advantage which the intracavity approach bestows. It is clear that in mobile, perhaps battery operated applications where diode-laser primary pump power might be limited to ~3W, the device, when optimised in this way, would operate at >2 times threshold delivering an idler power of 30mW. This level is more than adequate for a host of spectroscopic applications. When pumped at the maximum diode-laser pump power of 8.5W, the extracted idler power was 95mW. Hence the calculated total down-converted power, calculated in the previously described manner (equation 13), was 581mW. At this pumping level, the OPO was operating 6 times above threshold and 4.3 times greater than the optimum (in terms of down-conversion efficiency) pumping level of 1.96W, as determined by (1).

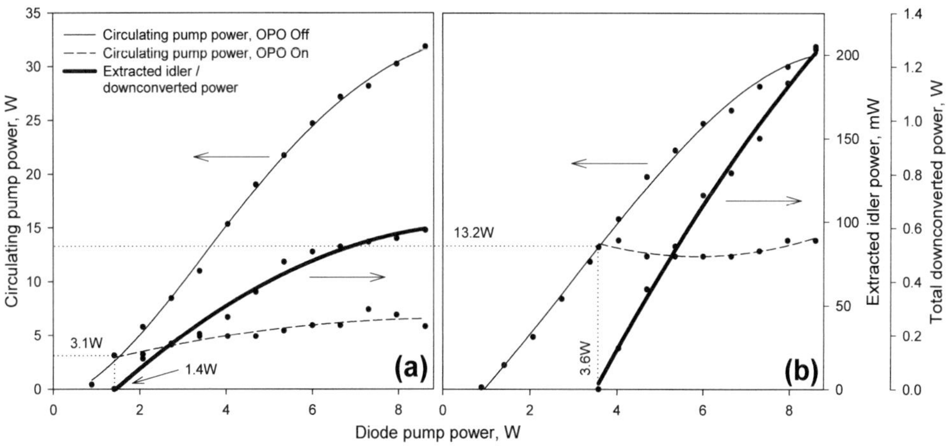

Fig. 26. Downconversion power characteristics of the VECSEL ICOPO when (a) optimised for lowest threshold and (b) optimised for maximum idler output power. In both cases the threshold of the pump laser occurred at a diode pump power of 1W, and the pump, signal and idler wavelengths were 1050nm, 1600nm and 3055nm respectively.

In order to optimise the down-conversion efficiency of the device when operating at the maximum available primary pumping power, and hence maximise the extracted idler power, it was necessary to increase the threshold of the OPO (but leave the laser threshold unchanged). This was achieved by increasing the mode size of the pump and signal fields within the nonlinear crystal. The resulting curve is shown in Fig. 26(b). The threshold of the OPO, in terms of diode pump power, increased from 1.4W (in the case of the low-threshold device) to 3.6W, at which point the circulating pump field was 13.2W. This threshold pump power is some 700mW greater than the calculated threshold of 2.9W for optimal down-conversion which (1) predicts; possibly a result of non-confocal focusing within the PPLN crystal. The extracted idler power at this pumping level was 205mW, corresponding to a total down-converted power of 1.25W. With the mirror M2 replaced with an optimal output coupler (but the cavity otherwise left unchanged), 1.5W power at the pump wavelength was

obtained. This implies that the 1.25W converted through the parametric process from pump into signal and idler power equates to a down-conversion efficiency of 83.3%.
This system, then, exhibits high-power, broadly tunable output with extremely high efficiency in a compact, robust mechanical design. The synergy of the intracavity technique and emergent semiconductor laser gain media is the culmination of the different topics discussed in this chapter, both in terms of exploiting the benefits of, and managing the issues associated with, the cw-ICOPO technique.

6. Conclusions and outlook

In this chapter we have examined various aspects of the cw-ICOPO in the context of realising low threshold, high efficiency and broadly tunable sources of mid-infrared radiation. We have seen that through an understanding of their origin, steps can be taken in order to either manage or eliminate the relaxation oscillations which have to date barred the cw-ICOPO from use in many of the applications to which it is otherwise ideally suited.
Due to the constraints of space we have unfortunately had to neglect many of the other design aspects of these devices; most notably spectral line narrowing, mode-hop free tuning mechanisms and output-coupling of the resonant down-converted field. It is hoped that armed with the knowledge provided in this chapter, the reader will have the necessary expertise and confidence to design and construct a cw-ICOPO and then be able to use well established laser frequency control and output coupling techniques, which are in the main entirely portable from their use in the parent pump laser to the resonant down-converted wave, in order to adapt their system to meet his or her precise needs.
Neither have we explored the operation of the ICOPO in the Q-switched, pulsed regime: yet another solution to the problem of relaxation oscillations. Here, the long upper-state (storage) lifetime of Nd gain media is turned to our advantage. Whilst the resulting very high intensity peak powers present within the pumping cavity allow the ICOPO to operate over a wider portion of its phase-matching bandwidth, and therefore increases the linewidth of the down converted waves, the system becomes very much less susceptible of the effects of pump and, in particular, signal cavity round-trip loss. This significantly relaxes the tolerance placed upon cavity finesse and therefore allows the tuning bandwidth of the device to be substantially expanded. Through careful gating of the pump cavity Q and taking advantage of high efficiency cavity-dumping of the pump field into signal and idler through the parametric process (Debuisschert, Raffy et. al. 1996), very high repetition rates (100's kHz) are achievable with mean-power efficiencies significantly higher than those normally associated with Q-switched devices (Stothard, Rae et. al. 2009).
The system we considered in section 5.2 is a particularly exciting stepping stone on our road to realising a truly utile cw-ICOPO ideally suited to high-resolution spectroscopic applications. As we saw, the use of state-of-the-art, semiconductor-based laser gain media eliminates the problems associated with relaxation oscillations, therefore releasing the potential of the cw-ICOPO. By utilising enhanced, ring-resonator designs coupled with existing well-established tuning mechanisms (such as cavity length control, the use of frequency-selective elements, etc.), very broad tunability with single-frequency, mode-hop free tuning over 100's GHz becomes possible. That the gain bandwidth of VECSEL gain media is dictated by its internal structure, rather than a particular electronic transition, opens up the particularly exciting possibility of realising cw devices capable of penetrating much further into the infrared spectral region. This band is currently inaccessible (to cw

devices) due to the limited transparency of established nonlinear materials beyond ~4μm, therefore mandating the use of semiconductor-based nonlinear optical crystals. These, however, suffer from very poor transparency below ~2μm. The combination of high power, long wavelength (>2μm) circulating pumping fields available within VECSEL-based lasers and very recent advances in low loss periodically-polled semiconductor nonlinear crystals fabricated from GaAs (Schunemann, Pomeranz et. al. 2009) could enable the kind of rapid and exciting developments in the field of mid- to deep- infrared parametric devices that PPLN did for shorter wavelength cw devices in the 1990's. Such developments would confirm the continuing importance of ICOPO technology and ensure that, in the absence of radical developments in laser gain media operating directly in the required spectral region, ICOPOs will continue to play an important role in extending the utility and spectral range of laser-based spectroscopic devices for many years to come.

7. References

Bosenberg, W. R., Drobshoff, A., et al. (1996). 93% pump depletion, 3.5-W continuous-wave, singly resonant optical parametric oscillator. *Optics Letters,* 21(17): 1336-1338.

Colville, F. G., Dunn, M. D., et al. (1997). Continuous-wave, singly resonant, intracavity parametric oscillator. *Optics Letters,* 22 (2), p75.

Debuisschert, T., Raffy, J., et. al. (1996) Intracavity optical parametric oscillator: study of the dynamics in the pulsed regime. *Journal of the Optical Society of America B - Optical Physics,* 12 (7), pp1569-1587.

Ebrahimzadeh, M. and Dunn, M. H., (1998). Optical Parametric Oscillators, In: *Handbook of Optics IV - Fiber and Nonlinear Optics,* Bass, M., and Enoch, J. M., (Eds.), pp22.1-22.72, Optical Society of America, ISBN 0-07-136456-0

Edwards, T. J., Turnbull, G. T. A., et al. (1998). High-power, continuous-wave, singly resonant, intracavity optical parametric oscillator. *Applied Physics Letters,* 72 (13), pp1527-1529.

Jundt, D. H. (1997). Temperature-dependent Sellmeier equation for the index of refraction ne in congruent lithium niobate. *Optics Letters,* 22, p1553.

Kim, J. Y., Cho, S., et al. (2007). Highly efficient green VECSEL with intra-cavity diamond heat spreader. *Electronics Letters,* 43 (2), pp105-107.

Kuznetsov, M., Hakimi, F., et al. (1999). Design and characteristics of high-power (> 0.5-W CW) diode-pumped vertical-external-cavity surface-emitting semiconductor lasers with circular TEM_{00} beams. *IEEE Journal of Selected Topics in Quantum Electronics,* 5 (3), pp561-573.

Maclean, A. J., Kemp, A. J., et al. (2008). Continuous tuning and efficient intracavity second-harmonic generation in a semiconductor disk laser with an intracavity diamond heatspreader. *IEEE Journal of Quantum Electronics,* 44 (3-4), pp216-225.

Oshman, M. K. and Harris, S. E., (1968). Theory of Optical Parametric Ocillation Internal to the Laser Cavity. *IEEE Journal of Quantum Electronics,* QE-4 (8), pp491-502.

Schunemann, P. G., Pomeranz, L. A., et. al. (2009). Recent advances in all-epitaxial growth and properties of opientation-patterned gallium arsenide (OP-GaAs). *Conference on Lasers and Electro-Optics.* Baltimore, Optical Society of America, Paper CWJ5.

Stothard, D. J. M. and Dunn, M. H., (2009). Relaxation-oscillation suppression in intracavity optical parametric oscillators. *Submitted to Optics Express,* September 2009.

Stothard, D. J. M., Ebrahimzadeh, M., et al. (1998). Low pump threshold, continuous-wave, singly resonant, optical parametric oscillator. *Optics Letters*, 23, p1895.

Stothard, D. J. M., Rae, C. F., and Dunn, M. H., (2009). An intracavity optical parametric oscillator with very high repetition rate and broad tunability based upon room temperature periodically-poled $MgO:LiNbO_3$ with fanned grating design. *IEEE Journal of Quantum Electronics*, 45 (3), pp256-263.

Stothard, D. J. M., Hopkins, J. M., et al. (2009). Stable, continuous-wave, intracavity optical parametric oscillator pumped by a semiconductor disk laser (VECSEL). *Optics Express*, 17 (13), pp10648-10658.

Tropper, A. C., Foreman H. D., et al. (2004). Vertical-external-cavity semiconductor lasers. *Journal of Physics D - Applied Physics* 37 (9), R75-R85.

Turnbull, G. A., Dunn, M. H., et al. (1998). Continuous-wave, intracavity optical parametric oscillators: an analysis of power characteristics. *Applied Physics B*, 66, p701.

Turnbull, G. A., Edwards, T. J., et al. (1997). Continuous-wave singly-resonant intracavity optical parametric oscillator based on periodically-poled $LiNbO_3$. *Electronics Letters*, 33 (21), pp1817-1818.

Vodopyanov, K. L., (2003) Pulsed Mid-IR Optical Parametric Oscillators, In: *Solid-state, mid-infrared laser sources*, Sorokina, I. T., Vodopyanov, K. L., (Eds.) p141-174, Springer, ISBN 978-3-540-00621-3

17

Ultra-Sensitive Optical Atomic Magnetometers and Their Applications

Igor Savukov
Los Alamos National Laboratory
USA

1. Introduction

In this chapter, we overview the most sensitive contemporary atomic magnetometers (AM) that are based on high-density alkali-metal vapors. These magnetometers are considered in a broader content of other magnetometers and their applications. The principles of the operation of the AMs are explained for better understanding of this topic. One point of focus in this chapter which establishes the connection to the title of this book is about the relation between lasers and most sensitive atomic magnetometers. The chapter is organized in the following way. After general introduction to the AMs and the applications of magnetometers, the principles of the operation of optical atomic magnetometers are given. With this background information, next so-called SERF magnetometers and their features are discussed. Then, the discussion continues to the topic about the operation of "SERF magnetometers" in the non-SERF regime. Finally, after covering the principles and theory, we return to some most notable applications of atomic magnetometers.

Since their discovery, lasers have revolutionized many fields - the field of AMs or magnetometers in general is no exception. Before the advent of lasers, AMs were based on discharge lamps which though relatively simple and inexpensive light sources did not provide enough power and had some other drawbacks for the realization of maximum sensitivity that can be achieved with atomic magnetometers. In a comparison of sensitivity of the state-of-the-art Cs magnetometers based on a Cs discharge lamp and a semiconductor laser made in Ref. (Groeger et al., 2005), the lamp Cs magnetometer had sensitivity of 25 $fT/Hz^{1/2}$ that was lower than that of the same but laser-based magnetometer, 15 $fT/Hz^{1/2}$. For most sensitive magnetometers the difference is expected much more significant, although there is no investigation of this question in the literature. From Ref.(Groeger et al., 2005), we can estimate that the intensity of the lamp light in the Cs absorption band was below 1 mW, and such power would be suboptimal in most sensitive magnetometer based on high-density vapors, such as spin-exchange relaxation free (SERF) magnetometer (Allred et al., 2002). The absence of lasers is one of the factors that initially atomic magnetometers had been far behind superconducting quantum interference devices (SQUIDs) in sensitivity.

Although the introduction of lasers into magnetometry improved sensitivity, laser-based magnetometers are not yet commercially available. However, this can change in the near future. Diode lasers of high quality are becoming less expensive, and some lasers such as vertical cavity surface-emitting lasers (VCSELs) in addition to extremely low price allow integration into microfabricated packages. Such packages will be not only inexpensive and easy to batch produce they will also have lower power consumption, light weight and unimpeded mobility (Knappe et al., 2006).

Sensitive magnetic field measurements, for which AMs can be used, are important owing to many existing and potential applications. Magnetometers have been in wide use, for example, in geology, military, biomagnetism, space magnetic field measurements. AMs have been used because they are both relatively sensitive compared to conventional inexpensive magnetometers, such as fluxgates, and more convenient compared to SQUIDs that require cryogenic cooling. For a long time low-T_c SQUIDs had been by far the most sensitive magnetometers at low frequency. However, this has changed with the demonstration with AMs of record 0.5 $fT/Hz^{1/2}$ sensitivity (Kominis et al., 2003).

High-sensitivity AMs are based on spin-polarized atomic gases or vapors, and AM research is closely related to the atomic physics subfield dedicated to the investigation of spin interactions in such media. This subfield includes the research on optical pumping and related topics such as atomic clocks, masers, hyperpolarized gases, spin-based spectroscopy, and some others. A comprehensive review of optical pumping experiments before 1972 and theory of optical pumping and various relaxation mechanisms are given by William Happer (Happer, 1972). This theory, which still stands the test of time, includes the formulation of density matrix equations, which can be directly applied to the analysis of magnetic resonances of vapors on which atomic magnetometers are based. Atomic magnetometers find many applications as other magnetometers, but they are most useful when high sensitivity and non-cryogenic operation are required. For example, AMs have been developed for submarine detection, and many other military applications are possible on similar principles. Some applications of Rb-vapor or ^{3}He magnetometers in geophysics and space physics were mentioned as early as in 1961 by Bloom (Bloom, 1961) who did some pioneering work on the high-sensitivity measurements of magnetic fields and theory for the operation of atomic magnetometers. Theoretical analysis of early atomic magnetometers was also done by Dehmelt (Dehmelt, 1957) and by Bell and Bloom (Bell & Bloom, 1957). Although the history of atomic magnetometers is quite fascinating and they played an important role in applications where simpler and less expensive methods did not provide sufficient sensitivity, the atomic magnetometers of the past were much less sensitive than the modern ones. Most exciting results appeared only recently with the various demonstrations of performance of high-density AMs, which we would like to focus on in this chapter.

2. Principles of operation of sensitive atomic magnetometers

2.1 The interaction of spins with magnetic field

First of all, AMs are based on atomic spins which due to their magnetic moment interact with magnetic field. The Hamiltonian related to this interaction in the formalism of quantum mechanics is

$$H = \gamma_e \mathbf{J} \cdot \mathbf{B} + \gamma_N \mathbf{I} \cdot \mathbf{B} + a_{hf} \mathbf{J} \cdot \mathbf{I}, \qquad (1)$$

where $\gamma_e = g_J \mu_B / \hbar$, $\gamma_N = g_I \mu_B / \hbar$, μ_B is the Bohr magneton, g_J, g_I are electron's and nuclear g-factors, $\mathbf{J}$ is the total angular momentum of the electron, the sum of the spin and the orbital momentum, $\mathbf{J} = \mathbf{S} + \mathbf{L}$; $\mathbf{I}$ is the nuclear angular momentum, and a_{hf} is the hyperfine constant. This Hamiltonian is responsible for the splitting of hyperfine sublevels in magnetic field, called the Zeeman splitting. Because in small external field the third term dominates, it is convenient to employ the basis of total momentum of the atom $\mathbf{F} = \mathbf{I} + \mathbf{J}$ for the classification of the hyperfine sublevels. In zero external field the total angular momentum F and its projection M are "good" quantum numbers, but by continuity we can use this classification when $|F, M\rangle$ states are no longer eigenstates of the Hamiltonian (1). The solution of the Schrödinger equation in the presence of magnetic field and hyperfine interaction for the case $\mathbf{J} = \mathbf{S} = 1/2$ is known as the Breit-Rabi equation:

$$W(F, M_F) = -\frac{\Delta W}{2(2I+1)} - \frac{\mu_I}{I} B M_F \pm \frac{\Delta W}{2}\sqrt{1 + \frac{4M_F}{2I+1}x + x^2}\,, \tag{2}$$

where $\Delta W = a_{hf}[F_2(F_2+1) - F_1(F_1+1)]/2$ is the hyperfine splitting between $F_2 = I + 1/2$ and $F_2 = I - 1/2$ states at zero field, $x = (g_J - g_I)\mu_B B / \Delta W$. Fig. 1 shows the dependence of the energy of hyperfine sublevels on applied DC magnetic field. The transitions between magnetic sublevels $M \to M \pm 1$ can be induced by time-varying magnetic interaction $H = \gamma_e \mathbf{J} \cdot \mathbf{B}(t) + \gamma_N \mathbf{I} \cdot \mathbf{B}(t)$ (we can neglect the second term, which is about 3 orders of magnitude smaller). If oscillating field is applied, by scanning its frequency, the Zeeman resonances can be observed, which are also called magnetic resonances. The resonance frequency at small field is directly proportional to the applied DC field. It is also possible to induce transitions between hyperfine levels $F_2 = I + 1/2$ and $F_2 = I - 1/2$ separated by the ΔW gap. These transitions are used in atomic clocks and masers. Because the splitting is quite stable, essentially it is dependent on constant internal field produced by the nucleus, high-quality clocks can be built. One figure of merit for clocks is the Q factor of resonance, and it is very high in the case of hyperfine transitions. Even in atomic vapors, where various collisions damp hyperfine resonances, not to mention traps and atomic fountains, where the collision effects are much smaller, Q factors can exceed 1 million. The Zeeman transitions can also have high Q factors at high frequencies because the resonance width, which is on the order of 1 kHz at high density of vapor and a few Hz at low density in cells with coating, does not change over large frequency range as long as gradients arising from the applied field are not very strong and the field is stable. In comparison to other resonance systems such as mechanical oscillators and LC-resonance circuits, the Zeeman resonances of atomic

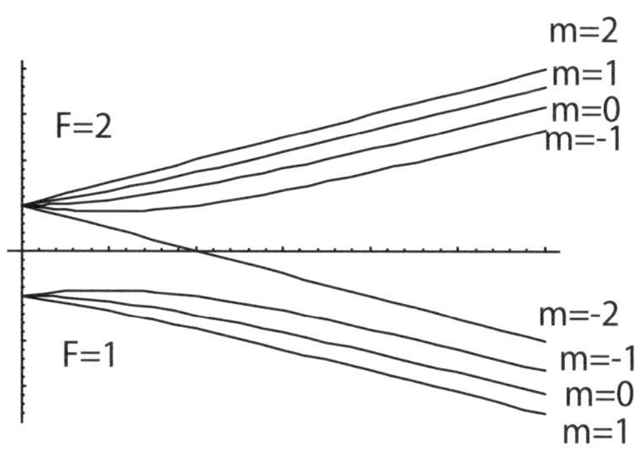

Fig. 1. Dependence of Zeeman sublevels on magnetic field for the case of I=3/2.

magnetometers at sufficiently high frequency can be considered of very high quality. Apart from direct applications in the magnetic field measurements, these resonance properties can be also useful for building radio receivers and filters.

The Zeeman level splitting and transitions between the Zeeman levels under the action of field is the quantum-mechanical picture of the interaction of the field with the spins, which is most appropriate for quantum objects such as atomic spins. However, in practice, the classical picture can be more convenient to use. In the classical picture, the magnetic field causes torque on the spins, and their behavior is described by the Bloch equations:

$$\begin{aligned} dS_x/dt &= \gamma S_y B_z - \gamma S_z B_y - S_x/T_2 \\ dS_y/dt &= -\gamma S_x B_z + \gamma S_z B_x - S_y/T_2 \\ dS_z/dt &= \gamma S_x B_y - \gamma S_y B_x + (S_0 - S_z)/T_1 \end{aligned} \tag{3}$$

Here γ is the gyromagnetic ratio of atomic spins (the slope of the energy-field curve in Fig. 1). In order to account for various relaxation mechanisms, the terms with phenomenological relaxation times T_1 and T_2 are added. T_1 is the longitudinal relaxation time, which shows how fast the spin ensemble reaches equilibrium when spins are oriented along the field, and T_2 is the transverse relaxation time, which characterizes the decay of transverse excitation of spins. The use of the Bloch equations can be justified in certain cases when the number of spins N is so large that quantum fluctuations of spin projections that scale as $1/N^{1/2}$ can be neglected and when resonance frequencies of multiple Zeeman transitions are the same so the majority of spins of various hyperfine levels precess together as the whole. Though very small at these conditions, quantum fluctuations can be added to the Bloch equations for the analysis of fundamental noise of the atomic magnetometer (Savukov et al., 2005). The conditions of collective precession of spins with the same frequency and relaxation times for all Zeeman sublevels are not always satisfied, but in some practically important regimes of operation, the Bloch equations give quite adequate explanations of observed effects. The Bloch equations are also convenient that they provide analogy with NMR, where they are routinely used for the analysis of various schemes for manipulating nuclear spins. NMR-like effects, such as the free-induction decay, spin echo, rf broadening, gradient broadening, exist in atomic magnetometers and can be used in applications. Even when the Bloch equations are not rigorously justified, they can still provide qualitative description of many experiments with atomic magnetometers.

Alternatively, if accurate description is desirable, the density matrix (DM) equation

$$\begin{aligned} \frac{d\rho}{dt} &= a_{hf}\frac{[\mathbf{I}\cdot\mathbf{S},\rho]}{i\hbar} + \mu_B g_S \frac{[\mathbf{B}\cdot\mathbf{S},\rho]}{i\hbar} + \frac{\varphi(1+4\langle\mathbf{S}\rangle\cdot\mathbf{S})-\rho}{T_{SE}} + \\ &+ \frac{\varphi-\rho}{T_{SD}} + R[\varphi(1+2\mathbf{s}\cdot\mathbf{S})-\rho] + D\nabla^2\rho \end{aligned} \tag{4}$$

can be used (Happer, 1972; Appelt et al., 1998; Alred et al., 2002; Savukov & Romalis, 2005). Here ρ is the density matrix, which has dimension of the number of hyperfine states; $\varphi = \rho/4 + \mathbf{S}\cdot\rho\mathbf{S}$ is the pure part of the density matrix, $\langle\mathbf{S}\rangle = Tr(\rho\mathbf{S})$, T_{SE} is the spin-exchange (SE) collision time, T_{SD} is the spin-destruction time, R is the pumping rate, and $\mathbf{s}$ is the optical pumping vector which is oriented in parallel with the direction of the pump beam and its magnitude and sing depends on the degree of circular polarization. A DM equation is the generalization of Schrödinger equation, normally applicable to pure states, for the case when states are mixed due to collisions between atoms, which cannot be ignored. Some terms (the first and the second) can be directly obtained from the Hamiltonian of the Schrödinger equation via the Von Neumann equation, $i\hbar\frac{d\rho}{dt}=[H,\rho]$, but others require some non-trivial theoretical derivation. Unlike the Bloch equations where the relaxation and equilibrium polarization were introduced phenomenologically, the DM equation contains explicitly relaxation terms and optical pumping terms that determine the equilibrium polarization. The solution of the DM equation can be used to explain many observed effects in atomic magnetometers, including precession frequency of spins and their relaxation rates in a wide range of experimental conditions, and is considered the most appropriate theoretical framework. Unfortunately, the DM equation has to be solved numerically, and the solutions are cumbersome for the analysis. The solution also takes significant computational time due to non-linear nature of the equation arising from the SE term, $\frac{\varphi(1+4\langle\mathbf{S}\rangle\cdot\mathbf{S})-\rho}{T_{SE}}$. Still in several limits the DM equation can be simplified and some intuitive pictures of spin dynamics can be obtained using for example the method of separation on spin subsystems and perturbation theory. Quite useful separation is on the upper ($F=I+1/2$) and the lower ($F=I-1/2$) hyperfine manifold subsystems of the ground state. In the presence of magnetic field, the upper manifold Zeeman components precess in one direction and the lower manifold components precess in the opposite direction. Thus instead of a single set of Bloch equations (2) two sets of the Bloch equations can be applied separately to these systems of precessing spins:

$$\begin{aligned}\frac{d\mathbf{S}_{up}}{dt} &= \gamma\mathbf{B}\times\mathbf{S}_{up}\\ \frac{d\mathbf{S}_{down}}{dt} &= -\gamma\mathbf{B}\times\mathbf{S}_{down}\end{aligned} \tag{5}$$

A coupling between the oppositely rotating spins exists due to SE collisions, which can lead either to full alignment of two subsystems if the precession frequency is much lower than the rate of SE collisions or to relaxation otherwise. In the former case, a single set of the Bloch equations can be used to describe the precession of the spins.

Another important concept that allows us greatly simplify the analysis is the spin-temperature (ST) distribution. A ST density matrix is $\rho_{ST}=k_n\exp(\boldsymbol{\beta}\mathbf{F})$, where $\boldsymbol{\beta}$ is the ST parameter, k_n is the normalization factor, and $\mathbf{F}$ is the total angular momentum vector. The ST density matrix can be characterized by a single vector – the total spin of the system or polarization vector. The ST distribution for $I=3/2$ is illustrated in Fig. 2. By substituting the ST distribution into the DM equation, Eq. 4, we find that it is a solution of this equation. The ST distribution is maintained in the SERF regime in the static and the rotating frame if spins change their orientations slow compared to the SE rate. When the DM has a ST distribution, the SE term does not lead to any relaxation and the spin dynamics is very similar to that described with the Bloch equations in the static or rotating frames. However, beyond the SERF regime, the ST distribution is not strictly valid. Still when the deviation from the SE distribution is small, perturbation theory can be effectively used to account for this small deviation. One practically important example is the case of small excitation by time-varying magnetic field. The theory of so-called rf magnetometer is based on this approach (Savukov et al., 2005). [See also (Appelt et al., 1998) where detailed solution of the DM equation is given.] It is also possible to apply perturbation theory in some cases of large excitation amplitudes. For example, when spin polarization is large, under condition of strong excitation most spins follow the ST distribution in the rotating frame, and perturbation theory can be used to account for small deviations from this pattern.

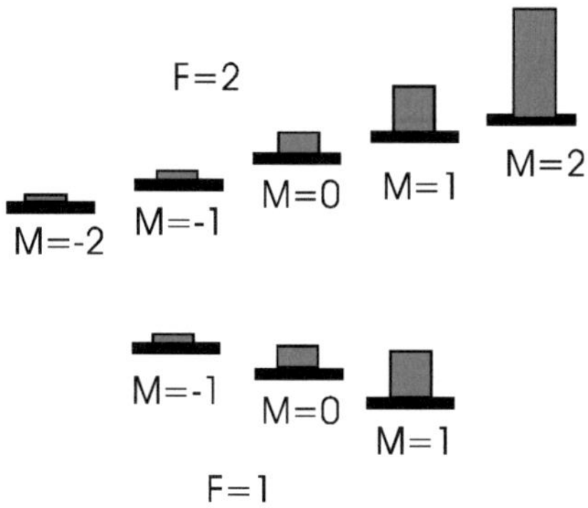

Fig. 2. A spin-temperature distribution

2.2 The interaction of spins with light

The interaction of spins with light leads to a number of phenomena such as optical pumping, dependence of optical properties on spin states, light shift, light-induced spin-destruction, and light narrowing of magnetic resonances that are encountered in atomic magnetometers. Optical pumping, in particular, is an essential feature of high sensitivity atomic magnetometers. In general optical pumping, quite common process in laser physics, leads to redistribution of atomic levels. For example, many lasers are based on population inversion that is created by irradiating laser medium with light. More specifically, in the context of AMs, optical pumping means the redistribution of magnetic sublevels due to absorption of light. Optical pumping can change the total spin of initially disoriented spins and can lead to build up of spin polarization. Although magnetic field prepolarization can be in principle used to create the preferential spin orientation and non-zero magnetometer signal, the optical pumping is much more efficient. It can increase polarization by many orders of magnitude compared to thermal equilibrium values even in strong field. With sufficient pumping power (about 10 mW), the expectation value of spin can reach almost maximum value, which is ½ in the case of S=1/2. For comparison, in NMR, such levels can not be reached with the strongest polarization magnets. Only the combination of high field (10 T) and low, liquid-helium temperature (4 K) can produce similar polarization of electron spins, but this method is not practical especially for alkali-metal vapors that have to be kept at much higher temperature.

Optical pumping of atomic spins can be illustrated and estimated in the case of circularly polarizing light by using the fundamental law of conservation of angular momentum. Circularly polarized photons have spin 1. In the act of absorption, according to this law the photons transfer their spins to atoms. The selection rules is another way to understand this process: the magnetic sublevel M changes by 1 in a transition to the excited state induced by circularly polarized light, so the expectation value of the atomic spin, which depends on M, changes. From excited states, atoms decay, either spontaneously or through collisions with other atoms, to the ground state. The overall cycle results in the change of angular momentum or spin of the ground state. The efficiency of pumping can be quite high - typically in atomic magnetometers on average only 1.5 photons are required to polarize one alkali-metal atom in the ground state. The described above pumping process is called depopulation pumping, because it is arranged that the photons preferentially depopulate the atomic states of some M with higher probability than others. It is possible to arrange other schemes of optical pumping. Some of them are analyzed in the review paper [Happer 1972]. In optical pumping not only the expectation value of spin (vector) can be changed, but also the expectation values of multipoles of higher order, if the state has angular momentum greater than ½. The terms orientation and alignment are used to differentiate odd and even multipoles or just dipole and quadrupole moments [(Budker et al., 2004), appendix about atomic polarization moments]. For example, if M=-1,0,1 levels are populated in the proportion 1:0:1, the system will have alignment but no orientation. Alignment, as orientation, precesses around magnetic field and can be used for magnetic-field measurements. Sensitive magnetometers based on various multipoles have been explored extensively by D. Budker group at UC Berkeley.

To build a sensitive magnetometer such as SERF, in addition to optical pumping it is necessary to utilize optical probing. Optical probing is a high-sensitivity method to detect the states of atomic spins based on strong spin-dependent interaction of light with polarized atoms. Alternatively, a pick-up coil can be used in some cases, but its sensitivity is low at low frequency. For example, the SERF magnetometer has only frequencies below a few hundred Hz range and a coil will not be very sensitive. The optical probing signal, on the other hand, does not depend much on frequency, and the optical probing can be used for detection of DC fields. The only problem could be 1/f noise, that exists at very low frequencies owing to various reasons. To reduce this technical noise, some methods of modulation can be implemented. For example, a polarization modulator can be inserted into the probe beam path to shift the low frequency AM signal to frequencies of a few kHz.

The high sensitivity of optical detection is due to both strong interaction of light with spins and high sensitivity of polarization angle measurements (or absorption measurements) that can reach quantum limit of photon fluctuations. This limit is extremely small, on the order of nrad/$Hz^{1/2}$. Note that one nrad is the angular size of a 1-mm object at the distance 1000 km! The interaction of light with atoms is strong because atoms, especially in gas phase, have very narrow optical absorption resonances and large transition amplitudes. For example, alkali-metal atoms of concentration of 10^{14} atoms/cc can absorb resonant light in the path length on the order of 1 mm. Quantitatively, the absorption coefficient α , or the inverse absorption length, can be found from the oscillator strength (f) , the density of atoms (n), and the width of absorption profile (γ). In the case when the buffer gas pressure is sufficiently high (which is the case in SERF magnetometers), the hyperfine splitting and Doppler broadening can be neglected, and the absorption coefficient of a specific line, such as D1, becomes a single Lorentzian:

$$\alpha = ncr_e f \frac{\gamma}{(\nu - \nu_0)^2 + \gamma^2} . \tag{6}$$

In the center of the resonance, the absorption coefficient $\alpha = ncr_e f / \gamma$. The absorption coefficient depends on optical linewidth, which in turn depends on the gas composition of the magnetometer cell, buffer gas pressure, if present, and temperature. Buffer gases are frequently added to reduce diffusion to the wall and for other functions, for example, to achieve high spatial resolution of AM measurements. When the potassium cell is filled with He, the linewidth is about 7 GHz (HWHM) or 0.014 nm per 1 amg (1 amg is the density of the gas for 1 atm at normal conditions). This line width at He density on the order of 1 amg exceeds the hyperfine spitting of K (I=3/2), equal to 462 MHz, and Doppler width HWHM=0.5 GHz. In heavier alkali-metal atoms the hyperfine splitting, which is 3036 MHz in Rb (I=5/2) and 9192 MHz in Cs (I=7/2), can become comparable to the buffer gas broadening for the pressures about 1 atm. Thus especially in the case of Cs two-component absorption profile will appear. The absorption coefficient is strong for all buffer gas densities used in SERF magnetometers, up to 10 amg. The optical probing sensitivity is also high in this range of densities.

Both absorption and light polarization rotation (Faraday effect) can be used for detection of spins. In most sensitive magnetometers, such as SERF, light polarization rotation measurements were chosen over absorption measurements. One drawback of the absorption method is that the probe laser has to be tuned close to the center of absorption line, and this leads to stronger spin-destruction as well as to strong attenuation of the probe beam, especially in optically thick high-density vapors. In the Faraday detection method, on the other hand, the probe laser is tuned away from the resonance, which facilitates the propagation of light through optically thick medium and reduces the spin destruction.

According to the rules of optics, the plane of polarization of linearly polarized light will be rotated by non-zero angle

$$\theta = \frac{\pi(n_- - n_+)l}{\lambda} \tag{7}$$

if the refractive indices for right and left circularly polarized light components n_+ and n_- are not equal, where λ is the wavelength and l is the pathlength. This is possible when spin-up and spin-down ground-state levels are unequally populated for quantization axis chosen along the direction of the light propagation. Large rotation of light polarization in optically pumped vapors is due to the strong dependence of refractive index on atomic spin orientation. It can be derived from Eqs.6 and 7 and the Kramers-Kronig relations that the rotation angle by alkali-metal atoms is

$$\theta = \pm\frac{1}{2} l r_e c f n P_x D(\nu - \nu_0), \tag{8}$$

where $D(\nu)$ is Lorentzian dispersion profile, l is the path length, and r_e is the classical electron radius. This is the expression for the D1and D2 lines, with opposite signs for their contributions.

The rotation of polarization by optically-pumped vapors, which can be evaluated with Eq. 8, exceeds by many orders of magnitude usual Faraday rotation in other substances.

In theory of AMs, the questions about spin-destruction and light shifts by pump and probe beams also arise. The mechanism of light-induced spin-destruction is similar to that of spin pumping: when light is absorbed it changes the spin states, or in other words perturbs the spins. In the light-induced spin destruction the change in the spin state leads to the loss of coherence and longitudinal polarization. Both circularly polarized and linear polarized light can induce spin destruction, but only circularly polarized component of light builds up the orientation of spins. Hence if the degree of circular polarization is smaller than 1, the maximum polarization level will be less than 1 for arbitrary light intensity. This result can be written as $P = sR/(R + R_{SD})$, where s is the degree of circular polarization and $R_{SD} = 1/T_{SD}$. This coefficient also includes the reduction owing to possibly non-zero angle between the beam direction and the spin orientation, which can be forced away from the beam direction by an applied magnetic field. Scattering of pump light by atoms can also lead to the reduction of the ultimate polarization. However, this effect is minimized in the SERF atomic cells that contain nitrogen buffer gas, which quenches excited states so atoms do not re-emit photons in random directions after the absorption of the pump light.

Light shift is the AC Stark effect, i.e. the shift of atomic energy levels in AC electric field produced by light. An unperturbed atom has zero electric moment (extremely small electric moment might exist, but its effect is hardly detectable), so Stark effect appears only in the second order of perturbation theory in dE term. In other words, electric field polarizes the atom and interacts with the dipole moment of the polarized atom. When light has wavelength in the vicinity of absorption resonance, significant enhancement of AC Stark effect appears due to the reduction of denominator in the perturbation theory. Minimal denominator value is determined by the width of the excited state. On the other hand, AC Stark shifts between Zeeman levels have significant cancellations so they are quite small, but observable since Zeeman resonances are narrow, reaching a few Hz in SERF magnetometers. It also can be shown that light shift [see for detail (Appelt et al., 1998)] is the imaginary part of the complex optical pumping, which can be introduced by replacing absorption Lorentzian with the complex Lorentzian.

Thus the maximum magnitude of light shift is on the order of the pumping rate that will be obtained at the center of the absorption line. Light shift follows dispersion Lorentzian, while the pumping rate follows absorption Lorentzian, both having the same prefactor. Light shift can be expressed in the units of frequency as the pumping rate, but by dividing light-shift frequency by gyromagnetic coefficient, it can also be expressed in the units of magnetic field. Actually, the effect of light shift on the spins is equivalent to that of a magnetic field and it can be included into the Bloch equations or in the density matrix equation on the equal footing as usual magnetic field. However, if there is more than one type of atom in the cell, the light shift "field" will be different for different atoms. The direction of the light-shift "field" is along the direction of the beam and the sign depend on the sign of circular polarization. Circularly polarized light creates light shift, but linear polarization does not except for very small light-shift noise arising from fundamental fluctuations in the difference of the number of photons of two circularly polarized components of which the linearly polarized light is composed.

Light shift is a parasitic effect in AMs which can add noise to the AM signal and lead to the broadening of magnetic resonances. Because light shift depends on the wavelength as the dispersion Lorentzian, it can be minimized by tuning the laser to the center of absorption resonance. However, this cannot be done for the probe beam, which is deliberately tuned off the resonance to avoid strong absorption. Although the probe beam is linearly polarized, due to imperfections, for example birefringence of glass cell walls, light-shift from the probe beam is always present. By minimizing its intensity, stabilizing wavelength, light shift can be made small and quite constant, so it won't lead to a large noise in the magnetometer.

As we mentioned above, optical pumping and optical probing are essential features of most sensitive AMs. Although it is possible to use very simple light sources for pumping and probing such as discharge lamps, their intensity over the absorption spectrum of atoms used in AMs is not sufficient for reaching best sensitivity and lasers have to be used. A question arises, then, what are requirements for the lasers to be good candidates for AMs? The primary parameter is the wavelength. The wavelength selection depends on the atoms that are used in the magnetometer. Ultra-sensitive magnetometers in order of their sensitivity are based on K, Rb, and Cs vapors. Usually D1 lines of these atoms are preferable, but D2 lines or other lines, which are less convenient from point of view of wavelength availability, in principle can be used. The D1 line ($ns_{1/2}$-$np_{1/2}$, where n=4,5,6 for K, Rb, Cs, respectively) has the advantages over D2 line ($ns_{1/2}$-$np_{3/2}$, the same n) that the pumping on the D1 line by circularly polarized light makes the vapor transparent to this light, so the pump beam can propagate over distances greatly exceeding the (low-intensity) absorption length. The intensity propagation equation for D1 line is $dI / dz = -\alpha(1 - P_z)I$, where I is the intensity and P_z is the polarization projection along the propagation direction z. When P_z is close to 1, the absorption coefficient will be multiplied by a small factor (1- P_z) and hence will be significantly reduced. In addition, as it follows from the solution of the propagation equation, the intensity will be attenuated linearly rather than exponentially, which allows to create a more uniform AM sensitivity across the cell. This is especially important in SERF magnetometers with optical densities exceeding 10. Although it is possible to tune the laser away from the resonance to reduce absorption and allow penetration through the cell for the D2 line, this method will have a drawback of large light shift, discussed earlier, which can introduce noise into the magnetometer signal and broadening owing to non-uniformity of the light shift across the cell. Another advantage of the D1 lines is that optimized magnetometer signal (when the probing is based on the Faraday effect) is two times larger. This advantage is not crucial, and D2 lines still can be used for probing spins especially if the laser of higher quality is easier to find for this line. The optimization of wavelength for each line is done experimentally by tuning the wavelength to maximize the signal or even better SNR. Because the magnetometer signal is the product of the Faraday rotation, which depends on wavelength as the dispersion Lorentzian, and intensity, which is reduced exponentially with absorption coefficient that depend on wavelength as absorption Lorentizian, the maximum of the signal is located a few linewdths from the center of the line. To reduce also spin destruction by the probe light, which has the same wavelength dependence as the absorption coefficient and is proportional to the intensity, the probe laser can be detuned further away than the wavelength of the maximum of the signal and the beam can be expanded to fill out the atomic cell. This can lead to the improvement of SNR. However, if probe light is tuned too far away from the resonance, the opposite rotations by D1 and D2 lines (Eq. 9) can become comparable and cancel each other. This precludes sensitive detection and effective pumping, for which similar cancelation occurs, with broad-spectrum light. If we choose to use the D1 lines because of the described advantages, then following wavelengths of the pump and probe lasers will be needed: 700 nm for K , 795 nm for Rb , 894.6 for Cs, with some tuning range 0.2 nm allowed for optimization. Many diode lasers of these wavelengths are available, making an AM a quite inexpensive device easy to operate.

In addition to the wavelength selection, the requirement for power is also important to consider. For relatively small cells employed in single-channel magnetometers, the power on the order of 50 mW is sufficient. The pump power requirements can be simply estimated from the analysis of the magnetometer sensitivity as the function of the pumping rate. For example, the signal of the SERF magnetometer scales as $R/(R+R_{SD})^2$ (it is directly proportional to polarization level, which is the ratio $R/(R+R_{SD})$, and inversely to magnetic resonance width or $R+R_{SD}$)and hence the optimal value of the pumping rate is $R=R_{SD}$. If the pumping rate is below this value, the signal linearly decreases and so SNR if it is limited by technical, probe beam noise (see below), which is often the case. The pumping rate is related to the pumping power. If we assume that 1.5 photons are required for pumping one atom (see previous discussion), for a cell containing N atoms, the power needed to achieve a given pumping rate R is: $P=1.5\hbar\omega NR$. For the specific case $R=R_{SD}$ for a typical spin-destruction rate of 45 s^{-1} [see Alred et al 2002] in a cell containing $N=8\times10^{14}$ atoms, $P=14$ mW. Various losses, including the reflection by the cell walls and optics, and some pumping efficiency factors would set the requirement above 20 mW for the considered case. If we have the power below this value the magnetometer signal will be below optimal value. (Here is one reason that lamps, which for a needed band have power below 1 mW, are not optimal for pumping SERF magnetometers.) In some applications the bandwidth has to be increased, so even higher pump power will be needed. To have some room for optimization for a single cell of the size 8 cc power on the order of 50 mW is needed. However, this is only a rough estimate, and each situation has to be analyzed separately, and it can be done using similar arguments. For example, the high-density rf magnetometer which relies on light narrowing and the condition $P_z\to1$ (to be discussed later), requires much higher power. The optimization condition is $R\sim\sqrt{R_{SD}R_{SE}}$, which in experiment of (Savukov et al., 2005) was about 2000 s^{-1}. This value is much larger than the value 45 s^{-1} of the optimized SERF magnetometer.

Many semiconductor lasers can provide enough power for the SERF magnetometers. However, multi-channel operation, important for magnetoencephalography (MEG) and magnetocardiography (MCG) applications, requires much higher power, on the order of 1W or more. This can be achieved with systems that have laser amplifiers or arrays of laser diodes.

While any diode laser with the required wavelength and power can be used, for best sensitivity additional requirements such as laser stability, single-frequency operation, and the absence of mode-hops have to be satisfied. In experiments with atomic magnetometers, it was discovered that so-called (distributed feedback) DFB lasers are almost ideal. These lasers are quite a novelty, with US patent issued on June 22, 2004. They work without mode hops and have low intensity noise. However, one problem exists that these lasers need additional cooling below freezing point for reaching the D1 line wavelength of K, 770 nm. In this respect, Rb magnetometers might have advantage for which the DFB laser is available for the D1 line without cooling. Alternative lasers based on tuning by mechanical rotation of a diffraction grating are quite unstable: they can change wavelength during experiments, can suddenly increase their intensity noise to high level, when modes jump. Another interesting laser type is VCSEL. These lasers are quite inexpensive and can perform well if the current is stable. Contrary to edge-emitter, VCSEL can operate only in a single mode. It is also possible to use apparently inexpensive lasers that are commonly found in CD players that have wavelength matching the Rb D2 line (about 780 nm).

The central question in the theory of atomic magnetometers is about noise and sensitivity. There are many sources of noise, some of which are of fundamental nature and cannot be avoided, and other technical noises that can be reduces by careful choice of components and experimental arrangement. Fundamental sources of noise are spin projection noise, photon-counting noise, and light-shift noise. Technical noise arises due to probe beam intensity fluctuations, environment magnetic field, pump intensity and wavelength fluctuations, vibration, air flow, etc. Some sources of noise can be tested separately: the environment field noise and noise from the pump beam can be turned off by blocking the pump beam. The residual noise in the magnetometer signal is frequently called the probe beam noise. The probe beam noise is due to intensity fluctuations of the probe beam, as well as due to wavelength instabilities and beam steering, which can occur due to vibration, temperature drifts, or air flow. This noise directly penetrates the probe beam detection photodiodes. The intensity fluctuation noise can be reduced using either a polarizing beam splitter (PBS) or with modulation methods. The PBS if well balanced has equal intensity fluctuations in its two channel and they can be removed from the signal. The subtraction level of 100 can be achieved, which is often sufficient to reach photon shot noise performance with a good laser. The alternative technique is based on the modulation of light polarization orientation, which is especially useful for detecting low-frequency (a few Hz) magnetic field. When the probe beam passes through one polarizer, a polarization angle modulator, an atomic vapor cell, and the second polarizer with polarization axis at 90 degree angle to the light polarization, the signal will be proportional to $[\theta_{\text{mod}}\cos(\omega t)+\theta_{atoms}]^2$, where ω is the frequency of modulation. With a lock-in amplifier, the first harmonic of ω can be extracted, which amplitude is proportional to the light polarization rotation by atoms θ_{atoms}, i.e. the magnetometer signal. At high enough frequency of modulation, the intensity fluctuation noise from the laser or beam steering falls down, and low frequency field can be measured with high sensitivity as if the signal occurred at high frequency. With a high-quality laser that has a noise level at modulation frequency close to the photon shot noise, it can be possible to achieve shot-noise sensitivity for the detection of polarization rotation at low frequency. One drawback of the modulation technique when it is based on Faraday modulators is the requirement for oscillating field of fairly high strength, on the order of 0.01 T. This makes the design cumbersome. Other options based on a photo-elastic modulator can be chosen to avoid this problem. It is also possible to use internal properties of the atomic vapors to modulate the AM signal.

Noise due to a pump laser is also important to consider and minimize. The intensity and wavelength fluctuations of the pump laser can penetrate to the atomic magnetometer signal due to various coupling mechanisms. One such a mechanism is light-shift. Light shift in an atomic magnetometer affects its signal similarly to the applied magnetic field, as we discussed earlier. The effect of light shift can be modeled with the Bloch equations which contain effective light-shift "magnetic field." The circularly polarized light of pump beam, when it is not exactly tuned to the center of the line, creates light shift field in the direction of its propagation. The change in the intensity as well as in wavelength on which light shift depends will result in the change of amplitude of the effective field, resulting in magnetometer noise. One solution to this problem developed in SERF magnetometers, which is based on the SERF signal output equation

$$S_x = S_0 \frac{\gamma B_y T_2 - (\gamma T_2)^2 B_x B_z}{1 + (\gamma T_2)^2 (B_x^2 + B_y^2 + B_z^2)}, \tag{9}$$

is to zero the x component of the field. Here x is the direction of the probe beam, z is the direction of the pump beam, and y is normal to x and z. Because the z component and the x component enter this equation as the product, zeroing of the x -component will make the magnetometer signal insensitive to the z component and hence the pump beam light shift.

At high frequency the magnetometer signal is not described by Eq. 9. Instead, the response of the magnetometer is linear to small oscillating B_x and B_y fields and the magnetometer exhibits resonance which is the function of the applied B_z field. Light shift from the pump laser, does not enter directly into the signal, but merely changes resonance frequency. Thus if there are no large oscillating fields the rf magnetometer will be immune to light shift fluctuations of pump beam. However, it is not immune to the probe beam light shift fluctuations, which is equivalent to B_x noise. Thus apart from noise that can be directly seen in the detection system arising from probe beam intensity fluctuation, addition noise with characteristic ringing of AM magnetic resonance (because it is capable of exciting the spin resonances), can be present due to these fluctuations.

Environment field noise can be reduced by magnetic shielding and with gradiometers. Vibration noise can be minimized by building magnetometes on optical tables that are floated. To achieve best sensitivity, many noise reduction strategies are combined, and the magnetometers become state-of-the-art.

3. Spin-exchange free atomic magnetometer

Among various atomic magnetometers, the so-called spin-exchange-relaxation-free (SERF) magmetometer introduced by Princeton group have the highest sensitivity at low frequency for a small cell size. (In general, for comparison, it necessary to specify the size, because many magnetometers can improve their sensitivity with the size as $1/V^{1/2}$, where V is the active volume of the magnetometer.) The absence of relaxation from SE collisions is the key to the high sensitivity of SERF magnetometers. In non-SERF magnetometers when the density of vapor is increased to increase the number of atoms and naively reduce fundamental spin-projection noise, which scales as $1/\sqrt{NT_2}$ or other noises, the width of magnetic resonance $1/T_2$ dominated by SE collisions also increases and sensitivity does not improve. However, in the SERF regime, the opposite happens: with the density increase, the resonance width falls down dramatically and reaches a very small value, which is determined by spin-destruction collisions rather than by SE collisions. The situation looks like SE is completely turned off. Because SE cross sections in many alkali-metal atoms are quite large (about 10^{-14} cm^{-2} for K, Rb, Cs) and greatly exceed those of spin-destruction collisions (by 4 orders in K, 3 orders in Rb, and 2 orders in Cs), the suppression of SE broadening leads to dramatic decrease of resonance width and improvement in sensitivity.

Although the SERF regime can be achieved at low densitities, the most beneficial densities are on the order of 10^{14} cm^{-3}. To achieve such densities the heating of alkali cells to relatively high temperatures is needed. Potassium cells require heating to about 180°C, Rb to 150°C, and Cs to 120°C. Thus one drawback of the SERF regime is relatively high temperature of operation requiring building ovens. Initially ovens of SERF magnetometers were based on hot air flow to avoid any magnetic field noise, and their design was quite cumbersome. The power efficiency of heating is also very low, with power on the order of 1 kW required. Apart from this, the air flow is arranged with an inconvenient hook-up to compressed air outlet. More recently, electrical oven designs were introduced for SERF magnetometers. While electrical current produces significant magnetic noise and can disorient atomic spins in the magnetometer, there are several ways to deal with it. The simplest method to reduce this noise is to turn off the power during measurements. While the noise becomes quite small with power off, the measurement is not continuous, which is probably the major drawback of the method. The interruptions also lead to periodic temperature variations which cause correlated variations of the AM signal. The heater element has to be chosen carefully to avoid ferromagnetic and highly conductive materials, which can cause field distortions and noise. In general, with the solution of these problems the electric oven design will be invaluable in out-of-the-lab applications where power consumption and portability are at premium.

Potassium SERF magnetometers set the record of the sensitivity, but from the comparative analysis of the noise (practical and fundamental) of SERF magnetometers for different alkali-metal atoms, which is dominated often by technical noise, it can be found that they can be used with similar results. The ratios of fundamental noise that scale as $1/Rsd^{1/2}$ in the sequence of K,Rb, Cs are about 1:3:10 for the corresponding approximate ratios of the SD rates R_{SD} 1:10:100, and even the least sensitive Cs SERF magnetometer has fundamental noise on the order of 0.1 $fT/Hz^{1/2}$, much lower than the best sensitivity level 0.5 $fT/Hz^{1/2}$ demonstrated. While the Cs SERF magnetometer is less sensitive than K fundamentally, Cs SERF requires lowest temperature of the operation and in some applications where the distance between the cell and the measured object has to be minimized, the Cs AM magnetometer will have great advantage. It can operate in the temperature range 100-140°C, and it is much easier to implement thermal insulation for these temperatures than those used in the potassium SERF magnetometer. The sensitivity of Cs SERF magnetometer has been measured in Ref. (Ledbetter et al., 2008a), and the advantage of low temperature was used for the detection of NMR in a microfluidic channel with a Cs SERF based on a microfabricated cell in Ref.(Ledbetter et al., 2008).

One important motivation for developing atomic magnetometers is owing to applications which have to be done outside the lab. In such applications, portability, small weight, low-power consumption, and vibration stability are highly desirable. The most sensitive SERF magnetometer (Kominis et al., 2003) was implemented on a special non-magnetic optical table with a multi-layer mu-metal shield reducing the ambient magnetic field by a factor of 1 million and due to the complexity of experimental arrangement and high price, such magnetometers would have only limited use, in the lab with the aim to demonstrate highest possible sensitivity or in fundamental experiments. For external applications the design has to be simplified and miniaturized, and for successful commercialization, the price also has to be greatly reduced.

With the goal of commercialization of AMs in mind, a NIST group has been working on the micro-fabrication of miniaturized atomic vapor cells motivated also by miniature atomic clock applications (Knappe et al., 2005). The NIST group showed that the clock package can be adapted to magnetic field measurements with sensitivity of 50 $pT/Hz^{1/2}$ at 10 Hz. The clocks or the magnetometer modules consisted of many layers of various functional components: lasers, filters, lenses, quartz waveplates, ITO heaters, atomic cells, and photodiodes. The components, thin wafers, were stacked on the top of each other to form a compact assembly. Although the demonstration of a miniature AM was a real breakthrough in the magnetometer technology, the performance of the first microfabricated magnetometer was not optimal. One reason of low sensitivity was that this magnetometer was not configured as the SERF magnetometer. However, in a following experiment, a microfabricated atomic cell was tested in the SERF configuration in the SERF regime, and dramatic improvement in sensitivity, almost 1000 times, to the level of 65 $fT/Hz^{1/2}$ was achieved [Shah et al 2007]. According to fundamental noise analysis, even higher sensitivity should be possible. One problem with microfabricated cells is that they have significant spin-destruction rate due to diffusion to the walls, so linewidth is much greater than in larger-cell SERF magnetometers. Perhaps in future if high-temperature coating is developed, the diffusion to the wall can be reduced and sensitivity of microfabricated SERF magnetometer can receive further boost.

In parallel, at Princeton small-scale magnetometers (quite larger than the microchip type) have been created and sensitivity on the order of a few $fT/Hz^{1/2}$ has been demonstrated (result is not yet published). Thus probably by converging these two approaches of NIST and Princeton groups, both inexpensive and highly sensitive magnetometers can emerge soon. The commercialization of this magnetometer will be important in many applications based on sensitive and portable magnetic-field measurements.

4. "SERF magnetometer" in non-SERF regime

The SERF magnetometer is a great advance in magnetometer technology. However, the operation in the SERF regime is limited in a frequency range and in the range of ambient fields. Thus a question arises about the performance of the "SERF magnetometer" - or high-density magnetometer with the arrangement of pumping and probing as well as of other elements similar to the SERF magnetometer– outside the SERF regime; in particular, about how sensitivity changes with a frequency and applied field. The investigation of the non-SERF regime of "the SERF magnetometer" was done in Ref. (Savukov & Romalis, 2005), which resulted later in discovery of rf magnetometer (Savukov et al., 2005) and rf-based scalar magnetometer (Smullin et al., 2009). One characteristic feature in operation outside the SERF regime is the effect of SE collisions on the magnetic resonance of the magnetometer. As we mentioned SE collisions have much higher cross section than SD collisions, and the broadening due to SE collisions can be on the order several kHz for typical temperatures of vapors used in SERF magnetometers, exceeding orders of magnitude a typical SERF bandwidth of several Hz. Because the bandwidth and the signal amplitude are inversely related in the AM, the bandwidth investigation is very important for the analysis of the sensitivity. The bandwidth of high-density magnetometers and the broadening due to SE was investigated in detail (Savukov & Romalis, 2005) experimentally and numerically by solving the DM equation.

Typically, the SERF magnetometer is operated with all fields close to zero, and the magnetometer has its frequency sensitivity profile similar to that of the first-order low-pass filter, with the bandwidth equal to the width of the magnetic resonance. The profile has the shape of Lorentzian centered on zero frequency. When the frequency of the measured field f is outside the bandwidth of the magnetometer, the signal decreases as 1/f. So SERF magnetometer loses its sensitivity very quickly beyond several bandwidths, 10-100 Hz. The sensitivity can be partially restored if a bias field is applied to tune the Zeeman resonance of the magnetometer to the applied field frequency. However, the applied bias field leads to the additional broadening arising from SE collisions. This also means that the operation goes outside the SERF regime. SE broadening increases first quadratically with the bias field, and then the dependence becomes more complicated, although monotonous. Eventually the resonance width reaches the maximum at high field at low pumping rate, which is $R_{SE}/8$ for K and or on this order for other alkali-metal atoms. The SE broadening depends not only on the bias field but also on the spin polarization or pumping rate. With high pumping rate, it is possible to increase polarization and significantly suppress SE broadening via the process of light narrowing. Because pumping leads to additional spin-destruction, the pumping rate cannot infinitely reduce the width of the magnetic resonance, and the trade-off between SE broadening and pump spin-destruction broadening exists. Light narrowing in more detail will be discussed later with regard to the operation of rf magnetometer where this effect is essential for improving sensitivity.

The transitions between Zeeman levels are excited with the change of magnetic number by 1, which is due to the term $\gamma \mathbf{J} \cdot \mathbf{B}$. Although there are many hyperfine sublevels, at low field the splitting is the same between them and multiple magnetic resonances collapse into a single resonance. In this case the Bloch equations can be a good description for spin dynamics (in the SERF regime and outside the SERF regime when excitation field is small) because the behavior of the system is similar to that of electron ½ spin. The only difference is in the presence of the so-called slow-down factor, which depends on the hyperfine level distribution (spin-polarization) and density of atoms. It was studied in detail in Ref. (Savukov & Romalis, 2005). In the K atomic vapor of high density at low field, the slow-down factor ranges from 4 to 6, which can be shown analytically using the assumption of the ST distribution. The maximum slow-down factor is reached for equal distribution over all hyperfine sublevels (small polarization limit, infinite ST), and the minimum factor is reached when only the stretched state M=I+1/2 is occupied (100% polarization, or zero ST). When the vapor density is not very high, the ST distribution is no longer enforced by SE collisions, so it is possible to create an arbitrary distribution of states; then slowing down factor can be made arbitrary large, though it is not of practical interest for magnetometers.

The case when the SE rate dominates other rates in the atomic spin system, including the rate of spin precession, is most simple for analysis. However, in practical situations Zeeman precession rate can exceed the SE rate. One consequence of this is that different Zeeman components can decouple in their motion. In particular the lower hyperfine components and the higher hyperfine components can evolve independently. Then SE collisions will "try" to bring these components into coherent motion causing relaxation.

For strong enough field, the Zeeman splitting between different levels can also become substantially unequal, and multiple magnetic resonances can be observed if the distances between them exceed the resonance widths. This can happen in the Earth field in atomic cells with low pressure of buffer gases and anti-relaxation coating. One consequence of these multiple resonances is that a magnetometer based on measurements of the position of this resonances (the field and the position are almost linearly related) will have heading error, that is the signal at a given modulation frequency will depend not only on the field strength as would be expected for a true scalar magnetometer but also on the direction. Thus the signal of the AM exposed to a large ambient field, such the Earth field, will fluctuate when its orientation changes and this can be a problem in mobile applications.

In the non-SERF regime, the SE broadening can reach levels of several kHz for typical SERF magnetometer operating temperatures. Good understanding of SE effects is essential for designing sensitive magnetometers at arbitrary frequency. For example, the SE broadening can be suppressed with light narrowing. Light narrowing was discovered and explained in Ref.(Appelt et al, 1998 & 1999). In very simple terms, light narrowing can be explained as follows. The SE broadening occurs due to the collisions between oppositely precessing spins of F=I+1/2 and F=I-1/2 hyperfine levels. Strong pumping populates the majority of atoms into the stretched state (F=I+1/2, M=F) and the number of atoms in the lower manifold (F=I-1/2) is small. Thus there will be not many SE collisions between oppositely precessing groups and relaxation due to SE will be suppressed. More detailed explanation is provided in Refs.(Appelt et al., 1998; Savukov & Romalis, 2005; Savukov et al., 2005) where also equations are given for calculations of light narrowing. In magnetometer experiments, light narrowing leads to more than 10 time reduction in bandwidth and similar improvement in magnetic field sensitivity because in practice it is limited by probe beam noise. Although high pump rate can suppress SE broadening completely, it also broadens resonance, more or less linearly with power, so there is an optimal rate to minimize the bandwidth and to maximize sensitivity. In Ref. (Savukov et al., 2005) it was found that the transverse relaxation rate or bandwidth are related to the magnetometer parameters as

$$T_2^{-1} = \frac{R}{4} + \frac{R_{SE} R_{SD}}{R} G(\omega_0, R_{SE}) \tag{10a}$$

$$G(\omega_0, R_{SE}) = \mathrm{Re}\left[\frac{R_{SE} + 4i\omega_0^2 / \pi\nu_{HF}}{5R_{SE} + 8i\omega_0^2 / \pi\nu_{HF}}\right] \tag{10b}$$

Here ω_0 is the spin precession frequency and ν_{HF} is hyperfine frequency. This equation is derived for atoms with I=3/2. In the case of precession frequency below the MHz range, $T_2^{-1} = \frac{R}{4} + \frac{R_{SE} R_{SD}}{5R}$ and the optimized pumping rate leads to the following minimal bandwidth: $(1/T_2)_{\min} = (R_{SE} R_{SD} / 5)^{1/2}$. This width is much smaller than spin-exchange broadening in no light narrowing regime, $R_{SE} / 8$, because $R_{SD} << R_{SE}$, about 10,000 times in potassium. The light-narrowing factor, which is the ratio of the minimal width for the optimal pumping rate and the maximum width without light narrowing, is $K = (5R_{SE} / R_{SD})^{1/2} / 8$. If the SD rate is dominated by K-K collisions, the condition that can always be achieved at high enough density of alkali-metal atoms, then $K = (5\sigma_{SE} / \sigma_{SD})^{1/2} / 8$, where σ_{SE} and σ_{SD} are spin-exchange and spin-destruction cross sections. Potassium has $\sigma_{SE} = 1.8 \times 10^{-14}$ cm^2 and $\sigma_{SD} = 1 \times 10^{-18}$ cm^2, so the maximum light-narrowing factor is $K_{max} \approx 37$.

Tuning to resonance and light narrowing are two main features of high-density rf atomic magnetometers. Another interesting feature is that laser noise, as well as many other technical noises, goes down with frequency, so the rf AM can be more sensitive than the SERF magnetometer, at least in terms of real experimental noise performance. Fundamental limit of the SERF might be by several orders better, but the rf magnetometer can approach its fundamental limit closer while SERF will be by far dominated by technical noise. The fundamental noise of the rf magnetometer has been investigated in Ref. (Savukov et al., 2005). After optimization of various parameters, such as the pumping rate and the probe laser intensity, this noise can be expressed in terms of fundamental quantities of atomic vapors, such as SE and SD cross sections:

$$\delta B_{min} = \frac{2}{\gamma}\sqrt{\frac{\bar{v}[\sigma_{SE}\sigma_{SD} / 5]^{1/2}}{V}(1 + \frac{1}{4\sqrt{\eta}})}, \qquad (11)$$

where $\bar{v}$ is the mean thermal velocity of K-K collisions. For a typical photodiode quantum efficiency $\eta = 50$ % and a cell active volume $V = 1$ cm^3 cell, the optimized fundamental magnetic field sensitivity is about 0.1 fT/Hz$^{1/2}$.

Because the rf magnetometer sensitivity exhibits resonance behavior with resonance frequency being a function of the applied DC magnetic field, this magnetometer can be converted to a scalar magnetometer by applying an rf modulation field near resonance frequency. One advantage of this approach is that the magnetometer can be used in the Earth-field environment, without mu-metal shielding or field compensation, unlike SERF. In the Earth field, the resonance frequency is about 350 kHz (I=3/2 atoms). Small variations in the Earth field can be readily observed as the shifts in the resonance. The in-phase and out-of-phase responses near the magnetic resonance have absorption and dispersion Lorenzian dependencies on frequency. It is convenient to use the dispersion component. Then the signal of the scalar magnetometer is proportional to the deviation of the field from the resonance condition. The lock-in amplifier can be used to convert DC magnetic field changes to the high-frequency rf magnetometer signal. The sensitivity to the DC field is determined by the slope of the dispersive component. The slope of the rf magnetometer was investigated in upcoming paper about scalar magnetometers (Smullin et al., 2009). Because the signal initially grows with the rf field excitation amplitude and then falls off, the optimal excitation amplitude exists. The fall off happens due to broadening of magnetic resonances, which consists of the conventional broadening that can be explained with the Bloch equations and the broadening due to SE collisions. The fundamental limit of the sensitivity of the scalar magnemeter can be derived from that of rf magnetometer in which the effects of large-excitation amplitude broadening are incorporated. Due to the additional SE broadening at the large excitation amplitude required for maximum sensitive, the fundamental noise of the scalar magnetometer has different dependence on the SE and SD rates than the rf magnetometer. The scalar magnetometer fundamental noise is mostly determined by the SE rate, while the rf magnetometer has $[\sigma_{SE}\sigma_{SD} / 5]^{1/4}$ dependence. As the result, the performance of the scalar magnetometer is expected to be quite similar for different alkali-metal atoms. Thus it can be beneficial to use other than K alkali-metal atoms requiring lower cell temperatures.

5. Applications of ultra-sensitive magnetometers

Research on atomic magnetometers is strongly motivated by many current and potential future applications. Among such applications, MEG is probably the most invaluable because no other device than the atomic magnetometer can rival low-Tc SQUIDs in sensitivity at low frequency, in the range of interest to MEG:

- MEG

History of MEG begins in 1968 when the first magnetic recordings of brain activity were registered with a Faraday coils (Cohen, 1968). Although the sensitivity of the coils in the initial demonstration was quite low, soon significant improvement was achieved when the coil was replaced with a SQUID magnetometer (Cohen, 1972). Later on, multi-channel systems have been developed to enable practical MEG source localization. Such systems became the basis for MEG research and clinical applications. However, the cost of multi-channel systems, of their maintenance, and of magnetically shielded rooms required for MEG measurements has been very high, restricting the clinical and research applications of this exciting technology. Consequently, some research has been done in the direction of cost reduction. For example, SQUID gradiometers were tested that could reduce ambient noise 1000 times to eliminate the requirement of expensive high-quality multi-layer shielded rooms. Some other ideas were tested, such as noise suppression based on open superconducting shield and reference channels (Volegov et al., 2004). However, until recently all practical MEG systems had been based on SQUIDs that required liquid helium supply, which is the major drawback.

Alternatively, MEG systems can be based on atomic magnetometers and requirements for cryogens can be eliminated. Recently, it was demonstrated that a SERF magnetometer can be successfully used for the detection and imaging of brain activities (Xia et al., 2006). Moreover, it was argued that a commercial multi-channel system can be built at a fraction of cost of a multi-channel SQUID system, so not only the AM MEG system would be more convenient in operation it would be less expensive as well. An inexpensive multi-channel operation is possible because a large atomic cell filled with a buffer gas can independenly measure magnetic field in different locations. Thus instead of building many separate magnetometers, it is sufficient to build a few large atomic cells to realize hundreds of channels. In Ref. (Xia et al., 2006) it was also demonstrated that a low cost shield can be designed for lying position consisting of mu-metal cylinders. With all these features, overall cost reduction is expected quite significant to make AM MEG system commercially viable and to extend the applications of the MEG method in research and hospitals. However, the currently demonstrated design is not suitable for building a full-head MEG system and requires further investigation.

Owing to rapid progress in AM technology and its novelty, it has to be noted that some even recent review books contain outdated notions about AMs with regard to MEG applications. For example, it is stated [see p. 273, (Clarke & Braginski, 2006) or (Wikswo, 2004), the original source] that one of the difficulties with atomic magnetometers in MEG applications is that they must be operated with a shielding factor about 10^4 times larger than that of magnetically shielded rooms currently used for biomagnetic measurements. This conclusion was drawn from the fact that shielding factor in Ref. (Kominis et al., 2003) was on the order of 10^6, but this was not essential. Actually, it was shown in a different paper (Seltzer & Romalis, 2004) that the SERF magnetometer can operate even in unshielded environment, if due field compensation is provided with a system of coils. Moreover, it was shown that with the gradiometric noise reduction technique based on multi-channel detection a relatively inexpensive shield can be used to conduct MEG experiments (Xia et al., 2006). Another outdated statement was about greater difficulty in thermal insulation of the hot oven, heated to 180°C, from the head compared to that of cryogenic systems such as SQUIDs. However, this problem was solved as well in Ref. (Xia et al., 2006). In addition to effective thermal insulation, a water cooling pad was added to make the subject comfortable during MEG sessions. The thermal insulation measures did not increase much distance from the sensor to the head or sources of the brain activity compared to that of the MEG SQUID Dewar. With regard to greater magnetic noise of the hot oven compared to that of the cold Dewar (Wikswo, 2004), it is the fact that the hot oven made of non conductive and non magnetic material does not create any magnetic noise, while in helium Dewars conductive materials are used for reflecting IR radiation to achieve minimal helium consumption, and such conductive materials produce significant noise.

- Fundamental experiments

Another possible application of AMs where high sensitivity is in demand is in fundamental physics. An example of such an application is the measurement of electric dipole moments (EDM) of atoms. There are several schemes for EDM experiments. The basic idea is to apply a strong electric field and to measure with high sensitivity a weak magnetic field arising due to EDM. Because the hypothetical atomic EDMs are extremely small, it is necessary to use sensors of highest sensitivity. Some schemes are based on unique properties of atomic spins and such experiments cannot be done with arbitrary magnetic sensors. Others do not necessarily need atomic magnetometers, and low-T_c SQUIDs can be used as well. A comprehensive review of EDM research is given in a book by Khriplovich and Lamoreaux (Khriplovich & Lamoreaux, 1997). Atomic magnetometers can be also used in other fundamental experiments such as the setting limits on CPT violation (Kornack & Romalis, 2002).

- NMR and MRI

High sensitivity of atomic magnetometers can be important for applications in unconventional low and ultra-low field (ULF) NMR and MRI. One motivation for exploring ULF MRI is that it is not based on bulky and expensive superconducting or permanent magnets and many applications supplemental to conventional MRI can be developed. For example, it is possible to combine MEG and MRI (Zotev et al., 2008) to reduce the corregistration error, or make a portable and inexpensive MRI scanner at a fraction of the

cost of conventional MRI machines. Such scanners can lead to wider spread of MRI diagnostics around the world. Normally in NMR/MRI simple pick-up coils are used, but the coils, which output signal according to Faraday's law is the time derivative of the magnetic flux, loose sensitivity at low frequency and do not perform well in the ULF regime. Apart from this, the polarization of NMR and MRI spins is also weak in this regime. On the other hand the standards of MR imaging are set very high with resolution on the order of 1 mm and SNR on the order of 30, setting demands on the sensitivity.
To enhance a weak NMR signal in the ULF regime, the method of prepolarization was proposed (Macovski & Conolly, 1993). In this method, relatively large field (whichever is practical to generate with a coil) is applied to polarize protons or other nuclear spins and is turned off during the NMR/MRI measurements. The process is repeated many times. With prepolarization method, the NMR signal is enhanced more than 1000 times compared to that would be generated in microTesla fields at which MRI enconding and readout are actually carried out. One advantage of the ULF readout and encoding is that gradients arising from the ULF coil are quite small and no shimming is necessary. Still even with prepolarization enhancement the SNR and resolution are quite inediquate if simple pick-up coils are used at low frequency, and some solution of this problem is necessary. One solution is to replace the coil with SQUIDs or AMs to achive additional gain in sensitivity. ULF MRI with SQUIDs is now a conventional way to do ULF MRI. For example, recently an airport security scanner have been built and tested based on a multi-channel SQUID detector. However, in all low-Tc SQUID applications the main drawback the requirement for cryogenics exists. The alternative solution to avoid cryogenics can be an atomic magnetometer.
The most potentially useful AM magnetometer for MRI applications (Savukov et al., 2007) is the high-density rf atomic magnetometer (Savukov et al., 2005) discussed in this chapter which not only has very high sensitivity (fundamental limit about 0.1 $fT/Hz^{1/2}$ for 1 cm^3 cell and practical noise was demonstrated 0.2 $fT/Hz^{1/2}$) but also has sufficient bandwidth, on the order of 1 kHz which is needed in MRI detection. The minimum bandwidth for efficient scanner is estimated as the product of tissue relaxation rate and the number of pixels in readout direction.
The demonstration of MRI with AM suitable for in situ imaging is the important step in the direction of developing non-cryogenic ULF-MRI system, which was done in Ref. (Savukov et al., 2009). The achieved sensitivity was on the order of 10 $fT/Hz^{1/2}$, but further significant improvement of the sensitivity is possible. Even without much modification of the system with which the demonstration was done the sensitivity on the order of 1 $fT/Hz^{1/2}$ can be achieved just by raising frequency from 3 kHz to 30 kHz, and with ultimate optimization the sensitivity limits on the order of 0.1 $fT/Hz^{1/2}$ are possible, which would make the MRI system suitable for clinical applications. This can be inferred from the scaling arguments given in Ref. (Savukov et al., 2009).

- Other potential applications based on similar sensitivity as low-Tc SQUIDs

There are many other potential application of AMs which can be developed following pioneering work on low-Tc SQUID applications. In biomedical imaging AMs can be applied to multi-channel MCG imaging. Since heart anomalies are among leading causes of death, their diagnostic is extremely important, and AM MCG can become a invaluable tool for saving millions of lives. Multi-channel MCG provides reach information on electrical activities in the heart non-invasively, and hence this modality can be crucial for revealing heart anomalies and the analysis of their localization. In addition to biomedical applications, AMs can be also used in submarine detection, geology, archeology, military applications, and many other fields as we have already discussed in the introduction. However, currently, high-sensitivity AMs are not available commercially, and this is the main impediment for applications. We expect that in the near future the situation will dramatically change, so all these applications will receive a significant boost.

6. Conclusion

We have considered the most sensitive atomic magnetometers based on high-density alkali-metal vapors. This chapter covered the principles of the operation of ultra-sensitive magnetometers and their applications. Among applications MEG and ULF MRI have been considered in some detail. Because low-T_c SQIUDs have been known as the most sensitive magnetometers for a long time and are still considered such by many researchers, the important conclusion from this chapter should be that atomic magnetometers can provide similar sensitivity and can be used instead of SQUIDs in their applications.

7. Literature

Allred, J.; Lyman, R.; Kornack, T. & Romalis, M. (2002). A high-sensitivity atomic magnetometer unaffected by spin-exchange relaxation. *Phys. Rev. Lett.* vol. 89, 130801.

Appelt, S.; Ben-Amar Baranga, A.; C. J. Erickson, C. J.; Romalis, M. V.; Young, A. R. & Happer, W. (1998). Theory of spin-exchange optical pumping of ^{3}He and ^{129}Xe. *Phys. Rev. A* vol. 58, p. 1412.

Appelt, S.; Ben-Amar Baranga, A.; Young, A. R. & Happer, W. (1999). Light narrowing of rubidium magnetic-resonance lines in high-pressure optical-pumping cells. *Phys. Rev. A* vol. 59, pp. 2078 – 2084.

Bell, W. E. & Bloom, A. L. (1957). *Phys. Rev.* Vol. 107, p. 1559.

Bloom, A. L. (1962). Principles of Operation of the Rubidium Vapor Magnetometers. *Appl. Optics* Vol. 1, pp. 61-68.

Budker, D.; Kimball, D. F. & D. P. DeMille, D. P. (2004). Atomic physics, an exploration through problems and solutions. Oxford Univ. Press.

Clarke, J. &Braginski, A. I. (Eds.) (2006). The SQUID Handbook. Wiley-VCH.

Cohen, D. (1972). Magnetoencephalography: detection of the brain's electrical activity with a superconducting magnetometer. *Science* vol. 175, pp. 664-666.

Cohen, D. (1968). Magnetoecephalography: evidence of magnetic field produced by alpha-rhythm current. *Science* vol. 161, pp. 784-786.

Dehmelt, H. G. (1957). *Phys. Rev.* Vol. 105, p. 1487, 1924.

Happer, W. (1972). Optical Pumping. *Rev. Mod. Phys.* Vol. 44, pp. 169-250.

Groeger, S.; Pazgalev, A. S. & Weis, A. (2005). Comparison of discharge lamp and laser pumped cesium magnetometers. *Appl Phy. B* Vol. 80, pp. 645-654.

Knappe, S.; Schwindt, P. D. D.; Gerginov, V., Shah, V.; Liew, L.; Moreland, J.; Robinson, H. G.; Hollberg, L. & Kitching, J. (2006). Microfabricated atomic clocks and magnetometers. *J. Opt. A: pure Appl. Opt.* pp. S318-S322.

Kornack, T. W. & Romalis, M. V. (2002). Dynamics of Two Overlapping Spin Ensembles Interacting by Spin Exchange. *Phys. Rev. Lett.* Vol. 89, p. 253002.

Ledbetter, M. P.; Savukov, I. M.; Acosta, V. M. & Budker, D. (2008a) Spin-exchange relaxation free magnetometry with Cs vapor. *Phys. Rev. A* vol. 77, p. 033408.

Ledbetter, M. P.; Savukov, I. M.; Budker, D.; Shah, V.; Knappe, S; J. Kitching, J.; Michalak, D. J.; Xu, S. & Pines, A. (2008b). Zero-field remote detection of NMR with a microfabricated atomic magnetometer. *Proc. Natl. Acad. Sci. USA* vol. 105, p. 2286.

Khriplovich, I. B. & Lamoreaux, S. K. (1997). CP Violation Without Strangeness: Electric Dipole Moments of Particles, Atoms, and Molecules. Springer-Verlag.

Kominis, I. K.; T. W. Kornack, T. W.; Allred, J. C.& Romalis, M. V. (2003). A subfemtotesla multichannel atomic magnetometer. *Nature* vol. 422, p. 596.

Macovski, A. & Conolly, S. (1993). Novel approaches to low-cost MRI. *Magn. Reson. Med.* Vol. 40, pp. 221-230.

McDermott, R.; Trabesinger, A. H.; Mück, M.; Hahn, E. L.; Pines, A. & Clarke, J. (2002). Liquid-State NMR and Scalar Couplings in Microtesla Magnetic Fields. *Science* vol. 295, pp. 2247-2249.

Savukov, I. M. & Romalis, M. V. (2005). Effects of spin-exchange collisions in a high-density alkali-metal vapor in low magnetic fields. *Phys. Rev. A* vol. 71, p. 023405.

Savukov, I. M. & Romalis, M. V. (2005). NMR detection with an atomic magnetometer. *Phys. Rev. Lett.* vol. 94, p. 123001.

Savukov, I. M.; Seltzer, S. J.; Romalis, M. V. & and Sauer, K. L. (2005). Tunable Atomic Magnetometer for Detection of Radio-Frequency Magnetic Fields. *Phys. Rev. Lett.* vol. 95, 063004.

Savukov, I. M.; Seltzer, S. J. & Romalis, M. V. (2007). Detection of NMR signals with a radio-frequency atomic magnetometer. *JMR* Vol. 185, p. 214.

Savukov, I. M.; Zotev, V. S.; Volegov, P. L.; Espy, M. A.; Matlashov, A. N.; Gomez, J. J.; & Kraus, R. H., Jr. (2009). MRI with an atomic magnetometer suitable for practical imaging applications. *JMR* 199, pp. 188-191.

Shah, V.; Knappe, S.; Schwindt, P. D. D. & J. Kitching, J. (2007). *Nat. Photonics* vol. 1, p. 649.

Seltzer, S. J. & Romalis, M. V. (2004). Unshielded three-axis vector operation of a spin-exchange-relaxation-free atomic magnetometer." *Appl. Phys. Lett.* Vol. 85(20), p. 4804.

Smullin, S. J.; Savukov, I. M.; Vasilakis, G.; Ghosh, R. K. & Romalis, M. V. (2009). A Low-Noise High-Density Alkali Metal Scalar Magnetometer. *Archives: physics/0611085, accepted to publication to PRA.*

Volegov, P.; Matlachov, A., Mosher, J., Espy, M. A. & Kraus, R. H., Jr. (2004). Noise-free magnetoencephalography recordings of brain function. *Phys. Med. Biol.* Vol. 49, pp. 2117-2128.

Wikswo, J. P. (2004). SQUIDs Remain Best Tools for Measuring Brain's Magnetic Field. Phys. Today Vol. 57, pp. 15-17.

Xia, H.; Ben-Amar Baranga, A.; Hoffman, D. & Romalis, M. V. (2006). Magnetoencephalography with an atomic magnetometer. *Appl. Phys. Lett.* vol. 89, p. 211104.

Zotev, V. S.; Matlashov, A. N.; Volegov, P. L.; Savukov, I. M.; Espy, M. A.; Mosher, J. C.; Gomez, J. J. & Kraus, R. H., Jr. (2008). Microtesla MRI of the human brain combined with MEG. *JMR* Vol. 194, pp. 115-120.